Designing Scientific Applications on GPUs

CHAPMAN & HALL/CRC
Numerical Analysis and Scientific Computing

Aims and scope:
Scientific computing and numerical analysis provide invaluable tools for the sciences and engineering. This series aims to capture new developments and summarize state-of-the-art methods over the whole spectrum of these fields. It will include a broad range of textbooks, monographs, and handbooks. Volumes in theory, including discretisation techniques, numerical algorithms, multiscale techniques, parallel and distributed algorithms, as well as applications of these methods in multi-disciplinary fields, are welcome. The inclusion of concrete real-world examples is highly encouraged. This series is meant to appeal to students and researchers in mathematics, engineering, and computational science.

Editors

Choi-Hong Lai
*School of Computing and
Mathematical Sciences
University of Greenwich*

Frédéric Magoulès
*Applied Mathematics and
Systems Laboratory
Ecole Centrale Paris*

Editorial Advisory Board

Mark Ainsworth
*Mathematics Department
Strathclyde University*

Todd Arbogast
*Institute for Computational
Engineering and Sciences
The University of Texas at Austin*

Craig C. Douglas
*Computer Science Department
University of Kentucky*

Ivan Graham
*Department of Mathematical Sciences
University of Bath*

Peter Jimack
*School of Computing
University of Leeds*

Takashi Kako
*Department of Computer Science
The University of Electro-Communications*

Peter Monk
*Department of Mathematical Sciences
University of Delaware*

Francois-Xavier Roux
ONERA

Arthur E.P. Veldman
*Institute of Mathematics and Computing Science
University of Groningen*

Proposals for the series should be submitted to one of the series editors above or directly to:
CRC Press, Taylor & Francis Group

Published Titles

Classical and Modern Numerical Analysis: Theory, Methods and Practice
Azmy S. Ackleh, Edward James Allen, Ralph Baker Kearfott, and
Padmanabhan Seshaiyer

Cloud Computing: Data-Intensive Computing and Scheduling
Frédéric Magoulès, Jie Pan, and Fei Teng

Computational Fluid Dynamics
Frédéric Magoulès

A Concise Introduction to Image Processing using C++
Meiqing Wang and Choi-Hong Lai

Decomposition Methods for Differential Equations: Theory and Applications
Juergen Geiser

Designing Scientific Applications on GPUs
Raphaël Couturier

Desktop Grid Computing
Christophe Cérin and Gilles Fedak

Discrete Dynamical Systems and Chaotic Machines: Theory and Applications
Jacques M. Bahi and Christophe Guyeux

Discrete Variational Derivative Method: A Structure-Preserving Numerical Method for Partial Differential Equations
Daisuke Furihata and Takayasu Matsuo

Grid Resource Management: Toward Virtual and Services Compliant Grid Computing
Frédéric Magoulès, Thi-Mai-Huong Nguyen, and Lei Yu

Fundamentals of Grid Computing: Theory, Algorithms and Technologies
Frédéric Magoulès

Handbook of Sinc Numerical Methods
Frank Stenger

Introduction to Grid Computing
Frédéric Magoulès, Jie Pan, Kiat-An Tan, and Abhinit Kumar

Iterative Splitting Methods for Differential Equations
Juergen Geiser

Mathematical Objects in C++: Computational Tools in a Unified Object-Oriented Approach
Yair Shapira

Numerical Linear Approximation in C
Nabih N. Abdelmalek and William A. Malek

Numerical Techniques for Direct and Large-Eddy Simulations
Xi Jiang and Choi-Hong Lai

Parallel Algorithms
Henri Casanova, Arnaud Legrand, and Yves Robert

Parallel Iterative Algorithms: From Sequential to Grid Computing
Jacques M. Bahi, Sylvain Contassot-Vivier, and Raphaël Couturier

Particle Swarm Optimisation: Classical and Quantum Perspectives
Jun Sun, Choi-Hong Lai, and Xiao-Jun Wu

XML in Scientific Computing
C. Pozrikidis

Designing Scientific Applications on GPUs

Edited by
Raphaël Couturier
University of Franche-Comte
Belfort, France

CRC Press
Taylor & Francis Group
Boca Raton London New York

CRC Press is an imprint of the
Taylor & Francis Group, an **informa** business

A CHAPMAN & HALL BOOK

First published in paperback 2024

First published 2014
by CRC Press
2385 NW Executive Center Drive, Suite 420, Boca Raton FL 33431

and by CRC Press
4 Park Square, Milton Park, Abingdon, Oxon, OX14 4RN

CRC Press is an imprint of Taylor & Francis Group, LLC

© 2014, 2024 Taylor & Francis Group, LLC

Reasonable efforts have been made to publish reliable data and information, but the author and publisher cannot assume responsibility for the validity of all materials or the consequences of their use. The authors and publishers have attempted to trace the copyright holders of all material reproduced in this publication and apologize to copyright holders if permission to publish in this form has not been obtained. If any copyright material has not been acknowledged please write and let us know so we may rectify in any future reprint.

Except as permitted under U.S. Copyright Law, no part of this book may be reprinted, reproduced, transmitted, or utilized in any form by any electronic, mechanical, or other means, now known or hereafter invented, including photocopying, microfilming, and recording, or in any information storage or retrieval system, without written permission from the publishers.

For permission to photocopy or use material electronically from this work, access www.copyright.com or contact the Copyright Clearance Center, Inc. (CCC), 222 Rosewood Drive, Danvers, MA 01923, 978-750-8400. For works that are not available on CCC please contact mpkbookspermissions@tandf.co.uk

Trademark notice: Product or corporate names may be trademarks or registered trademarks and are used only for identification and explanation without intent to infringe.

Publisher's Note
The publisher has gone to great lengths to ensure the quality of this reprint but points out that some imperfections in the original copies may be apparent.

Library of Congress Cataloging-in-Publication Data

Designing scientific applications on GPUs / editor, Raphael Couturier.
    pages cm. -- (Chapman & Hall/CRC numerical analysis and scientific computing series ; 21)
    Includes bibliographical references and index.
    ISBN 978-1-4665-7162-4 (hbk. : alk. paper) 1. Parallel programming (Computer science) 2. Graphics processing units--Programming. 3. Science--Data processing. 4. Numerical analysis--Computer programs. 5. Application software--Development. I. Couturier, Raphael, editor.

QA76.76.A65D474 2014
006.6'63--dc23   2013032465

ISBN: 978-1-4665-7162-4 (hbk)
ISBN: 978-1-03-291926-3 (pbk)
ISBN: 978-0-429-10085-7 (ebk)

DOI: 10.1201/b16051

Visit the Taylor & Francis Web site at
http://www.taylorandfrancis.com

and the CRC Press Web site at
http://www.crcpress.com

Contents

List of Figures   xi

List of Tables   xvii

Preface   xxi

I   Presentation of GPUs   1

1 Presentation of the GPU architecture and of the CUDA environment   3
Raphaël Couturier

2 Introduction to CUDA   13
Raphaël Couturier

II   Image processing   23

3 Setting up the environment   25
Gilles Perrot

4 Implementing a fast median filter   31
Gilles Perrot

5 Implementing an efficient convolution operation on GPU   53
Gilles Perrot

III   Software development   71

6 Development of software components for heterogeneous many-core architectures   73
Stefan L. Glimberg, Allan P. Engsig-Karup, Allan S. Nielsen, and Bernd Dammann

7 Development methodologies for GPU and cluster of GPUs   105
    Sylvain Contassot-Vivier, Stephane Vialle, and Jens Gustedt

IV Optimization   151

8 GPU-accelerated tree-based exact optimization methods   153
    Imen Chakroun and Nouredine Melab

9 Parallel GPU-accelerated metaheuristics   183
    Malika Mehdi and Ahcène Bendjoudi, Lakhdar Loukil, and
    Nouredine Melab

10 Linear programming on a GPU: a case study   215
    Xavier Meyer and Bastien Chopard and Paul Albuquerque

V Numerical applications   249

11 Fast hydrodynamics on heterogeneous many-core hardware   251
    Allan P. Engsig-Karup, Stefan L. Glimberg, Allan S. Nielsen, and
    Ole Lindberg

12 Parallel monotone spline interpolation and approximation on GPUs   295
    Gleb Beliakov and Shaowu Liu

13 Solving sparse linear systems with GMRES and CG methods on GPU clusters   311
    Lilia Ziane Khodja, Raphaël Couturier, and Jacques Bahi

14 Solving sparse nonlinear systems of obstacle problems on GPU clusters   331
    Lilia Ziane Khodja, Raphaël Couturier, and Jacques Bahi, Ming
    Chau, and Pierre Spitéri

15 Ludwig: multiple GPUs for a complex fluid lattice Boltzmann application   355
    Alan Gray and Kevin Stratford

16 Numerical validation and performance optimization on GPUs of an application in atomic physics   371
    Rachid Habel, Pierre Fortin, Fabienne Jézéquel, and Jean-Luc
    Lamotte, and Stan Scott

17 A GPU-accelerated envelope-following method for switching power converter simulation 395
Xuexin Liu, Sheldon Xiang-Dong Tan, Hai Wang, and Hao Yu

VI Other 413

18 Implementing multi-agent systems on GPU 415
Guillaume Laville, Christophe Lang, Bénédicte Herrmann, and Laurent Philippe, Kamel Mazouzi, and Nicolas Marilleau

19 Pseudorandom number generator on GPU 441
Raphaël Couturier and Christophe Guyeux

20 Solving large sparse linear systems for integer factorization on GPUs 453
Bertil Schmidt and Hoang-Vu Dang

Index 473

List of Figures

1.1	Comparison of number of cores in a CPU and in a GPU.	6
1.2	Comparison of low latency of a CPU and high throughput of a GPU.	7
1.3	Scalability of GPU.	9
1.4	Memory hierarchy of a GPU.	10
4.1	Example of 5x5 median filtering.	32
4.2	Illustration of window overlapping in 5x5 median filtering.	34
4.3	Example of median filtering, applied to salt and pepper noise reduction.	36
4.4	Comparison of pixel throughputs for CPU generic median, CPU 3×3 median register-only with bubble sort, GPU generic median, GPU 3×3 median register-only with bubble sort, and GPU libJacket.	39
4.5	Forgetful selection with the minimal element register count. Illustration for 3×3 pixel window represented in a row and supposedly sorted.	40
4.6	Determination of the median value by the *forgetful selection* process, applied to a 3×3 neighborhood window.	41
4.7	First iteration of the 5×5 selection process, with $k_{25} = 14$, which shows how Instruction Level Parallelism is maximized by the use of an incomplete sorting network.	41
4.8	Illustration of how window overlapping is used to combine 2 pixel selections in a 3×3 median kernel.	43
4.9	Comparison of pixel throughput on GPU C2070 for the different 3×3 median kernels.	45
4.10	Reducing register count in a 5×5 register-only median kernel outputting 2 pixels simultaneously.	45
4.11	Example of separable median filtering (smoother), applied to salt and pepper noise reduction.	49
5.1	Principle of a generic convolution implementation.	55
5.2	Mask window overlapping when processing a packet of 8 pixels per thread.	60
5.3	Organization of the prefetching stage of data, for a 5×5 mask and a thread block size of 8×4.	63

List of Figures

6.1 Schematic representation of the five main components, their type definitions, and member functions. 78
6.2 Discrete solution, at times $t = 0s$ and $t = 0.05s$, using (6.5) as the initial condition and a small 20×20 numerical grid. . 82
6.3 Single- and double-precision floating point operations per second for a two-dimensional stencil operator on a numerical grid of size 4096^2. 87
6.4 Algorithmic performance for the conjugate gradient, multigrid, and defect correction methods, measured in terms of the relative residual per iteration. 92
6.5 Message passing between two GPUs involves several memory transfers across lower bandwidth connections. 92
6.6 Domain distribution of a two-dimensional grid into three subdomains. 94
6.7 Performance timings for distributed stencil operations, including communication and computation times. Executed on test environment 3. 95
6.8 Time domain decomposition. 96
6.9 Schematic visualization of a fully distributed work scheduling model for the parareal algorithm as proposed by Aubanel. . 98
6.10 Parareal convergence properties as a function of R and number of GPUs used. 100
6.11 Parareal performance properties as a function of R and number of GPUs used. 100

7.1 Native overlap of internode CPU communications with GPU computations. 108
7.2 Overlap of internode CPU communications with a sequence of CPU/GPU data transfers and GPU computations. 110
7.3 Overlap of internode CPU communications with a streamed sequence of CPU/GPU data transfers and GPU computations. 113
7.4 Complete overlap of internode CPU communications, CPU/GPU data transfers, and GPU computations, interleaving computation-communication iterations. 116
7.5 Experimental performances of different synchronous algorithms computing a dense matrix product. 119
7.6 Computation times of the test application in synchronous and asynchronous modes. 138
7.7 Computation times with or without overlap of Jacobian updatings in asynchronous mode. 139

8.1 Illustration of the parallel tree exploration model. 158
8.2 Illustration of the parallel evaluation of bounds model. ... 159
8.3 Flow-shop problem instance with 3 jobs and 6 machines. .. 160

List of Figures

8.4	The lag l_j of a job J_j for a couple (k,l) of machines is the sum of the processing times of the job on all the machines between k and l.	161
8.5	The overall architecture of the parallel tree exploration-based GPU-accelerated branch-and-bound algorithm.	162
8.6	The overall architecture of the GPU-accelerated branch-and-bound algorithm based on the parallel evaluation of bounds.	163
9.1	Parallel models for metaheuristics.	186
9.2	A two level classification of state-of-the-art GPU-based parallel metaheuristics.	199
9.3	The skeleton of the ParadisEO-MO-GPU.	203
10.1	Solving an ILP problem using a branch-and-bound algorithm.	225
10.2	Example of a parallel reduction at block level. (Courtesy NVIDIA).	229
10.3	Communications between CPU and GPU.	232
10.4	Performance model and measurements comparison.	242
10.5	Time required to solve problems of Table 10.1.	243
10.6	Time required to solve problems of Table 10.2.	244
10.7	Time required to solve problems of Table 10.3.	245
11.1	Snapshot of steady state wave field generated by a Series 60 ship hull.	252
11.2	Numerical experiments to assess stability properties of numerical wave model.	265
11.3	Snapshots at intervals $T/8$ over one wave period in time.	268
11.4	Performance timings per PDC iteration as a function of increasing problem size N, for single, mixed, and double precision arithmetics.	270
11.5	Domain decomposition performance on multi-GPU systems.	272
11.6	The accuracy in phase celerity c determined by (11.43a) for small-amplitude (linear) wave.	277
11.7	Assessment of kinematic error is presented in terms of the depth-averaged error.	278
11.8	Comparison between convergence histories for single- and double-precision computations using a PDC method for the solution of the transformed Laplace problem.	279
11.9	Comparison of accuracy as a function of time for double-precision calculations vs. single-precision with and without filtering.	282
11.10	Harmonic analysis for the experiment of Whalin for $T = 1, 2, 3$ s.	283
11.11	Parareal absolute timings and parareal speedup.	284
11.12	Parallel time integration using the parareal method.	285

11.13 Computed results. Comparison with experiments for hydrodynamics force calculations confirming engineering accuracy for low Froude numbers. 288

12.1 Cubic spline (solid) and monotone quadratic spline (dashed) interpolating monotone data 296
12.2 Hermite cubic spline (solid) and Hermite rational spline interpolating monotone data 296

13.1 A data partitioning of the sparse matrix A, the solution vector x, and the right-hand side b into four portions. 318
13.2 Data exchanges between *Node 1* and its neighbors *Node 0*, *Node 2*, and *Node 3*. 320
13.3 Columns reordering of a sparse submatrix. 321
13.4 General scheme of the GPU cluster of tests composed of six machines, each with two GPUs. 322
13.5 Sketches of sparse matrices chosen from the University of Florida collection. 323
13.6 Parallel generation of a large sparse matrix by four computing nodes. 326

14.1 Data partitioning of a problem to be solved among $S = 3 \times 4$ computing nodes. 338
14.2 Decomposition of a subproblem in a GPU into nz slices. . . 340
14.3 Matrix constant coefficients in a three-dimensional domain. 342
14.4 Computation of a vector element with the projected Richardson method. 344
14.5 Red-black ordering for computing the iterate vector elements in a three-dimensional space. 349
14.6 Weak scaling of both synchronous and asynchronous algorithms of the projected Richardson method using red-black ordering technique. 352

15.1 The lattice is decomposed between MPI tasks. 358
15.2 The weak (top) and strong (bottom) scaling of *Ludwig*. . . . 363
15.3 A two-dimensional schematic picture of spherical particles on the lattice. 364

16.1 Subdivision of the configuration space (r_1,r_2) into a set of connected sectors. 374
16.2 Propagation of the R-matrix from domain D to domain D'. 375
16.3 Error distribution for medium case in single precision. . . . 378
16.4 1s2p cross-section, 10 sectors. 380
16.5 1s4d cross-section, 10 sectors. 381
16.6 1s2p cross-section, threshold = 10^{-4}, 210 sectors. 382
16.7 1s4d cross-section, threshold = 10^{-4}, 210 sectors. 382

List of Figures

16.8	The six steps of an off-diagonal sector evaluation.	383
16.9	Constructing the local R-matrix R34 from the j amplitude array associated with edge 4 and the i amplitude array associated with edge 3.	384
16.10	Compute and I/O times for the GPU V3 on one C1060.	386
16.11	Speedup of the successive GPU versions.	388
16.12	CPU (1 core) execution times for the off-diagonal sectors of the large case.	389
16.13	GPU execution times for the off-diagonal sectors of the large case.	390
17.1	Transient envelope-following analysis.	397
17.2	The flow of envelope-following method.	401
17.3	GPU parallel solver for envelope-following update.	403
17.4	Diagram of a zero-voltage quasi-resonant flyback converter.	407
17.5	Illustration of power/ground network model.	407
17.6	Flyback converter solution calculated by envelope-following.	408
17.7	Buck converter solution calculated by envelope-following.	409
18.1	Evolution algorithm of the Collembola model.	421
18.2	Performance of the Collembola model on CPU and GPU.	424
18.3	Consolidation of multiple simulations in one OpenCL kernel execution.	426
18.4	Compact representation of the topology of a MIOR simulation.	428
18.5	CPU and GPU performance on a Tesla C1060 node.	432
18.6	CPU and GPU performance on a personal computer with a Geforce 8800GT.	432
18.7	Execution time of one multi-simulation kernel on the Tesla platform.	433
18.8	Total execution time for 1000 simulations on the Tesla platform, while varying the number of simulations for each kernel.	433
19.1	Quantity of pseudorandom numbers generated per second with the xorlike-based PRNG.	449
20.1	An example square matrix of size 6×6 (zero values are not shown).	456
20.2	Partitioning of a row-sorted NFS matrix into four formats.	459
20.3	Example of the memory access pattern for a 6×6 matrix stored in sliced COO format (slice size = 3 rows).	461
20.4	Partitioning of a row-sorted floating-point matrix into SCOO format.	465
20.5	Performance comparison of SCOO and other GPU formats for each test matrix on a Fermi Tesla C2075 (ECC disabled).	468
20.6	Visualization of *nlpkkt120*, *relat9*, and *GL7d19* matrix.	469

20.7 Performance of the SCOO on a GTX-580 and a CPU implementation using MKL performed on a Core-i7 2700K using 8 threads. 470

List of Tables

4.1 Performance results of `kernel medianR`. 35
4.2 Performance of various 5×5 median kernel implementations, applied on 4096×4096 pixel image with C2070 GPU card. . 47
4.3 Measured performance of one generic pseudo-separable median kernel applied to 4096×4096 pixel image with various window sizes. 48

5.1 Timings (time) and throughput values (TP in MP/s) of one register-only nonseparable convolution kernel, for small mask sizes of 3×3, 5×5, and 7×7 pixels, on a C2070 card. . . . 57
5.2 Timings (time) and throughput values (TP in MP/s) of one register-only nonseparable convolution kernel, for small mask sizes of 3×3, 5×5, and 7×7 pixels, on a GTX280. 57
5.3 Time cost of data transfers between CPU and GPU memories, on C2070 and GTX280 cards (in milliseconds). 58
5.4 Timings (time) and throughput values (TP in MP/s) of our generic fixed mask size convolution kernel run on a C2070 card. 60
5.5 Performances, in milliseconds, of our generic 8 pixels per thread kernel using shared memory, run on a C2070 card. Data transfers duration are not included. 62
5.6 Throughput values, in MegaPixel per second, of our generic 8 pixels per thread kernel using shared memory, run on a C2070 card. 63
5.7 Performances, in milliseconds, of our generic 8 pixels per thread 1D convolution kernels using shared memory, run on a C2070 card. 66
5.8 Throughput values, in megapixel per second, of our generic 8 pixels per thread 1D convolution kernel using shared memory, run on a C2070 card. 66
5.9 Time cost of data copy between the vertical and the horizontal 1D convolution stages, on a C2070 cards (in milliseconds). . 66

8.1 The different data structures of the LB algorithm and their associated complexities in memory size and numbers of accesses. 170

8.2	The sizes of each data structure for the different experimented problem instances.	170
8.3	The sequential resolution time of each instance according to its number of jobs and machines.	174
8.4	Speedups for different problem instances and pool sizes with the GPU-PTE-BB approach.	174
8.5	Speedups for different problem instances and pool sizes with the GPU-PEB-BB approach.	175
8.6	Speedups for different instances and pool sizes using thread divergence management.	176
8.7	Speedup for different FSP instances and pool sizes obtained with data access optimization.	177
8.8	Speedup for different FSP instances and pool sizes obtained with data access optimization.	177
8.9	Speedup for different FSP instances and pool sizes obtained with data access optimization.	177
9.1	Results of the GPU-based iterated tabu search for different Q3AP instances.	209
10.1	NETLIB problems solved in less than 1 second.	243
10.2	NETLIB problems solved in the range of 1 to 4 seconds.	244
10.3	NETLIB problems solved in more than 5 seconds.	245
12.1	The average CPU time (sec) of the serial PAVA, MLS, and parallel MLS algorithms.	307
13.1	Main characteristics of sparse matrices chosen from the University of Florida collection.	323
13.2	Performances of the parallel CG method on a cluster of 24 CPU cores vs. on a cluster of 12 GPUs.	324
13.3	Performances of the parallel GMRES method on a cluster 24 CPU cores vs. on cluster of 12 GPUs.	324
13.4	Main characteristics of sparse banded matrices generated from those of the University of Florida collection.	326
13.5	Performances of the parallel CG method for solving linear systems associated to sparse banded matrices on a cluster of 24 CPU cores vs. on a cluster of 12 GPUs.	327
13.6	Performances of the parallel GMRES method for solving linear systems associated to sparse banded matrices on a cluster of 24 CPU cores vs. on a cluster of 12 GPUs.	327
14.1	Execution times in seconds of the parallel projected Richardson method implemented on a cluster of 24 CPU cores.	347
14.2	Execution times in seconds of the parallel projected Richardson method implemented on a cluster of 12 GPUs.	347

List of Tables

14.3	Execution times in seconds of the parallel projected Richardson method using red-black ordering technique implemented on a cluster of 12 GPUs.	351
16.1	Characteristics of four data sets.	375
16.2	Impact on FARM of the single precision version of PROP.	379
16.3	Execution time of PROP on CPU and GPU.	387
16.4	Performance results with multiple concurrent energies on one C2070 GPU.	391
17.1	CPU and GPU time comparisons (in seconds) for solving Newton update equation with the proposed Gear-2 sensitivity.	409
20.1	Properties of some NFS matrices.	458
20.2	Sliced COO subformat comparison (# rows per slices is based on $n = 64$).	463
20.3	Overview of hardware used in the experiments.	466
20.4	Performance of SpMV on RSA-170 matrix.	467
20.5	Performance for each of the four subformat partitions of the RSA-170 matrix on a C2075.	467
20.6	Overview of sparse matrices used for performance evaluation.	468

Preface

This book is intended to present the design of significant scientific applications on GPUs. Scientific applications require more and more computational power in a large variety of fields: biology, physics, chemistry, phenomon model and prediction, simulation, mathematics, etc.

In order to be able to handle more complex applications, the use of parallel architectures is the solution to decrease the execution times of these applications. Using many computing cores simulataneously can significantly speed up the processing time.

Nevertheless using parallel architectures is not so easy and has always required an endeavor to parallelize an application. Nowadays with general purpose graphics processing units (GPGPU), it is possible to use either general graphic cards or dedicated graphic cards to benefit from the computational power of all the cores available inside these cards. The NVIDIA company introduced Compute Unified Device Architecture (CUDA) in 2007 to unify the programming model to use their video card. CUDA is currently the most used environment for designing GPU applications although some alternatives are available, such as Open Computing Language (OpenCL). According to applications and the GPU considered, a speed up from 5 up to 50, or even more can be expected using a GPU over computing with a CPU.

The programming model of GPU is quite different from the one of CPU. It is well adapted to data parallelism applications. Several books present the CUDA programming models and multi-core applications design. This book is only focused on scientific applications on GPUs. It contains 20 chapters gathered in 6 parts.

The first part presents the GPUs. The second part focuses on two significant image processing applications on GPUs. Part three presents two general methodologies for software development on GPUs. Part four describes three optimization problems on GPUs. The fifth part, the longest one, presents seven numerical applications. Finally part six illustrates three other applications that are not included in the previous parts.

Some codes presented in this book are available online on my webpage: http://members.femto-st.fr/raphael-couturier/en/gpu-book/

Part I

Presentation of GPUs

Chapter 1

Presentation of the GPU architecture and of the CUDA environment

Raphaël Couturier

Femto-ST Institute, University of Franche-Comte, France

1.1	Introduction	3
1.2	Brief history of the video card	4
1.3	GPGPU	4
1.4	Architecture of current GPUs	5
1.5	Kinds of parallelism	7
1.6	CUDA multithreading	8
1.7	Memory hierarchy	9
1.8	Conclusion	11
	Bibliography	11

1.1 Introduction

This chapter introduces the Graphics Processing Unit (GPU) architecture and all the concepts needed to understand how GPUs work and can be used to speed up the execution of some algorithms. First of all this chapter gives a brief history of the development of the graphics cards up to the point when they started being used in order to perform general purpose computations. Then the architecture of a GPU is illustrated. There are many fundamental differences between a GPU and a traditional processor. In order to benefit from the power of a GPU, a CUDA programmer needs to use threads. They have some particularities which enable the CUDA model to be efficient and scalable when some constraints are addressed.

1.2 Brief history of the video card

Video cards or graphics cards have been introduced in personal computers to produce high quality graphics faster than classical Central Processing Units (CPU) and to free the CPU from this task. In general, display tasks are very repetitive and very specific. Hence, some manufacturers have produced more and more sophisticated video cards, providing 2D accelerations, then 3D accelerations, then some light transforms. Video cards own their own memory to perform their computations. For at least two decades, every personal computer has had a video card which is simple for desktop computers or which provides many accelerations for game and/or graphic-oriented computers. In the latter case, graphics cards may be more expensive than a CPU.

Since 2000, video cards have allowed users to apply arithmetic operations simultaneously on a sequence of pixels, later called stream processing. In this case, the information of the pixels (color, location and other information) is combined in order to produce a pixel color that can be displayed on a screen. Simultaneous computations are provided by shaders which calculate rendering effects on graphics hardware with a high degree of flexibility. These shaders handle the stream data with pipelines.

Some researchers tried to apply those operations on other data, representing something different from pixels, and consequently this resulted in the first uses of video cards for performing general purpose computations. The programming model was not easy to use at all and was very dependent on the hardware constraints. More precisely it consisted in using either DirectX of OpenGL functions providing an interface to some classical operations for videos operations (memory transfers, texture manipulation, etc.). Floating point operations were most of the time unimaginable. Obviously when something went wrong, programmers had no way (and no tools) to detect it.

1.3 GPGPU

In order to benefit from the computing power of more recent video cards, CUDA was first proposed in 2007 by NVIDIA. It unifies the programming model for some of their most efficient video cards. CUDA [3] has quickly been considered by the scientific community as a great advance for general purpose graphics processing unit (GPGPU) computing. Of course other programming models have been proposed. The other well-known alternative is OpenCL which aims at proposing an alternative to CUDA and which is multiplatform and portable. This is a great advantage since it is even possible to execute OpenCL programs on traditional CPUs. The main drawback is that it is less close to the hardware and, consequently, it sometimes provides less efficient programs. Moreover, CUDA benefits from more mature compilation and optimization procedures. Other less known environments have been proposed, but most of them have been discontinued, such as FireStream by ATI, which is not maintained anymore and has been replaced by OpenCL and BrookGPU by Stanford University [1]. Another environment based on pragma (insertion of pragma directives inside the code to help the compiler to generate efficient code) is called OpenACC. For a comparison with OpenCL, interested readers may refer to [2].

1.4 Architecture of current GPUs

The architecture of current GPUs is constantly evolving. Nevertheless some trends remain constant throughout this evolution. Processing units composing a GPU are far simpler than a traditional CPU and it is much easier to integrate many computing units inside a GPU card than to do so with many cores inside a CPU. In 2012, the most powerful GPUs contained more than 500 cores and the most powerful CPUs had 8 cores. Figure 1.1 shows the number of cores inside a CPU and inside a GPU. In fact, in a current NVIDIA GPU, there are multiprocessors which have 32 cores (for example, on Fermi cards). The core clock of a CPU is generally around 3GHz and the one of a GPU is about 1.5GHz. Although the core clock of GPU cores is slower, the number of cores inside a GPU provides more computational power. This measure is commonly represented by the number of floating point operation per seconds. Nowadays the most powerful GPUs provide more than 1TFlops, i.e., 10^{12} floating point operations per second. Nevertheless GPUs are very efficient at executing repetitive work in which only the data change. It is important to keep in mind that multiprocessors inside a GPU have 32 cores. Later we will see that these 32 cores need to do the same work to get maximum performance.

On the most powerful GPU cards, called Fermi, multiprocessors are called streaming multiprocessors (SMs). Each SM contains 32 cores and is able to perform 32 floating points or integer operations per clock on 32-bit numbers or 16 floating points per clock on 64-bit numbers. SMs have their own registers, execution pipelines and caches. On Fermi architecture, there are 64Kb shared memory plus L1 cache and 32,536 32-bit registers per SM. More precisely the programmer can decide what amounts of shared memory and L1 cache SM are to be used. The constraint is that the sum of both amounts should be less than or equal to 64Kb.

Threads are used to benefit from the large number of cores of a GPU. These threads are different from traditional threads for a CPU. In Chapter 2, some examples of GPU programming will explain the details of the GPU threads.

 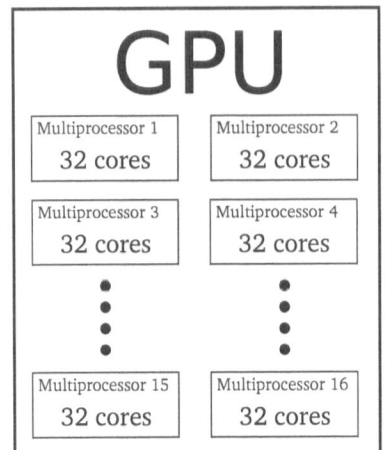

FIGURE 1.1. Comparison of number of cores in a CPU and in a GPU.

Threads are gathered into blocks of 32 threads, called "warps". These warps are important when designing an algorithm for GPU.

Another big difference between a CPU and a GPU is the latency of memory. In a CPU, everything is optimized to obtain a low latency architecture. This is possible through the use of cache memories. Moreover, nowadays CPUs carry out many performance optimizations such as speculative execution which roughly speaking consists of executing a small part of the code in advance even if later this work reveals itself to be useless. GPUs do not have low latency memory. In comparison GPUs have small cache memories; nevertheless the architecture of GPUs is optimized for throughput computation and it takes into account the memory latency.

Figure 1.2 illustrates the main difference of memory latency between a CPU and a GPU. In a CPU, tasks "ti" are executed one by one with a short memory latency to get the data to process. After some tasks, there is a context switch that allows the CPU to run concurrent applications and/or multi-threaded applications. Memory latencies are longer in a GPU. The principle to obtain a high throughput is to have many tasks to compute. Later we will see that these tasks are called threads with CUDA. With this principle, as soon as a task is finished the next one is ready to be executed while the wait for data for the previous task is overlapped by the computation of other tasks.

CPU: optimized for low latency

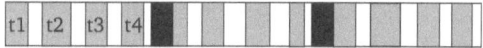

GPU: optimized for high throughput

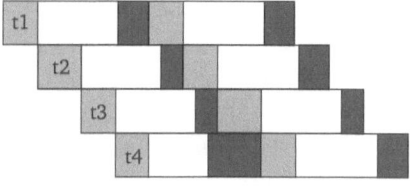

FIGURE 1.2. Comparison of low latency of a CPU and high throughput of a GPU.

1.5 Kinds of parallelism

Many kinds of parallelism are available according to the type of hardware. Roughly speaking, there are three classes of parallelism: instruction-level parallelism, data parallelism, and task parallelism.

Instruction-level parallelism consists in reordering some instructions in order to execute some of them in parallel without changing the result of the code. In modern CPUs, instruction pipelines allow the processor to execute instructions faster. With a pipeline a processor can execute multiple instructions simultaneously because the output of a task is the input of the next one.

Data parallelism consists in executing the same program with different data on different computing units. Of course, no dependency should exist among the data. For example, it is easy to parallelize loops without dependency using the data parallelism paradigm. This paradigm is linked with the Single Instructions Multiple Data (SIMD) architecture. This is the kind of parallelism provided by GPUs.

Task parallelism is the common parallelism achieved on clusters and grids and high performance architectures where different tasks are executed by different computing units.

1.6 CUDA multithreading

The data parallelism of CUDA is more precisely based on the Single Instruction Multiple Thread (SIMT) model, because a programmer accesses the cores by the intermediate of threads. In the CUDA model, all cores execute the same set of instructions but with different data. This model has similarities with the vector programming model proposed for vector machines through the 1970s and into the 90s, notably the various Cray platforms. On the CUDA architecture, the performance is led by the use of a huge number of threads (from thousands up to millions). The particularity of the model is that there is no context switching as in CPUs and each thread has its own registers. In practice, threads are executed by SM and gathered into groups of 32 threads, called warps. Each SM alternatively executes active warps and warps becoming temporarily inactive due to waiting of data (as shown in Figure 1.2).

The key to scalability in the CUDA model is the use of a huge number of threads. In practice, threads are gathered not only in warps but also in thread blocks. A thread block is executed by only one SM and it cannot migrate. The typical size of a thread block is a power of two (for example, 64, 128, 256, or 512).

In this case, without changing anything inside a CUDA code, it is possible to run code with a small CUDA device or the best performing Tesla CUDA cards. Blocks are executed in any order depending on the number of SMs available. So the programmer must conceive code having this issue in mind. This independence between thread blocks provides the scalability of CUDA codes.

A kernel is a function which contains a block of instructions that are executed by the threads of a GPU. When the problem considered is a two-dimensional or three-dimensional problem, it is possible to group thread blocks into a grid. In practice, the number of thread blocks and the size of thread blocks are given as parameters to each kernel. Figure 1.3 illustrates an example of a kernel composed of 8 thread blocks. Then this kernel is executed on a small device containing only 2 SMs. So in this case, blocks are executed 2 by 2 in any order. If the kernel is executed on a larger CUDA device containing 4 SMs, blocks are executed 4 by 4 simultaneously. The execution times should be approximately twice as fast in the latter case. Of course, that depends on other parameters that will be described later (in this chapter and other chapters).

Thread blocks provide a way to cooperate in the sense that threads of the same block cooperatively load and store blocks of memory they all use. Synchronizations of threads in the same block are possible (but not between threads of different blocks). Threads of the same block can also share results in order to compute a single result. In Chapter 2, some examples will explain that.

Presentation of the GPU architecture and of the CUDA environment   9

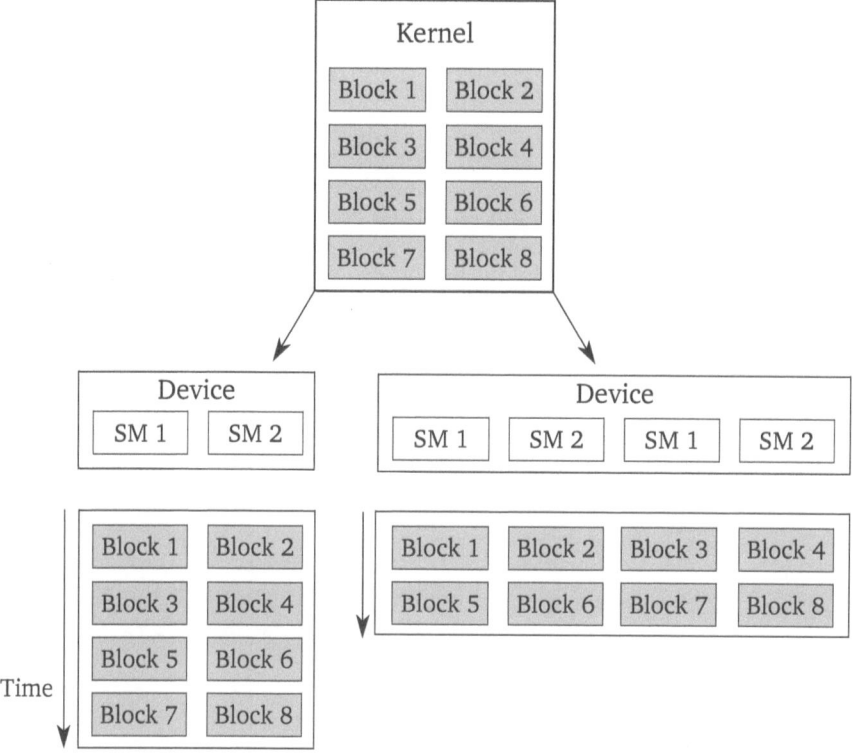

FIGURE 1.3. Scalability of GPU.

1.7 Memory hierarchy

The memory hierarchy of GPUs is different from that of CPUs. In practice, there are registers, local memory, shared memory, cache memory, and global memory.

As previously mentioned each thread can access its own registers. It is important to keep in mind that the number of registers per block is limited. On recent cards, this number is limited to 64Kb per SM. Access to registers is very fast, so it is a good idea to use them whenever possible.

Likewise each thread can access local memory which, in practice, is much slower than registers. Local memory is automatically used by the compiler when all the registers are occupied, so the best idea is to optimize the use of registers even if this involves reducing the number of threads per block.

Shared memory allows cooperation between threads of the same block. This kind of memory is fast but it needs to be manipulated manually and its size is limited. It is accessible during the execution of a kernel. So the idea is to fill the shared memory at the start of the kernel with global data that are used very frequently, then threads can access it for their computation. Threads can obviously change the content of this shared memory either with computation or by loading other data and they can store its content in the global memory. So shared memory can be seen as a cache memory, which is manually managed. This obviously requires effort from the programmer.

On recent cards, the programmer may decide what amount of cache memory and shared memory is attributed to a kernel. The cache memory is an L1 cache which is directly managed by the GPU. Sometimes, this cache provides very efficient result and sometimes the use of shared memory is a better solution.

Figure 1.4 illustrates the memory hierarchy of a GPU. Threads are represented on the top of the figure. They can have access to their own registers and their local memory. Threads of the same block can access the shared memory of that block. The cache memory is not represented here but it is local to a thread. Then each block can access the global memory of the GPU.

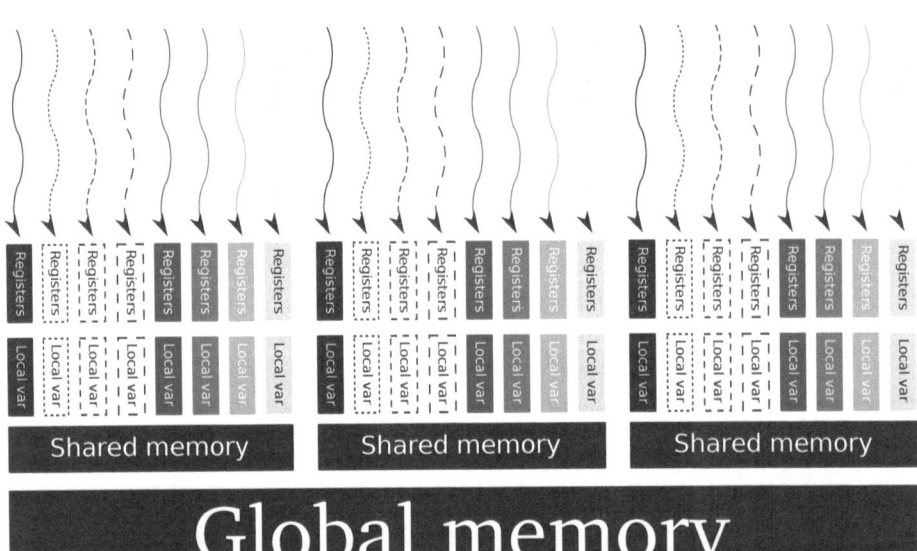

FIGURE 1.4. Memory hierarchy of a GPU.

1.8 Conclusion

In this chapter, a brief presentation of the video card, which has later been used to perform computation, has been given. The architecture of a GPU has been illustrated focusing on the particularity of GPUs in terms of parallelism, memory latency, and threads. In order to design an efficient algorithm for GPU, it is essential to keep all these parameters in mind.

Bibliography

[1] I. Buck, T. Foley, D. Horn, J. Sugerman, K. Fatahalian, M. Houston, and P. Hanrahan. Brook for GPUs: stream computing on graphics hardware. *ACM Transactions on Graphics*, 23(3):777–786, August 2004.

[2] P. Dua, R. Webera, P. Luszczeka, S. Tomova, G. Petersona, and J. Dongarra. From CUDA to OpenCL: Towards a performance-portable solution for multi-platform GPU programming. *Parallel Computing*, 38(8):391–407, 2012.

[3] NVIDIA Corporation. NVIDIA CUDA C Programming Guide, 2011. Version 4.0.

Chapter 2

Introduction to CUDA

Raphaël Couturier

Femto-ST Institute, University of Franche-Comte, France

2.1	Introduction	13
2.2	First example	13
2.3	Second example: using CUBLAS	16
2.4	Third example: matrix-matrix multiplication	18
2.5	Conclusion	21
	Bibliography	21

2.1 Introduction

In this chapter we give some simple examples of CUDA programming. The goal is not to provide an exhaustive presentation of all the functionalities of CUDA but rather to give some basic elements. Of course, readers who do not know CUDA are invited to read other books that are specialized on CUDA programming (for example, [2]).

2.2 First example

This first example is intented to show how to build a very simple program with CUDA. Its goal is to perform the sum of two arrays and put the result into a third array. A CUDA program consists in a C code which calls CUDA kernels that are executed on a GPU. This code is in Listing 2.1.

As GPUs have their own memory, the first step consists of allocating memory on the GPU. A call to `cudaMalloc` allocates memory on the GPU. The first parameter of this function is a pointer on a memory on the device, i.e., the GPU. The second parameter represents the size of the allocated variables; this size is expressed in bits.

Listing 2.1. simple example

```c
#include <stdlib.h>
#include <stdio.h>
#include <string.h>
#include <math.h>
#include <assert.h>
#include "cutil_inline.h"

const int nbThreadsPerBloc=256;

__global__
void addition(int size, int *d_C, int *d_A, int *d_B) {
    int tid = blockIdx.x * blockDim.x + threadIdx.x;
    if(tid<size) {
        d_C[tid]=d_A[tid]+d_B[tid];
    }
}

int main( int argc, char** argv)
{
    if(argc!=2) {
        printf("usage: ex1 nb_components\n");
        exit(0);
    }
    int size=atoi(argv[1]);
    int i;
    int *h_arrayA=(int*)malloc(size*sizeof(int));
    int *h_arrayB=(int*)malloc(size*sizeof(int));
    int *h_arrayC=(int*)malloc(size*sizeof(int));
    int *h_arrayCgpu=(int*)malloc(size*sizeof(int));
    int *d_arrayA, *d_arrayB, *d_arrayC;

    cudaMalloc((void**)&d_arrayA,size*sizeof(int));
    cudaMalloc((void**)&d_arrayB,size*sizeof(int));
    cudaMalloc((void**)&d_arrayC,size*sizeof(int));

    for(i=0;i<size;i++) {
        h_arrayA[i]=i;
        h_arrayB[i]=2*i;
    }

    unsigned int timer_cpu = 0;
    cutilCheckError(cutCreateTimer(&timer_cpu));
    cutilCheckError(cutStartTimer(timer_cpu));
    for(i=0;i<size;i++) {
        h_arrayC[i]=h_arrayA[i]+h_arrayB[i];
    }
    cutilCheckError(cutStopTimer(timer_cpu));
    printf("CPU processing time : %f (ms) \n", cutGetTimerValue(
        timer_cpu));
    cutDeleteTimer(timer_cpu);

    unsigned int timer_gpu = 0;
    cutilCheckError(cutCreateTimer(&timer_gpu));
    cutilCheckError(cutStartTimer(timer_gpu));
    cudaMemcpy(d_arrayA,h_arrayA, size * sizeof(int),
        cudaMemcpyHostToDevice);
    cudaMemcpy(d_arrayB,h_arrayB, size * sizeof(int),
        cudaMemcpyHostToDevice);

    int nbBlocs=(size+nbThreadsPerBloc-1)/nbThreadsPerBloc;
    addition<<<nbBlocs,nbThreadsPerBloc>>>(size,d_arrayC,d_arrayA,
        d_arrayB);
```

```
        cudaMemcpy(h_arrayCgpu,d_arrayC, size * sizeof(int),
            cudaMemcpyDeviceToHost);

        cutilCheckError(cutStopTimer(timer_gpu));
        printf("GPU processing time : %f (ms)\n", cutGetTimerValue(
            timer_gpu));
65      cutDeleteTimer(timer_gpu);

        for(i=0;i<size;i++) {
            assert(h_arrayC[i]==h_arrayCgpu[i]);
        }
70      cudaFree(d_arrayA);
        cudaFree(d_arrayB);
        cudaFree(d_arrayC);
        free(h_arrayA);
        free(h_arrayB);
75      free(h_arrayC);
        return 0;
}
```

In this example, we want to compare the execution time of the additions of two arrays in CPU and GPU. So for both these operations, a timer is created to measure the time. CUDA manipulates timers quite easily. The first step is to create the timer, then to start it, and at the end to stop it. For each of these operations a dedicated function is used.

In order to compute the same sum with a GPU, the first step consists of transferring the data from the CPU (considered as the host with CUDA) to the GPU (considered as the device with CUDA). A call to cudaMemcpy copies the content of an array allocated in the host to the device when the fourth parameter is set to cudaMemcpyHostToDevice. The first parameter of the function is the destination array, the second is the source array, and the third is the number of elements to copy (expressed in bytes).

Now the GPU contains the data needed to perform the addition. In sequential programming, such addition is achieved with a loop on all the elements. With a GPU, it is possible to perform the addition of all the elements of the two arrays in parallel (if the number of blocks and threads per blocks is sufficient). In Listing 2.1 at the beginning, a simple kernel, called addition is defined to compute in parallel the summation of the two arrays. With CUDA, a kernel starts with the keyword __global__ which indicates that this kernel can be called from the C code. The first instruction in this kernel is used to compute the variable tid which represents the thread index. This thread index is computed according to the values of the block index (called blockIdx in CUDA) and of the thread index (called threadIdx in CUDA). Blocks of threads and thread indexes can be decomposed into 1 dimension, 2 dimensions, or 3 dimensions. According to the dimension of manipulated data, the dimension of blocks of threads must be chosen carefully. In our example, only one dimension is used. Then using the notation .x, we can access the first dimension (.y and .z, respectively allow access to the second and third dimensions). The variable blockDim gives the size of each block.

2.3 Second example: using CUBLAS

The Basic Linear Algebra Subprograms (BLAS) allow programmers to use efficient routines for basic linear operations. Those routines are heavily used in many scientific applications and are optimized for vector operations, matrix-vector operations, and matrix-matrix operations [1]. Some of those operations seem to be easy to implement with CUDA; however, as soon as a reduction is needed, implementing an efficient reduction routine with CUDA is far from being simple. Roughly speaking, a reduction operation is an operation which combines all the elements of an array and extracts a number computed from all the elements. For example, a sum, a maximum, or a dot product are reduction operations.

In this second example, we have two vectors A and B. First of all, we want to compute the sum of both vectors and store the result in a vector C. Then we want to compute the scalar product between $1/C$ and $1/A$. This is just an example which has no direct interest except to show how to program it with CUDA.

Listing 2.2 shows this example with CUDA. The first kernel for the addition of two arrays is exactly the same as the one described in the previous example.

The kernel to compute the inverse of the elements of an array is very simple. For each thread index, the inverse of the array replaces the initial array.

In the main function, the beginning is very similar to the one in the previous example. First, the user is asked to define the number of elements. Then a call to `cublasCreate` initializes the CUBLAS library. It creates a handle. Then all the arrays are allocated in the host and the device, as in the previous example. Both arrays A and B are initialized. The CPU computation is performed and the time for this is measured. In order to compute the same result for the GPU, first of all, data from the CPU need to be copied into the memory of the GPU. For that, it is possible to use CUBLAS function `cublasSetVector`. This function has several arguments. More precisely, the first argument represents the number of elements to transfer, the second arguments is the size of each element, the third element represents the source of the array to transfer (in the GPU), the fourth is an offset between each element of the source (usually this value is set to 1), the fifth is the destination (in the GPU), and the last is an offset between each element of the destination. Then we call the kernel `addition` which computes the sum of all elements of arrays A and B. The `inverse` kernel is called twice, once to inverse elements of array C and once for A. Finally, we call the function `cublasDdot` which computes the dot product of two vectors. To use this routine, we must specify the handle initialized by CUDA, the number of elements to consider, then each vector is followed by the offset between every element. After the GPU computation, it is possible to check that both computations produce the same result.

Listing 2.2. simple example with CUBLAS

```
#include <stdlib.h>
#include <stdio.h>
#include <string.h>
#include <math.h>
#include <assert.h>
#include "cutil_inline.h"
#include <cublas_v2.h>

const int nbThreadsPerBloc=256;

__global__
void addition(int size, double *d_C, double *d_A, double *d_B) {
    int tid = blockIdx.x * blockDim.x + threadIdx.x;
    if(tid<size) {
        d_C[tid]=d_A[tid]+d_B[tid];
    }
}

__global__
void inverse(int size, double *d_x) {
    int tid = blockIdx.x * blockDim.x + threadIdx.x;
    if(tid<size) {
        d_x[tid]=1./d_x[tid];
    }
}


int main( int argc, char** argv)
{
    if(argc!=2) {
        printf("usage: ex2 nb_components\n");
        exit(0);
    }

    int size=atoi(argv[1]);
    cublasStatus_t stat;
    cublasHandle_t handle;
    stat=cublasCreate(&handle);
    int i;
    double *h_arrayA=(double*)malloc(size*sizeof(double));
    double *h_arrayB=(double*)malloc(size*sizeof(double));
    double *h_arrayC=(double*)malloc(size*sizeof(double));
    double *h_arrayCgpu=(double*)malloc(size*sizeof(double));
    double *d_arrayA, *d_arrayB, *d_arrayC;

    cudaMalloc((void**)&d_arrayA, size*sizeof(double));
    cudaMalloc((void**)&d_arrayB, size*sizeof(double));
    cudaMalloc((void**)&d_arrayC, size*sizeof(double));

    for(i=0;i<size;i++) {
        h_arrayA[i]=i+1;
        h_arrayB[i]=2*(i+1);
    }

    unsigned int timer_cpu = 0;
    cutilCheckError(cutCreateTimer(&timer_cpu));
    cutilCheckError(cutStartTimer(timer_cpu));
    double dot=0;
    for(i=0;i<size;i++) {
        h_arrayC[i]=h_arrayA[i]+h_arrayB[i];
        dot+=(1./h_arrayC[i])*(1./h_arrayA[i]);
```

```
         }
         cutilCheckError(cutStopTimer(timer_cpu));
         printf("CPU_processing_time_:_%f_(ms)_\n", cutGetTimerValue(
             timer_cpu));
65       cutDeleteTimer(timer_cpu);

         unsigned int timer_gpu = 0;
         cutilCheckError(cutCreateTimer(&timer_gpu));
         cutilCheckError(cutStartTimer(timer_gpu));
70       stat = cublasSetVector(size, sizeof(double), h_arrayA, 1, d_arrayA, 1);
         stat = cublasSetVector(size, sizeof(double), h_arrayB, 1, d_arrayB, 1);
         int nbBlocs=(size+nbThreadsPerBloc-1)/nbThreadsPerBloc;

         addition<<<nbBlocs,nbThreadsPerBloc>>>(size,d_arrayC,d_arrayA,
             d_arrayB);
75       inverse<<<nbBlocs,nbThreadsPerBloc>>>(size,d_arrayC);
         inverse<<<nbBlocs,nbThreadsPerBloc>>>(size,d_arrayA);
         double dot_gpu=0;
         stat = cublasDdot(handle, size, d_arrayC,1,d_arrayA,1,&dot_gpu);

80       cutilCheckError(cutStopTimer(timer_gpu));
         printf("GPU_processing_time_:_%f_(ms)_\n", cutGetTimerValue(
             timer_gpu));
         cutDeleteTimer(timer_gpu);
         printf("cpu_dot_%e_---_gpu_dot_%e\n",dot,dot_gpu);

85       cudaFree(d_arrayA);
         cudaFree(d_arrayB);
         cudaFree(d_arrayC);
         free(h_arrayA);
         free(h_arrayB);
90       free(h_arrayC);
         free(h_arrayCgpu);
         cublasDestroy(handle);
         return 0;
}
```

2.4 Third example: matrix-matrix multiplication

Matrix-matrix multiplication is an operation which is quite easy to parallelize with a GPU. If we consider that a matrix is represented using a two-dimensional array, $A[i][j]$ represents the element of the i row and of the j column. In many cases, it is easier to manipulate a one-dimentional (1D) array rather than a 2D array. With CUDA, even if it is possible to manipulate 2D arrays, in the following we present an example based on a 1D array. For the sake of simplicity, we consider we have a square matrix of size `size`. So with a 1D array, `A[i*size+j]` allows us to have access to the element of the i row and of the j column.

With sequential programming, the matrix-matrix multiplication is performed using three loops. We assume that A, B represent two square matrices and the result of the multiplication of $A \times B$ is C. The element `C[i*size+j]` is computed as follows:

$$C[size*i+j] = \sum_{k=0}^{size-1} A[size*i+k]*B[size*k+j]. \qquad (2.1)$$

In Listing 2.3, the CPU computation is performed using 3 loops, one for i, one for j, and one for k. In order to perform the same computation on a GPU, a naive solution consists of considering that the matrix C is split into 2-dimensional blocks. The size of each block must be chosen such that the number of threads per block is less than 1,024.

In Listing 2.3, we consider that a block contains 16 threads in each dimension, the variable width is used for that. The variable nbTh represents the number of threads per block. So, to compute the matrix-matrix product on a GPU, each block of threads is assigned to compute the result of the product of the elements of that block. The main part of the code is quite similar to the previous code. Arrays are allocated in the CPU and the GPU. Matrices A and B are randomly initialized. Then arrays are transferred to the GPU memory with call to cudaMemcpy. So the first step for each thread of a block is to compute the corresponding row and column. With a 2-dimensional decomposition, int i= blockIdx.y*blockDim.y+ threadIdx.y; allows us to compute the corresponding line and int j= blockIdx.x*blockDim.x+ threadIdx.x; the corresponding column. Then each thread has to compute the sum of the product of the row of A by the column of B. In order to use a register, the kernel matmul uses a variable called sum to compute the sum. Then the result is set into the matrix at the right place. The computation of CPU matrix-matrix multiplication is performed as described previously. A timer measures the time. In order to use 2-dimensional blocks, dim3 dimGrid(size/width,size/width); allows us to create size/width blocks in each dimension. Likewise, dim3 dimBlock(width,width); is used to create width thread in each dimension. After that, the kernel for the matrix multiplication is called. At the end of the listing, the matrix C computed by the GPU is transferred back into the CPU and we check that both matrices C computed by the CPU and the GPU are identical with a precision of 10^{-4}.

With $1,024 \times 1,024$ matrices, on a C2070M Tesla card, this code takes 37.68ms to perform the multiplication. With an Intel Xeon E31245 at 3.30GHz, it takes 2465ms without any parallelization (using only one core). Consequently the speed up between the CPU and GPU version is about 65 which is very good considering the difficulty of parallelizing this code.

Listing 2.3. simple matrix-matrix multiplication with cuda

```
#include <stdlib.h>
#include <stdio.h>
#include <string.h>
#include <math.h>
#include <assert.h>
#include "cutil_inline.h"
```

```
#include <cublas_v2.h>

const int width=16;
const int nbTh=width*width;

const int size=1024;
const     int sizeMat=size*size;

__global__
void matmul(float *d_A, float *d_B, float *d_C) {
   int i= blockIdx.y*blockDim.y+ threadIdx.y;
   int j= blockIdx.x*blockDim.x+ threadIdx.x;

   float sum=0;
   for(int k=0;k<size;k++) {
      sum+=d_A[i*size+k]*d_B[k*size+j];
   }
   d_C[i*size+j]=sum;
}

int main( int argc, char** argv)
{
   float *h_arrayA=(float*)malloc(sizeMat*sizeof(float));
   float *h_arrayB=(float*)malloc(sizeMat*sizeof(float));
   float *h_arrayC=(float*)malloc(sizeMat*sizeof(float));
   float *h_arrayCgpu=(float*)malloc(sizeMat*sizeof(float));

   float *d_arrayA, *d_arrayB, *d_arrayC;

   cudaMalloc((void**)&d_arrayA,sizeMat*sizeof(float));
   cudaMalloc((void**)&d_arrayB,sizeMat*sizeof(float));
   cudaMalloc((void**)&d_arrayC,sizeMat*sizeof(float));

   srand48(32);
   for(int i=0;i<sizeMat;i++) {
      h_arrayA[i]=drand48();
      h_arrayB[i]=drand48();
      h_arrayC[i]=0;
      h_arrayCgpu[i]=0;

   }
   cudaMemcpy(d_arrayA,h_arrayA, sizeMat * sizeof(float),
       cudaMemcpyHostToDevice);
   cudaMemcpy(d_arrayB,h_arrayB, sizeMat * sizeof(float),
       cudaMemcpyHostToDevice);
   cudaMemcpy(d_arrayC,h_arrayC, sizeMat * sizeof(float),
       cudaMemcpyHostToDevice);

   unsigned int timer_cpu = 0;
   cutilCheckError(cutCreateTimer(&timer_cpu));
   cutilCheckError(cutStartTimer(timer_cpu));
   int sum=0;
   for(int i=0;i<size;i++) {
      for(int j=0;j<size;j++) {
         for(int k=0;k<size;k++) {
            h_arrayC[size*i+j]+=h_arrayA[size*i+k]*h_arrayB[size*k+j];
         }
      }
   }
   cutilCheckError(cutStopTimer(timer_cpu));
   printf("CPU processing time : %f (ms) \n", cutGetTimerValue(
       timer_cpu));
   cutDeleteTimer(timer_cpu);

   unsigned int timer_gpu = 0;
   cutilCheckError(cutCreateTimer(&timer_gpu));
```

```
    cutilCheckError(cutStartTimer(timer_gpu));

    dim3 dimGrid(size/width,size/width);
    dim3 dimBlock(width,width);

    matmul<<<dimGrid,dimBlock>>>(d_arrayA,d_arrayB,d_arrayC);
    cudaThreadSynchronize();

    cutilCheckError(cutStopTimer(timer_gpu));
    printf("GPU_processing_time_:_%f_(ms)_\n", cutGetTimerValue(
        timer_gpu));
    cutDeleteTimer(timer_gpu);

    cudaMemcpy(h_arrayCgpu,d_arrayC, sizeMat * sizeof(float),
        cudaMemcpyDeviceToHost);

    for(int i=0;i<sizeMat;i++)
        if (fabs(h_arrayC[i]-h_arrayCgpu[i])>1e-4)
            printf("%f_%f\n",h_arrayC[i],h_arrayCgpu[i]);

    cudaFree(d_arrayA);
    cudaFree(d_arrayB);
    cudaFree(d_arrayC);
    free(h_arrayA);
    free(h_arrayB);
    free(h_arrayC);
    free(h_arrayCgpu);
    return 0;
}
```

2.5 Conclusion

In this chapter, three simple CUDA examples have been presented. As we cannot present all the possibilities of the CUDA programming, interested readers are invited to consult CUDA programming introduction books if some issues regarding the CUDA programming are not clear.

Bibliography

[1] J. Dongarra. Basic linear algebra subprograms technical (BLAST) forum standard (1). *International Journal of High Performance Computing Applications*, 16(1):1–111, 2002.

[2] J. Sanders and E. Kandrot. *CUDA by example: An Introduction To General-Purpose GPU Programming.* Addison-Wesley, Upper Saddle River, NJ, 2010.

Part II

Image processing

Chapter 3

Setting up the environment

Gilles Perrot
Femto-ST Institute, University of Franche-Comte, France

3.1 Data transfers, memory management 25
3.2 Performance measurements 28

Image processing using a GPU often means using it as a general purpose computing processor, which soon brings up the issue of data transfers, especially when kernel runtime is fast and/or when large data sets are processed. The truth is that, in certain cases, data transfers between GPU and CPU are slower than the actual computation on GPU. It remains that global runtime can still be faster than similar processes run on CPU. Therefore, to fully optimize global runtimes, it is important to pay attention to how memory transfers are done. This leads us to propose, in the following section, an overall code structure to be used with all our kernel examples.

Obviously, our code originally accepts various image dimensions and can process color images when an extrapolated definition of the median filter is choosen. However, so as to propose concise and more readable code, we will assume the following limitations: 16 bit-coded gray-level input images whose dimensions $H \times W$ are multiples of 512 pixels.

3.1   Data transfers, memory management

This section deals with the following issues:

1. Data transfer from CPU memory to GPU global memory: several GPU memory areas are available as destination memory but the 2D caching mechanism of texture memory, specifically designed for fetching neighboring pixels, is currently the fastest way to fetch gray-level pixel values inside a kernel computation. This has led us to choose **texture memory** as primary GPU memory area for input images.

2. Data fetching from GPU global memory to kernel local memory: as

said above, we use texture memory. Depending on which process is run, texture data is used either by direct fetching in kernel local memory or through a prefetching in shared memory.

3. Data outputting from kernels to GPU memory: there is actually no alternative to global memory, as kernels cannot directly write into texture memory and as copying from texture to CPU memory would not be faster than from simple global memory.

4. Data transfer from GPU global memory to CPU memory: it can be drastically accelerated by use of **pinned memory**, keeping in mind it has to be used sparingly.

Algorithm 1 summarizes all the above considerations and describes how data are handled in our examples. For more information on how to handle the different types of GPU memory, we suggest referring to the CUDA programmer's guide.

Algorithm 1: global memory management on CPU and GPU sides
---
1 allocate and populate CPU memory **h_in**;
2 allocate CPU pinned-memory **h_out**;
3 allocate GPU global memory **d_out**;
4 declare GPU texture reference **tex_img_in**;
5 allocate GPU array in global memory **array_img_in**;
6 bind GPU array **array_img_in** to texture **tex_img_in**;
7 copy data from **h_in** to **array_img_in**;
8 kernel<<< gridDim,blockDim>>>()   /* outputs to d_out */;
9 copy data from **d_out** to **h_out** ;

At debug stage, for simplicity's sake, we use the **cutil** library supplied by the NVIDIA software development kit (SDK). Thus, in order to easily implement our examples, we suggest readers download and install the latest NVIDIA-SDK (ours is SDK4.0), create a new directory *SDK-root-dir/C/src/fast_kernels* and adapt the generic *Makefile* that can be found in each subdirectory of *SDK-root-dir/C/src/*. Then, only two more files will be needed to have a fully operational environnement: *main.cu* and *fast_kernels.cu*. Listings 3.1, 3.2 and 3.3 implement all the above considerations minimally, while remaining functional.

The main file of Listing 3.1 is a simplified version of our actual main file. It has to be noted that functions `cutLoadPGMi` and `cutSavePGMi` of the **cutil** library operate only on unsigned integer data. As data is coded in short integer format for performance reasons, the use of these functions involves one data cast after loading and before saving. This may be overcome by use of a different library. Actually, our choice was to modify the above mentioned cutil functions.

Setting up the environment

Listing 3.2 gives a minimal kernel skeleton that will serve as the basis for all other kernels. Lines 5 and 6 determine the coordinates (i,j) of the pixel to be processed, each pixel being associated to one thread. The instruction in line 8 combines writing the output gray-level value into global memory and fetching the input gray-level value from 2D texture memory. The Makefile given in Listing 3.3 shows how to adapt examples given in SDK.

Listing 3.1. generic main.cu file used to launch CUDA kernels

```
#include <cuda_runtime.h>
#include <cutil_inline.h>
#include "fast_kernels.cu"

int main(int argc, char **argv){
    cudaSetDevice( 0 );          // select first GPU
    char filename[80] = "image.pgm" ;
    short *h_in , *h_out, *d_out ;
    int size , bsx=16, bsy=16 ;
    dim3 dimBlock, dimGrid ;
    cudaChannelFormatDesc channelD=cudaCreateChannelDesc<short>();
    cudaArray * array_img_in ;
    /*..................... load image and cast ...........*/
    unsigned int * h_img = NULL ;
    unsigned int *h_outui, H, L ;
    cutilCheckError( cutLoadPGMi(filename , &h_img, &L, &H));
    size = H * L * sizeof( short );
    h_in = new short[H*L] ;
    for (int k=0; k<H*L ; k++)
        h_in[k] = (short)h_img[k] ;
    /*..................... end of image load ............*/
    cudaHostAlloc((void**)&h_out , size , cudaHostAllocDefault) ;
    cudaMalloc((void**) &d_out, size);
    cudaMallocArray( &array_img_in , &channelD, W, H );
    cudaBindTextureToArray( tex_img_in , array_img_in , channelD);
    cudaMemcpyToArray( array_img_in , 0, 0, h_in , size ,
        cudaMemcpyHostToDevice) ;
    dimBlock = dim3(bsx,bsy,1) ;
    dimGrid = dim3( L/dimBlock.x, H/dimBlock.y, 1) ;

    kernel_test <<< dimGrid, dimBlock, 0>>>(d_out, W, H) ;

    cutilSafeCall( cudaMemcpy(h_out , d_out, size ,
        cudaMemcpyDeviceToHost) ) ;
    /*............ cast and  save output image (optional) */
    h_outui = new unsigned int[H*L] ;
    for (int k=0; k<H*L ; k++)
        h_outui[k] = (unsigned int)h_out[k] ;
    cutilCheckError( cutSavePGMi("image_out.pgm", h_outui, L, H) ) ;

    cudaFreeHost(h_out) ;
    cudaFree(d_out);
    return 0;
}
```

Listing 3.2. fast_kernels.cu file featuring one kernel skeleton

```
texture<short, 2, cudaReadModeElementType> tex_img_in ;

__global__ void kernel_ident( short *output, int w)
{
    int j = __mul24(blockIdx.x,blockDim.x) + threadIdx.x ;
    int i = __mul24( blockIdx.y, blockDim.y) + threadIdx.y ;
```

```
   output[ __mul24(i, w) + j ] = tex2D(tex_img_in, j, i);
10 }
```

Listing 3.3. generic makefile based on those provided by NVIDIA SDK
```
EXECUTABLE   := fast_median
CUFILES      := main.cu

include ../../common/common.mk

NVCCFLAGS    += -arch=sm_20
NVCCFLAGS    += --ptxas-options=-v
```

3.2 Performance measurements

As our goal is to design very fast implementations of basic image processing algorithms, we need to make quite accurate time-measurements, within the order of magnitude of 0.01 ms. Again, the easiest way of doing so is to use the helper functions of the **cutil** library. As usual, because the durations we are measuring are short and possibly subject to nonnegligible variations, a good practice is to measure multiple executions and report the mean runtime. All time results given in this chapter have been obtained through 1000 calls to each kernel.

Listing 3.4 shows how to use the dedicated **cutil** functions. Timer declaration and creation need to be performed only once while reset, start and stop functions can be used as often as necessary. Synchronization is mandatory before stopping the timer (Line 7), to avoid runtime measurement being biased.

Listing 3.4. Time measurement technique using cutil functions
```
unsigned int timer ;
cutilCheckError( cutCreateTimer(&timer) );
cutilCheckError( cutResetTimer(timer)   );
cutilCheckError( cutStartTimer(timer)   );
for (int ct=0; ct<1000 ; ct++)
   kernel_ident<<< dimGrid, dimBlock, 0>>>(d_out, W);
cudaThreadSynchronize() ;
cutilCheckError( cutStopTimer(timer)    );
cutilCheckError( cutStopTimer(timer)    );
printf("Mean runtime: %f ms.\n", cutGetTimerValue(timer)/1000);
```

In an attempt to provide relevant speedup values, we either implemented CPU versions of the algorithms studied or used the values found in existing literature. Still, the large number and diversity of hardware platforms and GPU cards makes it impossible to benchmark every possible combination and significant differences may occur between the speedups we report and those obtained with different devices. As a reference, our developing platform details as follows:

- CPU codes run on
    - **Xeon**: a recent and very efficient Quad Core Xeon E31245 at 3.3GHz-8GByte RAM running Linux kernel 3.2.
- GPU codes run on
    - **C2070**: NVIDIA Tesla C2070 hosted by a PC QuadCore Xeon E5620 at 2.4GHz-12GByte RAM, running Linux kernel 2.6.18
    - **GTX280**: NVIDIA GeForce GTX 280 hosted by a PC QuadCore Xeon X5482 at 3.20GHz-4GByte RAM, running Linux kernel 2.6.32

All kernels have also been tested with various image sizes from 512×512 to 4096×4096 pixels. This allows estimating runtime dependancy over image size.

Last, like many authors, we chose to use the pixel throughput value of each process in Mega Pixels per second (MP/s) as a performance indicator, including data transfers and kernel runtimes. In order to estimate the potential for improvement of each kernel, a reference throughput measurement, involving the identity kernel of Listing 3.2, was performed. As this kernel only fetches input values from texture memory and outputs them to global memory without doing any computation, it represents the smallest, thus fastest, possible process and is taken as the reference throughput value (100%). The same measurement was performed on CPU, with a maximum effective pixel throughput of 130 MP/s. On GPU, depending on grid parameters this measurement was 800 MP/s on GTX280 and 1300 MP/s on C2070.

Chapter 4

Implementing a fast median filter

Gilles Perrot

Femto-ST Institute, University of Franche-Comte, France

4.1	Introduction		31
4.2	Median filtering		32
	4.2.1	Basic principles	32
	4.2.2	A naive implementation	33
4.3	NVIDIA GPU tuning recipes		35
4.4	A 3×3 median filter: using registers		37
	4.4.1	The simplest way	37
	4.4.2	Further optimization	38
		4.4.2.1 Reducing register count	39
		4.4.2.2 More data output per thread	42
4.5	A 5×5 and more median filter		44
	4.5.1	A register-only 5×5 median filter	44
	4.5.2	Fast approximated $n \times n$ median filter	47
	Bibliography		51

4.1 Introduction

Median filtering is a well-known method used in a wide range of application frameworks as well as a standalone filter especially for *salt and pepper* denoising. It is able to greatly reduce the power of noise without blurring edges too much. That is actually why we originally focused on this filtering technique as a preprocessing stage when we were in the process of designing a GPU implementation of one region-based image segmentation algorithm [8].

First introduced by Tukey in [10], it has been widely studied since then, and many researchers have proposed efficient implementations of it, adapted to various hypotheses, architectures and processors. Originally, its main drawbacks were its compute complexity, its nonlinearity and its data-dependent runtime. Several researchers have addressed these issues and designed, for example, efficient histogram-based median filters with predictible runtimes [3, 11].

More recently, the advent of GPUs opened new perspectives in terms of image processing performance, and some researchers managed to take advantage of the new graphics capabilities: in that respect, we can cite the Branchless Vectorized Median (BVM) filter [2, 4] which allows very interesting runtimes on CUDA-enabled devices but, as far as we know, the fastest implementation to date is the histogram-based Parallel Ccdf-based Median Filter (PCMF) [9] where Ccdf means Complementary Cumulative Distribution Function.

Some of the following implementations feature very fast runtimes. They are targeted on NVIDIA Tesla GPU (Fermi architecture, compute capability 2.x) but may easily be adapted to other models, e.g., those of compute capability 1.3.

The fastest ones are based on one efficient parallel implementation of the BVM algorithm described in [5], improving its performance through fine tuning of its implementation as presented in [7] and detailed in the following sections.

4.2 Median filtering

4.2.1 Basic principles

Designing a 2D median filter basically consists of defining a square window $H(i,j)$ for each pixel $I(i,j)$ of the input image, containing $n \times n$ pixels and centered on $I(i,j)$. The output value $I'(i,j)$ is the median value of the gray-level values of the $n \times n$ pixels of $H(i,j)$. Figure 4.1 illustrates this principle with an example of a 5x5 median filter applied on pixel $I(5,6)$. The output value is the median value of the 25 values of the dark gray window centered on pixel $I(5,6)$. Figure 4.3 shows an example of a 512×512 pixel image, corrupted by a *salt and pepper* noise and the denoised versions, output respectively by a 3×3, a 5×5, and 2 iterations of a 3×3 median filter.

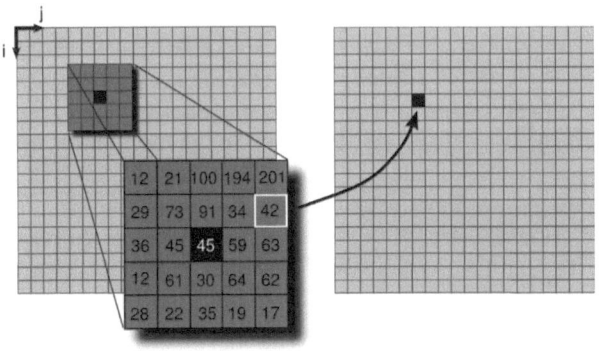

FIGURE 4.1. Example of 5x5 median filtering.

The generic filtering method is given by Algorithm 2. After the data transfer stage of the first line, which copies data from CPU memory to GPU texture memory, the actual median computing occurs, before the final transfer which copies data back to CPU memory at the last line. Obviously, one key issue is the selection method that identifies the median value. But, as shown in Figure 4.2, since two neighboring pixels share part of the values to be sorted, a second key issue is how to rule redundancy between consecutive positions of the running window $H(i,j)$. As mentioned earlier, the selection of the median value

---

Algorithm 2: generic n×n median filter
---
1 copy data from CPU to GPU texture memory;
2 **foreach** *pixel at position (x,y)* **do**   /* in parallel */
3 | Read gray-level values of the n×n neighborhood;
4 | Selects the median value among those n×n values;
5 | Outputs the new gray-level value ;
6 **end**
7 copy data from GPU global memory to CPU memory;

---

can be performed by more than one technique, using either histogram-based or sorting methods, each having its own benefits and drawbacks as will be discussed further down.

4.2.2 A naive implementation

As a reference, Listing 4.1 gives a simple, not to say simplistic, implementation of a CUDA kernel (kernel_medianR) achieving generic $n \times n$ histogram-based median filtering. Its runtime has a very low data dependency, but this implementation does not suit GPU architecture very well. Each pixel loads the whole of its $n \times n$ neighborhood, meaning that one pixel is loaded multiple times inside one single thread block, and even more time-consuming, the use of a local vector (histogram[]) considerably downgrades performance, as the compiler automatically stores such vectors in local memory (slow) .

Table 4.1 displays measured runtimes of kernel_medianR and pixel throughputs for each GPU version (C2070 and GTX480 targets) and for both CPU and GPU implementations. Usual window sizes of 3×3, 5×5, and 7×7 are shown. Though some specific applications require larger window sizes and dedicated algorithms, such small square window sizes are most widely used in general purpose image processing. GPU runtimes have been obtained with a grid of 64-thread blocks.

The first observation to make when analysing results of Table 4.1 is that, on CPU, window size has almost no influence on the effective pixel throughput. Since inner loops that fill the histogram vector contain very few fetching instructions (from 9 to 49, depending on the window size), it is not surprising to note their negligible impact compared to outer loops that fetch image pixels (from 256k to 16M instructions). One could be tempted to claim that CPU has no chance to win, which is not so obvious as it highly depends on what kind of algorithm is run and, above all, how it is implemented. To illustrate this, we can observe that, despite a maximum effective throughput potential that is almost five times higher, measured GTX280 throughput values sometimes prove slower than CPU values, as shown in Table 4.1.

On the GPU's side, we note high dependence on window size due to the redundancy induced by the multiple fetches of each pixel inside each block, becoming higher with the window size. Figure 4.2 shows for example that two 5 × 5 windows, centered on two neighbor pixels share at least 16 pixels. On a C2070 card, thanks to a more efficient caching mechanism, this effect is less. On GPUs, dependency on image size is low, and due to slightly more efficient data transfers when copying larger data amounts, pixel throughputs increases with image size. As an example, transferring a 4096×4096 pixel image (32 MBytes) is a bit faster than transferring a 512×512 pixel image (0.5 MBytes) 64 times.

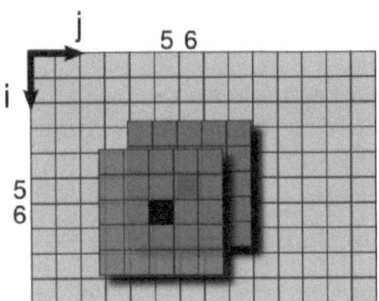

FIGURE 4.2. Illustration of window overlapping in 5x5 median filtering.

Listing 4.1. generic CUDA kernel achieving median filtering

```
__global__ void kernel_medianR ( short *output,
                                int i_dim, int j_dim, int r)
{
    // absolute coordinates of the center pixel
    int j = __mul24(blockIdx.x,blockDim.x) + threadIdx.x ;
    int i = __mul24(blockIdx.y,blockDim.y) + threadIdx.y ;

    short cpt, ic, jc ;
    short histogram[256] ; // 8 bit image
    // zeroing histogram data
    for (ic =0; ic<256; ic++) histogram[ ic ]=0 ;
    // histogram filling
    for(ic=i-r; ic<=i+r; ic++ )
        for(jc=j-r; jc<=j+r; jc++)
            histogram[ tex2D(tex_img_ins, jc, ic) ]++ ;
    // histogram parsing
    cpt = 0 ;
    for(ic=0; ic<256; ic++)
    {
        cpt += histogram[ ic ] ;
        // selection of the median value
```

```
            if ( cpt > ((2*r+1)*(2*r+1))>>1 ) break ;
       }
       output[ __mul24(i, j_dim) +j ] = ic ;
}
```

Processor		GTX280			C2070			CPU (Xeon)		
Performances→ sizes (pixels)↓		t (ms)	output (MP/s)	rate %	t (ms)	output (MP/s)	rate %	t (ms)	output (MP/s)	rate %
512^2	3×3	11.50	22	2.2	7.58	33	3.4	19.25	14	11
	5×5	19.10	14	1.3	8.60	30	3.0	18.49	14	11
	7×7	31.30	8	0.8	10.60	24	2.5	20.27	13	10
1024^2	3×3	44.50	23	2.3	29.60	34	3.5	75.49	14	11
	5×5	71.10	14	1.4	33.00	31	3.2	73.88	14	11
	7×7	114.50	9	0.9	39.10	26	2.7	77.40	13	10
2048^2	3×3	166.00	24	2.4	115.20	36	3.6	296.18	14	11
	5×5	261.00	16	1.5	128.20	32	3.3	294.55	14	11
	7×7	411.90	10	1.0	143.30	28	2.8	303.48	14	11
4096^2	3×3	523.80	31	3.0	435.00	38	3.9	1184.16	14	11
	5×5	654.10	25	2.4	460.20	36	3.7	1158.26	14	11
	7×7	951.30	17	1.7	509.60	32	3.3	1213.55	14	11

TABLE 4.1. Performance results of `kernel medianR`.

4.3 NVIDIA GPU tuning recipes

When designing GPU code, besides thinking of the actual data computing process, one must choose the memory type in which to store temporary data. Three types of GPU memory are available:

1. **Global memory, the most versatile:**
   Offers the largest storing space and global scope but is the slowest (400 to 800 clock cycles latency). **Texture memory** is physically included in it, but allows access through an efficient 2D caching mechanism.

2. **Registers, the fastest:**
   Allow access without latency, but only 63 registers are available per thread (thread scope), with a maximum of 32K per Streaming Multiprocessor (SM).

3. **Shared memory, a complex compromise:**
   All threads in one block can access 48 $KBytes$ of shared memory, which is faster than global memory (20 clock cycles latency) but slower than registers. However, bank conflicts can occur if two threads of a warp try to access data stored in one single memory bank. In such cases, the parallel process is serialized which may cause significant performance

(a) Airplane image, corrupted by salt and pepper noise of density 0.25

(b) Image denoised by a 3 × 3 median filter

(c) Image denoised by a 5 × 5 median filter

(d) Image denoised by 2 iterations of a 3 × 3 median filter

FIGURE 4.3. Example of median filtering, applied to salt and pepper noise reduction.

decrease. One easy way to avoid this is to ensure that two consecutive threads in one block always access 32-bit data at two consecutive addresses.

As observed earlier, designing a median filter GPU implementation using only global memory is fairly straightforward, but its performance remains quite low even if it is faster than CPU. To overcome this, the most frequent choice made in efficient implementations found in literature is to use shared memory. Such option implies prefetching data prior to doing the actual computations, a relevant choice, as each pixel of an image belongs to n^2 different neighborhoods. Thus, it can be expected that fetching each gray-level value from global memory only once should be more efficient than doing it each time it is required. One of the most efficient implementations using shared memory is presented in [4]. In the case of the generic kernel of Listing 4.1, using shared memory without further optimization would not bring valuable speedup because that would just move redundancy from texture to shared memory fetching and would generate bank conflicts. For information, we wrote such a version of the generic median kernel and our measurements showed a speedup of around 3% (as an example, 32 ms for 5×5 median on a 1024^2 pixel image, i.e., 33 MP/s).

As for registers, designing a generic median filter that would only use that type of memory seems difficult, due to the above mentioned 63 register-per-thread limitation. Yet, nothing forbids us to design fixed-size filters, each of them specific to one of the most popular window sizes. It might be worth the effort as dramatic increase in performance could be expected.

Another track to follow in order to improve performance of GPU implementations consists of hiding latencies generated by arithmetic instruction calls and memory accesses. Both can be partially hidden by introducing Instruction-Level Parallelism (ILP) and by increasing the data count outputted by each thread. Though such techniques may seem to break the NVIDIA occupancy paradigm, they can lead to dramatically higher data throughput values. The following sections illustrate these ideas and detail the design of the fastest CUDA median filter known to date.

4.4 A 3×3 median filter: using registers

Designing a median filter dedicated to the smallest possible square window size is a good challenge to start using registers. One first issue is that the exclusive use of registers forbids us to implement a naive histogram-based method. In a *8-bit gray-level pixel per thread* rule, each histogram requires one 256-element vector to store its values, i.e., more than four times the maximum register count allowed per thread (63). Considering that a 3×3 median filter involves only 9 pixel values per thread, it seem obvious they can be sorted within the 63-register limit.

4.4.1 The simplest way

In the case of a 3×3 median filter, the simplest solution consists of associating one register to each gray-level value, then sorting those 9 values and selecting the fifth one, i.e., the median value. For such a small amount of data to sort, a simple selection method is well indicated. As shown in Listing 4.2 (`kernel_Median3RegSort9()`), the constraint of only using registers forces the adoption of an unusual manner of coding. However, results are persuasive: runtimes are divided by around 120 on GTX280 and 80 on C2070, while only reduced by a 3.5 factor on CPU (CPU median3 bubble sort). The diagram of Figure 4.4 summarizes these first results for C2070, obtained with a block size of 256 threads, and Xeon CPU. We included the maximum effective pixel throughput in order to see the improvement potential of the different implementations. We also introduced throughput achieved by libJacket, a commercial implementation, as it was the fastest known implementation of a 3×3 median filter to date, as illustrated in [2]. One of the authors of libJacket kindly posted the CUDA code of its 3×3 median filter, which we inserted into our own coding structure. The algorithm itself is quite similar to ours, but running it in our own environment produced higher throughput values than those published in [2], not due to different hardware capabilities between our GTX280 and the GTX260 those authors used, but due to the way we perform memory transfers and our register-only method of storing temporary data.

Listing 4.2. 3×3 median filter kernel using one register per neighborhood pixel and bubble sort

```
__global__ void kernel_Median3RegSort9( short *output,
                                int i_dim, int j_dim)
{
    int j = __mul24(blockIdx.x,blockDim.x) + threadIdx.x ;
    int i = __mul24(blockIdx.y,blockDim.y) + threadIdx.y ;
    int a0, a1, a2, a3, a4, a5, a6, a7, a8 ;  // 1 register per pixel

    a0 = tex2D(tex_img_ins, j-1, i-1) ;   // fetching values
    a1 = tex2D(tex_img_ins, j  , i-1) ;
    a2 = tex2D(tex_img_ins, j+1, i-1) ;
    a3 = tex2D(tex_img_ins, j-1, i  ) ;
    a4 = tex2D(tex_img_ins, j  , i  ) ;
    a5 = tex2D(tex_img_ins, j+1, i  ) ;
    a6 = tex2D(tex_img_ins, j-1, i+1) ;
    a7 = tex2D(tex_img_ins, j  , i+1) ;
    a8 = tex2D(tex_img_ins, j+1, i+1) ;

    bubReg9(&a0,&a1,&a2,&a3,&a4,&a5,&a6,&a7,&a8) ;   // bubble sort

    output[ __mul24(i, j_dim) +j ] = a4 ;   // median at the middle
}
```

4.4.2 Further optimization

Running the above register-only 3×3 median filter through the NVIDIA CUDA profiler teaches us that the memory throughput achieved by the kernel remains quite low. To improve this, two methods can be used:

- increasing the number of concurrent threads, which can be achieved by reducing the number of registers used by each thread.

- having each thread process more data which can be achieved at thread level by processing and outputting the gray-level value of two pixels or more.

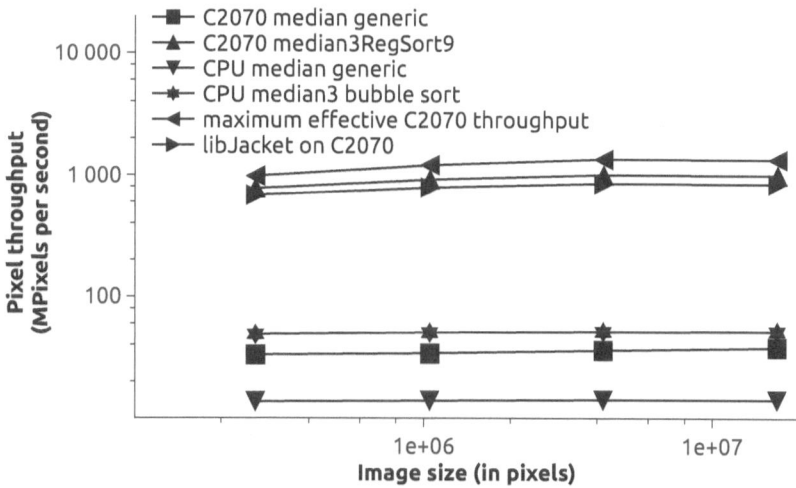

FIGURE 4.4. Comparison of pixel throughputs for CPU generic median, CPU 3×3 median register-only with bubble sort, GPU generic median, GPU 3×3 median register-only with bubble sort, and GPU libJacket. The GPU is the C2070 card and the CPU is the Xeon processor. The maximum effective C2070 throughput is also shown.

4.4.2.1 Reducing register count

Our current kernel (kernel_Median3RegSort9) uses one register per gray-level value, which amounts to 9 registers for the entire 3×3 window. This count can be reduced by use of an iterative sorting process called *forgetful selection*, where both *extrema* are eliminated at each sorting stage, until only 3 elements remain. The question is to learn the minimal register count k_{n^2} that allows the selection of the median amoung n^2 values. The answer can be evaluated considering that, when eliminating the maximum and the minimum values, one has to make sure not to eliminate the global median value. Such a situation is illustrated in Figure 4.5 for a 3 × 3 median filter. For better comprehension, the 9 elements of the 3×3 pixel window have been represented in a row.

We must remember that by definition, in the fully sorted vector, the median value will have the middle index, i.e., $\lfloor n^2/2 \rfloor$. Moreover, assuming that both *extrema* are eliminated from the first k elements and that the global median is one of them would mean that

- if the global median was the minimum among the k elements, then at least $k-1$ elements would have a higher index. Considering the above median definition, at least $k-1$ elements should also have a lower index in the entire vector.

- if the global median was the maximum among the k elements, then at least $k-1$ elements would have a lower index. Considering the above median definition, at least $k-1$ elements should also have a higher index in the entire vector.

Therefore, the number k of elements that are part of the first selection stage can be defined by the condition

$$n^2 - k \le \lfloor \frac{n^2}{2} \rfloor - 1$$

which leads to

$$k_{n^2} = \lceil \frac{n^2}{2} \rceil + 1$$

This rule can be applied to the first eliminating stage and remains true with the next ones as each stage suppresses exactly two values, one above and one below the median value. In our 3×3 pixel window example, the minimum register count becomes $k_9 = \lceil 9/2 \rceil + 1 = 6$. This iterative process is illustrated in Figure 4.6, where it achieves one entire 3×3 median selection, beginning with $k_9 = 6$ elements.

The *forgetful selection* method, used in [5], does not imply full sorting of values, but only selecting minimum and maximum values, which, at the price of a few iteration steps $(n^2 - k)$, reduces arithmetic complexity. Listing 4.3 details this process where forgetful selection is achieved by use of simple 2-value swapping function ($s()$, lines 1 to 5) that swaps input values if necessary, so as to achieve the first steps of an incomplete sorting network [1]. Moreover, whenever possible, in order to increase the ILP, successive calls to $s()$ are done with independant elements as arguments. This is illustrated by the macro definitions of lines 7 to 12 and by Figure 4.7 which details the first iteration of the 5×5 selection, starting with $k_{25} = 14$ elements.

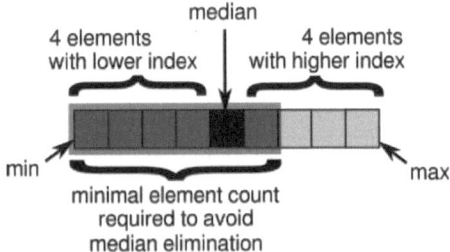

FIGURE 4.5. Forgetful selection with the minimal element register count. Illustration for 3×3 pixel window represented in a row and supposedly sorted.

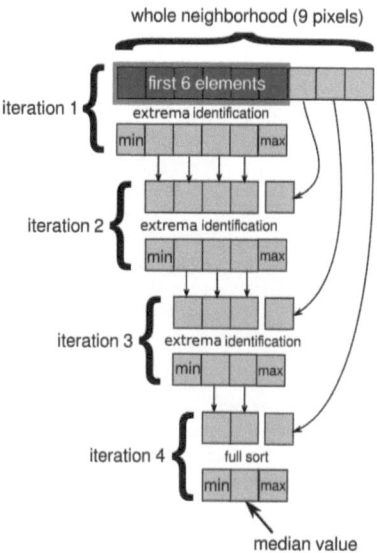

FIGURE 4.6. Determination of the median value by the *forgetful selection* process, applied to a 3 × 3 neighborhood window.

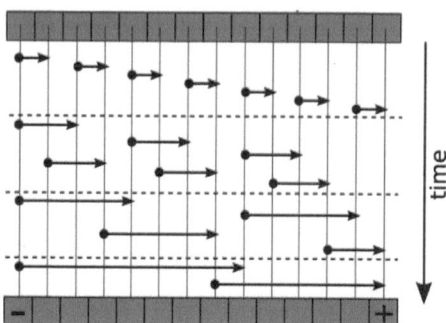

FIGURE 4.7. First iteration of the 5 × 5 selection process, with $k_{25} = 14$, which shows how Instruction Level Parallelism is maximized by the use of an incomplete sorting network. Arrows represent the result of the swapping function, with the lower value at the starting point and the higher value at the end point.

Listing 4.3. 3×3 median filter kernel using the minimum register count of 6 to find the median value by forgetful selection method. The optimal thread block size is 128 on GTX280 and 256 on C2070

```
__device__ inline void s(int* a, int* b)
{
  int tmp ;
  if (*a > *b) { tmp = *b; *b = *a; *a = tmp;}
}

#define min3(a, b, c)  s(a, b); s(a, c);
#define max3(a, b, c)  s(b, c); s(a, c);
#define minmax3(a, b, c)  max3(a, b, c); s(a, b);
#define minmax4(a, b, c, d)  s(a, b); s(c, d); s(a, c); s(b, d);
#define minmax5(a, b, c, d, e)  s(a, b); s(c, d); min3(a, c, e); max3(b
    , d, e);
#define minmax6(a, b, c, d, e, f)  s(a,d); s(b, e); s(c, f); min3(a, b,
    c); max3(d, e, f);

__global__ void kernel_medianForget1pix3( short *output, int i_dim,
    int j_dim)
{
  int j = __mul24(blockIdx.x,blockDim.x) + threadIdx.x ;
  int i = __mul24(blockIdx.y,blockDim.y) + threadIdx.y ;
  int a0, a1, a2, a3, a4, a5 ;

  a0 = tex2D(tex_img_ins, j-1, i-1) ; // first 6 values
  a1 = tex2D(tex_img_ins, j,   i-1) ;
  a2 = tex2D(tex_img_ins, j+1, i-1) ;
  a3 = tex2D(tex_img_ins, j-1, i) ;
  a4 = tex2D(tex_img_ins, j,   i) ;
  a5 = tex2D(tex_img_ins, j+1, i) ;

  minmax6(&a0, &a1, &a2, &a3, &a4, &a5);//min->a0 max->a5
  a5 = tex2D(tex_img_in, j-1, i+1) ;       //next value in a5
  minmax5(&a1, &a2, &a3, &a4, &a5) ;        //min->a1 max->a5
  a5 = tex2D(tex_img_ins, j, i+1) ;        //next value in a5
  minmax4(&a2, &a3, &a4, &a5) ;             //min->a1 max->a5
  a5 = tex2D(tex_img_ins, j+1, i+1) ;      //next value in a5
  minmax3(&a3, &a4, &a5) ;                  //min->a1 max->a5

  output[ __mul24(i, j_dim) +j ] = a4 ; //middle value
}
```

Our such modified kernel provides significantly improved runtimes: an average speedup of 16% is obtained, and pixel throughput reaches around 1000 MP/s on C2070.

4.4.2.2 More data output per thread

In the case of a kernel achieving an effective memory throughput value far from the GPU peak value, and if enough threads are run, another technique may help with hiding memory latency and thus leverage performance: making sure that each thread generates multiple pixel outputs.

Attentive readers could remark that it would increase the register count per thread, which can be compensated by dividing thread block size accordingly, thus keeping the same register count per block. Moreover, it is now possible to take advantage of window overlapping, first illustrated in Figure 4.2, and further detailed in Figure 4.8. As the selection is first processed on the first

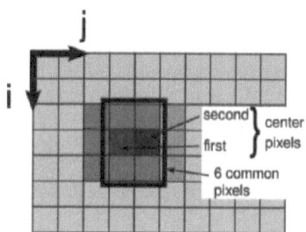

FIGURE 4.8. Illustration of how window overlapping is used to combine 2 pixel selections in a 3 × 3 median kernel.

6 gray-level values, i.e., exactly the number of pixels that overlap between the neighborhoods of two adjacent center pixels, 6 texture fetches, and one `minmax6` selection per thread can be saved. There again, some speedup can be expected through our modified kernel source code presented in Listing 4.4. One important difference from previous versions lies in the way pixel coordinates are computed from thread indexes. As each thread has to process two pixels, the number of threads in each block is divided by 2, while the grid size remains unchanged. Consequently, in our kernel code, each thread whose block-related coordinates are (tx, ty) will be in charge of processing pixels of block-related coordinates $(2tx, ty)$ and $(2tx + 1, ty)$; lines 5 and 6 implement this.

Listing 4.4. 3 × 3 median filter kernel processing 2 output pixel values per thread using combined forgetful selection

```
__global__ void kernel_median3_2pix( short *output,
                       int i_dim, int j_dim)
{
              // j base coordinate = 2*(thread index)
  int j= __mul24(__mul24(blockIdx.x,blockDim.x) + threadIdx.x,2) ;
  int i= __mul24(blockIdx.y,blockDim.y) + threadIdx.y ;
  int a0, a1, a2, a3, a4, a5 ;     // for left window
  int b0, b1, b2, b3, b4, b5 ;     // for right window

  a0 = tex2D(tex_img_ins, j   , i-1); // 6 common pixels
  a1 = tex2D(tex_img_ins, j+1, i-1);
  a2 = tex2D(tex_img_ins, j   , i  );
  a3 = tex2D(tex_img_ins, j+1, i  );
  a4 = tex2D(tex_img_ins, j   , i+1);
  a5 = tex2D(tex_img_ins, j+1, i+1);

  minmax6(&a0, &a1, &a2, &a3, &a4, &a5);// common minmax
  b0=a0; b1=a1; b2=a2; b3=a3; b4=a4; b5=a5;// separation

  a5 = tex2D(tex_img_ins, j-1, i);   //separate processes
  b5 = tex2D(tex_img_ins, j+2, i);
  minmax5(&a1, &a2, &a3, &a4, &a5);
  minmax5(&b1, &b2, &b3, &b4, &b5);
  a5 = tex2D(tex_img_ins, j-1, i-1);
  b5 = tex2D(tex_img_ins, j+2, i-1);
  minmax4(&a2, &a3, &a4, &a5);
  minmax4(&b2, &b3, &b4, &b5);
  a5 = tex2D(tex_img_ins, j-1, i+1);
  b5 = tex2D(tex_img_ins, j+2, i+1);
  minmax3(&a3, &a4, &a5);
```

```
    minmax3(&b3, &b4, &b5);

    output[ __mul24(i, j_dim) +j   ] = a4 ;      //2 outputs
    output[ __mul24(i, j_dim) +j+1 ] = b4 ;
35 }

//grid dimensions to be set in main.cu file
 dimGrid = dim3( (W/dimBlock.x)/2, H/dimBlock.y, 1 ) ;
```

Running this 3 × 3 kernel saves another 10% runtime, as shown in Figure 4.9 and provides the best peak pixel throughput value known so far on the C2070: 1155 MP/s which is 86% of the maximum effective throughput.

4.5 A 5×5 and more median filter

Considering the maximum register count allowed per thread (63) and trying to push this technique to its limit potentially allows designing up to 9×9 median filters. Such maximum would actually use $k_{81} = \lceil 81/2 \rceil + 1 = 42$ registers per thread plus 9, used by the compiler to complete arithmetic operations, and 9 more when outputting 2 pixels per thread. This leads to a total register count of 60, which would limit the number of concurrent threads per block. As for larger window sizes, one option could be using shared memory. The next two sections will first detail the particular case of the 5×5 median through register-only method and eventually a generic kernel for larger window sizes.

4.5.1 A register-only 5×5 median filter

The minimum register count required to apply the forgetful selection method to a 5×5 median filter is $k_{25} = \lceil 25/2 \rceil + 1 = 14$. Moreover, two adjacent overlapping windows share 20 pixels ($n^2 - one\_column$) so that, when processing 2 pixels simultaneously, a count of 7 common selection stages can be carried out from the first selection stage with 14 common values to the processing of the last common value. This allows limiting the register count to 22 per thread. Figure 4.10 describes the distribution of overlapping pixels, implemented in Listing 4.5: common selection stages take place from line 25 to line 37, while the remaining separate selection stages occur between lines 45 and 62 after the separation of line 40.

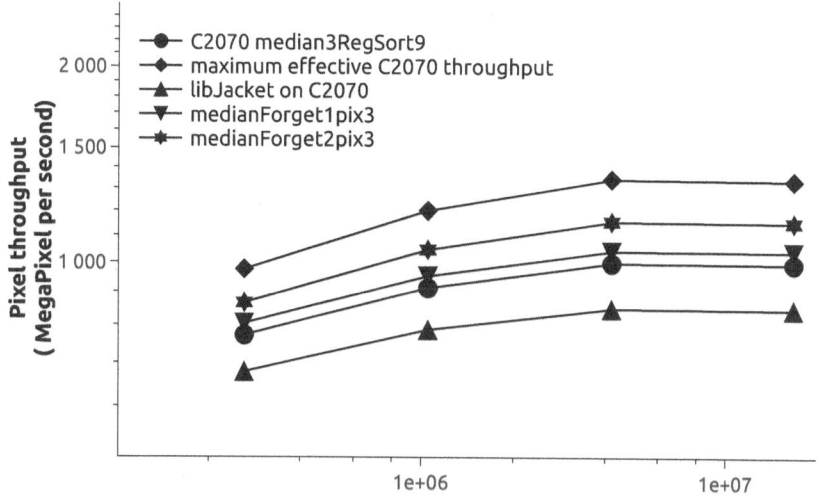

FIGURE 4.9. Comparison of pixel throughput on GPU C2070 for the different 3×3 median kernels.

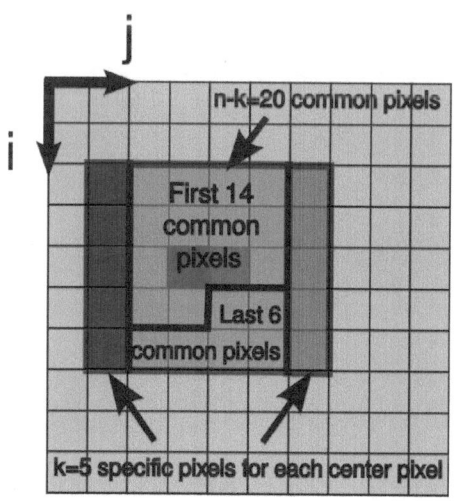

FIGURE 4.10. Reducing register count in a 5×5 register-only median kernel outputting 2 pixels simultaneously. The first 7 forgetful selection stages are common to both processed center pixels. Only the last 5 selections have to be done separately.

Listing 4.5. kernel 5×5 median filter processing 2 output pixel values per thread by a combined forgetfull selection

```
__global__ void kernel_median5_2pix( short *output,
                        int i_dim, int j_dim)
{
  int j= __mul24(__mul24(blockIdx.x,blockDim.x) + threadIdx.x,2);
  int i= __mul24(blockIdx.y,blockDim.y) + threadIdx.y;
  int a0,a1,a2,a3,a4,a5,a6,a7,a8,a9,a10,a11,a12,a13;//left window
  int b7,b8,b9,b10,b11,b12,b13 ;                    //right window
  //first 14 common pixels
  a0  = tex2D(tex_img_ins, j-1, i-2) ;   // first line
  a1  = tex2D(tex_img_ins, j  , i-2) ;
  a2  = tex2D(tex_img_ins, j+1, i-2) ;
  a3  = tex2D(tex_img_ins, j+2, i-2) ;
  a4  = tex2D(tex_img_ins, j-1, i-1) ;   //seconde line
  a5  = tex2D(tex_img_ins, j  , i-1) ;
  a6  = tex2D(tex_img_ins, j+1, i-1) ;
  a7  = tex2D(tex_img_ins, j+2, i-1) ;
  a8  = tex2D(tex_img_ins, j-1, i) ;     // third line
  a9  = tex2D(tex_img_ins, j  , i) ;
  a10 = tex2D(tex_img_ins, j+1, i) ;
  a11 = tex2D(tex_img_ins, j+2, i) ;     // first 2 of fourth line
  a12 = tex2D(tex_img_ins, j-1, i+1) ;
  a13 = tex2D(tex_img_ins, j  , i+1) ;

  //common selection
  minmax14(&a0,&a1,&a2,&a3,&a4,&a5,&a6,&a7,&a8,&a9,&a10,&a11,&a12,&a13
          );
  a13 = tex2D(tex_img_ins, j+1, i+1);
  minmax13(&a1,&a2,&a3,&a4,&a5,&a6,&a7,&a8,&a9,&a10,&a11,&a12,&a13);
  a13 = tex2D(tex_img_ins, j+2, i+1);
  minmax12(&a2,&a3,&a4,&a5,&a6,&a7,&a8,&a9,&a10,&a11,&a12,&a13);
  a13 = tex2D(tex_img_ins, j-1, i+2);
  minmax11(&a3,&a4,&a5,&a6,&a7,&a8,&a9,&a10,&a11,&a12,&a13);
  a13 = tex2D(tex_img_ins, j  , i+2);
  minmax10(&a4,&a5,&a6,&a7,&a8,&a9,&a10,&a11,&a12,&a13);
  a13 = tex2D(tex_img_ins, j+1, i+2);
  minmax9(&a5,&a6,&a7,&a8,&a9,&a10,&a11,&a12,&a13);
  a13 = tex2D(tex_img_ins, j+2, i+2);
  minmax8(&a6,&a7,&a8,&a9,&a10,&a11,&a12,&a13);

  // separation
  b7=a7; b8=a8; b9=a9; b10=a10; b11=a11; b12=a12; b13=a13;

  // separate selections: 5 remaining pixels in both windows
  a13 = tex2D(tex_img_ins, j-2, i-2);
  b13 = tex2D(tex_img_ins, j+3, i-2);
  minmax7(&a7,&a8,&a9,&a10,&a11,&a12,&a13);
  minmax7(&b7,&b8,&b9,&b10,&b11,&b12,&b13);
  a13 = tex2D(tex_img_ins, j-2, i-1);
  b13 = tex2D(tex_img_ins, j+3, i-1);
  minmax6(&a8,&a9,&a10,&a11,&a12,&a13);
  minmax6(&b8,&b9,&b10,&b11,&b12,&b13);
  a13 = tex2D(tex_img_ins, j-2, i  );
  b13 = tex2D(tex_img_ins, j+3, i  );
  minmax5(&a9,&a10,&a11,&a12,&a13);
  minmax5(&b9,&b10,&b11,&b12,&b13);
  a13 = tex2D(tex_img_ins, j-2, i+1);
  b13 = tex2D(tex_img_ins, j+3, i+1);
  minmax4(&a10,&a11,&a12,&a13);
  minmax4(&b10,&b11,&b12,&b13);
  a13 = tex2D(tex_img_ins, j-2, i+2);
  b13 = tex2D(tex_img_ins, j+3, i+2);
  minmax3(&a11,&a12,&a13);
  minmax3(&b11,&b12,&b13);
```

```
output[ __mul24(i, j_dim) +j   ] = a12 ; //middle values
output[ __mul24(i, j_dim) +j+1 ] = b12 ;
}
```

Timing results follow the same variations with image size as in previously presented kernels. That is why Table 4.2 shows only throughput values obtained for C2070 card and 4096×4096 pixel image.

Implementation	registers only 1 pix/thread	registers only 2 pix/thread	libJacket (interpolated)	shared mem
Throughput (MP/s)	551	738	152	540

TABLE 4.2. Performance of various 5×5 median kernel implementations, applied on 4096×4096 pixel image with C2070 GPU card.

4.5.2 Fast approximated $n \times n$ median filter

Large window median filters are less widespread but are used in more specific fields, such as digital microscopy where, for example, background estimation of images is achieved through 64×64 or 128×128 median filters [12]. In such cases, a possible technique is to split median selection into two separate 1D stages: one in the vertical direction and the other in the horizontal direction. Image processing specialists may object that this method does not select the actual median value. This is true but, in the case of large window sizes and *real-life* images, the value selected in this manner is statistically near the actual median value and often represents an acceptable approximation. Such a filter is sometimes called a *smoother*.

As explained earlier in this section, the use of large window median filters rules out register-only implementation, which favors the use of shared memory. The 1D operation almost completely avoids bank conflicts in shared memory accesses. Furthermore, the above-described forgetful selection method cannot be used anymore, as too many registers would be required. Instead, the Torben Morgensen sorting algorithm is used, as its required register count is both low and constant, and avoids the use of a local vector, unlike histogram-based methods.

Listing 4.6 presents a kernel code that implements the above considerations and achieves a 1D vertical $n \times 1$ median filter. The shared memory vector is declared as `extern` (Line 16) as its size is determined at runtime and passed to the kernel call as an argument. Lines 20 to 29 perform data prefetching, including the $2n$-row halo (n at the bottom and n at the top of each block). Then one synchronization barrier is mandatory (line 31) to ensure that all needed data is ready prior to its use by the different threads. Torben Morgensen sorting takes place between lines 37 and 66 and eventually, the transposed output value is stored in global memory at line 69. Outputting the transposed image in global memory saves time and allows the reuse of the same kernel to achieve the second step, e.g 1D horizontal $n \times 1$ median filtering. It has to be noticed that this smoother, unlike the technique we proposed for fixed-size median filters, cannot be considered as a state-of-the-art technique as, for example, the one presented in [6]. However, it may be considered as a good, easy to use and efficient alternative as confirmed by the results presented in Table 4.3. Pixel throughput values achieved by our kernel, though not constant with window size, remain very competitive if window size is kept under 120×120 pixels, especially when outputting 2 pixels per thread (in [6], pixel throughput is around 7MP/s). Figure 4.11 shows an example of a 512×512 pixel image, corrupted by a *salt and pepper* noise, and the denoised versions, outputted respectively by a 3×3, a 5×5, and a 55×55 separable smoother.

Window edge size (in pixels)	41	81	111	121
Throughput (MP/s)	54	27	20	18

TABLE 4.3. Measured performance of one generic pseudo-separable median kernel applied to 4096×4096 pixel image with various window sizes.

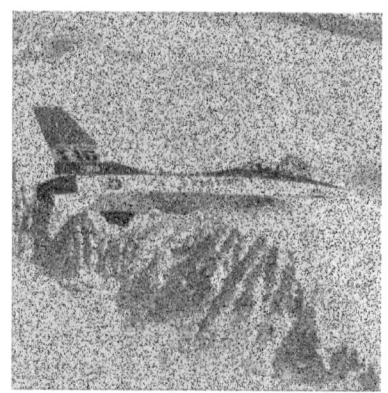

(a) Airplane image, corrupted with by salt and pepper noise of density 0.25

(b) Image denoised by a 3 × 3 separable smoother

(c) Image denoised by a 5 × 5 separable smoother

(d) Image background estimation by a 55 × 55 separable smoother

FIGURE 4.11. Example of separable median filtering (smoother), applied to salt and pepper noise reduction.

Listing 4.6. generic pseudo median kernel.

```
__global__ void kernel_medianV_sh( short *output, int i_dim, int j_dim
    , int r)
{
    int idc, val, min, max, inf, equal, sup, mxinf, minsup, estim ;
    //coordinates in the block
    int ib = threadIdx.y ;
    int jb = threadIdx.x ;
    int idx_h = __mul24(ib+r,blockDim.x) +jb; // base pixel index
    int offset = __mul24(blockDim.x,r) ;

    int j = __mul24(blockIdx.x,blockDim.x) + jb ;
    int i = __mul24(blockIdx.y,blockDim.y) + ib ;

    //       DATA PREFETCHING INTO SHARED MEM
    extern __shared__ int buff[] ;
    buff[ idx_h ] = tex2D(tex_img_ins , j, i) ;

    if (ib < r)
    {
        buff[ idx_h - offset ]    = tex2D(tex_img_ins, j, i-r) ;
    } else
    if (ib >= (blockDim.y-r))
    {
        buff[ idx_h + offset ]    = tex2D(tex_img_ins, j, i+r) ;
    }

    __syncthreads() ;

    //       TORBEN MOGENSEN SORTING
    min = max = buff[ ib*blockDim.x +jb] ;

    for (idc= 0 ; idc< 2*r+1 ; idc++ )
    {
        val = buff[ __mul24(ib+idc, blockDim.x) +jb ] ;
        if ( val < min ) min = val ;
        if ( val > max ) max = val ;
    }

    while (1)
    {
        estim = (min+max)/2 ;
        inf = sup = equal = 0  ;
        mxinf = min ;
        minsup= max ;
        for (idc =0; idc< 2*r+1 ; idc++)
        {
            val = buff[ __mul24(ib+idc, blockDim.x) +jb ] ;
            if( val < estim  )
            {
                inf++;
                if( val > mxinf) mxinf = val ;
            } else if ( val > estim)
            {
                sup++;
                if( val < minsup) minsup = val ;
            } else equal++ ;
        }
        if ( (inf <= (r+1))&&(sup <=(r+1)) ) break ;
        else if (inf>sup) max = mxinf ;
        else min = minsup ;
    }

    if ( inf >= r+1 ) val = mxinf ;
```

```
65     else if (inf+equal >= r+1) val = estim ;
       else val = minsup ;

       output[ __mul24(j, i_dim) +i ] = val ;
   }
```

Bibliography

[1] K. E. Batcher. Sorting networks and their applications. In *Proceedings of the April 30–May 2, 1968, Spring Joint Computer Conference*, AFIPS '68 (Spring), pages 307–314, New York, NY, USA, 1968. ACM.

[2] W. Chen, M. Beister, Y. Kyriakou, and M. Kachelries. High performance median filtering using commodity graphics hardware. In *Nuclear Science Symposium Conference Record (NSS/MIC), 2009 IEEE*, pages 4142–4147, November 2009.

[3] T. S. Huang. *Two-Dimensional Digital Signal Processing II: Transforms and Median Filters*. Springer-Verlag New York, Inc., Secaucus, NJ, USA, 1981.

[4] M. Kachelriess. Branchless vectorized median filtering. In *Nuclear Science Symposium Conference Record (NSS/MIC), 2009 IEEE*, pages 4099 – 4105, November 2009.

[5] M. McGuire. A fast, small-radius gpu median filter. In *ShaderX6*, February 2008.

[6] S. Perreault and P. Hebert. Median filtering in constant time. *IEEE Transactions on Image Processing*, 16(9):2389–2394, Sept. 2007.

[7] G. Perrot, S. Domas, and R. Couturier. Fine-tuned high-speed implementation of a gpu-based median filter. *Journal of Signal Processing Systems*, pages 1–6, 2013.

[8] G. Perrot, S. Domas, R. Couturier, and N. Bertaux. Gpu implementation of a region based algorithm for large images segmentation. In *2011 IEEE 11th International Conference on Computer and Information Technology (CIT)*, pages 291–298, Sept. 2011.

[9] R. M. Sánchez and P. A. Rodríguez. Highly parallelable bidimensional median filter for modern parallel programming models. *Journal of Signal Processing Systems*, 71(3):221–235, 2012.

[10] J. W. Tukey. *Exploratory Data Analysis*. Addison-Wesley, 1977.

[11] B Weiss. Fast median and bilateral filtering. In *ACM SIGGRAPH 2006 Papers*, SIGGRAPH '06, pages 519–526, New York, NY, USA, 2006. ACM.

[12] Y. Wu, M. Eghbali, J. Ou, R. Lu, L. Toro, and E. Stefani. Quantitative determination of spatial protein-protein correlations in fluorescence confocal microscopy. *Biophysical Journal*, 98(3):493–504, Feb. 2010.

Chapter 5

Implementing an efficient convolution operation on GPU

Gilles Perrot

Femto-ST Institute, University of Franche-Comte, France

5.1	Overview		53
5.2	Definition		54
5.3	Implementation		54
	5.3.1	First test implementation	55
	5.3.2	Using parameterizable masks	58
	5.3.3	Increasing the number of pixels processed by each thread	59
	5.3.4	Using shared memory to store prefetched data	62
5.4	Separable convolution		65
5.5	Conclusion		69
	Bibliography		69

5.1 Overview

In this chapter, after dealing with GPU median filter implementations, we propose to explore how convolutions can be implemented on modern GPUs. Widely used in digital image processing filters, the *convolution operation* basically consists of taking the sum of products of elements from two 2D functions, letting one of the two functions move over every element of the other, producing a third function that is typically viewed as a modified version of one of the original functions. To begin with, we shall examine nonseparable or generic convolutions, before addressing the matter of separable convolutions. We shall refer to I as an $H \times L$ pixel gray-level image and to $I(x,y)$ as the gray-level value of each pixel of coordinates (x,y).

5.2 Definition

Within a digital image I, the convolution operation is performed between image I and convolution mask h (To avoid confusion with other GPU functions referred to as kernels, we shall use *convolution mask* instead of *convolution kernel*) is defined by

$$I'(x,y) = (I * h) = \sum_{(i<H)} \sum_{(j<L)} I(x-j, y-j) h(j,i) \quad (5.1)$$

While processing an image, function h is often bounded by a square window of size $k = 2r + 1$, i.e., an uneven number, to ensure there is a center. We shall also point out that, as stated earlier, the square shape is not a limiting factor to the process, as any shape can be inscribed into a square. In the case of a more complex shape, the remaining space is filled by null values (padding).

5.3 Implementation

The basic principle of computing a convolution between one I picture and one h convolution mask defined on domain Ω is given by Algorithm 3 and illustrated by Figure 5.1, which mainly shows how gray-level values of the center pixel's neighborhood are combined with the convolution mask values to compute the output value. For more readability, only part of the connecting lines are shown.

Algorithm 3: generic convolution

1 **foreach** *pixel at position* (x,y) **do**
2   Read all gray-level values $I(x,y)$ in the neighborhood;
3   Compute the weighted sum $I_\Omega = \sum_{(j,i) \in \Omega} I(x-j, y-j) h(j,i)$;
4   Normalize $I'(x,y)$ value;
5   Output the new gray-level value
6 **end**

The gray-level value of each pixel of output image I' is the weighted sum of pixels included in the neighborhood defined by Ω around the corresponding pixel in the input image. It has to be noted that, in case the sum S of all coefficients in the mask is not 1, the original brightness of the image will be altered and a normalization stage has to take place, as, for example, in the case of an 8-bit coded image:

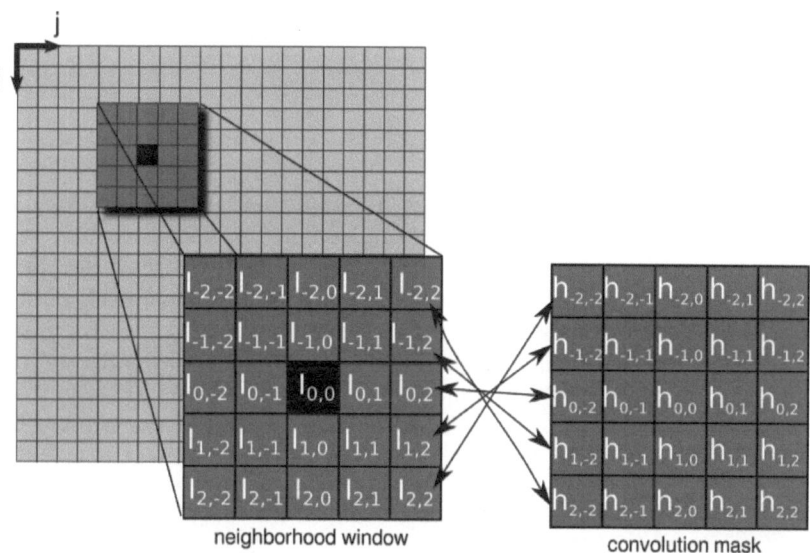

FIGURE 5.1. Principle of a generic convolution implementation. The center pixel is represented with a black background and the pixels of its neighborhood are denoted $I_{p,q}$ where (p,q) is the relative position of the neighbor pixel. Elements $h_{t,u}$ are the values of the convolution mask.

1. if $S > 0$ then $I' = I_\Omega / S$
2. if $S = 0$ then $I' = I_\Omega + 128$
3. if $S < 0$ then $I' = I_\Omega + 255$

In case one, normalizing means performing a division operation for each pixel, which will be quite time-costly when performed on a GPU. A simple workaround is to normalize mask values before using them in GPU kernels.

5.3.1 First test implementation

This first implementation consists of a rather naive application to convolutions of the techniques applied to median filters in the previous chapter, as a reminder: texture memory used with incoming data, pinned memory with output data, optimized use of registers while processing data and multiple output per thread. One significant difference lies in the fact that the median filter uses only one parameter, the size of the window mask, which can be hard-coded, while a convolution mask requires referring to several parameters; hard-coding the elements of the mask would lead to severe lack of flexibility (one function per filter, no external settings) so we will just use it as a starting point in our approach.

Let us assume that we are planning to implement the convolution defined by the following 3 × 3 mask (low-pass filter or averaging filter):

$$h = \frac{1}{9} \begin{bmatrix} 1 & 1 & 1 \\ 1 & 1 & 1 \\ 1 & 1 & 1 \end{bmatrix}$$

The kernel code presented in Listing 5.1 implements the convolution operation and applies all above optimizations except, for clarity reasons, multiple outputs per thread. In the particular case of a generic convolution, it is important to note how mask coefficients are applied to image pixels in order to fit the definition of equation 5.1: if the coordinates of the center pixel had been set to (0,0), then the gray-level value of pixel of coordinates (i, j) would have been multiplied by the element $(-i, -j)$ of the mask, which, transposed in our kernel code, leads to multiplying the pth pixel of the window by the $(n-p)^{th}$ element of the convolution mask.

Listing 5.1. generic CUDA kernel achieving a convolution operation with hard-coded mask values

```
__global__ void kernel_convoGene3Reg8( unsigned char *output, int
    j_dim)
{
  float outval0=0.0 ;
  float n0,n1,n2,n3,n4,n5,n6,n7,n8 ;
  // convolution mask values
  n0 = (1.0/9) ;
  n1 = (1.0/9) ;
  n2 = (1.0/9) ;
  n3 = (1.0/9) ;
  n4 = (1.0/9) ;
  n5 = (1.0/9) ;
  n6 = (1.0/9) ;
  n7 = (1.0/9) ;
  n8 = (1.0/9) ;

  // absolute base point coordinates
  int j = __mul24(blockIdx.x, blockDim.x) + threadIdx.x ;
  int i = __mul24(blockIdx.y, blockDim.y) + threadIdx.y ;
  // weighted sum
  outval0 = n8*tex2D(tex_img_inc , j-1, i-1 )
          + n7*tex2D(tex_img_inc , j  , i-1 )
          + n6*tex2D(tex_img_inc , j+1, i-1 )
          + n5*tex2D(tex_img_inc , j-1, i   )
          + n4*tex2D(tex_img_inc , j  , i   )
          + n3*tex2D(tex_img_inc , j+1, i   )
          + n2*tex2D(tex_img_inc , j-1, i+1 )
          + n1*tex2D(tex_img_inc , j  , i+1 )
          + n0*tex2D(tex_img_inc , j+1, i+1 ) ;

  output[ __mul24(i, j_dim) + j ] = (unsigned char) outval0 ;
}
```

Table 5.1 shows kernel timings and throughput values for such a low-pass filter extended to 5 × 5 and 7 × 7 masks applied on 8-bit coded gray-level images of sizes 512 × 512, 1024 × 1024, 2048 × 2048, and 4096 × 4096 run on a C2070 card with 32 × 8 thread blocks.

Table 5.2 shows timings and global throughput values achieved by those convolution masks on an NVIDIA GT200 Tesla architecture (GTX280 card)

Mask size→ Image size↓	3 × 3		5 × 5		7 × 7	
	time (ms)	TP	time (ms)	TP	time (ms)	TP
512 × 512	0.077	1165	0.209	559	0.407	472
1024 × 1024	0.297	1432	0.820	836	1.603	515
2048 × 2048	1.178	1549	**3.265**	**875**	6.398	529
4096 × 4096	4.700	1585	13.05	533	25.56	533

TABLE 5.1. Timings (time) and throughput values (TP in MPx/s) of one register-only nonseparable convolution kernel, for small mask sizes of 3 × 3, 5 × 5, and 7 × 7 pixels, on a C2070 card (fermi architecture). Data transfer duration are those of Table 5.3. The bold value points out the result obtained in the reference situation.

with 16 × 8 thread blocks. This measurement has been done in order to make a relevant comparison with a reference given by NVIDIA in [1] in which they state that their fastest kernel achieves a 5×5 convolution of an 8-bit 2048×2048 pixel image in 1.4 ms, leading to a throughput value of 945 MP/s. In all the result tables, the values associated to this reference will be presented in boldface. Our current value of 802 MP/s, though not unsatisfactory, remains lower to the one reached by the manufacturer's own coding. Tested in the same conditions, the newer Fermi architecture of NVIDIA's GPUs proved slower (3.3 ms, see Table 5.1) due to the lower maximum register count allowed (63 as opposed to 128 for Tesla GT200).

Mask size→ Image size↓	3 × 3		5 × 5		7 × 7	
	time (ms)	TP	time (ms)	TP	time(ms)	TP
512 × 512	0.060	1186	0.148	848	0.280	594
1024 × 1024	0.209	1407	0.556	960	1.080	649
2048 × 2048	0.801	1092	**2.189**	**802**	4.278	573
4096 × 4096	3.171	1075	8.720	793	17.076	569

TABLE 5.2. Timings (time) and throughput values (TP in MP/s) of one register-only nonseparable convolution kernel, for small mask sizes of 3 × 3, 5 × 5, and 7 × 7 pixels, on a GTX280 (GT200 architecture). Data transfer duration are those of Table 5.3. The bold value points out the result obtained in the reference situation.

It is interesting to note that, as long as each thread processes one single pixel, kernel execution time is ruled in proportion with the number of pixels in the image multiplied by that of the mask. The proportionality factor, that we call *slope*, is $3.14 \cdot 10^{-8}$ ms/pix on C2070 in this first implementation. As a reminder, Table 5.3 details the data transfer costs that helped in computing throughput values.

GPU card→ Image size↓	C2070	GTX280
512 × 512	0.148	0.161
1024 × 1024	0.435	0.536
2048 × 2048	1.530	3.039
4096 × 4096	5.882	12.431

TABLE 5.3. Time cost of data transfers between CPU and GPU memories, on C2070 and GTX280 cards (in milliseconds).

5.3.2 Using parameterizable masks

To further improve the above implementation, it becomes necessary to free ourselves from the hard-coding constraint. To achieve this, as was the case with input image storing, several memory options are available, but, since the amount of data involved in processing a mask is quite small and constant, we considered it relevant to copy data into *symbol memory*. Listing 5.2 details this process, involving the CUDA function *cudaMemcpyToSymbol()*.

Listing 5.2. code snippet showing how to setup a mask in GPU symbol memory

```
// on GPU side
__device__ __constant__ float d_mask[256] ;
// on CPU side
float * h_mask = new float[(2*r+1)*(2*r+1)] ;
for (int i=0; i<(2*r+1)*(2*r+1); i++)
   h_mask[i]= 1.0/((2*r+1)*(2*r+1)) ;

cudaMemcpyToSymbol( d_mask, h_mask,
                    (2*r+1)*(2*r+1)*sizeof(float), 0) ;
```

In parallel, giving up the register-only constraint allows a more conventional coding practice (loops). Listing 5.3 presents a generic convolution kernel, whose code immediately appears both simple and concise. Its global time performance, however, is comparatively lower than the register-only process, due to the use of constant memory and of the r parameter (radius of the mask). The average slope amounts to $3.81.10^{-8}$ ms/pix on C2070, which means a time-cost increase of around 20 %.

Listing 5.3. generic CUDA kernel achieving a convolution operation with the mask in symbol memory and its radius passed as a parameter

```
__global__ void kernel_convoGene8r( unsigned char *output, int j_dim,
    int r)
{
  int ic, jc ;
  int k=2*r+1 ;
  float outval0=0.0 ;

  // absolute coordinates of base point
  int j = __umul24( blockIdx.x, blockDim.x ) + threadIdx.x ;
  int i = __umul24( blockIdx.y, blockDim.y) + threadIdx.y ;
```

```
       // convolution computation
       for (ic=-r ; ic<=r ; ic++)
       for (jc=-r ; jc<=r ; jc++)
           outval0 += mask[ __umul24(ic,k)+jc+r ]
15                    *tex2D(tex_img_inc, j+jc, i+ic) ;

       output[ __umul24(i, j_dim) + j ] = outval0 ;
   }
```

5.3.3 Increasing the number of pixels processed by each thread

Much in the same way as we did with the Median Filter, we shall now attempt to reduce the average latency due to writes into global memory by having each thread process more than one output value. As the basic structure of the above GPU kernel uses only 14 registers per thread, regardless of the size of the convolution mask, one can envisage processing 2 or more pixels per thread while keeping safely within the 63-per-thread rule.

However, when doing so, e.g., processing what we shall call a *packet* of pixels, window mask overlapping has to be taken into account to avoid multiple texture fetches of each pixel's gray-level value, while benefiting from the 2D cache. In that case, both mask size and pixel packet shape determine the number of texture fetches to be performed for each pixel value. Figure 5.2 illustrates two different situations: (a) a mask of radius 1 (3×3) applied to a packet of 8 pixels in a row; (b) a mask of radius 2 (5×5). The dark gray pixels are the center pixels (pixels of the packet), while light gray pixels belong to the halo around the packet. The number in each pixel box corresponds to the convolution count in which it is involved. There would be little interest in using different *packet* shapes, as the final global memory writes would not be coalescent, generating multiple latencies.

Although we actually have written GPU kernels able to process 2, 4, 8, and 16 pixels per thread, only the one that processes 8 pixels per thread is presented below, as it proved to be the fastest one. Listing 5.4 reproduces the source code of the kernel for 3×3 masks. The bottom line is that each thread is associated with one base pixel of coordinates (x, y) which is the first, in the packet, to be processed, the last one being $(x + 7, y)$.

In this particular case of a 3×3 mask, each pixel value is used in 3 different convolution sums, except for pixels located near both ends of the packet, whose values are used in fewer sums. The general rule, when performing an $n \times n$ convolution (radius k) by 8-pixel packets is that each of the $(8 - 2k).(2k + 1)$ *center* pixels of the halo is used in k sums, while the $4k.(2k + 1)$ remaining pixels, located around the ends of the packet, are used in fewer sums, from $k - 1$ to 1 ($2(2k + 1)$ pixels each).

Timing results and throughput values are shown in Table 5.4, and show that this solution now outperforms NVIDIA references. It is important to remember that the above kernels have been optimized for the Fermi architecture, unlike those mentioned earlier, which were more efficient on the GT200

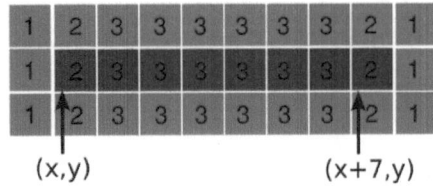

(a) 3 × 3 mask: there are 18 pixels (out of 30) involved in 3 computations.

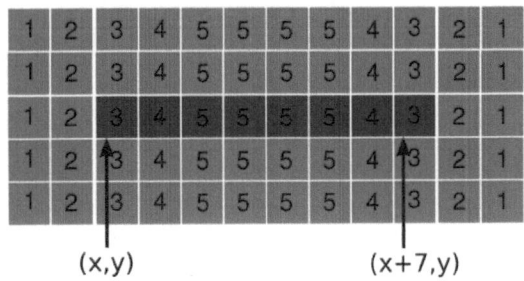

(b) 5 × 5 mask: only 20 pixels (out of 60) are involved in 5 computations.

FIGURE 5.2. Mask window overlapping when processing a packet of 8 pixels per thread. The dark gray pixels are the center pixels, while light gray pixels belong to the halo. The number in each pixel box is the convolution count in which it is involved. (a) 3 × 3 mask; (b) 5 × 5 mask.

Mask size→ Image size↓	3 × 3 time (ms)	TP	5 × 5 time (ms)	TP	7 × 7 time (ms)	TP
512 × 512	0.036	1425	0.069	1208	0.110	1016
1024 × 1024	0.128	1862	0.253	1524	0.413	1237
2048 × 2048	0.495	2071	**0.987**	1666	1.615	1334
4096 × 4096	1.964	2138	3.926	1711	6.416	1364

TABLE 5.4. Timings (time) and throughput values (TP in MP/s) of our generic fixed mask size convolution kernel run on a C2070 card. Data transfer durations are those of Table 5.3. The bold value points out the result obtained in the reference situation.

architecture. However, our technique requires writing one kernel per mask size, which can be seen as a major constraint. To make it easier to use this method, we are working on a kernel code generator that is currently under development and will be made available in the near future.

Listing 5.4. CUDA kernel achieving a 3 × 3 convolution operation with the mask in symbol memory and direct data fetches in texture memory

```
__global__ void kernel_convoGene8x8pL3( unsigned char *output, int
    j_dim )
{
  int ic, jc ;
  const int k=3 ;
  unsigned char pix ;
  float outval0=0.0, outval1=0.0, outval2=0.0, outval3=0.0 ;
  float outval4=0.0, outval5=0.0, outval6=0.0, outval7=0.0 ;

  // coordonnees absolues du point de base en haut a gauche
  int j = ( __umul24( blockIdx.x, blockDim.x) + threadIdx.x)<< 3 ;
  int i = ( __umul24( blockIdx.y, blockDim.y) + threadIdx.y) ;

  // center pixels
  for (ic=0 ; ic<k ; ic++)
    {
      pix = tex2D(tex_img_inc, j+1, i-1+ic) ;
      outval0 += mask[ __umul24(ic,k) +2 ]*pix ;
      outval1 += mask[ __umul24(ic,k) +1 ]*pix ;
      outval2 += mask[ __umul24(ic,k)    ]*pix ;
      pix = tex2D(tex_img_inc, j+2, i-1+ic) ;
      outval1 += mask[ __umul24(ic,k) +2 ]*pix ;
      outval2 += mask[ __umul24(ic,k) +1 ]*pix ;
      outval3 += mask[ __umul24(ic,k)    ]*pix ;
      pix = tex2D(tex_img_inc, j+3, i-1+ic) ;
      outval2 += mask[ __umul24(ic,k) +2 ]*pix ;
      outval3 += mask[ __umul24(ic,k) +1 ]*pix ;
      outval4 += mask[ __umul24(ic,k)    ]*pix ;
      pix = tex2D(tex_img_inc, j+4, i-1+ic) ;
      outval3 += mask[ __umul24(ic,k) +2 ]*pix ;
      outval4 += mask[ __umul24(ic,k) +1 ]*pix ;
      outval5 += mask[ __umul24(ic,k)    ]*pix ;
      pix = tex2D(tex_img_inc, j+5, i-1+ic) ;
      outval4 += mask[ __umul24(ic,k) +2 ]*pix ;
      outval5 += mask[ __umul24(ic,k) +1 ]*pix ;
      outval6 += mask[ __umul24(ic,k)    ]*pix ;
      pix = tex2D(tex_img_inc, j+6, i-1+ic) ;
      outval5 += mask[ __umul24(ic,k) +2 ]*pix ;
      outval6 += mask[ __umul24(ic,k) +1 ]*pix ;
      outval7 += mask[ __umul24(ic,k)    ]*pix ;
      // end zones
      pix = tex2D(tex_img_inc, j,   i-1+ic) ;
      outval0 += mask[ __umul24(ic,k) +1 ]*pix ;
      outval1 += mask[ __umul24(ic,k)    ]*pix ;
      pix = tex2D(tex_img_inc, j-1, i-1+ic) ;
      outval0 += mask[ __umul24(ic,k)    ]*pix ;

      pix = tex2D(tex_img_inc, j+7, i-1+ic) ;
      outval6 += mask[ __umul24(ic,k) +2 ]*pix ;
      outval7 += mask[ __umul24(ic,k) +1 ]*pix ;
      pix = tex2D(tex_img_inc, j+8, i-1+ic) ;
      outval7 += mask[ __umul24(ic,k) +2 ]*pix ;
    }
  // multiple output
  output[ __umul24(i, j_dim) + j   ] = outval0 ;
  output[ __umul24(i, j_dim) + j+1 ] = outval1 ;
  output[ __umul24(i, j_dim) + j+2 ] = outval2 ;
```

```
output[ __umul24(i, j_dim) + j+3 ] = outval3 ;
output[ __umul24(i, j_dim) + j+4 ] = outval4 ;
output[ __umul24(i, j_dim) + j+5 ] = outval5 ;
output[ __umul24(i, j_dim) + j+6 ] = outval6 ;
output[ __umul24(i, j_dim) + j+7 ] = outval7 ;
}
```

5.3.4 Using shared memory to store prefetched data

A more convenient way of coding a convolution kernel is to use shared memory to perform a prefetching stage of the whole halo before computing the convolution sums. This proves to be quite efficient and more versatile, but it obviously generates some overhead because

- Each pixel value has to be read at least twice, first from texture memory into shared memory and then one or several more times from shared memory to be used in convolution computations.

- Reducing the number of times a single pixel value is read from shared memory is bound to generate bank conflicts, hence once again performance loss.

Still, we also implemented this method, in a similar manner as NVIDIA did in its SDK sample code. Some improvement has been obtained by increasing the number of pixels processed by each thread, to an optimum 8 pixels per thread. The principle is to prefetch all pixel values involved in the computations performed by all threads of a block, including 8 pixels per thread plus the halo of radius r (the radius of the convolution mask). As this obviously represents more values than the thread count in one block, some threads have to load more than one value. The general organization is reproduced in Figure 5.3 for 5×5 mask and a 8×4 thread block, while Listing 5.5 gives the details of the implementation with its two distinct code blocks: preload in shared memory (Lines 20 to 42) and convolution computations (Lines 45 to 57). Tables 5.5 and 5.6 detail timing results and throughput values of this implementation (16×8 threads/block), up to 13×13 masks, that will serve as a reference in the next section, devoted to separable convolution.

Mask size→ Image size↓	3×3	5×5	7×7	9×9	11×11	13×13
512×512	0.040	0.075	0.141	0.243	0.314	0.402
1024×1024	0.141	0.307	0.524	0.917	1.192	1.535
2048×2048	0.543	1.115	2.048	3.598	4.678	6.037
4096×4096	2.146	4.364	8.156	14.341	18.652	24.020

TABLE 5.5. Performances, in milliseconds, of our generic 8 pixels per thread kernel using shared memory, run on a C2070 card. Data transfers duration are not included.

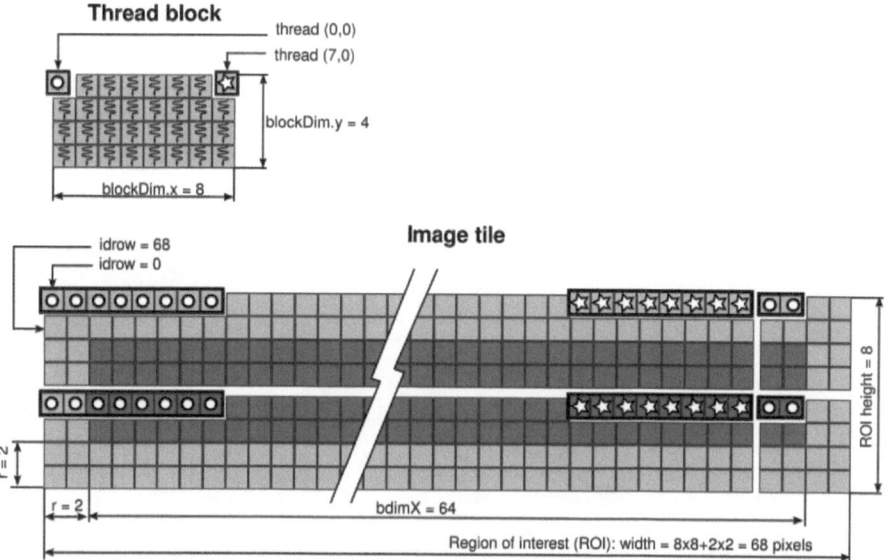

FIGURE 5.3. Organization of the prefetching stage of data, for a 5 × 5 mask and a thread block size of 8 × 4. Threads in both top corners of the top figure are identified either by a circle or by a star symbol. The image tile, loaded into shared memory, includes the pixels to be updated by the threads of the block, as well as its 2-pixel wide halo. Here, circle and star symbols in the image tile show which pixels are actually loaded into one shared memory vector by its corresponding thread.

Mask size→ Image size↓	3 × 3	5 × 5	7 × 7	9 × 9	11 × 11	13 × 13
512 × 512	1394	1176	907	670	567	477
1024 × 1024	1820	1413	1093	776	644	532
2048 × 2048	2023	1586	1172	818	676	554
4096 × 4096	2090	1637	1195	830	684	561

TABLE 5.6. Throughput values, in MegaPixel per second, of our generic 8 pixels per thread kernel using shared memory, run on a C2070 card. Data transfer durations are those of Table 5.3.

Listing 5.5. CUDA kernel achieving a generic convolution operation after a preloading of data in shared memory

```
__global__ void kernel_convoNonSepSh_8p(unsigned char *output, int
    j_dim, int r)
{
  int ic, jc;
  int k = 2*r+1 ;
  float outval0=0.0, outval1=0.0, outval2=0.0, outval3=0.0, outval4
    =0.0, outval5=0.0, outval6=0.0, outval7=0.0 ;
  int bdimX = blockDim.x<<3 ;
  int tidX = threadIdx.x<<3 ;

  // absolute coordinates of packet's base point
  int j = (__umul24(blockIdx.x,blockDim.x) + threadIdx.x) << 3 ;
  int i = __umul24( blockIdx.y, blockDim.y) + threadIdx.y ;
  int j0= __umul24(blockIdx.x,blockDim.x)<<3 ;   // block's base point
  int idx = __umul24(i,j_dim) + j ;              // absolute index
  int idrow = threadIdx.y*(bdimX+k-1) ;    // line's offset in sh mem

  // shared memory declaration
  extern __shared__ unsigned char roi8p[];

  // top left
  for (int p=0; p<8; p++)
    roi8p[ idrow + tidX +p ] = tex2D(tex_img_inc, j-r+p  , i-r) ;

  // top right
  if ( threadIdx.x < r )
  {
    roi8p[ idrow + bdimX + threadIdx.x    ] = tex2D( tex_img_inc, j0-
      r +bdimX+threadIdx.x , i-r ) ;
    roi8p[ idrow + bdimX + threadIdx.x +r ] = tex2D( tex_img_inc, j0
      +bdimX+threadIdx.x  , i-r ) ;
  }
  // bottom left
  if ( threadIdx.y < k-1 )
  {
    idrow = (threadIdx.y+blockDim.y)*(bdimX+k-1) ;
    for (int p=0; p<8; p++)
      roi8p[ idrow + tidX +p ] = tex2D( tex_img_inc, j-r+p  , i+
        blockDim.y-r ) ;
    // bottom right
    if ( threadIdx.x < r )
    {
      roi8p[ idrow + bdimX +threadIdx.x    ] = tex2D( tex_img_inc, j0
        -r +bdimX +threadIdx.x, i+blockDim.y-r ) ;
      roi8p[ idrow + bdimX +threadIdx.x +r] = tex2D( tex_img_inc, j0
        +bdimX +threadIdx.x, i+blockDim.y-r ) ;
    }
  }
  __syncthreads();

  // computations
  for (ic=0 ; ic<k ; ic++)
    for( jc=0 ; jc<k ; jc++)
    {
      int baseRoi = __umul24(ic+threadIdx.y,(bdimX+k-1)) + jc+tidX ;
      float valMask = mask[ __umul24(ic,k)+jc ] ;
      outval0 += valMask*roi8p[ baseRoi    ] ;
      outval1 += valMask*roi8p[ baseRoi +1 ] ;
      outval2 += valMask*roi8p[ baseRoi +2 ] ;
      outval3 += valMask*roi8p[ baseRoi +3 ] ;
      outval4 += valMask*roi8p[ baseRoi +4 ] ;
      outval5 += valMask*roi8p[ baseRoi +5 ] ;
      outval6 += valMask*roi8p[ baseRoi +6 ] ;
      outval7 += valMask*roi8p[ baseRoi +7 ] ;
```

```
                  }
60      // multiple output ---> global mem
        output[ idx++ ] = outval0 ;
        output[ idx++ ] = outval1 ;
        output[ idx++ ] = outval2 ;
        output[ idx++ ] = outval3 ;
65      output[ idx++ ] = outval4 ;
        output[ idx++ ] = outval5 ;
        output[ idx++ ] = outval6 ;
        output[ idx   ] = outval7 ;
     }
```

5.4 Separable convolution

A convolution operation is said separable when its masks h is the product of 2 vectors h_v and h_h, as is the case in the following example:

$$h = h_v \times h_h = \begin{bmatrix} 1 \\ 2 \\ 1 \end{bmatrix} \times \begin{bmatrix} -1 & 2 & -1 \end{bmatrix} = \begin{bmatrix} -1 & 2 & -1 \\ -2 & 4 & -2 \\ -1 & 2 & -1 \end{bmatrix}$$

Such a mask allows us to replace a generic 2D convolution operation by two consecutive stages of a 1D convolution operation: a vertical of mask h_v and a horizontal of mask h_h. This saves a lot of arithmetic operations, as a generic $n \times n$ convolution applied on an $H \times L$ image basically represents HLn^2 multiplications and as many additions, while two consecutive $n \times 1$ convolutions represents only $2HLn$ of each, e.g., 60% operations are saved per pixel of the image for a 5×5 mask.

However, besides reducing the operation count, performing a separable convolution also means writing an intermediate image into global memory. CPU implementations of separable convolutions often use a single function to perform both 1D convolution stages. To do so, this function reads the input image and actually ouputs the transposed filtered image. Applying this principle to GPUs is not efficient, as outputting the transposed image means noncoalescent writes into global memory, generating severe performance loss. Hence the idea of developing two different kernels, one for each of the vertical and horizontal convolutions.

Here, the use of shared memory is the best choice, as there is no overlapping between neighbor windows and thus no possible optimization. Moreover, to ensure efficiency, it is important to read the input image from texture memory, which implies an internal GPU data copy between both 1D convolution stages. This, even if it is faster than CPU/GPU data transfer, makes separable convolutions slower than generic convolutions for small mask sizes. On C2070, the lower limit is 7×7 pixels (9×9 for 512×512 images).

Both vertical and horizontal kernels feature similar runtimes: Table 5.7

contains only their average execution time, including the internal data copy stage, while Table 5.8 shows the achieved global throughput values. Timings of the data copy stage are given in Table 5.9. Listings 5.7 and 5.8 detail the implementation of both 1D kernels, while Listing 5.6 shows how to use them in addition with the data copy function in order to achieve a whole separable convolution. The shared memory size is dynamically passed as a parameter at kernel call time. Its expression is given in both Listings (5.7 and 5.8), in the comment lines before its declaration.

Mask size→ Image size↓	3 × 3	5 × 5	7 × 7	9 × 9	11 × 11	13 × 13
512 × 512	0.080	0.087	0.095	**0.108**	**0.115**	**0.126**
1024 × 1024	0.306	0.333	**0.333**	**0.378**	**0.404**	**0.468**
2048 × 2048	1.094	1.191	**1.260**	**1.444**	**1.545**	**1.722**
4096 × 4096	4.262	4.631	**5.000**	**5.676**	**6.105**	**6.736**

TABLE 5.7. Performances, in milliseconds, of our generic 8 pixels per thread 1D convolution kernels using shared memory, run on a C2070 card. Timings include data copy. Bold values correspond to situations where separable-convolution kernels run faster than nonseparable ones.

Mask size→ Image size↓	3 × 3	5 × 5	7 × 7	9 × 9	11 × 11	13 × 13
512 × 512	1150	1116	1079	**1024**	**997**	**957**
1024 × 1024	1415	1365	**1365**	**1290**	**1250**	**1169**
2048 × 2048	1598	1541	**1503**	**1410**	**1364**	**1290**
4096 × 4096	1654	1596	**1542**	**1452**	**1400**	**1330**

TABLE 5.8. Throughput values, in MegaPixel per second, of our generic 8 pixels per thread 1D convolution kernel using shared memory, run on a C2070 card. Bold values correspond to situations where separable-convolution kernels run faster than nonseparable ones (data transfer durations are those of Table 5.3).

Image size	C2070
512 × 512	0.029
1024 × 1024	0.101
2048 × 2048	0.387
4096 × 4096	1.533

TABLE 5.9. Time cost of data copy between the vertical and the horizontal 1D convolution stages, on a C2070 cards (in milliseconds).

Implementing an efficient convolution operation on GPU

Listing 5.6. data copy between the calls to 1D convolution kernels achieving a 2D separable convolution operation

```
dimGrid = dim3( (L/dimBlock.x)/8, H/dimBlock.y, 1 ) ;

kernel_convoSepShx8pV<<< dimGrid, dimBlock, 8*dimBlock.x*(dimBlock.y
   +2*r)*sizeof(char) >>>(d_outc, L, r );
cudaMemcpyToArray( array_img_inc , 0, 0, d_outc, H*L*sizeof(char) ,
   cudaMemcpyDeviceToDevice) ;
kernel_convoSepShx8pH<<< dimGrid, dimBlock, 8*(dimBlock.x+2*r)*
   dimBLock.y*sizeof(char) >>>(d_outc, L, r );
```

Listing 5.7. CUDA kernel achieving a horizontal 1D convolution operation after a preloading of data into shared memory

```
__global__ void kernel_convoSepShx8pV(unsigned char *output, int j_dim
    , int r )
{
  int ic, jc, p;
  int k = 2*r+1 ;
  float outval0=0.0, outval1=0.0, outval2=0.0, outval3=0.0 ;
  float outval4=0.0, outval5=0.0, outval6=0.0, outval7=0.0 ;
  int bdimX = blockDim.x<<3 ; // all packets width
  int tidX = threadIdx.x<<3 ; // one packet offset

  // absolute coordinates of the base point
  int j = (__umul24(blockIdx.x,blockDim.x) + threadIdx.x)<<3 ;
  int i = __umul24( blockIdx.y, blockDim.y) + threadIdx.y ;
  // absolute index in the image
  int idx = __umul24(i,j_dim) + j ;
  // offset of one ROI row in shared memory
  int idrow = threadIdx.y*bdimX ;

  extern __shared__ unsigned char roi8p [];

  // top block
  for (p=0; p<8; p++)
    roi8p[ idrow + tidX +p ] = tex2D(tex_img_inc, j+p , i-r) ;

  // bottom block
  if ( threadIdx.y < k-1 )
    {
      idrow = (threadIdx.y+blockDim.y)*bdimX ;
      for (int p=0; p<8; p++)
        roi8p[ idrow + tidX +p ] = tex2D( tex_img_inc, j+p , i+
          blockDim.y-r ) ;
    }
  __syncthreads();

  // vertical convolution
  for (ic=0 ; ic<k ; ic++)
    {
      int baseRoi = __umul24(ic+threadIdx.y,bdimX) + tidX ;
      float valMask = mask[ ic ] ;
      outval0 += valMask*roi8p[ baseRoi    ] ;
      outval1 += valMask*roi8p[ baseRoi +1 ] ;
      outval2 += valMask*roi8p[ baseRoi +2 ] ;
      outval3 += valMask*roi8p[ baseRoi +3 ] ;
      outval4 += valMask*roi8p[ baseRoi +4 ] ;
      outval5 += valMask*roi8p[ baseRoi +5 ] ;
      outval6 += valMask*roi8p[ baseRoi +6 ] ;
      outval7 += valMask*roi8p[ baseRoi +7 ] ;
    }

  // 8 pixels per thread --> global mem
  output[ idx++ ] = outval0 ;
```

```
50    output[ idx++ ] = outval1 ;
      output[ idx++ ] = outval2 ;
      output[ idx++ ] = outval3 ;
      output[ idx++ ] = outval4 ;
      output[ idx++ ] = outval5 ;
55    output[ idx++ ] = outval6 ;
      output[ idx   ] = outval7 ;
   }
```

Listing 5.8. CUDA kernel achieving a vertical 1D convolution operation after a preloading of data into shared memory

```
   __global__ void kernel_convoSepShx8pH(unsigned char *output, int j_dim
       , int r)
   {
     int ic, jc, p;
     int k = 2*r+1 ;
5    float outval0=0.0, outval1=0.0, outval2=0.0, outval3=0.0 ;
     float outval4=0.0, outval5=0.0, outval6=0.0, outval7=0.0 ;
     int bdimX = blockDim.x<<3 ; // all packets width
     int tidX  = threadIdx.x<<3 ; // one packet offset

10   // absolute coordinates of one packet base point
     int j = (__umul24(blockIdx.x,blockDim.x) + threadIdx.x)<<3 ;
     int i = __umul24( blockIdx.y, blockDim.y) + threadIdx.y ;
     int j0= __umul24(blockIdx.x,blockDim.x)<<3 ;
     // absolute index in the image
15   int idx = __umul24(i,j_dim) + j ;

     // offset of one ROI row in shared memory
     int idrow = threadIdx.y*(bdimX+k-1) ;

20   extern __shared__ unsigned char roi8p[];

     // top left block
     for (p=0; p<8; p++)
       roi8p[ idrow + tidX +p ] = tex2D(tex_img_inc, j-r+p  , i) ;
25   // top right block
     if ( threadIdx.x < r  )
       {
         roi8p[ idrow + bdimX + threadIdx.x    ] = tex2D( tex_img_inc, j0
             -r +bdimX+threadIdx.x   , i  ) ;
         roi8p[ idrow + bdimX + threadIdx.x +r ] = tex2D( tex_img_inc, j0
             +bdimX+threadIdx.x   , i  ) ;
30     }

     __syncthreads() ;

     // horizontal convolution
35   for (jc=0 ; jc<k ;  jc++)
       {
         int baseRoi = idrow + tidX +jc ;
         float valMask = mask[ jc ]  ;
         outval0 += valMask*roi8p[  baseRoi       ] ;
40       outval1 += valMask*roi8p[  baseRoi +1  ] ;
         outval2 += valMask*roi8p[  baseRoi +2  ] ;
         outval3 += valMask*roi8p[  baseRoi +3  ] ;
         outval4 += valMask*roi8p[  baseRoi +4  ] ;
         outval5 += valMask*roi8p[  baseRoi +5  ] ;
45       outval6 += valMask*roi8p[  baseRoi +6  ] ;
         outval7 += valMask*roi8p[  baseRoi +7  ] ;
       }

     // 8 pixels per thread --> global mem
50   output[ idx++ ] = outval0 ;
     output[ idx++ ] = outval1 ;
```

```
    output[ idx++ ] = outval2 ;
    output[ idx++ ] = outval3 ;
    output[ idx++ ] = outval4 ;
55  output[ idx++ ] = outval5 ;
    output[ idx++ ] = outval6 ;
    output[ idx   ] = outval7 ;
}
```

5.5 Conclusion

Extensively detailing the various techniques that may be applied when designing a median or a convolution operation on GPU has enabled us determine that

- the use of registers with direct data fetching from texture often allows kernels to run faster than those which use the more conventionnal way of prefetching data from texture memory and storing them in shared memory.

- increasing the pixel count processed by each thread brings important speedups. In this case, if neighboring windows overlap, optimized direct data fetching from texture will likely outperform the shared memory prefetching technique. This is the case for generic convolution kernels.

- coding such optimized data fetching is not straightforward. Consequently, we are currently developing a kernel code generator that will make our kernels more accessible by GPU users.

The presented kernels, optimized for a C2070 card, achieve up to 2138 MP/s including data transfers, which comes close to the absolute maximum throughput value allowed by the Fermi architecture. The next GPU generation (called Kepler) may allow us not only to benefit from new dynamic parallelism capability to increase kernel paralelism level, but also to take advantage of an increase in the register count allowed per thread block which would allow us, for example, to extend our register-only median filter technique to larger mask sizes.

Bibliography

[1] J. Stam. Convolution soup. In *GPU Technology Conference*, Aug. 2010.

Part III

Software development

Chapter 6

Development of software components for heterogeneous many-core architectures

Stefan L. Glimberg, Allan P. Engsig-Karup, Allan S. Nielsen, and Bernd Dammann

Technical University of Denmark

6.1	Software development for heterogeneous architectures		74
	6.1.1	Test environments	76
6.2	Heterogeneous library design for PDE solvers		76
	6.2.1	Component and concept design	77
	6.2.2	A matrix-free finite difference component	77
6.3	Model problems		81
	6.3.1	Heat conduction equation	82
		6.3.1.1 Assembling the heat conduction solver	84
		6.3.1.2 Numerical solutions to the heat conduction problem	86
	6.3.2	Poisson equation	87
		6.3.2.1 Assembling the Poisson solver	89
		6.3.2.2 Numerical solutions to the Poisson problem	91
6.4	Optimization strategies for multi-GPU systems		91
	6.4.1	Spatial domain decomposition	93
	6.4.2	Parareal–parallel time integration	95
		6.4.2.1 The parareal algorithm	96
		6.4.2.2 Computational complexity	98
6.5	Conclusion and outlook		100
	Acknowledgments		101
	Bibliography		102

6.1 Software development for heterogeneous architectures

Massively parallel processors, such as graphical processing units (GPUs), have in recent years proven to be effective for a vast amount of scientific applications. Today, most desktop computers are equipped with one or more powerful GPUs, offering heterogeneous high-performance computing to a broad range of scientific researchers and software developers. Though GPUs are now programmable and can be highly effective computing units, they still pose challenges for software developers to fully utilize their efficiency. Sequential legacy codes are not always easily parallelized, and the time spent on conversion might not pay off in the end. This is particular true for heterogeneous computers, where the architectural differences between the main and coprocessor can be so significant that they require completely different optimization strategies. The cache hierarchy management of CPUs and GPUs are an evident example hereof. In the past, industrial companies were able to boost application performance solely by upgrading their hardware systems, with an overt balance between investment and performance speedup. Today, the picture is different; not only do they have to invest in new hardware, but they also must account for the adaption and training of their software developers. What traditionally used to be a hardware problem, addressed by the chip manufacturers, has now become a software problem for application developers.

Software libraries can be a tremendous help for developers as they make it easier to implement an application, without requiring special knowledge of the underlying computer architecture and hardware. A library may be referred to as *opaque* when it automatically utilizes the available resources, without requiring specific details from the developer [1]. The ultimate goal for a successful library is to simplify the process of writing new software and thus to increase developer productivity. Since programmable heterogeneous CPU/GPU systems are a rather new phenomenon, there is a limited number of established software libraries that take full advantage of such heterogeneous high performance systems, and there are no de facto design standards for such systems either. Some existing libraries for conventional homogeneous systems have already added support for offloading computationally intense operations onto coprocessing GPUs. However, this approach comes at the cost of frequent memory transfers across the low bandwidth PCIe bus.

In this chapter, we focus on the use of a software library to help application developers achieve their goals without spending an immense amount of time on optimization details, while still offering close-to-optimal performance. A good library provides performance-portable implementations with intuitive interfaces, that hide the complexity of underlaying hardware optimizations. Unfortunately, opaqueness sometimes comes at a price, as one does not necessarily get the best performance when the architectural details are not *visible* to the programmer [1]. If, however, the library is flexible enough and permits developers to supply their own low-level implementations as well, this does not need to be an issue. These are some of the considerations library developers should take into account, and what we will try to address in this chapter.

For demonstrative purposes we present details from an in-house generic CUDA-based C++ library for fast assembling of partial differential equation (PDE) solvers, utilizing the computational resources of GPUs. This library has been developed as part of research activities associated with the GPUlab, at the Technical University of Denmark and, therefore, is referred to as the *GPUlab library*. It falls into the category of computational libraries, as categorized by Hoefler and Snir [12]. Memory allocation and basic algebraic operations are supported via object-oriented components, without the user having to write CUDA specific kernels. As a back-end vector class, the parallel CUDA Thrust template-based library is used, enabling easy memory allocation and a high-level interface for vector manipulation [6]. Inspirations for *good library design*, some of which we will present in this chapter, originate from guidelines proposed throughout the literature [10, 12, 24]. An identification of desirable properties, which any library should strive to achieve, is pointed out by Korson and McGregor [16]. In particular we mention being easy-to-use, extensible, and intuitive.

The library is designed to be effective and scalable for fast prototyping of PDE solvers, (primarily) based on matrix-free implementations of finite difference (stencil) approximations on logically structured grids. It offers functionalities that will help assemble PDE solvers that automatically exploit heterogeneous architectures much faster than manually having to manage GPU memory allocation, memory transfers, kernel launching, etc.

In the following sections we demonstrate how software components that play important roles in scientific applications can be designed to fit a simple framework that will run efficiently on heterogeneous systems. One example is finite difference approximations, commonly used to find numerical solutions to differential equations. Matrix-free implementations minimize both memory consumption and memory access, two important features for efficient GPU utilization and for enabling the solution of large-scale problems. The bottleneck problem for many PDE applications is to solve large sparse linear systems, arising from the discretization. In order to help solve these systems, the library includes a set of iterative solvers. All iterative solvers are template-based, such that vector and matrix classes, along with their underlying implementations, can be freely interchanged. New solvers can also be implemented without much coding effort. The generic nature of the library, along with a predefined set of interface rules, allows assembling components into PDE solvers. The use of parameterized-type binding allows the user to assemble PDE solvers at a high abstraction level, without having to change the remaining implementation.

Since this chapter is mostly dedicated to the discussion of software development for high performance heterogeneous systems, the focus will be more on the development and usage of an in-house GPU-based library, than on specific scientific applications. We demonstrate how to use the library on two elementary model problems and refer the reader to Chapter 11 for a detailed description of an advanced application tool for free surface water wave simulations. These examples are assembled using the library components presented in this chapter.

6.1.1 Test environments

Throughout the chapter we use three different test environments: two high-end desktop computers located at the GPUlab–Technical University of Denmark, and a GPU cluster located at the Center for Computing and Visualization, Brown University, USA. Hardware details for the two systems are as follows:

Test environment 1. Desktop computer, Linux Ubuntu, Intel Xeon E5620 (2.4GHz) quad-core Westmere processor, 12GB of DDR-3 memory (1066 MHz), 2x NVIDIA GeForce GTX590 GPU with 3GB DDR5 memory, PCIe 2.0.

Test environment 2. Desktop computer, Linux Ubuntu, Intel Core i7-3820 (3.60GHz) Sandy Bridge processors, 32GB RAM, 2x NVIDIA Tesla K20 GPUs with 5GB DDR5 memory, PCIe 2.0.

Test environment 3. GPU cluster, Linux, up to 44 compute nodes based on dual Intel Xeon E5540 (2.53GHz) quad-core Nehalem processors, 24 GB of DDR-3 memory (1333MHz), 2x NVIDIA Tesla M2050 GPUs with 3GB GDDR5 memory, 40 Gb/s Quad-Data-Rate (QDR) InfiniBand interconnect.

6.2 Heterogeneous library design for PDE solvers

A generic CUDA-based C++ library has been developed to ease the assembling of PDE solvers. The template-based design allows users to assemble solver parts easily and to supply their own implementation of problem specific parts. In the following, we present an overview of the library and the supported features, introduce the concepts of the library components, and give short code examples to ease understanding. The library is a starting point for fast assembling of GPU-based PDE solvers, developed mainly to support finite difference operations on regular grids. However, this is not a limitation, since existing vector objects could be used as base classes for extending to other discretization methods or grid types as well.

6.2.1 Component and concept design

The library is grouped into component classes. Each component should fulfill a set of simple interface and template rules, called concepts, in order to guarantee compatibility with the rest of the library. In the context of PDE solving, we present five component classes: vectors, matrices, iterative solvers for linear system of equations, preconditioners for the iterative solvers, and time integrators. Figure 6.1 lists the five components along with a subset of the type definitions they should provide and the methods they should implement. It is possible to extend the implementation of these components with more functionality that relate to specific problems, but this is the minimum requirement for compatibility with the remaining library. With these concept rules fulfilled, components can rely on other components to have their respective functions implemented.

A component is implemented as a generic C++ class, and normally takes as a template arguments the same types that it offers through type definitions: a matrix takes a vector as template argument, and a vector takes the working precision type. The matrix can then access the working precision through the vector class. Components that rely on multiple template arguments can combine these arguments via type binders to reduce the number of arguments and maintain code simplicity. We will demonstrate use of such type binders in the model problem examples. A thorough introduction to template-based programming in C++ can be found in [27].

The generic configuration allows the developer to define and assemble solver parts at the very beginning of the program using type definitions. Changing PDE parts at a later time is then only a matter of changing type definitions. We will give two model examples of how to assemble PDE solvers in Section 6.3.

6.2.2 A matrix-free finite difference component

Common vector operations, such as memory allocation, element-wise assignments, and basic algebraic transformations, require many lines of codes for a purely CUDA-based implementation. These CUDA-specific operations and kernels are hidden from the user behind library implementations, to ensure a high abstraction level. The vector class inherits from the CUDA-based Thrust library and therefore offer the same level of abstraction that enhances developer productivity and enables performance portability. Creating and allocating device (GPU) memory for two vectors can be done in a simple and intuitive way using the GPUlab library, as shown in Listing 6.1 where two vectors are added together.

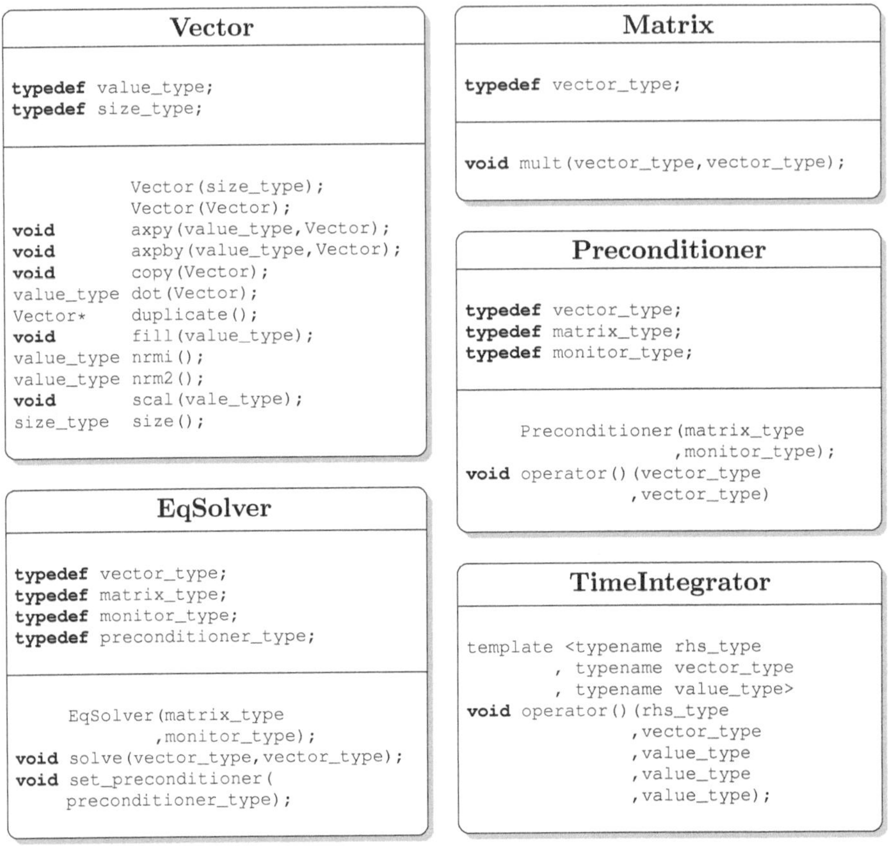

Vector
typedef value_type; **typedef** size_type;

Vector(size_type);
          Vector(Vector);
void   axpy(value_type,Vector);
void   axpby(value_type,Vector);
void   copy(Vector);
value_type dot(Vector);
Vector*   duplicate();
void   fill(value_type);
value_type nrm1();
value_type nrm2();
void   scal(vale_type);
size_type size();

Matrix
typedef vector_type;
void mult(vector_type,vector_type);

Preconditioner
typedef vector_type; **typedef** matrix_type; **typedef** monitor_type;

Preconditioner(matrix_type
                  ,monitor_type);
void operator()(vector_type
                    ,vector_type)

EqSolver
typedef vector_type; **typedef** matrix_type; **typedef** monitor_type; **typedef** preconditioner_type;

EqSolver(matrix_type
         ,monitor_type);
void solve(vector_type,vector_type);
void set_preconditioner(
    preconditioner_type);

TimeIntegrator

template <typename rhs_type
       , typename vector_type
       , typename value_type>
void operator()(rhs_type
              ,vector_type
              ,value_type
              ,value_type
              ,value_type);

FIGURE 6.1. Schematic representation of the five main components, their type definitions, and member functions. Because components are template based, the argument types cannot be known beforehand. The concepts ensure compliance among components.

Listing 6.1. allocating, initializing, and adding together two vectors on the GPU: first example uses pure CUDA C; second example uses the built-in library template-based vector class

```
#include <gpulab/vector.h>

__global__ void add(double* a, double const* b, int N)
{
    int i = blockDim.x*blockIdx.x + threadIdx.x;
    if(i<N)
        a[i] += b[i];
}

int main(int argc, char *argv[])
{
    int N = 1000;
```

```cpp
// Basic CUDA example
double *a1, *b1;
cudaMalloc((void**)&a1, N*sizeof(double));
cudaMalloc((void**)&b1, N*sizeof(double));
cudaMemset(a1, 2.0, N);
cudaMemset(b1, 3.0, N);
int blocksize = 128;
add<<<(N+blocksize-1)/blocksize, blocksize>>>(a1, b1, N);

// gpulab example
gpulab::vector<double, gpulab::device_memory> a2(N, 2.0);
gpulab::vector<double, gpulab::device_memory> b2(N, 3.0);
a2.axpy(1.0, b2);   // BLAS1: a2 = 1*b2 + a2

return 0;
}
```

The vector class (and derived classes hereof) is compliant with the rest of the library components. Matrix-vector multiplications are usually what makes PDE-based applications different from each other, and the need to write a user specific implementation of the matrix-vector product is essential when solving specific PDE problems. The PDE and the choice of discretization method determine the structure and sparsity of the resulting matrix. Spatial discretization is supported by the library with finite difference approximations, and it offers an efficient low-storage (matrix-free) flexible order implementation to help developers tailor their custom codes. These matrix-free operators are feasible for problems where the matrix structure is known in advance and can be exploited, such that the matrix values can be either precomputed or computed on the fly. Furthermore, the low constant memory requirement makes them perfect in the context of solving large scale problems, whereas traditional sparse matrix formats require increasingly more memory, see e.g., [5] for details on GPU sparse matrix formats.

Finite differences approximate the derivative of some function $u(x)$ as a weighted sum of neighboring elements. In compact notation we write

$$\frac{\partial^q u(x_i)}{\partial x^q} \approx \sum_{n=-\alpha}^{\beta} c_n u(x_{i+n}), \qquad (6.1)$$

where q is the order of the derivative, c_n is a set of finite difference coefficients, and α plus β define the number of coefficients that are used for the approximation. The total set of contributing elements is called the stencil, and the size of the stencil is called the rank, given as $\alpha+\beta+1$. The stencil coefficients c_n can be derived from a Taylor expansion based on the values of α and β, and q, using the method of undetermined coefficients [17]. An example of a three-point finite difference matrix that approximates the first ($q=1$) or second ($q=2$) derivative of a one-dimensional uniformly distributed vector u of length 8 is given here:

$$\begin{bmatrix} c_{00} & c_{01} & c_{02} & 0 & 0 & 0 & 0 & 0 \\ c_{10} & c_{11} & c_{12} & 0 & 0 & 0 & 0 & 0 \\ 0 & c_{10} & c_{11} & c_{12} & 0 & 0 & 0 & 0 \\ 0 & 0 & c_{10} & c_{11} & c_{12} & 0 & 0 & 0 \\ 0 & 0 & 0 & c_{10} & c_{11} & c_{12} & 0 & 0 \\ 0 & 0 & 0 & 0 & c_{10} & c_{11} & c_{12} & 0 \\ 0 & 0 & 0 & 0 & 0 & c_{10} & c_{11} & c_{12} \\ 0 & 0 & 0 & 0 & 0 & c_{20} & c_{21} & c_{22} \end{bmatrix} \begin{bmatrix} u_0 \\ u_1 \\ u_2 \\ u_3 \\ u_4 \\ u_5 \\ u_6 \\ u_7 \end{bmatrix} \approx \begin{bmatrix} u_0^{(q)} \\ u_1^{(q)} \\ u_2^{(q)} \\ u_3^{(q)} \\ u_4^{(q)} \\ u_5^{(q)} \\ u_6^{(q)} \\ u_7^{(q)} \end{bmatrix}. \quad (6.2)$$

It is clear from this example that the matrix is sparse and that the same coefficients are repeated for all centered rows. The coefficients differ only near the boundaries, where off-centered stencils are used. It is natural to pack this information into a stencil operator that stores only the unique set of coefficients:

$$\mathbf{c} = \begin{bmatrix} c_{00} & c_{01} & c_{02} \\ c_{10} & c_{11} & c_{12} \\ c_{20} & c_{21} & c_{22} \end{bmatrix}. \quad (6.3)$$

Matrix components precompute these compact stencil coefficients and provides member functions that computes the finite difference approximation of input vectors. Unit scaled coefficients (assuming grid spacing is one) are computed and stored to be accessible via both CPU and GPU memory. On the GPU, the constant memory space is used for faster memory access [21]. In order to apply a stencil on a nonunit-spaced grid, with grid space Δx, the scale factor $1/(\Delta x)^q$ will have to be multiplied by the finite difference sum, i.e., $(c_{00}u_0 + c_{01}u_1 + c_{02}u_2)/(\Delta x)^q \approx u_0^{(q)}$ as in the first row of (6.2).

Setting up a two-dimensional grid of size $N_x \times N_y$ in the unit square and computing the first derivative hereof is illustrated in Listing 6.2. The grid is a vector component, derived from the vector class. It is by default treated as a device object and memory is automatically allocated on the device to fit the grid size. The finite difference approximation as in (6.1), is performed via a CUDA kernel behind the scenes during the calls to `mult` and `diff_x`, utilizing the memory hierarchy as the CUDA guidelines prescribe [21, 22]. To increase developer productivity, kernel launch configurations have default settings, based on CUDA guidelines, principles, and experiences from performance testings, such that the user does not have to explicitly specify them. For problem-specific finite difference approximations, where the built-in stencil operators are insufficient, a pointer to the coefficient matrix (6.3) can be accessed as demonstrated in Listing 6.2 and passed to customized kernels.

Listing 6.2. two-dimensional finite difference stencil example: computing the first derivative using five points ($\alpha = \beta = 2$) per dimension, a total nine-point stencil

```
#include <gpulab/grid.h>
#include <gpulab/FD/stencil.h>

int main(int argc, char *argv[])
{
    // Initialize grid dimensions
    unsigned int Nx = 10, Ny = 10;
    gpulab::grid_dim<unsigned int> dim(Nx,Ny);
    gpulab::grid<double> u(dim);
    gpulab::grid<double> ux(u);
    gpulab::grid<double> uxy(u);

    // Put meaningful values into u here ...

    // Stencil size, alpha=beta=2, 9pt 2D stencil
    int alpha = 2;
    // 1st order derivative
    gpulab::FD::stencil_2d<double> stencil(1, alpha);
    // Calculate uxy = du/dx + du/dy
    stencil.mult(u,uxy);
    // Calculate ux = du/dx
    stencil.diff_x(u,ux);
    // Host and device pointers to stencil coeffs
    double const* hc = stencil.coeffs_host();
    double const* dc = stencil.coeffs_device();

    return 0;
}
```

In the following sections we demonstrate how to go from an initial value problem (IVP) or a boundary value problem (BVP) to a working application solver by combining existing library components along with new custom-tailored components. We also demonstrate how to apply spatial and temporal domain decomposition strategies that can make existing solvers take advantage of systems equipped with multiple GPUs. Next section demonstrates how to rapidly assemble a PDE solver using library components.

6.3 Model problems

We present two elementary PDE model problems, to demonstrate how to assemble PDE solvers, using library components that follow the guidelines described above. The first model problem is the unsteady parabolic heat conduction equation; the second model problem is the elliptic Poisson equation. The two model problems consist of elements that play important roles in solving a broad range of more advanced PDE problems.

We refer the reader to Chapter 11 for an example of a scientific application relevant for coastal and maritime engineering analysis that has been assembled using customized library components similar to those presented in the following.

6.3.1 Heat conduction equation

First, we consider a two-dimensional heat conduction problem defined on a unit square. The heat conduction equation is a parabolic partial differential diffusion equation, including both spatial and temporal derivatives. It describes how the diffusion of heat in a medium changes with time. Diffusion equations are of great importance in many fields of sciences, e.g., fluid dynamics, where the fluid motion is uniquely described by the Navier-Stokes equations, which include a diffusive viscous term [7, 9].

The heat problem is an IVP, it describes how the heat distribution evolves from a specified initial state. Together with homogeneous Dirichlet boundary conditions, the heat problem in the unit square is given as

$$\frac{\partial u}{\partial t} - \kappa \nabla^2 u = 0, \quad (x, y) \in \Omega([0,1] \times [0,1]), \quad t \geq 0, \quad \text{(6.4a)}$$

$$u = 0, \quad (x, y) \in \partial\Omega, \quad \text{(6.4b)}$$

where $u(x, y, t)$ is the unknown heat distribution defined within the domain Ω, t is the time, κ is a heat conductivity constant (let $\kappa = 1$), and ∇^2 is the two-dimensional Laplace differential operator $(\partial_{xx} + \partial_{yy})$. We use the following initial condition:

$$u(x, y, t_0) = \sin(\pi x) \sin(\pi y), \quad (x, y) \in \Omega, \quad \text{(6.5)}$$

because it has a known analytic solution over the entire time span, and it satisfies the homogeneous boundary condition given by (6.4b). An illustrative example of the numerical solution to the heat problem, using (6.5) as the initial condition, is given in Figure 6.2.

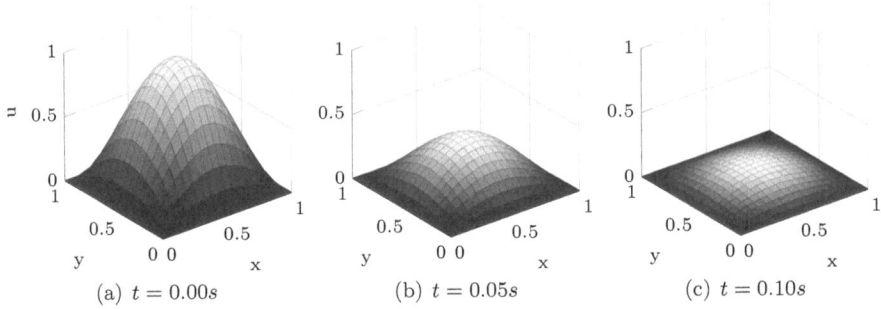

FIGURE 6.2. Discrete solution, at times $t = 0s$ and $t = 0.05s$, using (6.5) as the initial condition and a small 20×20 numerical grid.

We use a Method of Lines (MoL) approach to solve (6.4). Thus, the spatial derivatives are replaced with finite difference approximations, leaving only the temporal derivative as unknown. The spatial derivatives are approximated from $\mathbf{u}^n$, where $\mathbf{u}^n$ represents the approximate solution to $u(t_n)$ at a given time t_n with time step size δt such that $t_n = n\delta t$ for $n = 0, 1, \ldots$ The finite difference approximation can be interpreted as a matrix-vector product as sketched in (6.2), and so the semi-discrete heat conduction problem becomes

$$\frac{\partial u}{\partial t} = \mathcal{A}\mathbf{u}, \qquad \mathcal{A} \in \mathbb{R}^{N \times N}, \quad \mathbf{u} \in \mathbb{R}^N, \tag{6.6}$$

where $\mathcal{A}$ is the sparse finite difference matrix and N is the number of unknowns in the discrete system. The temporal derivative is now free to be approximated by any suitable choice of a time-integration method. The most simple integration scheme would be the first-order accurate explicit forward Euler method,

$$\mathbf{u}^{n+1} = \mathbf{u}^n + \delta t \,\mathcal{A}\mathbf{u}^n, \tag{6.7}$$

where $n + 1$ refers to the solution at the next time step. The forward Euler method can be exchanged with alternative high-order accurate time integration methods, such as Runge-Kutta methods or linear multistep methods, if numerical instability becomes an issue, see, e.g., [17] for details on numerical stability analysis. For demonstrative purpose, we simply use conservative time step sizes to avoid stability issues. However, the component-based library design provides exactly the flexibility for the application developer to select or change PDE solver parts, such as the time integrator, with little coding effort. A generic implementation of the forward Euler method that satisfies the library concept rules is illustrated in Listing 6.3. According to the component guidelines in Figure 6.1, a time integrator is basically a functor, which means that it implements the parenthesis operator, taking five template arguments: a right hand side operator, the state vector, integration start time, integration end time, and a time step size. The method takes as many time steps necessary to integrate from the start to the end, continuously updating the state vector according to (6.7). Notice, that nothing in Listing 6.3 indicates wether GPUs are used or not. However, it is likely that the underlying implementation of the right hand side functor and the axpy vector function, do rely on fast GPU kernels. However, it is not something that the developer of the component has to account for. For this reason, the template-based approach, along with simple interface concepts, make it easy to create new components that will fit well into a generic library.

The basic numerical approach to solve the heat conduction problem has now been outlined, and we are ready to assemble the PDE solver.

Listing 6.3. generic implementation of explicit first-order forward Euler integration

```
struct forward_euler
{
    template <typename F, typename T, typename V>
    void operator()(F fun, V& x, T t, T tend, T dt)
    {
        V rhs(x);                    // Initialize RHS vector
        while(t < tend)
        {
            if(tend-t < dt)
                dt = tend-t;         // Adjust dt for last time step

            (*fun)(t, x, rhs);       // Apply rhs function
            x.axpy(dt,rhs);          // Update stage
            t += dt;                 // Next time step
        }
    }
}
```

6.3.1.1 Assembling the heat conduction solver

Before we are able to numerically solve the discrete heat conduction problem (6.4), we need implementations to handle the the following items:

Grid – A discrete numerical grid to represent the two-dimensional heat distribution domain and the arithmetical working precision (32-bit single-precision or 64-bit double-precision).

RHS – A right-hand side operator for (6.6) that approximates the second-order spatial derivatives (matrix-vector product).

Boundary conditions – A strategy that ensures that the Dirichlet conditions are satisfied on the boundary.

Time integrator – A time integration scheme, that approximates the time derivative from (6.6).

All items are either directly available in the library or can be designed from components herein. The built-in stencil operator may assist in implementing the matrix-vector product, but we need to explicitly ensure that the Dirichlet boundary conditions are satisfied. We demonstrated in Listing 6.2 how to approximate the derivative using flexible order finite difference stencils. However, from (6.4b) we know that boundary values are zero. Therefore, we extend the stencil operator with a simple kernel call that assigns zero to the entire boundary. Listing 6.4 shows the code for the two-dimensional Laplace right-hand side operator. The constructor takes as an argument the stencil half size α and assumes $\alpha = \beta$. Thus, the total two-dimensional stencil rank will be $4\alpha + 1$. For simplicity we also assume that the grid is uniformly distributed, $N_x = N_y$. Performance optimizations for the stencil kernel, such as shared memory utilization, are handled in the underlying implementation, accordingly to CUDA guidelines [21,22]. The macros, BLOCK1D and GRID1D, are used to help set up kernel configurations based on grid sizes and RAW_PTR is used to cast the vector object to a valid device memory pointer.

Listing 6.4. the right-hand side Laplace operator: the built-in stencil approximates the two-dimensional spatial derivatives, while the custom set_dirichlet_bc kernel takes care of satisfying boundary conditions

```
template <typename T>
__global__ void set_dirichlet_bc(T* u, int Nx)
{
    int i = blockDim.x*blockIdx.x+threadIdx.x;
    if(i<Nx)
    {
        u[i]            = 0.0;
        u[(Nx-1)*Nx+i]  = 0.0;
        u[i*Nx]         = 0.0;
        u[i*Nx+Nx-1]    = 0.0;
    }
};

template <typename T>
struct laplacian
{
    gpulab::FD::stencil_2d<T> m_stencil;

    laplacian(int alpha) : m_stencil(2,alpha) {}

    template <typename V>
    void operator()(T t, V const& u, V & rhs) const
    {
        m_stencil.mult(u,rhs); // rhs = du/dxx + du/dyy

        // Make sure bc is correct
        dim3 block = BLOCK1D(rhs.Nx());
        dim3 grid  = GRID1D(rhs.Nx());
        set_dirichlet_bc<<<grid,block>>>(RAW_PTR(rhs),rhs.Nx());
    }
};
```

With the right-hand side operator in place, we are ready to implement the solver. For this simple PDE problem we compute all necessary initial data in the body of the main function and use the forward Euler time integrator to compute the solution until $t = t_{end}$. For more advanced solvers, a built-in ode_solver class is defined that helps take care of initialization and storage of multiple state variables. Declaring type definitions for all components at the beginning of the main file gives a good overview of the solver composition. In this way, it will be easy to control or change solver components at later times. Listing 6.5 lists the type definitions that are used to assemble the heat conduction solver.

Listing 6.5. type definitions for all the heat conduction solver components used throughout the remaining code

```
typedef double                              value_type;
typedef laplacian<value_type>               rhs_type;
typedef gpulab::grid<value_type>            vector_type;
typedef vector_type::property_type          property_type;
typedef gpulab::integration::forward_euler  time_integrator_type;
```

The grid is by default treated as a device object, and memory is allocated on the GPU upon initialization of the grid. Setting up the grid can be done via the property type class. The property class holds information about the discrete and physical dimensions, along with fictitious ghost (halo) layers and periodicity conditions. For the heat conduction problem we use a nonperiodic domain of size $N \times N$ within the unit square with no ghost layers. Listing 6.6 illustrates the grid assembly.

Listing 6.6. creating a two-dimensional grid of size N times N and physical dimension 0 to 1

```
// Setup discrete and physical dimensions
gpulab::grid_dim<int>         dim(N,N,1);
gpulab::grid_dim<value_type>  p0(0,0);
gpulab::grid_dim<value_type>  p1(1,1);
property_type                 props(dim,p0,p1);

// Initialize vector
vector_type                   u(props);
```

Hereafter the vector u can be initialized accordingly to (6.5). Finally we need to instantiate the right-hand side Laplacian operator from Listing 6.4 and the forward Euler time integrator in order to integrate from t_0 until t_{end}.

Listing 6.7. creating a time integrator and the right-hand side Laplacian operator

```
rhs_type rhs(alpha);          // Create right-hand side operator
time_integrator_type solver;  // Create time integrator
solver(&rhs,u,0.0f,tend,dt);  // Integrate from 0 to tend using dt
```

The last line invokes the forward Euler time integration scheme defined in Listing 6.5. If the developer decides to change the integrator into another explicit scheme, only the time integrator type definition in Listing 6.5 needs to be changed. The heat conduction solver is now complete.

6.3.1.2 Numerical solutions to the heat conduction problem

Solution time for the heat conduction problem is in itself not very interesting, as it is only a simple model problem. What is interesting for GPU kernels, such as the finite differences kernel, is that increased computational work often comes with a very small price, because the fast computations can be hidden by the relatively slower memory fetches. Therefore, we are able to improve the accuracy of the numerical solution via more accurate finite differences (larger stencil sizes), while improving the computational performance in terms of floating point operations per second (flops). Figure 6.3 confirms, that larger stencils improve the kernel performance. Notice that even though these performance results are favorable compared to single core systems ($\sim$ 10 GFlops double-precision on a 2.5-GHz processor), they are still far from their peak performance, e.g., $\sim$ 2.4 TFlops single-precision for the GeForce GTX590. The reason is that the kernel is bandwidth bound, i.e., performance is limited by the time it takes to move memory between the global GPU memory and the chip. The Tesla K20 performs better than the GeForce GTX590 because it obtains the highest bandwidth. Being bandwidth bound is a general limitation for matrix-vector-like operations that arise from the discretization of PDE problems. Only matrix-matrix multiplications, which have a high ratio of computations versus memory transactions, are able to reach near-optimal performance results [15]. These kinds of operators are, however, rarely used to solve PDE problems.

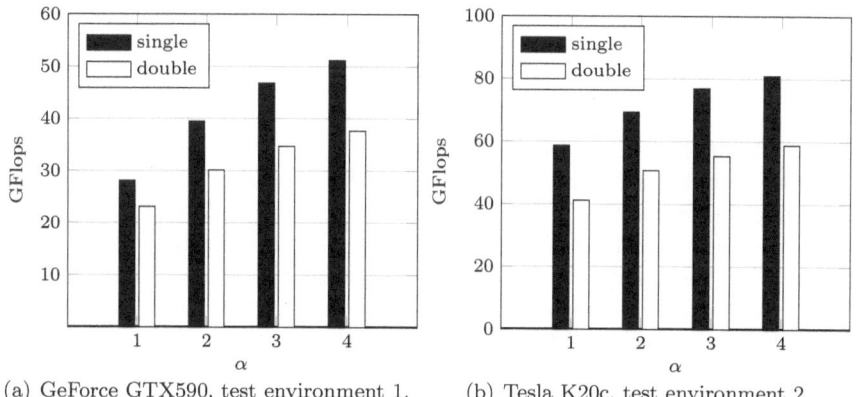

(a) GeForce GTX590, test environment 1. (b) Tesla K20c, test environment 2.

FIGURE 6.3. Single- and double-precision floating point operations per second for a two-dimensional stencil operator on a numerical grid of size 4096^2. Various stencil sizes are used $\alpha = 1, 2, 3, 4$, equivalent to 5pt, 9pt, 13pt, and 17pt stencils.

6.3.2 Poisson equation

The Poisson equation is a second-order elliptic differential equation, often encountered in applications within scientific fields such as electrostatics and mechanics. We consider the two-dimensional BVP defined in terms of Poisson's equation with homogeneous Dirichlet boundary conditions on the form

$$\nabla^2 u = f(x,y), \quad (x,y) \in \Omega([0,1] \times [0,1]), \tag{6.8a}$$
$$u = 0, \quad (x,y) \in \partial\Omega. \tag{6.8b}$$

Notice the similarities to the heat conduction equation (6.4). In fact, (6.8) could be a steady-state solution to the heat equation, when there is no temporal change $\frac{\partial u}{\partial t} = 0$, but a source term $f(x,y)$. Since the Laplace operator and the boundary conditions are the same for both problems, we are able to reuse the same implementation with few modifications.

Opposite to the heat equation, there are no initial conditions. Instead, we seek some $u(x,y)$ that satisfies (6.8), given a source term $f(x,y)$, on the right-hand side. For simplicity, assume that we know the exact solution, u_{true}, corresponding to (6.5). Then we use the method of manufactured solutions to derive an expression for the corresponding right-hand side $f(x,y)$:

$$f(x,y) = \nabla^2 u_{\text{true}} = -2\pi^2 \sin(\pi x) \sin(\pi y). \qquad (6.9)$$

The spatial derivative in (6.8) is again approximated with finite differences, similar to the example in (6.2), except boundary values are explicitly set to zero. The discrete form of the system can now be written as a sparse linear system of equations:

$$\mathcal{A}\mathbf{u} = \mathbf{f}, \quad \mathbf{u}, \mathbf{f} \in \mathbb{R}^N, \quad \mathcal{A} \in \mathbb{R}^{N \times N}, \qquad (6.10)$$

where $\mathcal{A}$ is the sparse matrix formed by finite difference coefficients, N is the number of unknowns, and $\mathbf{f}$ is given by (6.9). Equation (6.10) can be solved in numerous ways, but a few observations may help do it more efficiently. Direct solvers based on Gaussian elimination are accurate and use a finite number of operations for a constant problem size. However, the arithmetic complexity grows with the problem size by as much as $\mathcal{O}(N^3)$ and does not exploit the sparsity of $\mathcal{A}$. Direct solvers are therefore mostly feasible for dense systems of limited sizes. Sparse direct solvers exist, but they are often difficult to parallelize, or applicable for only certain types of matrices. Regardless of the discretization technique, the discretization of an elliptic PDE into a linear system as in (6.10) yields a very sparse matrix $\mathcal{A}$ when N is large. Iterative methods for solving large sparse linear systems find broad use in scientific applications, because they require only an implementation of the matrix-vector product, and they often use a limited amount of additional memory. Comprehensive introductions to iterative methods may be found in any of [4, 13, 23].

One benefit of the high abstraction level and the template-based library design is to allow developers to implement their own components, such as iterative methods for solving sparse linear systems. The library includes three popular iterative methods: conjugate gradient, defect correction, and geometric multigrid. The conjugate gradient method is applicable only to systems with symmetric positive definite matrices. This is true for the two-dimensional Poisson problem, when it is discretized with a five-point finite difference stencil, because then there will be no off-centered approximations near the boundary. For high-order approximations ($\alpha > 1$), we use the defect correction method with multigrid preconditioning. See e.g., [26] for details on multigrid methods.

We will not present the implementation details for all three methods but briefly demonstrate the simplicity of implementing the body of such an iterative solver, given a textbook recipe or mathematical formulation. The defect correction method iteratively improves the solution to $\mathcal{A}\mathbf{x} = \mathbf{b}$, given an initial start guess $\mathbf{x}^0$, by continuously solving a preconditioned error equation. The defect correction iteration can be written as

$$\mathbf{x}^{k+1} = \mathbf{x}^k + \mathcal{M}^{-1}(\mathbf{b} - \mathcal{A}\mathbf{x}^k), \quad \mathcal{A}, \mathcal{M} \in \mathbb{R}^{N \times N}, \quad \mathbf{x}, \mathbf{b} \in \mathbb{R}^N, \qquad (6.11)$$

where k is the iteration number and $\mathcal{M}$ is the preconditioner which should be an approximation to the original coefficient matrix $\mathcal{A}$. To achieve fast numerical convergence, applying the preconditioner should be a computationally inexpensive operation compared to solving the original system. How to implement (6.11) within the library context is illustrated in Listing 6.8. The host CPU traverses each line in Listing 6.8 and tests for convergence, while the computationally expensive matrix-vector operation and preconditioning, can be executed on the GPU, if GPU-based components are used. The defect correction method has two attractive properties. First, global reduction is required to monitor convergence only once per iteration during convergence evaluation, which reduces communication requirements and provides a basis for efficient and scalable parallelization. Second, it has a minimal constant memory footprint, making it a suitable method for solving very large systems.

Listing 6.8. main loop for the iterative defect correction solver: the solver is instantiated with template argument types for the matrix and vector classes, allowing underlying implementations to be based on GPU kernels

```
while(r.nrm2() > tol)
{
    // Calculate residual
    A.mult(x,r);
    r.axpby(1, -1, b);

    // Reset initial guess
    d.fill(0);

    // Solve M*d=r
    M(d,r);

    // Defect correction update
    x.axpy(1, d);
}
```

In the following section we demonstrate how to assemble a solver for the discrete Poisson problem, using one of the three iterative methods to efficiently solve (6.10).

6.3.2.1 Assembling the Poisson solver

Assembling the Poisson solver follows almost the same procedure as the heat conduction solver, except the time integration part is exchanged with an iterative method to solve the system of linear equations (6.10). For the discrete matrix-vector product we reuse the Laplace operator from the heat conduction problem in Listing 6.4 with few modifications. The Laplace operator is now a matrix component, so to be compatible with the component interface rules in Figure 6.1, a `mult` function taking two vector arguments is implemented instead of the parentheses operator. We leave out this code example as it almost identical to the one in Listing 6.4.

At the beginning of the solver implementation we list the type definitions for the Poisson solver that will be used throughout the implementation. Here we use a geometric multigrid method as a preconditioner for the defect correction method. Therefore the multigrid solver is assembled first, so that it can be used in the assembling of the defect correction solver. Listing 6.9 defines the types for the vector, the matrix, the multigrid preconditioner, and the defect correction solver. The geometric multigrid method needs two additional template arguments that are specific for multigrid, namely, a smoother and a grid restriction/interpolation operator. These arguments are free to be implemented and supplied by the developer if special care is required, e.g., for a custom grid structure. For the Poisson problem on a regular grid, the library contains built-in restriction and interpolation operators, and a red-black Gauss-Seidel smoother. We refer the reader to [26] for extensive details on multigrid methods. The monitor and config types that appear in Listing 6.9 are used for convergence monitoring within the iterative solver and to control runtime parameters, such as tolerances and iteration limits.

Listing 6.9. type definitions for the Laplacian matrix component and the multigrid preconditioned iterative defect correction solver

```
typedef double                                              value_type;
typedef gpulab::grid<value_type>                            vector_type;
typedef laplacian<vector_type>                              matrix_type;

// MULTIGRID solver types
typedef gpulab::solvers::multigrid_types<
    vector_type
    , matrix_type
    , gpulab::solvers::gauss_seidel_rb_2d
    , gpulab::solvers::grid_handler_2d>                     mg_types;
typedef gpulab::solvers::multigrid<mg_types>                mg_solver_type;
typedef mg_solver_type::monitor_type                        monitor_type;
typedef monitor_type::config_type                           config_type;

// DC solver types
typedef gpulab::solvers::defect_correction_types<
    vector_type
    , matrix_type
    , monitor_type
    , mg_solver_type>                                       dc_types;
typedef gpulab::solvers::defect_correction<dc_types>        dc_solver_type;
```

With the type definitions set up, the implementation for the Poisson solver follows in Listing 6.10. Some of the initializations are left out, as they follow the same procedure as for the heat conduction example. The defect correction and geometric multigrid solvers are initialized and then multigrid is set as a preconditioner to the defect correction method. Finally the system is solved via a call to solve().

Software components for heterogeneous many-core architectures   91

> Listing 6.10. initializing the preconditioned defect correction solver to approximate the solution to $\mathcal{A}u = f$

```
matrix_type A(alpha);                        // High-order matrix
matrix_type M(1);                            // Low-order matrix

/* Omitted: create and init vectors u, f */

config_type config;                          // Create configuration
config.set("iter",30);                       // Set max iteration count
config.set("rtol",1e-10);                    // Set relative tolerance
monitor_type monitor(config);                // Create monitor
dc_solver_type solver(A,monitor);            // Create DC solver
mg_solver_type precond(M,monitor);           // Create MG preconditioner
solver.set_preconditioner(precond);          // Set preconditioner
solver.solve(u,f);                           // Solve M^-1(Au = f)
if(monitor.converged())
    printf("SUCCESS\n");
```

6.3.2.2 Numerical solutions to the Poisson problem

The discrete Poisson problem (6.10) has been solved using the three iterative methods presented above. Convergence histories for the conjugate gradient method and geometric multigrid method, using two different resolutions, are illustrated in Figure 6.4(a). Multigrid methods are very robust and algorithmic efficient, independent of the problem size. Figure 6.4(a) confirms that the rate of convergence for the multigrid method is unchanged for both problem sizes. Only the attainable accuracy is slightly worsened, as a consequence of a more ill-conditioned system for large problem sizes.

Defect correction in combination with multigrid preconditioning enables efficient solution of high-order approximations of the Poisson problem, illustrated in Figure 6.4(b). The multigrid preconditioning matrix $\mathcal{M}$ is based on a low-order approximation to (6.10), whereas matrix $\mathcal{A}$ is a high-order approximation. When $\mathcal{M}$ is a close approximation to $\mathcal{A}$, defect correction converges most rapidly. This is the effect that can be seen between the three convergence lines in Figure 6.4(b).

6.4 Optimization strategies for multi-GPU systems

CUDA-enabled GPUs are optimized for high memory bandwidth and fast on-chip performance. However, the role as a separate coprocessor to the CPU can be a limiting factor for large scale scientific applications, because the GPU memory capacity is fixed and is only in the range of a few gigabytes. In comparison, it is not unusual for a high-end workstation to be equipped with $\sim 32\text{GB}$ of main memory, plus a terabyte hard disk capacity for secondary storage. Therefore, large scale scientific applications that process gigabytes of data, require distributed computations on multiple GPU devices. Multi-GPU

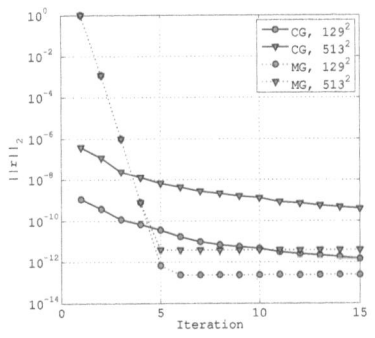 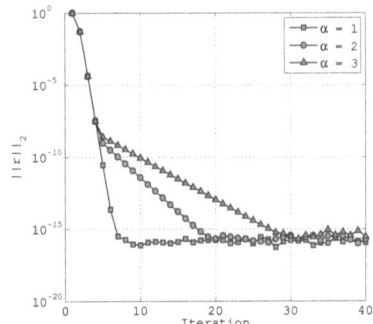

(a) Convergence histories for the conjugate gradient (CG) and multigrid (MG) methods, for two different problem sizes.

(b) Defect correction convergence history for three different stencil sizes.

FIGURE 6.4. Algorithmic performance for the conjugate gradient, multigrid, and defect correction methods, measured in terms of the relative residual per iteration.

desktop computers and clusters can have a very attractive peak performance, but the addition of multiple devices introduces the potential performance bottleneck of slow data transfers across PCIe busses and network interconnections, as illustrated in Figure 6.5. The ratio between data transfers and computational work has a significant impact on the possibility for latency hiding and thereby overall application performance.

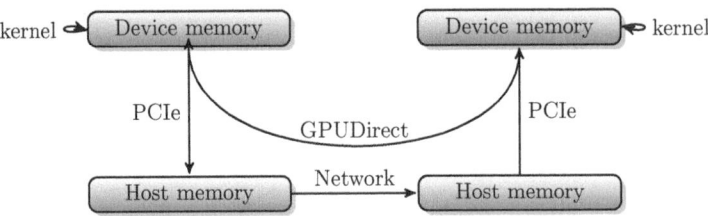

FIGURE 6.5. Message passing between two GPUs involves several memory transfers across lower bandwidth connections. The kernel call is required if the data is not already sequentially stored in device memory. Recent generations of NVIDIA GPUs, CUDA, and MPI support direct transfers without explicitly transfering data to the host first.

Developing applications that exploit the full computational capabilities of modern clusters–GPU-based or not–is no trivial matter. Developers are faced with the complexity of distributing and coordinating computations on nodes consisting of many-core CPUs, GPUs, and potentially other types of accelerators as well. These complexities give rise to challenges in finding numerical algorithms, that are well suited for such systems, forcing developers to search for novel methods that utilize concurrency.

To ease software development, we use MPI-2 for message passing and ensure a safe and private communication space by creation of a communicator private to the library during initialization, as recommended by Hoefler and Snir [12]. With the addition of remote direct memory access (RDMA) for GPUDirect it is possible to make direct memory transfers between recent generation of GPUs (Kepler), eliminating CPU overhead. Unfortunately there are some strict system and driver requirements to enable these features. Therefore, in the following examples, device memory is first transferred to the CPU main memory before invoking any MPI calls. The library provides device-to-device transfers via template-based routines that work directly with GPU vector objects. This hides the complexity of message passing from the developer and helps developers design new components for multi-GPU execution.

In the following sections we present two very different methods for distributed computing based on spatial and temporal decomposition. Each method has its own characteristic, which makes the method attractive for various types of PDE problems and for different problem sizes.

6.4.1 Spatial domain decomposition

Domain decomposition methods can be used for distributing computational work in the numerical solution of boundary value problems [25]. These methods add parallelism by splitting the spatial dimensions, on which the boundary values are defined, into a number of smaller boundary value problems and then coordinating the solution between adjacent subdomains. Domain decomposition techniques, such as the classical overlapping Schwarz methods, may be considered as preconditioners to the system of equations that arise from the discretization of PDEs, e.g., as for the Poisson problem in (6.10). The algorithmic efficiency of such methods depends on the size of the domain overlaps, while an additional coarse grid correction method is sometimes necessary to maintain fast global convergence.

An alternative to the preconditioning strategy is to have each subdomain query information from adjacent subdomains whenever needed. For PDEs that are discretized onto regular shaped grids, this can be an attractive strategy, as the decomposition of subdomains, and thereby the communication topology, is straightforward. The library supports decomposition of regular grids by either pre- or user-defined topologies. A topology in this context, is a description of connectivity between processors that share the same grid along with information about the local and global discretization. Layers of ghost points are inserted at the artificially introduced boundaries to account for grid points that reside in adjacent subdomains. The ghost point values can be updated upon request from the user. An illustrative example is given in Figure 6.6; the arrows indicate message passing between adjacent subdomains, when updating grid points within the ghost layers.

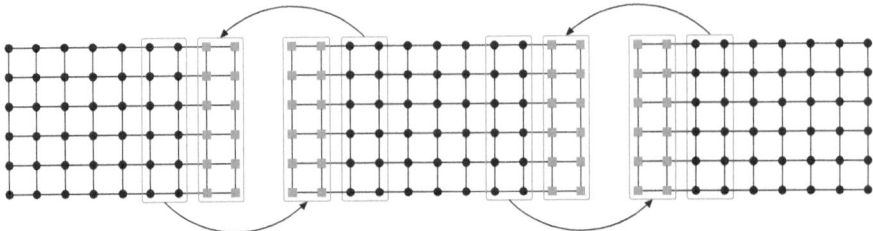

FIGURE 6.6. Domain distribution of a two-dimensional grid into three subdomains. • and ■ represent internal grid points and ghost points, respectively.

Topologies are introduced via an extra template argument to the grid class. A grid is by default not decomposed, because the default template argument is based on a nondistribution topology implementation. The grid class is extended with a new member function update(), which makes sure that all ghost points are updated according to the grid topology. The library contains topologies based on one-dimensional and two-dimensional distributions of the grid. The number of grid subdomains will be equal to the number of MPI processes executing the program.

If grid ghost layers are updated whenever information from adjacent subdomains is needed, e.g., before a stencil operation, all interior points will be exactly the same as they would be for the nondistributed setup. Therefore, one advantage of this approach is that the algorithmic efficiency of an application can be preserved, if grid updates are consistently invoked at the proper times.

Distributed performance for the finite difference stencil operation is illustrated in Figure 6.7. The timings include the compute time for the finite difference approximation and the time for updating ghost layers via message passing. It is obvious from Figure 6.7(a) that communication overhead dominates for the smallest problem sizes, where the nondistributed grid (1 GPU) is fastest. However, communication overhead does not grow as rapidly as computation times, due to the surface-to-volume ratio. Therefore message passing becomes less influential for large problems, where reasonable performance speedups are obtained. Figure 6.7(b) demonstrates how the computational performance on multi-GPU systems can be significantly improved for various stencil sizes. With this simple domain decomposition technique, developers are able to implement applications based on heterogeneous distributed computing, without explicitly dealing with message passing and it is still possible to provide user specific implementations of the topology class for customized grid updates.

Software components for heterogeneous many-core architectures 95

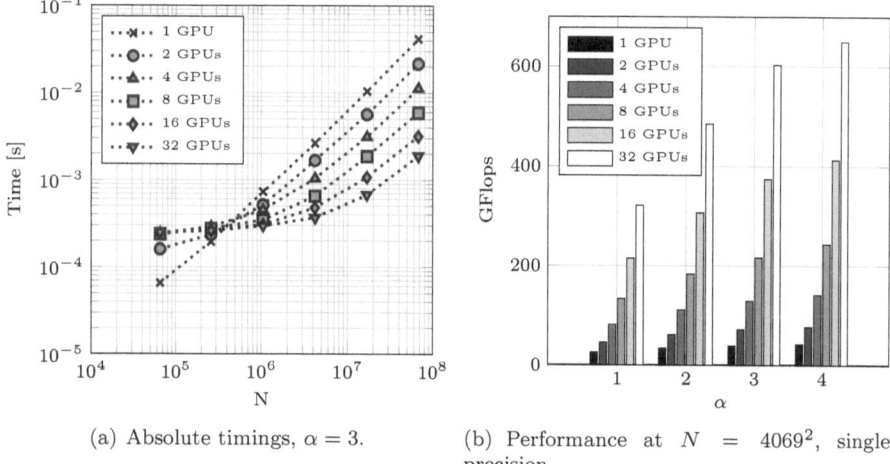

(a) Absolute timings, $\alpha = 3$.

(b) Performance at $N = 4069^2$, single-precision.

FIGURE 6.7. Performance timings for distributed stencil operations, including communication and computation times. Executed on test environment 3.

6.4.2 Parareal–parallel time integration

The use of spatial domain decomposition methods is widespread, as they have proven to be efficient on a wide range of problems. Unfortunately, applications that are concerned with limited numerical problem sizes can rapidly reach a speedup limit for a low number of processors due to scalability degradation when the number of processors becomes large, as this leads to an increasingly unfavourable communication-to-compute ratio. This issue is continuously worsened by the fact that communication speed has been increasing at a far slower pace than compute speed for the past several years, and this trend is expected to continue for years to come. It is often referred to as *the memory wall* [1], one of the grand challenges facing development and architectural design of future high-performance systems [8, 14]. Also, there are applications based on ordinary differential equations, where classical domain decomposition methods are not even applicable [19]. For these type of applications, a method of adding parallelism in the temporal integration is of great interest. Contrary to space however, time is–by its very nature–sequential, which precludes a straightforward implementation of a parallel approach.

One method that introduces concurrency to the solution of evolution problems is the parareal algorithm. Parareal is an iterative method imposed on a time decomposition. Gander and Vandewalle showed in [11] that the algorithm can be written both as a multiple shooting method and as a two-level multigrid-in-time approach, even though the leading idea came from spatial domain decomposition. The method has many exciting features: it is fault tolerant and has different communication characteristics than those of the classical domain decomposition methods. It has also been demonstrated to work effectively on a wide range of problems, and most importantly, once the proper distribution infrastructure is in place, it can easily be wrapped around any type of numerical integrator, for any type of initial value problem.

6.4.2.1 The parareal algorithm

The parareal algorithm was first presented in 2001, in a paper by Lions et al. [18], and later introduced in a slightly revised predictor-corrector form in 2002 by Baffico et al. [3]. The parareal-in-time approach proposes to break the global problem of time evolution into a series of independent evolution problems on smaller intervals, see Figure 6.8. Initial states for these problems

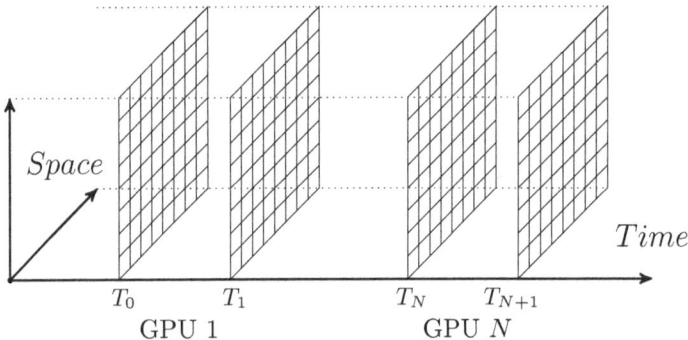

FIGURE 6.8. Time domain decomposition. A compute node is assigned to each individual time subdomain to compute the initial value problem. Consistency at the time subdomain boundaries is obtained with the application of a computationally cheap integrator in conjunction with the parareal iterative predictor-corrector algorithm.

are needed and supplied by a simple, less accurate, but computationally cheap sequential integrator. The smaller independent evolution problems can then be solved in parallel. The information, generated during the concurrent solution of the independent evolution problems with accurate propagators and inaccurate initial states, is used in a predictor-corrector fashion in conjunction with the coarse integrator to propagate the solution faster, now using the information generated in parallel. We define the decomposition into N intervals, that is,

$$T_0 < T_1 < \cdots < T_n = n\Delta T < T_{n+1} < T_N, \qquad (6.12)$$

where ΔT is the size of the time intervals and $n = 0, 1, \ldots, N$. The general initial value problem on the decomposed time domain is defined as

$$\frac{\partial u}{\partial t} + \mathcal{A}u = 0, \qquad u(T_0) = u^0, \quad t \in [T_0, T_N], \tag{6.13}$$

where $\mathcal{A}$ is an operator from one Hilbert space to another. To solve the differential problem (6.13) we define an operator $\mathcal{F}_{\Delta T}$ that operates on some initial state $U_n \approx u(T_n)$ and returns an approximate solution to (6.13), at time $T_n + \Delta T$. Such an operator is achieved by the implementation of a numerical time integrator, using some small time-step $\delta t \ll \Delta T$ in the integration. The numerical solution to (6.13) can then be obtained by applying the fine propagator sequentially for $n = 1, 2, \ldots, N$.

$$\hat{U}_n = \mathcal{F}_{\Delta T}\left(T_{n-1}, \hat{U}_{n-1}\right), \qquad \hat{U}_0 = u^0. \tag{6.14}$$

For the purpose of parallel acceleration of the otherwise purely sequential process of obtaining $\mathcal{F}_{\Delta T}^N u^0 \approx u(T_N)$, we define the coarse propagator $\mathcal{G}_{\Delta T}$. $\mathcal{G}_{\Delta T}$ also operates on some initial state U_n, propagating the solution over the time interval ΔT, but now using a time step δT. Typically $\delta t < \delta T < \Delta T$. For the parareal algorithm to be effective, the coarse propagator $\mathcal{G}_{\Delta T}$ has to be substantially faster to evaluate than the fine propagator $\mathcal{F}_{\Delta T}$. There are many ways of constructing the coarse propagator, the simplest one being to apply the same numerical integrator as for the fine propagator, but using a coarser time discretization. We refer the reader to [20] for an introduction to other methods. The coarse operator reads

$$\tilde{U}_n = \mathcal{G}_{\Delta T}\left(T_{n-1}, \tilde{U}_{n-1}\right), \qquad \tilde{U}_0 = u^0. \tag{6.15}$$

Using the defined $\mathcal{F}_{\Delta T}$ and $\mathcal{G}_{\Delta T}$ operators, the predictor-corrector form of the parareal algorithm can be written in a single line as

$$U_n^{k+1} = \mathcal{G}_{\Delta T}\left(U_{n-1}^{k+1}\right) + \mathcal{F}_{\Delta T}\left(U_{n-1}^k\right) - \mathcal{G}_{\Delta T}\left(U_{n-1}^k\right), \qquad U_0^k = u^0, \tag{6.16}$$

with the initial prediction $U_n^0 = \mathcal{G}_{\Delta T}^n u^0$ for $n = 1 \ldots N$ and $k = 1 \ldots K$. N being the number of time subdomains, while $K \geq 1$ is the number of predictor-corrector iterations applied. The parareal algorithm is implemented in the library as a separate time-integration component, using a fully distributed work scheduling model, as proposed by Aubanel [2]. The model is schematically presented in Figure 6.9. The parareal component hides all communication and work distribution from the application developer. It is defined such that a user only has to decide what coarse and fine propagators to use. Setting up the type definitions for parareal time-integration using forward Euler for coarse propagation and fourth order Runge-Kutta for fine propagation could then be defined as in Listing 6.11. The number of GPUs used for parallelization depends on the number of MPI processes executing the application.

Listing 6.11. assembling a parareal time integrator using forward Euler for coarse propagation and a Runge-Kutta method for fine propagation

```
typedef gpulab::integration::forward_euler         coarse;
typedef gpulab::integration::ERK4                  fine;
typedef gpulab::integration::parareal<coarse,fine> integrator;
```

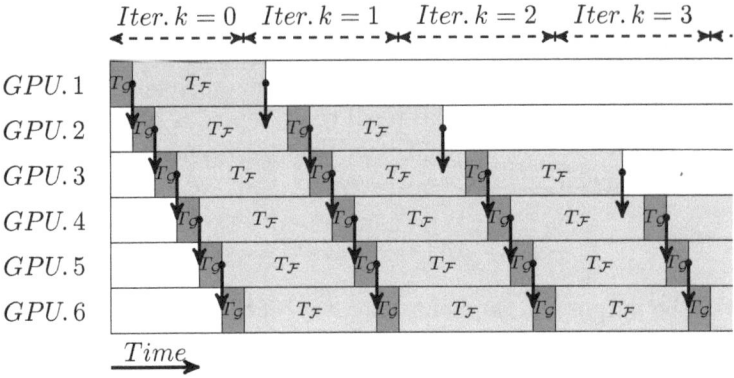

FIGURE 6.9. Schematic visualization of a fully distributed work scheduling model for the parareal algorithm as proposed by Aubanel [2]. Each GPU is responsible for computing the solution on a single time subdomain. The computation is initiated at rank 0 and cascades through to rank N where the final solution can be fetched.

6.4.2.2 Computational complexity

In the analysis of the computational complexity, we first recognize that both the coarse and the fine propagators, regardless of the type of discretization scheme, involve a complexity that is proportional to the number of time steps being used. Let us define two scalar values $\mathcal{C}_\mathcal{F}$ and $\mathcal{C}_\mathcal{G}$ as the computational cost of performing a single step with the fine and coarse propagators. The computational complexity of a propagator integrating over an interval ΔT is then given by $\mathcal{C}_\mathcal{F} \frac{\Delta T}{\delta t}$ and $\mathcal{C}_\mathcal{G} \frac{\Delta T}{\delta T}$, respectively. R is introduced as the relation between the two; that is, R is a measure of how much faster the coarse propagator is compared to the fine propagator in integrating the time interval ΔT. The total computational cost for parareal over N intervals is then proportional to

$$(k+1) N \mathcal{C}_\mathcal{G} \frac{\Delta T}{\delta T} + k N \mathcal{C}_\mathcal{F} \frac{\Delta T}{\delta t}. \qquad (6.17)$$

Recognizing that the second term can be distributed over N processors, we are left with

$$(k+1) N \mathcal{C}_\mathcal{G} \frac{\Delta T}{\delta T} + k \mathcal{C}_\mathcal{F} \frac{\Delta T}{\delta t}. \qquad (6.18)$$

The above should be compared to the computational complexity of a purely sequential propagation, using only the fine operator,

$$\frac{T_N - T_0}{\delta t} \mathcal{C}_\mathcal{F} = N \frac{\Delta T}{\delta t} \mathcal{C}_\mathcal{F}. \qquad (6.19)$$

We can now estimate the speedup, here denoted ψ, as the ratio between the computational complexity of the purely sequential solution, and the complexity of the solution obtained by the parareal algorithm (6.18). Neglecting the influence of communication speed and correction time, we are left with the estimate

$$\psi = \frac{N \frac{\Delta T}{\delta t} \mathcal{C}_\mathcal{F}}{(k+1) N \mathcal{C}_\mathcal{G} \frac{\Delta T}{\delta T} + k \mathcal{C}_\mathcal{F} \frac{\Delta T}{\delta t}} = \frac{N}{(k+1) N \frac{\mathcal{C}_\mathcal{G}}{\mathcal{C}_\mathcal{F}} \frac{\delta t}{\delta T} + k}. \qquad (6.20)$$

If we additionally assume that the time spent on coarse propagation is negligible compared to the time spent on the fine propagation, i.e., the limit $\frac{\mathcal{C}_\mathcal{G}}{\mathcal{C}_\mathcal{F}} \frac{\delta t}{\delta T} \to 0$, the estimate reduces to $\psi = \frac{N}{k}$. It is thus clear that the number of iterations k for the algorithm to converge poses an upper bound on obtainable parallel efficiency. The number of iterations needed for convergence is intimately coupled with the ratio R between the speed of the fine and the coarse integrators $\frac{\mathcal{C}_\mathcal{F}}{\mathcal{C}_\mathcal{G}} \frac{\delta T}{\delta t}$. Using a slow, but more accurate coarse integrator will lead to convergence in fewer iterations k, but at the same time it also makes R smaller. Ultimately, this will degrade the obtained speedup as can be deduced from (6.20), and by Amdahl's law it will also lower the upper bound on possible attainable speedup. Thus, R *cannot* be made arbitrarily large since the ratio is inversely proportional to the number of iterations k needed for convergence. This poses a challenge in obtaining speedup and is a trade-off between time spent on the fundamentally sequential part of the algorithm and the number of iterations needed for convergence. It is particularly important to consider this trade-off in the choice of stopping strategy; a more thorough discussion on this topic is available in [20] for the interested reader. Measurements on parallel efficiency are typically observed in the literature to be in the range of 20–50%, depending on the problem and the number of time subdomains, which is also confirmed by our measurements using GPUs. Here we include a demonstration of the obtained speedup of parareal applied to the two-dimensional heat problem (6.4). In Figure 6.10 the iterations needed for convergence using the forward Euler method for both fine and coarse integration are presented. R is regulated by changing the time step size for the coarse integrator. In Figure 6.11 speedup and parallel efficiency measurements are presented. Notice, when using many GPUs it is advantageous to use a faster, less accurate coarse propagator, despite it requires an extra parareal iteration that increases the total computational complexity.

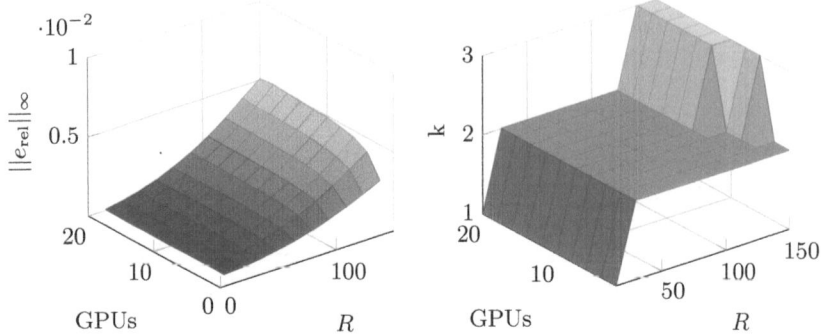

(a) The relative error after one parareal iteration ($k = 1$).

(b) Iterations K needed to obtain a relative error less than 10^{-5}.

FIGURE 6.10. Parareal convergence properties as a function of R and number of GPUs used. The error is measured as the relative difference between the purely sequential solution and the parareal solution.

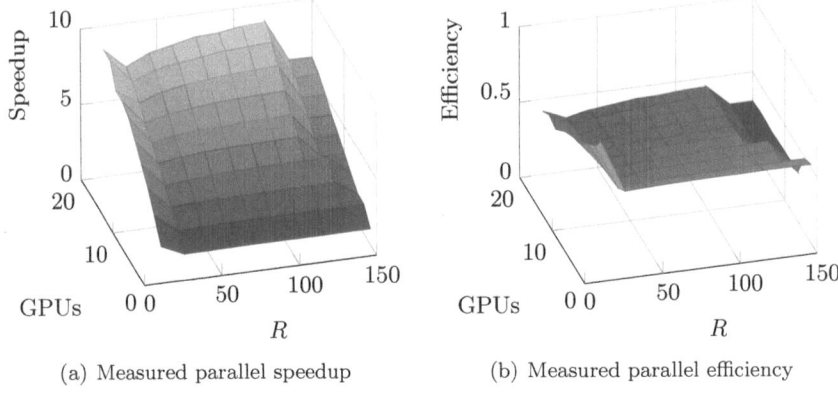

(a) Measured parallel speedup

(b) Measured parallel efficiency

FIGURE 6.11. Parareal performance properties as a function of R and number GPUs used. Notice how the obtained performance depends greatly on the choice of R as a function of the number of GPUs. Executed on test environment 3.

6.5 Conclusion and outlook

Massively parallel heterogeneous systems continue to enter the consumer market, and there has been no identification that this trend will stop for years to come. However, these parallel architectures require software vendors to adjust to new programming models and optimization strategies. Good software libraries are important tools for reducing the time and complexity of adjusting to new architectures, and they provide the user with an intuitive programming interface.

We have presented ideas for a generic GPU-based library for fast assembling of PDE solvers. A high-level interface and simple implementation concepts were created with the objective of enhancing developer productivity and to ensure performance portability. Two elementary PDE model problems were used to demonstrate assembling of both spatial and temporal approximation techniques. Results confirmed that numerical methods for solution of PDE systems are bandwidth limited on GPU systems. Therefore higher-order methods can be advantageous, when extra computational work can be performed during memory transfer idle times, leading to increased flop performance.

Decomposition strategies for spatial and temporal parallelization on multi-GPU systems have been presented, with reasonable performance speedups. Novel results for the parareal algorithm on multi-GPU systems have been presented, and an example of parallel efficiency for the heat conduction problem has been demonstrated. Furthermore, a domain decomposition technique that preserves algorithmic efficiency has been presented for the Poisson problem.

The library has already successfully been used for development of a fast tool intended for scientific applications within maritime engineering; see Chapter 11 for details. We intend to further extend the library, as we explore new techniques, suitable for parallelization on heterogeneous systems, that fit the scope of our applications.

For more information about the library and our work within the GPUlab group at the Technical University of Denmark, we encourage you to visit http://gpulab.imm.dtu.dk

Acknowledgments

This work have been supported by grant no. 09-070032 from the Danish Research Council for Technology and Production Sciences. A special thanks goes to Professor Jan S. Hesthaven for supporting parts of this work. Scalability and performance tests were done in the GPUlab at DTU Informatics, Technical University of Denmark and using the GPU-cluster at the Center for Computing and Visualization, Brown University, USA. The NVIDIA Corporation is acknowledged for generous hardware donations to facilities of the GPUlab.

Bibliography

[1] K. Asanovic, R. Bodik, B. C. Catanzaro, J. J. Gebis, P. Husbands, K. Keutzer, D. A. Patterson, W. L. Plishker, J. Shalf, S. W. Williams, and K. A. Yelick. The landscape of parallel computing research: A view from Berkeley. Technical Report UCB/EECS-2006-183, EECS Department, University of California, Berkeley, Dec 2006.

[2] E. Aubanel. Scheduling of tasks in the parareal algorithm. *Parallel Computing*, 37:172–182, 2010.

[3] L. Baffico, S. Bernard, Y. Maday, G. Turinici, and G. Zérah. Parallel-in-time molecular-dynamics simulations. *Phys. Rev. E*, 66:057701, Nov 2002.

[4] R. Barrett, M. Berry, T. F. Chan, J. Demmel, J. Donato, J. Dongarra, V. Eijkhout, R. Pozo, C. Romine, and H. Van der Vorst. *Templates for the Solution of Linear Systems: Building Blocks for Iterative Methods*. SIAM, Philadelphia, PA, 2nd edition, 1994.

[5] N. Bell and M. Garland. Implementing sparse matrix-vector multiplication on throughput-oriented processors. In *SC '09: Proceedings of the Conference on High Performance Computing Networking, Storage and Analysis*, pages 1–11, New York, NY, 2009. ACM.

[6] N. Bell and J. Hoberock. Thrust: A productivity-oriented library for CUDA. *In GPU Computing Gems, Jade Edition,* Edited by Wen-mei W. Hwu. Elsevier Science, 2:359–371, 2011.

[7] A. J. Chorin and J. E. Marsden. *A Mathematical Introduction to Fluid Mechanics*. Texts in Applied Mathematics. Springer, New York, NY, 3rd edition, 1993.

[8] P. Messina et. al D. L. Brown. Scientific grand challenges, crosscutting technologies for computing at the exascale. Technical Report, U.S. Department of Energy, Washington, D.C., February 2010.

[9] J.H. Ferziger and M. Perić. *Computational Methods for Fluid Dynamics*. Numerical Methods: Research and Development. Springer-Verlag, Berlin Heidelberg New York, 1996.

[10] E. Gamma, R. Helm, R. Johnson, and J. Vlissides. *Design Patterns– Elements of Reusable Object-Oriented Software*. Addison-Wesley Professional Computing Series, 1995.

[11] M. Gander and S. Vandewalle. Analysis of the parareal time-parallel time-integration method. *SIAM Journal of Scientific Computing*, 29(2):556–578, 2007.

[12] T. Hoefler and M. Snir. Writing parallel libraries with MPI–Common practice, issues, and extensions. In Y. Cotronis, A. Danalis, D. Nikolopoulos, and J. Dongarra, editors, *Recent Advances in the Message Passing Interface*, volume 6960 of *Lecture Notes in Computer Science*, pages 345–355. Springer, Berlin/Heidelberg, 2011.

[13] C. T. Kelley. *Iterative Methods for Linear and Nonlinear Equations*. Frontiers in Applied Mathematics Series. Society for Industrial and Applied Mathematics, Philadelphia, PA,, 1995.

[14] D. E. Keyes. Exaflop/s: The why and the how. *Comptes Rendus Mecanique*, 339:70–77, 2011.

[15] D. B. Kirk and W.-M. W. Hwu. *Programming Massively Parallel Processors: A Hands-on Approach*. Morgan Kaufmann Publishers Inc., San Francisco, CA, 2010.

[16] T. Korson and J. D. McGregor. Technical criteria for the specification and evaluation of object-oriented libraries. *Softw. Eng. J.*, 7(2):85–94, March 1992.

[17] R. J. LeVeque. *Finite Difference Methods for Ordinary and Partial Differential Equations–Steady-state and Time-dependent Problems*. SIAM, Philadelphia, PA, 2007.

[18] J.-L. Lions, Y. Maday, and G. Turinici. Résolution d'edp par un schéma en temps pararéel. *C.R. Acad. Sci. Paris Sér. I Math*, 332:661–668, 2001.

[19] Y. Maday. The parareal in time algorithm. Technical Report R08030, Universite Pierré et Marie Curie, Paris, 2008.

[20] A. S. Nielsen. Feasibility study of the parareal algorithm. Master thesis, Technical University of Denmark, Department of Informatics and Mathematical Modeling, Lyngby, 2012.

[21] NVIDIA Corporation. CUDA C Programming Guide, 2012.

[22] NVIDIA Corporation. CUDA C Best Practices Guide, 2012.

[23] Y. Saad. *Iterative Methods for Sparse Linear Systems*. Society for Industrial and Applied Mathematics, Philadelphia, PA, 2nd edition, 2003.

[24] A. Skjellum, N. E. Doss, and P. V. Bangaloret. Writing libraries in MPI. Technical Report, Department of Computer Science and NSF Engineering Research Center for Computational Fiels Simulation. Mississippi State University, 1994.

[25] B. F. Smith, P. E. Bjørstad, and W. D. Gropp. *Domain Decomposition: Parallel Multilevel Methods for Elliptic Partial Differential Equations*. Cambridge University Press, New York, 1996.

[26] U. Trottenberg, C. W. Oosterlee, and A. Schüller. *Multigrid*. Elsevier Academic Press, London, 2001.

[27] D. Vandevoorde and N. M. Josuttis. *C++ Templates: The Complete Guide*. Addison-Wesley Professional, November 2002.

Chapter 7

Development methodologies for GPU and cluster of GPUs

Sylvain Contassot-Vivier

Université Lorraine, Loria UMR 7503 & AlGorille INRIA Project Team, Nancy, France

Stephane Vialle

SUPELEC, UMI GT-CNRS 2958 & AlGorille INRIA Project Team, Metz, France

Jens Gustedt

INRIA Nancy–Grand Est, AlGorille INRIA Project Team, Strasbourg, France

7.1	Introduction	106
7.2	General scheme of synchronous code with computation/communication overlapping in GPU clusters	106
	7.2.1 Synchronous parallel algorithms on GPU clusters	106
	7.2.2 Native overlap of CPU communications and GPU computations	108
	7.2.3 Overlapping with sequences of transfers and computations	110
	7.2.4 Interleaved communications-transfers-computations overlapping	115
	7.2.5 Experimental validation	118
7.3	General scheme of asynchronous parallel code with computation/communication overlapping	120
	7.3.1 A basic asynchronous scheme	122
	7.3.2 Synchronization of the asynchronous scheme	126
	7.3.3 Asynchronous scheme using MPI, OpenMP, and CUDA	130
	7.3.4 Experimental validation	137
7.4	Perspective: a unifying programming model	140
	7.4.1 Resources	141
	7.4.2 Control	142
	7.4.3 Example: block-cyclic matrix multiplication (MM)	142
	7.4.4 Tasks and operations	145
7.5	Conclusion	146
7.6	Glossary	146

Bibliography 147

7.1 Introduction

This chapter proposes to draw upon several development methodologies to obtain efficient codes in classical scientific applications. Those methodologies are based on the feedback from several research works involving GPUs, either in a single machine or in a cluster of machines. Indeed, our past collaborations with industries have allowed us to point out that in their economical context, they can adopt a parallel technology only if its implementation and maintenance costs are small compared with the potential benefits (performance, accuracy, etc.). So, in such contexts, GPU programming is still regarded with some distance due to its specific field of applicability (SIMD/SIMT model: Single Instruction Multiple Data/Thread) and its still higher programming complexity and maintenance. In the academic domain, things are a bit different, but studies for efficiently integrating GPU computations in multicore clusters with maximal overlapping of computations with communications and/or other computations are still rare.

For these reasons, the major aim of that chapter is to propose general programming patterns, as simple as possible, that can be followed or adapted in practical implementations of parallel scientific applications. In addition, we propose a prospect analysis together with a particular programming tool that is intended to ease multicore GPU cluster programming.

7.2 General scheme of synchronous code with computation/communication overlapping in GPU clusters

7.2.1 Synchronous parallel algorithms on GPU clusters

Considered parallel algorithms and implementations

This section focuses on synchronous parallel algorithms implemented with overlapping computations and communications. Parallel synchronous algorithms are easier to implement, debug, and maintain than asynchronous ones (see Section 7.3). Usually, they follow a BSP-like parallel scheme (Bulk Synchronous Parallel model), alternating local computation steps and communication steps (see [19]). Their execution is usually deterministic, except for stochastic algorithms that contain random number generations. Even in this case, their execution can be controlled during debug steps, allowing the user to track and to fix bugs quickly.

However, depending on the properties of the algorithm, it is sometimes possible to overlap computations and communications. If processes exchange data that is not needed for the computation immediately following, it is possible to implement such an overlap. We have investigated the efficiency of this approach in previous works [14,20], using standard parallel programming tools to achieve the implementation.

The normalized and well-known Message Passing Interface (MPI) includes some asynchronous point-to-point communication routines, that should allow the implementation of some communication/computation overlap. However, current MPI implementations do not achieve this goal efficiently; effective overlapping with MPI requires a group of dedicated threads (in our case implemented with OpenMP) for the basic synchronous communications while another group of threads executes computations in parallel. Nevertheless, communication and computation are not completely independent on modern multicore architectures: they use shared hardware components such as the interconnection bus and the RAM. Therefore, this approach saves only up to 20% of the expected time on such a platform. This picture changes on clusters equipped with GPUs. Indeed, GPUs effectively allow independence of computations they perform and communications done on the mainboard (CPU, interconnection bus, RAM, network card). We save up to 100% of the expected time on our GPU cluster, as we will expose in the next section.

Specific interests in GPU clusters

In a computing node, a GPU is a kind of scientific coprocessor, usually located on an auxiliary board, with its own memory. So, once data are transferred from the CPU memory to the GPU memory, GPU computations can be achieved on the GPU board, totally in parallel with any CPU activities (such as internode cluster communications). The CPU and the GPU access their respective memories and do not interfere with each other, so they can achieve a very good overlap of their activities (better than two CPU cores).

But using a GPU on a computing node requires the transfer of data from the CPU to the GPU memory, as well as the transfer of the computation results back from the GPU to the CPU. Transfer times are not excessive, but depending on the application they still can be significant compared to the GPU computation times. So, sometimes it can be interesting to overlap the internode cluster communications with both the CPU/GPU data transfers and the GPU computations. We can identify four main parallel programming schemes on a GPU cluster:

1. parallelizing only internode CPU communications with GPU computations, and achieving CPU/GPU data transfers before and after this parallel step,

2. parallelizing internode CPU communications with a (sequential) sequence of CPU/GPU data transfers and GPU computations,

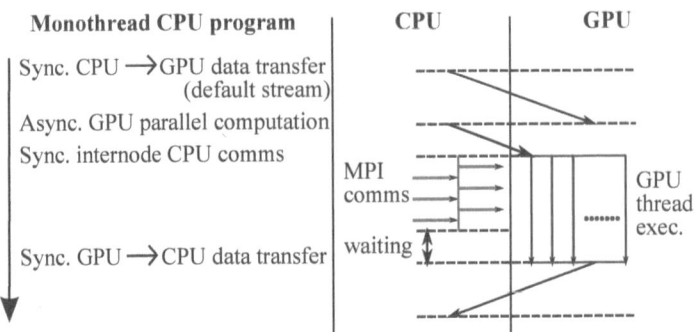

FIGURE 7.1. Native overlap of internode CPU communications with GPU computations.

3. parallelizing internode CPU communications with a streamed sequence of CPU/GPU data transfers and GPU computations,

4. parallelizing internode CPU communications with CPU/GPU data transfers and with GPU computations, interleaving computation-communication iterations.

7.2.2 Native overlap of CPU communications and GPU computations

Using CUDA, GPU kernel executions are nonblocking, and GPU/CPU data transfers are blocking or nonblocking operations. All GPU kernel executions and CPU/GPU data transfers are associated to "streams", and all operations on a same stream are serialized. When transferring data from the CPU to the GPU, then running GPU computations, and finally transferring results from the GPU to the CPU, there is a natural synchronization and serialization if these operations are achieved on the same stream. GPU developers can choose to use one (default) or several streams. In this first scheme of overlapping, we consider parallel codes using only one GPU stream.

"Nonblocking GPU kernel execution" means a CPU routine runs a parallel execution of a GPU computing kernel, and the CPU routine continues its execution (on the CPU) while the GPU kernel is running (on the GPU). Then the CPU routine can initiate some communications with some other CPUs, and so it automatically overlaps the internode CPU communications with the GPU computations (see Figure 7.1). This overlapping is natural when programming with CUDA and MPI: it is easy to deploy, but does not overlap the CPU/GPU data transfers.

Listing 7.1 introduces the generic code of a MPI+CUDA implementation, natively and implicitly overlapping MPI communications with CUDA GPU computations.

Development methodologies for GPU and cluster of GPUs

```
Listing 7.1. generic scheme implicitly overlapping MPI communications with
CUDA GPU computations

   // Input data and result variables and arrays (example with
   // float datatype, 1D input arrays, and scalar results)
   float *cpuInputTabAdr, *gpuInputTabAdr;
   float *cpuResTabAdr, *gpuResAdr;
5
   // CPU and GPU array allocations
   cpuInputTabAdr = malloc(sizeof(float)*N);
   cudaMalloc(&gpuInputTabAdr, sizeof(float)*N);
   cpuResTabAdr = malloc(sizeof(float)*NbIter);
10 cudaMalloc(&gpuResAdr, sizeof(float));

   // Definition of the grid of blocks of GPU threads
   dim3 Dg, Db;
   Dg.x = ...
15 ...

   // Indexes of source and destination MPI processes
   int dest = ...
   int src = ...
20
   // Computation loop (using the GPU)
   for (int i = 0; i < NbIter; i++) {
      cudaMemcpy(gpuInputTabAdr, cpuInputTabAdr,  // Data transfer:
                 sizeof(float)*N,                   // CPU —> GPU (sync. op)
25                cudaMemcpyHostToDevice);
      gpuKernel_k1<<<Dg,Db>>>();                  // GPU comp. (async. op)
      MPI_Sendrecv_replace(cpuInputTabAdr,        // MPI comms. (sync. op)
                 N, MPI_FLOAT,
                 dest, 0, src, 0, ...);
30    // IF there is (now) a result to transfer from the GPU to the CPU:
      cudaMemcpy(cpuResTabAdr + i, gpuResAdr,     // Data transfer:
                 sizeof(float),                     // GPU —> CPU (sync. op)
                 cudaMemcpyDeviceToHost);
   }
35 ...
```

Some input data and output results arrays and variables are declared and allocated from line 3 through 10, and a computation loop is implemented from line 22 through 34. At each iteration:

- cudaMemcpy on line 23 transfers data from the CPU memory to the GPU memory. This is a basic and synchronous data transfer.

- gpuKernel_k1<<<Dg,Db>>> on line 26 starts GPU computation (running a GPU kernel on the grid of blocks of threads defined in lines 13 to 15). This is a standard GPU kernel run; it is an asynchronous operation. The CPU can continue to run its code.

- MPI_Sendrecv_replace on line 27 achieves some blocking internode communications, overlapping GPU computations started just previously.

- If needed, cudaMemcpy on line 31 transfers the iteration result from one variable in the GPU memory to one array index in the CPU memory (in this example the CPU collects all iteration results in an array). This operation is started after the end of the MPI communication (previous instruction) and after the end of the GPU kernel execution. CUDA

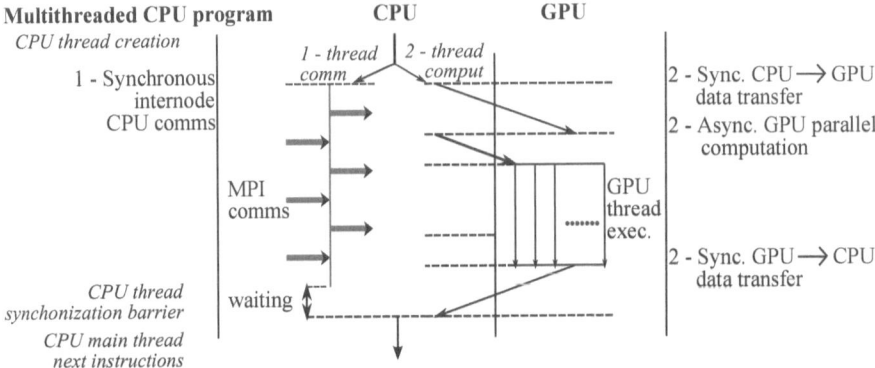

FIGURE 7.2. Overlap of internode CPU communications with a sequence of CPU/GPU data transfers and GPU computations.

insures an implicit synchronization of all operations involving the same GPU stream, like the default stream in this example. The transfer of the results has to wait until the GPU kernel execution is finished. If there is no results transfer implemented, the next operation on the GPU will wait until the GPU kernel execution has ended.

This implementation is the easiest one involving the GPU. It achieves an implicit overlap of internode communications and GPU computations with no explicit multithreading required on the CPU. However, CPU/GPU data transfers are achieved serially and not overlapped.

7.2.3 Overlapping with sequences of transfers and computations

Overlapping with a sequential GPU sequence

When CPU/GPU data transfers are not negligible compared to GPU computations, it can be interesting to overlap internode CPU computations with a *GPU sequence* including CPU/GPU data transfers and GPU computations (see Figure 7.2). Algorithmic issues of this approach are basic, but their implementation requires explicit CPU multithreading and synchronization, and CPU data buffer duplication. We need to implement two threads, one starting and achieving MPI communications and the other running the *GPU sequence*. OpenMP allows an easy and portable implementation of this overlapping strategy. However, it remains more complex to develop and to maintain than the previous strategy (overlapping only internode CPU communications and GPU computations) and should be adopted only when CPU/GPU data transfer times are not negligible.

Listing 7.2 introduces the generic code of a MPI+OpenMP+CUDA implementation, explicitly overlapping MPI communications with GPU sequences.

Listing 7.2. generic scheme explicitly overlapping MPI communications with sequences of CUDA CPU/GPU transfers and CUDA GPU computations

```c
// Input data and result variables and arrays (example with
// float datatype, 1D input arrays, and scalar results)
float *cpuInputTabAdrCurrent, *cpuInputTabAdrFuture, *gpuInputTabAdr;
float *cpuResTabAdr, *gpuResAdr;

// CPU and GPU array allocations
cpuInputTabAdrCurrent = malloc(sizeof(float)*N);
cpuInputTabAdrFuture = malloc(sizeof(float)*N);
cudaMalloc(&gpuInputTabAdr, sizeof(float)*N);
cpuResTabAdr = malloc(sizeof(float)*NbIter);
cudaMalloc(&gpuResAdr, sizeof(float));

// Definition of the grid of blocks of GPU threads
dim3 Dg, Db;
Dg.x = ...
...

// Indexes of source and destination MPI processes
int dest = ...
int src = ...

// Set the number of OpenMP threads (to create) to 2
omp_set_num_threads(2);
// Create threads and start the parallel OpenMP region
#pragma omp parallel
{
  // Buffer pointers (thread local variables)
  float *current = cpuInputTabAdrCurrent;
  float *future = cpuInputTabAdrFuture;
  float *tmp;

  // Computation loop (using the GPU)
  for (int i = 0; i < NbIter; i++) {

    // - Thread 0: achieves MPI communications
    if (omp_get_thread_num() == 0) {
      MPI_Sendrecv(current,                  // MPI comms. (sync. op)
                   N, MPI_FLOAT, dest, 0,
                   future,
                   N, MPI_FLOAT, dest, 0, ...);

    // - Thread 1: achieves the GPU sequence (GPU computations and
    //             CPU/GPU data transfers)
    } else if (omp_get_thread_num() == 1) {
      cudaMemcpy(gpuInputTabAdr, current,    // Data transfer:
                 sizeof(float)*N,            // CPU --> GPU (sync. op)
                 cudaMemcpyHostToDevice);
      gpuKernel_k1<<<Dg,Db>>>();             // GPU comp. (async. op)
    // IF there is (now) a result to transfer from the GPU to the CPU:
      cudaMemcpy(cpuResTabAdr + i, gpuResAdr,// Data transfer:
                 sizeof(float),              // GPU --> CPU (sync. op)
                 cudaMemcpyDeviceToHost);
    }

    // - Wait until both threads have achieved their iteration tasks
    #pragma omp barrier
    // - Each thread permutes its local buffer pointers
    tmp = current;
    current = future;
    future = tmp;
  } // End of computation loop
} // End of OpenMP parallel region
...
```

Lines 25–62 implement the OpenMP parallel region, around the computation loop (lines 33–61). For efficient performances it is important to create and destroy threads only one time (not at each iteration): the parallel region has to surround the computation loop. Lines 3–11 consist of declaration and allocation of input data arrays and result arrays and variables, as in the previous algorithm (Listing 7.1). However, we implement two input data buffers on the CPU (current and future version). As we aim to overlap internode MPI communications and GPU sequence, including CPU to GPU data transfer of current input data array, we need to store the received new input data array in a separate buffer. Then, the current input data array will be safely read on the CPU and copied into the GPU memory.

The thread creations are easily achieved with one OpenMP directive (line 25). Then each thread defines and initializes *its* local buffer pointers, and enters *its* computing loop (lines 28–33). Inside the computing loop, a test on the thread number makes it possible to run a different code in each thread. Lines 37–40 implement the MPI synchronous communication run by thread number 0. Lines 45–52 implement the GPU sequence run by thread 1: CPU to GPU data transfer, GPU computation, and GPU to CPU result transfer (if needed). Details of the three operations of this sequence have not changed from the previous overlapping strategy.

At the end of Listing 7.2, an OpenMP synchronization barrier on line 56 forces the OpenMP threads to wait until MPI communications and GPU sequence are achieved. Then, each thread permutes its local buffer pointers (lines 58–60), and is ready to enter the next iteration, processing the new current input array.

Overlapping with a streamed GPU sequence

Depending on the algorithm implemented, it is sometimes possible to split the GPU computation into several parts processing distinct data. Then, we can speedup the GPU sequence using several CUDA streams. The goal is to overlap CPU/GPU data transfers with GPU computations inside the GPU sequence. Compared to the previous overlapping strategy, we have to split the initial data transfer into a set of n asynchronous and smaller data transfers, and split the initial GPU kernel call into a set of n calls to the same GPU kernel. Usually, these smaller calls are deployed with fewer GPU threads (i.e., associated to a smaller grid of blocks of threads). Then, the first GPU computations can start as soon as the first data transfer has been achieved, and next transfers can be done in parallel with next GPU computations (see Figure 7.3).

NVIDIA advises starting all asynchronous CUDA data transfers and then calling all CUDA kernel executions, using up to 16 streams [16]. Then, the CUDA driver and runtime optimize the global execution of these operations. So, we accumulate two overlapping mechanisms. The former is controlled by CPU multithreading and overlaps MPI communications and the streamed GPU sequence. The latter is controlled by CUDA programming and overlaps CPU/GPU data transfers and GPU computations. Again, OpenMP allows

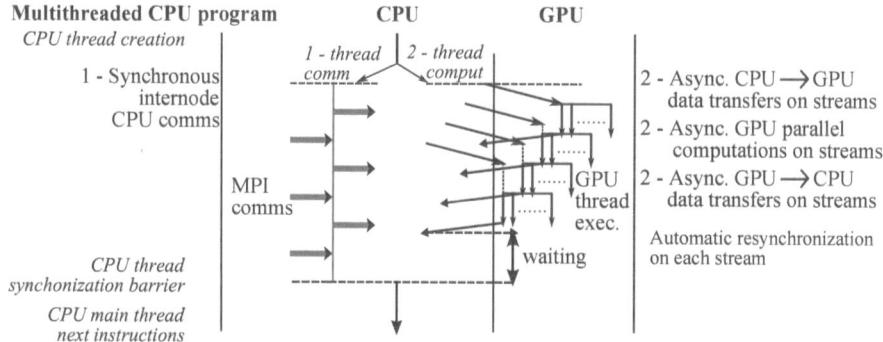

FIGURE 7.3. Overlap of internode CPU communications with a streamed sequence of CPU/GPU data transfers and GPU computations.

the easy implementation of the CPU multithreading and waiting for the end of both CPU threads before executing the next instructions of the code.

Listing 7.3 introduces the generic MPI+OpenMP+CUDA code, explicitly overlapping MPI communications with streamed GPU sequences.

Listing 7.3. generic scheme explicitly overlapping MPI communications with streamed sequences of CUDA CPU/GPU transfers and CUDA GPU computations

```
// Input data and result variables and arrays (example with
// float datatype, 1D input arrays, and scalar results)
float *cpuInputTabAdrCurrent, *cpuInputTabAdrFuture, *gpuInputTabAdr;
float *cpuResTabAdr, *gpuResAdr;
// CPU and GPU array allocations (allocates page-locked CPU memory)
cudaHostAlloc(&cpuInputTabAdrCurrent, sizeof(float)*N,
    cudaHostAllocDefault);
cudaHostAlloc(&cpuInputTabAdrFuture, sizeof(float)*N,
    cudaHostAllocDefault);
cudaMalloc(&gpuInputTabAdr, sizeof(float)*N);
cpuResTabAdr = malloc(sizeof(float)*NbIter);
cudaMalloc(&gpuResAdr, sizeof(float));
// Stream declaration and creation
cudaStream_t TabS[NbS];
for(int s = 0; s < NbS; s++)
    cudaStreamCreate(&TabS[s]);
// Definition of the grid of blocks of GPU threads
...
// Set the number of OpenMP threads (to create) to 2
omp_set_num_threads(2);
// Create threads and start the parallel OpenMP region
#pragma omp parallel
{
    // Buffer pointers (thread local variables)
    float *current = cpuInputTabAdrCurrent;
    float *future  = cpuInputTabAdrFuture;
    float *tmp;
    // Stride of data processed per stream
    int stride = N/NbS;
    // Computation loop (using the GPU)
    for (int i = 0; i < NbIter; i++) {
        // - Thread 0: achieves MPI communications
        if (omp_get_thread_num() == 0) {
```

```
            MPI_Sendrecv(current,              // MPI comms. (sync. op)
                        N, MPI_FLOAT, dest, 0,
                        future,
35                      N, MPI_FLOAT, dest, 0, ...);
       // - Thread 1: achieves the streamed GPU sequence (GPU
       //   computations and CPU/GPU data transfers)
       } else if (omp_get_thread_num() == 1) {
         for (int s = 0; s < NbS; s++) {    // Start all data transfers:
40         cudaMemcpyAsync(gpuInputTabAdr + s*stride,  // CPU --> GPU
                           current + s*stride,         // (async. ops)
                           sizeof(float)*stride,
                           cudaMemcpyHostToDevice,
                           TabS[s]);
45       }
         for (int s = 0; s < NbS; s++) { // Start all GPU comps. (async.)
           gpuKernel_k1<<<Dg, Db, 0, TabS[s]>>>(gpuInputTabAdr + s*stride
                        );
         }
         cudaThreadSynchronize();            // Wait all threads are ended
50       // IF there is (now) a result to transfer from the GPU to the CPU:
         cudaMemcpy(cpuResTabAdr,                // Data transfers:
                    gpuResAdr,                   // GPU --> CPU (sync. op)
                    sizeof(float),
                    cudaMemcpyDeviceToHost);
55     }
       // - Wait until both threads have achieved their iteration tasks
       #pragma omp barrier
       // - Each thread permutes its local buffer pointers
       tmp = current; current = future; future = tmp;
60     } // End of computation loop
     } // End of OpenMP parallel region
     ...
     // Destroy the streams
     for(int s = 0; s < NbS; s++)
65     cudaStreamDestroy(TabS[s]);
     ...
```

Efficient usage of CUDA streams requires executing asynchronous CPU/GPU data transfers, which implies reading page-locked data in CPU memory. So, CPU memory allocations on lines 6 and 7 are implemented with `cudaHostAlloc` instead of the basic `malloc` function. Then, NbS *streams* are created on lines 12–14. Usually we create 16 streams: the maximum number supported by CUDA.

An OpenMP parallel region including two threads is implemented on lines 18–61 of Listing 7.3, as in the previous algorithm (see Listing 7.2). Code of thread 0 achieving MPI communication is unchanged, but code of thread 1 is now using streams. Following NVIDIA recommendations, we first implement a loop starting NbS asynchronous data transfers (lines 39–45): transferring N/NbS data on each stream. Then we implement a second loop (lines 46–48), starting asynchronous executions of NbS grids of blocks of GPU threads (one per stream). Data transfers and kernel executions on the same stream are synchronized by CUDA and the GPU. So, each kernel execution will start after its data has been transferred into the GPU memory, and the GPU scheduler ensures the start of some kernel executions as soon as the first data transfers are achieved. Then, next data transfers will be overlapped with GPU computations. After the kernel calls, on the different streams, we wait for the end of all GPU threads previously run, calling an explicit synchronization function on line 49. This synchronization is not mandatory, but it will make the implementation more robust and will facilitate the debugging steps: all GPU computations run by the OpenMP thread number 1 will be achieved before this thread enters a new loop iteration, or before the computation loop has ended.

If a partial result has to be transferred from GPU to CPU memory at the end of each loop iteration (for example, the result of one *reduction* per iteration), this transfer is achieved synchronously on the default stream (no particular stream is specified) on lines 51–54. Availability of the result values is ensured by the synchronization implemented on line 49. However, if a partial result has to be transferred onto the CPU on each stream, then NbS asynchronous data transfers could be started in parallel (one per stream) and should be implemented before the synchronization operation on line 49. The end of the computation loop includes a synchronization barrier of the two OpenMP threads, waiting until they have finished accessing the different data buffers in the current iteration. Then, each OpenMP thread exchanges its local buffer pointers, as in the previous algorithm. After the computation loop, we have added the destruction of the CUDA streams (lines 64–65).

In conclusion, CUDA streams have been used to extend Listing 7.2 with respect to its global scheme. Listing 7.3 still creates an OpenMP parallel region, with two CPU threads, one in charge of MPI communications and the other managing data transfers and GPU computations. Unfortunately, using GPU streams requires the ability to split a GPU computation into independent subparts, working on independent subsets of data. Listing 7.3 is not so generic as Listing 7.2.

7.2.4 Interleaved communications-transfers-computations overlapping

Many algorithms do not support splitting data transfers and kernel calls, and cannot exploit CUDA streams, for example, when each GPU thread requires access to some data spread in the global set of transferred data. Then, it is possible to overlap internode CPU communications, CPU/GPU data transfers, and GPU computations, if the algorithm achieves *computation-communication iterations* and if we can interleave these iterations. At iteration k: CPUs exchange data D_k, each CPU/GPU couple transfers data D_k, and each GPU achieves computations on data D_{k-1} (see Figure 7.4). Compared to the previous strategies, this strategy requires twice as many CPU data buffers and twice as many GPU buffers.

Listing 7.4 introduces the generic code of a MPI+OpenMP+CUDA implementation, explicitly interleaving computation-communication iterations and overlapping MPI communications, CUDA CPU/GPU transfers, and CUDA GPU computations.

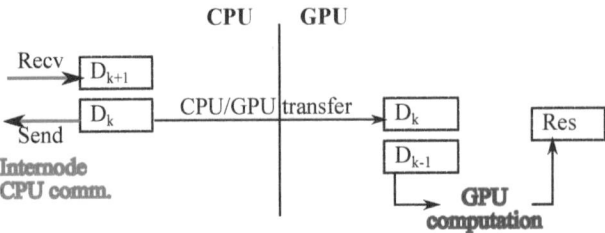

FIGURE 7.4. Complete overlap of internode CPU communications, CPU/GPU data transfers, and GPU computations, interleaving computation-communication iterations.

Listing 7.4. generic scheme explicitly overlapping MPI communications, CUDA CPU/GPU transfers, and CUDA GPU computations, interleaving computation-communication iterations

```
// Input data and result variables and arrays (example with
// float datatype, 1D input arrays, and scalar results)
float *cpuInputTabAdrCurrent, *cpuInputTabAdrFuture;
float *gpuInputTabAdrCurrent, *gpuInputTabAdrFuture;
float *cpuResTabAdr, *gpuResAdr;

// CPU and GPU array allocations
cpuInputTabAdrCurrent = malloc(sizeof(float)*N);
cpuInputTabAdrFuture  = malloc(sizeof(float)*N);
cudaMalloc(&gpuInputTabAdrCurrent, sizeof(float)*N);
cudaMalloc(&gpuInputTabAdrFuture,  sizeof(float)*N);
cpuResTabAdr = malloc(sizeof(float)*NbIter);
cudaMalloc(&gpuResAdr, sizeof(float));

// Definition of the grid of blocks of GPU threads
dim3 Dg, Db; Dg.x = ...
// Indexes of source and destination MPI processes
int dest, src; dest = ...

// Set the number of OpenMP threads (to create) to 2
omp_set_num_threads(3);
// Create threads and start the parallel OpenMP region
#pragma omp parallel
{
  // Buffer pointers (thread local variables)
  float *cpuCurrent = cpuInputTabAdrCurrent;
  float *cpuFuture  = cpuInputTabAdrFuture;
  float *gpuCurrent = gpuInputTabAdrCurrent;
  float *gpuFuture  = gpuInputTabAdrFuture;
  float *tmp;

  // Computation loop on NbIter + 1 iterations
  for (int i = 0; i < NbIter + 1; i++) {
    // - Thread 0: achieves MPI communications
    if (omp_get_thread_num() == 0) {
      if (i < NbIter) {
        MPI_Sendrecv(cpuCurrent,              // MPI comms. (sync. op)
                     N, MPI_FLOAT, dest, 0,
                     cpuFuture,
                     N, MPI_FLOAT, dest, 0, ...);
      }
    // - Thread 1: achieves the CPU/GPU data transfers
    } else if (omp_get_thread_num() == 1) {
      if (i < NbIter) {
```

```
            cudaMemcpy(gpuFuture, cpuCurrent,   // Data transfer:
                      sizeof(float)*N,          // CPU --> GPU (sync. op)
                      cudaMemcpyHostToDevice);
       }
      // - Thread 2: achieves the GPU computations and result transfer
       } else if (omp_get_thread_num() == 2) {
          if (i > 0) {
             gpuKernel_k1<<<Dg,Db>>>(gpuCurrent);// GPU comp. (async. op)
             // IF there is (now) a result to transfer from GPU to CPU:
             cudaMemcpy(cpuResTabAdr + (i-1),    // Data transfer:
                        gpuResAdr, sizeof(float),// GPU --> CPU (sync. op)
                        cudaMemcpyDeviceToHost);
          }
       }
       // - Wait until both threads have achieved their iteration tasks
       #pragma omp barrier
       // - Each thread permutes its local buffer pointers
       tmp = cpuCurrent; cpuCurrent = cpuFuture; cpuFuture = tmp;
       tmp = gpuCurrent; gpuCurrent = gpuFuture; gpuFuture = tmp;
    } // End of computation loop
 } // End of OpenMP parallel region
...
```

As in the previous algorithms, we declare two CPU input data arrays (current and future version) on line 3. However, in this version we also declare two GPU input data arrays on line 4. On lines 8–11, these four data arrays are allocated, using `malloc` and `cudaMalloc`. We do not need to allocate page-locked memory space. On lines 23–65 we create an OpenMP parallel region, configured to run three threads (see line 21). Lines 26–30 are declarations of thread local pointers on data arrays and variables (each thread will use its own pointers). On line 33, the three threads enter a computation loop of `NbIter + 1` iterations. We need to run one more iteration than with previous algorithms.

Lines 35 41 are the MPI communications, achieved by the thread number 0. They send the current CPU input data array to another CPU, and receive the future CPU input data array from another CPU, like in previous algorithms. But this thread achieves communications only during the *first* `NbIter` iterations. Lines 43–48 are the CPU to GPU input data transfers, achieved by thread number 1. These data transfers are run in parallel with MPI communications. They are run during the *first* `NbIter` iterations and transfer current CPU input data array into the future GPU data array. Lines 50–57 correspond to the code run by thread number 2. They start GPU computations, process the current GPU input data array, and if necessary transfer a GPU result to an index of the CPU result array. These GPU computations and result transfers are run during the *last* `NbIter` iterations: the GPU computations have to wait until the first data transfer is ended before starting to process any data and cannot run during the first iteration. So, the activity of the third thread is shifted by one iteration compared to the activities of the other threads. Moreover, the address of the current GPU input data array has to be passed as a parameter of the kernel call on line 52, in order for the GPU threads to access the right data array. As in previous algorithms the GPU result is copied to one index of the CPU result array, in lines 54–56, but due to the shift of the third thread activity this index is now (i - 1).

Line 60 is a synchronization barrier of the three OpenMP threads, followed by a pointer permutation of local pointers on current and future data arrays, on line 62 and 63. Each thread waits for the completion of other threads to use the data arrays, and then permutes its data array pointers before entering a new loop iteration.

This complete overlap of MPI communications, CPU/GPU data transfers, and GPU computations is not too complex to implement, and can be a solution when GPU computations are not adapted to use CUDA streams: when GPU computations cannot be split into subparts working on independent subsets of input data. However, this requires running one more iteration (a total of NbIter + 1 iterations). If the number of iterations is very small, it could be more interesting not to attempt to overlap CPU/GPU data transfers and GPU computations, and to implement Listing 7.2.

7.2.5 Experimental validation

Experimentation testbed

Two clusters located at SUPELEC in Metz (France) have been used for the entire set of experiments presented in this chapter:

- The first consists of 17 nodes with an Intel Nehalem quad-core processor at 2.67Ghz, 6 Gb RAM, and an NVIDIA GeForce GTX480 GPU, each.

- The second consists of 16 nodes with an Intel core2 dual-core processor at 2.67Ghz, 4 Gb RAM, and an NVIDIA GeForce GTX580 GPU, each.

Both clusters have a gigabit Ethernet interconnection network that is connected through a Dell Power Object 5324 switch. The two switches are linked twice, ensuring the interconnection of the two clusters. The software environment consists of a Linux Fedora 64bit OS (kernel v. 2.6.35), GNU C and C++ compilers (v. 4.5.1), and the CUDA library (v. 4.2).

Validation of the synchronous approach

We have tested our approach of synchronous parallel algorithms with a classic block cyclic algorithm for dense matrix multiplication. This problem requires splitting two input matrices (A and B) on a ring of computing nodes and establishing a circulation of the slices of A matrix on the ring (B matrix partition does not evolve during all the run). Compared to our generic algorithms, there is no partial result to transfer from GPU to CPU at the end of each computing iteration. The part of the result matrix computed on each GPU is transferred onto the CPU at the end of the computation loop.

We have first implemented a synchronous version without any overlap of MPI communications, CPU/GPU data transfers, and GPU computations. We

Development methodologies for GPU and cluster of GPUs

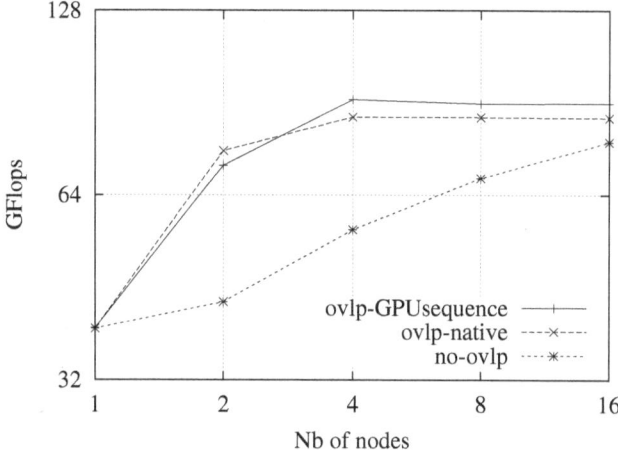

FIGURE 7.5. Experimental performances of different synchronous algorithms computing a dense matrix product.

have added some synchronizations in the native overlapping version in order to avoid any overlap. We have measured the performance achieved on our cluster with NVIDIA GTX480 GPUs and matrices sizes of 4096×4096, and we have obtained the curves in Figure 7.5 labeled *no-ovlp*. We observe that performance increases when the number of processor increases. Of course, there is a significant increase in cost when comparing a single node (without any MPI communication) with two nodes (starting to use MPI communications). But beyond two nodes we get a classical performance curve.

Then, we implemented and tested Listing 7.1, labeled *ovlp-native* in Figure 7.5. The native overlap of MPI communications with the asynchronous run of CUDA kernels appears efficient. When the number of nodes increases the ratio of the MPI communications increases a lot (because the computation times decrease a lot). So, there is not a lot of GPU computation time that remains to be overlapped, and both *no-ovlp* and *ovlp-native* tend to the same limit. Already, the native overlap performed in Listing 7.1 achieves a high level of performance very quickly, using only four nodes. Beyond four nodes, a faster interconnection network would be required for a performance increase.

Finally, we implemented Listing 7.2, overlapping MPI communications with a GPU sequence including both CPU/GPU data transfers and GPU computations, labeled *ovlp-GPUsequence* in Figure 7.5. From four up to sixteen nodes it achieves better performances than *ovlp-native*: the overlapping of MPI communications is wider and thus more efficient. However, this parallelization mechanism has more overhead: OpenMP threads have to be created and synchronized. With only two nodes it is less efficient than the native overlapping algorithm. Beyond two nodes, the CPU multithreading overhead seems compensated. Listing 7.2 requires more time for the implementation and

7.3 General scheme of asynchronous parallel code with computation/communication overlapping

In the previous section, we have seen how to efficiently implement overlap of computations (CPU and GPU) with communications (GPU transfers and internode communications). However, we have previously shown that for some parallel iterative algorithms, it is sometimes even more efficient to use an asynchronous scheme of iterations [3, 4, 11]. In that case, the nodes do not wait for each other but they perform their iterations using the last external data they have received from the other nodes, even if this data was produced *before* the previous iteration on the other nodes.

Formally, if we denote by $f = (f_1, ..., f_n)$ the function representing the iterative process and by $x^t = (x_1^t, ..., x_n^t)$ the values of the n elements of the system at iteration t, we pass from a synchronous iterative scheme of the form given in Algorithm 4

Algorithm 4: synchronous iterative scheme

---
1 $x^0 = (x_1^0, ..., x_n^0)$;
2 **for** $t = 0, 1, ...$ **do**
3 $\quad$ **for** $i = 1, ..., n$ **do**
4 $\quad\quad$ $x_i^{t+1} = f_i(x_1^t, ..., x_i^t, ..., x_n^t)$;
5 $\quad$ **end**
6 **end**

---

to an asynchronous iterative scheme of the form given in Algorithm 5.

Algorithm 5: asynchronous iterative scheme

---
1 $x^0 = (x_1^0, ..., x_n^0)$;
2 **for** $t = 0, 1, ...$ **do**
3 $\quad$ **for** $i = 1, ..., n$ **do**
4 $\quad\quad$ $x_i^{t+1} = \begin{cases} x_i^t & \text{if } i \text{ is } not \text{ updated at iteration } i \\ f_i(x_1^{s_1^i(t)}, ..., x_n^{s_n^i(t)}) & \text{if } i \text{ is updated at iteration } i \end{cases}$
5 $\quad$ **end**
6 **end**

In this scheme, $s_j^i(t)$ is the iteration number of the production of the value x_j of element j that is used on element i at iteration t (see, for example, [9, 12] for further details). Such schemes are called AIAC for *Asynchronous Iterations and Asynchronous Communications*. They combine two aspects that are respectively different computation speeds of the computing elements and communication delays between them.

The key feature of such algorithmic schemes is that they may be faster than their synchronous counterparts due to the implied total overlap of computations with communications: in fact, this scheme suppresses all the idle times induced by nodes synchronizations between each iteration.

However, the efficiency of such a scheme is directly linked to the frequency at which new data arrives on each node. Typically, if a node receives newer data only every four or five local iterations, it is strongly probable that the evolution of its local iterative process will be slower than if it receives data at every iteration. The key point here is that not only does this frequency depend on the hardware configuration of the parallel system but it also depends on the software that is used to implement the algorithmic scheme.

The impact of the programming environments used to implement asynchronous algorithms has already been investigated in [5]. Although the features required to efficiently implement asynchronous schemes have not changed, the available programming environments and computing hardware have evolved, in particular now that GPUs are available. So, there is a need to reconsider the implementation schemes of AIAC according to the new de facto standards for parallel programming (communications and threads) as well as the integration of the GPUs. One of the main objective here is to obtain a maximal overlap between the activities of the three types of devices: the CPU, the GPU, and the network. Moreover, another objective is to present what we think is the best compromise between the simplicity of the implementation and its maintainability on one side and its performance on the other side. This is especially important for industries where implementation and maintenance costs are strong constraints.

For the sake of clarity, we present the different algorithmic schemes in a progressive order of complexity, from the basic asynchronous scheme to the complete scheme with full overlap. Between these two extremes, we propose a synchronization mechanism on top of our asynchronous scheme that can be used either statically or dynamically during the application execution.

Although there exist several programming environments for internode communications, multithreading, and GPU programming, a few of them have become de facto standards, due to their good stability, their ease of use, and/or their wide adoption by the scientific community. Therefore, as in the previous section, all the schemes presented in the following use MPI [1], OpenMP [2], and CUDA [17]. However, there is no loss of generality as these schemes may easily be implemented with other libraries.

Finally, in order to stay as clear as possible, only the parts of code and variables related to the control of parallelism (communications, threads, etc.) are presented in our schemes. The inner organization of data is not detailed as it depends on the application. We only consider that we have two data arrays (previous version and current version) and communication buffers. However, in most of the cases, those buffers can correspond to the data arrays themselves to avoid data copies.

7.3.1 A basic asynchronous scheme

The first step toward our complete scheme is to implement a basic asynchronous scheme that includes an actual overlap of the communications with the computations. In order to ensure that the communications are actually performed in parallel with the computations, it is necessary to use different threads. It is important to remember that asynchronous communications provided in communication libraries such as MPI are not systematically performed in parallel with the computations [15, 20]. So, the logical and classical way to implement such an overlap is to use three threads: one for computing, one for sending, and one for receiving. Moreover, since the communication is performed by threads, blocking synchronous communications can be used without deteriorating the overall performance.

In this basic version, the termination of the global process is performed individually on each node according to its own termination. This can be guided by either a number of iterations or a local convergence detection. The important step at the end of the process is to perform the receptions of all pending communications in order to ensure the termination of the two communication threads.

So, the global organization of this scheme is set up in Listing 7.5.

Listing 7.5. initialization of the basic asynchronous scheme
```
// Variables declaration and initialization
// Controls the sendings from the computing thread
omp_lock_t lockSend;
// Ensures the initial reception of external data
omp_lock_t lockRec;
char Finished = 0; // Boolean indicating the end of the process
// Boolean indicating if previous data sendings are still in progress
char SendsInProgress = 0;
// Threshold of the residual for convergence detection
double Threshold;

// Parameters reading
...

// MPI initialization
MPI_Init_thread(argc, argv, MPI_THREAD_MULTIPLE, &provided);
MPI_Comm_size(MPI_COMM_WORLD, &nbP);
MPI_Comm_rank(MPI_COMM_WORLD, &numP);

// Data initialization and distribution among the nodes
...

// OpenMP initialization (mainly declarations and setting up of locks)
omp_set_num_threads(3);
omp_init_lock(&lockSend);
omp_set_lock(&lockSend);//Initially locked, unlocked to start sendings
omp_init_lock(&lockRec);
//Initially locked, unlocked when initial data are received
omp_set_lock(&lockRec);
```

```
      #pragma omp parallel
      {
        switch(omp_get_thread_num()){
          case COMPUTATION :
35          computations(... relevant parameters ...);
            break;

          case SENDINGS :
            sendings();
40          break;

          case RECEPTIONS :
            receptions();
            break;
45      }
      }

      // Cleaning of OpenMP locks
      omp_test_lock(&lockSend);
50    omp_unset_lock(&lockSend);
      omp_destroy_lock(&lockSend);
      omp_test_lock(&lockRec);
      omp_unset_lock(&lockRec);
      omp_destroy_lock(&lockRec);
55
      // MPI termination
      MPI_Finalize();
```

In this scheme, the `lockRec` mutex is not mandatory. It is only used to ensure that data dependencies are actually exchanged at the first iteration of the process. Data initialization and distribution (lines 20–21) are not detailed here because they are directly related to the application. The important point is that, in most cases, they should be done before the iterative process. The computing function is given in Listing 7.6.

Listing 7.6. computing function in the basic asynchronous scheme

```
    // Variables declaration and initialization
    int iter = 1;       // Number of the current iteration
    double difference;  // Variation of one element between two iterations
    double residual;    // Residual of the current iteration
5
    // Computation loop
    while(!Finished){
      // Sending of data dependencies if there is no previous sending
      // in progress
10    if(!SendsInProgress){
        // Potential copy of data to be sent into additional buffers
        ...
        // Change of sending state
        SendsInProgress = 1;
15      omp_unset_lock(&lockSend);
      }

      // Blocking receptions at the first iteration
      if(iter == 1){
20      omp_set_lock(&lockRec);
      }

      // Initialization of the residual
      residual = 0.0;
25    // Swapping of data arrays (current and previous)
      tmp = current;       // Pointers swapping to avoid
      current = previous;  // actual data copies between
```

```
      previous = tmp;      // the two data versions
      // Computation of current iteration over local data
30    for(ind=0; ind<localSize; ++ind){
         // Updating of current array using previous array
         ...
         // Updating of the residual
         // (max difference between two successive iterations)
35       difference = fabs(current[ind] - previous[ind]);
         if(difference > residual){
            residual = difference;
         }
      }
40
      // Checking of the end of the process (residual under threshold)
      // Other conditions can be added to the termination detection
      if(residual <= Threshold){
         Finished = 1;
45       omp_unset_lock(&lockSend);  // Activation of end messages sendings
         MPI_Ssend(&Finished, 1, MPI_CHAR, numP, tagEnd, MPI_COMM_WORLD);
      }

      // Updating of the iteration number
50    iter++;
   }
```

As mentioned above, it can be seen in lines 19–21 of Listing 7.6 that the `lockRec` mutex is used only at the first iteration to wait for the initial data dependencies before the computations. The residual, initialized in line 24 and computed in lines 35–38, is defined by the maximal difference between the elements from two consecutive iterations. It is classically used to detect the local convergence of the process on each node. In the more complete schemes presented in the sequel, a global termination detection that takes the states of all the nodes into account will be exhibited.

Finally, the local convergence is tested and updated when necessary. In line 45, the `lockSend` mutex is unlocked to allow the sending function to send final messages to the dependency nodes. Those messages are required to keep the reception function alive until all the final messages have been received. Otherwise, a node could stop its reception function while other nodes are still trying to communicate with it. Moreover, a local sending of a final message to the node itself is required (line 46) to ensure that the reception function will not stay blocked in a message probing (see Listing 7.8, line 12). This may happen if the node receives the final messages from its dependencies *before* reaching its own local convergence.

All the messages but this final local one are performed in the sending function described in Listing 7.7. The main loop is only conditioned by the end of the computing process (line 4). At each iteration, the thread waits for the permission from the computing thread (according to the `lockSend` mutex). Then, data are sent with blocking synchronous communications. The `SendsInProgress` boolean allows the computing thread to skip data sendings as long as a previous sending is in progress. This skip is possible due to the nature of asynchronous algorithms that allows such *message loss* or *message miss*. After the main loop, the final messages are sent to the dependencies of the node.

Listing 7.7. sending function in the basic asynchronous scheme

```
// Variables declaration and initialization
...

while(!Finished){
    omp_set_lock(&lockSend); // Waiting for signal from the comp. thread
    if(!Finished){
        // Blocking synchronous sends to all dependencies
        for(i=0; i<nbDeps; ++i){
            MPI_Ssend(&dataToSend[deps[i]], nb_data, type_of_data, deps[i],
                tagCom, MPI_COMM_WORLD);
        }
        SendsInProgress = 0; // Indicates that the sendings are done
    }
}
// At the end of the process, sendings of final messages
for(i=0; i<nbDeps; ++i){
    MPI_Ssend(&Finished, 1, MPI_CHAR, deps[i], tagEnd, MPI_COMM_WORLD);
}
```

The last function, detailed in Listing 7.8, does all the messages receptions.

Listing 7.8. Reception function in the basic asynchronous scheme

```
// Variables declaration and initialization
char countReceipts = 1; // Boolean indicating whether receptions are
                        // counted or not
int nbEndMsg = 0;       // Number of end messages received
int arrived = 0;        // Boolean indicating if a message has arrived
int srcNd;              // Source node of the message
int size;               // Message size

// Main loop of receptions
while(!Finished){
    // Waiting for an incoming message
    MPI_Probe(MPI_ANY_SOURCE, MPI_ANY_TAG, MPI_COMM_WORLD, &status);
    if(!Finished){
        // Management of data messages
        switch(status.MPI_TAG){
            case tagCom: // Management of data messages
                // Get the source node of the message
                srcNd = status.MPI_SOURCE;
                // Actual data reception in the corresponding buffer
                MPI_Recv(dataBufferOf(srcNd), nbDataOf(srcNd), dataTypeOf(srcNd
                    ), srcNd, tagCom, MPI_COMM_WORLD, &status);
                // Unlocking of the computing thread when data are received
                // from all dependencies
                if(countReceipts == 1 && ... receptions from ALL dependencies ...){
                    omp_unset_lock(&lockRec);
                    countReceipts = 0; // No more counting after first iteration
                }
                break;
            case tagEnd: // Management of end messages
                // Actual end message reception in dummy buffer
                MPI_Recv(dummyBuffer, 1, MPI_CHAR, status.MPI_SOURCE, tagEnd,
                    MPI_COMM_WORLD, &status);
                nbEndMsg++;
        }
    }
}
// Reception of pending messages and counting of end messages
do{ // Loop over the remaining incoming/end messages
    MPI_Probe(MPI_ANY_SOURCE, MPI_ANY_TAG, MPI_COMM_WORLD, &status);
    MPI_Get_count(&status, MPI_CHAR, &size);
    // Actual reception in dummy buffer
```

```
    MPI_Recv(dummyBuffer, size, MPI_CHAR, status.MPI_SOURCE, status.
       MPI_TAG, MPI_COMM_WORLD, &status);
    if(status.MPI_TAG == tagEnd){ // Counting of end messages
       nbEndMsg++;
    }
45  MPI_Iprobe(MPI_ANY_SOURCE, MPI_ANY_TAG, MPI_COMM_WORLD, &arrived, &
       status);
  }while(arrived == 1 || nbEndMsg < nbDeps + 1);
```

As in the sending function, the main loop of receptions is done while the iterative process is not Finished. In line 12, the thread waits until a message arrives on the node. Then, it performs the actual reception and the corresponding subsequent actions (potential data copies for data messages and counting for end messages). Lines 23–26 check, only at the first iteration of computations, that all data dependencies have been received before unlocking the lockRec mutex. Although this is not mandatory, it ensures that all data dependencies are received before starting the computations. Lines 28–31 are required to manage end messages that arrive on the node *before* it reaches its own termination process. As the nodes are *not* synchronized, this may happen. Finally, lines 37–46 perform the receptions of all pending communications, including the remaining end messages (at least the one from the node itself).

So, with these algorithms, we obtain a quite simple and efficient asynchronous iterative scheme. It is interesting to notice that GPU computing can be easily included in the computing thread. This will be fully addressed in Section 7.3.3. However, before presenting the complete asynchronous scheme with GPU computing, we have to detail how our initial scheme can be made synchronous.

7.3.2 Synchronization of the asynchronous scheme

The presence of synchronization in the previous scheme may seem contradictory to our goal, and obviously, it is neither the simplest way to obtain a synchronous scheme nor the most efficient (as presented in Section 7.2). However, it is necessary for our global convergence detection strategy. Recall that the global convergence is the extension of the local convergence concept to all the nodes. This implies that all the nodes have to be in local convergence at the same time to achieve global convergence. Typically, if we use the residual and a threshold to stop the iterative process, all the nodes have to continue their local iterative process until *all* of them obtain a residual under the threshold.

In our context, being able to dynamically change the operating mode (sync/async) during the process execution strongly simplifies the global convergence detection. In fact, our past experience in the design and implementation of global convergence detection in asynchronous algorithms [5–7] has led us to the conclusion that although a decentralized detection scheme is possible and may be more efficient in some situations, its much higher complexity is an obstacle to actual use in practice, especially in industrial contexts where implementation/maintenance costs are strong constraints. Moreover, although the decentralized scheme does not slow down the computations, it requires more iterations than a synchronous version and thus may induce longer detection times in some cases. So, the solution we present below is a good compromise between simplicity and efficiency. It consists in dynamically changing the operating mode between asynchronous and synchronous during the execution of the process in order to check the global convergence. This is why we need to synchronize our asynchronous scheme.

In each algorithm of the initial scheme, we only give the additional code required to change the operating mode.

Listing 7.9. initialization of the synchronized scheme

```
// Variables declarations and initialization
...
// Controls the synchronous exchange of local states
omp_lock_t lockStates;
// Controls the synchronization at the end of each iteration
omp_lock_t lockIter;
//Boolean indicating whether the local stabilization is reached or not
char localCV = 0;
// Number of other nodes being in local stabilization
int nbOtherCVs = 0;

// Parameters reading
...
// MPI initialization
...
// Data initialization and distribution among the nodes
...
// OpenMP initialization (mainly declarations and setting up of locks)
...
omp_init_lock(&lockStates);
// Initially locked, unlocked when all state messages are received
omp_set_lock(&lockStates);
omp_init_lock(&lockIter);
// Initially locked, unlocked when all "end of iteration" messages are
// received
omp_set_lock(&lockIter);

// Threads launching
#pragma omp parallel
{
   switch(omp_get_thread_num()){
      ...
   }
}

// Cleaning of OpenMP locks
...
omp_test_lock(&lockStates);
omp_unset_lock(&lockStates);
omp_destroy_lock(&lockStates);
omp_test_lock(&lockIter);
omp_unset_lock(&lockIter);
omp_destroy_lock(&lockIter);

// MPI termination
MPI_Finalize();
```

As can be seen in Listing 7.9, the synchronization implies two additional mutex. The `lockStates` mutex is used to wait for the receptions of all state messages coming from the other nodes. As shown in Listing 7.10, those messages contain only a boolean indicating for each node if it is in local convergence. So, once all the states are received on a node, it is possible to determine if all the nodes are in local convergence and, thus, to detect the global convergence. The `lockIter` mutex is used to synchronize all the nodes at the end of each iteration. There are also two new variables that represent the local state of the node (`localCV`) according to the iterative process (convergence) and the number of other nodes that are in local convergence (`nbOtherCVs`).

The computation thread is where most of the modifications take place, as shown in Listing 7.10.

Listing 7.10. computing function in the synchronized scheme
```
// Variables declarations and initialization
...

// Computation loop
while(!Finished){
  // Sending of data dependencies at each iteration
  // Potential copy of data to be sent in additional buffers
  ...
  omp_unset_lock(&lockSend);

  // Blocking receptions at each iteration
  omp_set_lock(&lockRec);

  // Local computation
  // (init of residual, arrays swapping and iteration computation)
  ...

  // Checking of the stabilization of the local process
  // Other conditions than the residual can be added
  if(residual <= Threshold){
    localCV = 1;
  }else{
    localCV = 0;
  }

  // Global exchange of local states of the nodes
  for(ind=0; ind<nbP; ++ind){
    if(ind != numP){
      MPI_Ssend(&localCV, 1, MPI_CHAR, ind, tagState, MPI_COMM_WORLD);
    }
  }

  // Waiting for the state messages receptions from the other nodes
  omp_set_lock(&lockStates);

  //Determination of global convergence (if all nodes are in local CV)
  if(localCV + nbOtherCVs == nbP){
    // Entering global CV state
    Finished = 1;
    // Unlocking of sending thread to start sendings of end messages
    omp_unset_lock(&lockSend);
    MPI_Ssend(&Finished, 1, MPI_CHAR, numP, tagEnd, MPI_COMM_WORLD);
  }else{
    // Resetting of information about the states of the other nodes
    ...
    // Global barrier at the end of each iteration during the process
    for(ind=0; ind<nbP; ++ind){
      if(ind != numP){
```

```
                MPI_Ssend(&Finished , 1, MPI_CHAR, ind , tagIter , MPI_COMM_WORLD
                    );
50          }
        }
        omp_set_lock(&lockIter);
    }

55
    // Updating of the iteration number
    iter++;
}
```

Most of the added code is related to the waiting for specific communications. Between lines 6 and 7, the use of the flag SendsInProgress is no longer needed since the sends are performed at each iteration. In line 12, the thread waits for the data receptions from its dependencies. In lines 27–34, the local states are determined and exchanged among all nodes. A new message tag (tagState) is required for identifying those messages. In line 37, the global termination state is determined. When it is reached, lines 39–42 change the Finished boolean to stop the iterative process and send the end messages. Otherwise each node resets its local state information about the other nodes and a global barrier is added between all the nodes at the end of each iteration with another new tag (tagIter). That barrier is needed to ensure that data messages from successive iterations are actually received during the *same* iteration on the destination nodes. Nevertheless, it is not useful at the termination of the global process as it is replaced by the global exchange of end messages.

There is no big modification induced by the synchronization in the sending function. The function stays almost the same as in Listing 7.7. The only change could be the suppression of line 11 that is not useful in this case.

In the reception function, given in Listing 7.11, there are mainly two insertions (in lines 19–31 and 32–42), corresponding to the additional types of messages to receive. There is also the insertion of three variables that are used for the receptions of the new message types. In lines 24–30 and 35–41 are located messages counting and mutex unlocking mechanisms that are used to block the computing thread at the corresponding steps of its execution. They are similar to the mechanism used for managing the end messages at the end of the entire process. Line 23 directly updates the number of other nodes that are in local convergence by adding the received state of the source node. This is possible due to the encoding that is used to represent the local convergence (1) and the nonconvergence (0).

Listing 7.11. reception function in the synchronized scheme
```
// Variables declarations and initialization
...
int nbStateMsg = 0;  // Number of local state messages received
int nbIterMsg  = 0;  // Number of "end of iteration" messages received
5 char recvdState;    // Received state from another node (0 or 1)

// Main loop of receptions
while(!Finished){
```

```
        // Waiting for an incoming message
10      MPI_Probe(MPI_ANY_SOURCE, MPI_ANY_TAG, MPI_COMM_WORLD, &status);
        if(!Finished){
          switch(status.MPI_TAG){ // Actions related to message type
            case tagCom: // Management of data messages
              ...
15            break;
            case tagEnd: // Management of termination messages
              ...
              break;
            case tagState: // Management of local state messages
20            // Actual reception of the message
              MPI_Recv(&recvdState, 1, MPI_CHAR, status.MPI_SOURCE, tagState,
                  MPI_COMM_WORLD, &status);
              // Updates of numbers of stabilized nodes and recvd state msgs
              nbOtherCVs += recvdState;
              nbStateMsg++;
25            // Unlocking of the computing thread when states of all other
              // nodes are received
              if(nbStateMsg == nbP-1){
                nbStateMsg = 0;
                omp_unset_lock(&lockStates);
30            }
              break;
            case tagIter: // Management of "end of iteration" messages
              // Actual reception of the message in dummy buffer
              MPI_Recv(dummyBuffer, 1, MPI_CHAR, status.MPI_SOURCE, tagIter,
                  MPI_COMM_WORLD, &status);
35            nbIterMsg++; // Update of the number of iteration messages
              // Unlocking of the computing thread when iteration messages
              // are received from all other nodes
              if(nbIterMsg == nbP - 1){
                nbIterMsg = 0;
40              omp_unset_lock(&lockIter);
              }
              break;
          }
        }
45      }
        // Reception of pending messages and counting of end messages
        do{ // Loop over the remaining incoming/end messages
          ...
50      }while(arrived == 1 || nbEndMsg < nbDeps + 1);
```

Now that we can synchronize our asynchronous scheme, the final step is to dynamically alternate the two operating modes in order to regularly check the global convergence of the iterative process. This is detailed in the following section together with the inclusion of GPU computing in the final asynchronous scheme.

7.3.3 Asynchronous scheme using MPI, OpenMP, and CUDA

As mentioned above, the strategy proposed to obtain a good compromise between simplicity and efficiency in the asynchronous scheme is to dynamically change the operating mode of the process. A good way to obtain a maximal simplification of the final scheme while preserving good performance is to perform local and global convergence detections only in synchronous mode. Moreover, as two successive iterations are sufficient in synchronous mode to detect local and global convergences, the key is to alternate some asynchronous iterations with two synchronous iterations until convergence.

The last problem is to decide *when* to switch from the asynchronous to the synchronous mode. Here again, for the sake of simplicity, any asynchronous mechanism for *detecting* such moment is avoided, and we prefer to use a mechanism that is local to each node. Obviously, that local system must rely neither on the number of local iterations done nor on the local convergence. The former would slow down the fastest nodes according to the slowest ones. The latter would provoke too much synchronization because the residuals on all nodes generally do not evolve in the same way, and in most cases, there is a convergence wave phenomenon throughout the elements. So, a good solution is to insert a local timer mechanism on each node with a given initial duration. Then, that duration may be modified during the execution according to the successive results of the synchronous sections.

Another problem induced by entering synchronous mode from the asynchronous one is the possibility of receiving some data messages from previous asynchronous iterations during synchronous iterations. This could lead to deadlocks. In order to avoid this, a wait for the end of previous send is added to the transition between the two modes. This is implemented by replacing the variable SendsInProgress with a mutex lockSendsDone which is unlocked once all the messages have been sent in the sending function. Moreover, it is also necessary to stamp data messages (by the function stampData) with a boolean indicating whether they have been sent during a synchronous or asynchronous iteration. Then, the lockRec mutex is unlocked only after to the complete reception of data messages from synchronous iterations. The message ordering of point-to-point communications in MPI and the barrier at the end of each iteration ensure two important properties of this mechanism. First, data messages from previous asynchronous iterations will be received but not taken into account during synchronous sections. Then, a data message from a given synchronous iteration cannot be received during another synchronous iteration. In the asynchronous sections, no additional mechanism is needed as there are no such constraints concerning the data receptions.

Finally, the required modifications of the previous scheme are mainly related to the computing thread. Small additions or modifications are also required in the main process and the other threads.

In the main process, two new variables are added to store the main operating mode of the iterative process (mainMode) and the duration of asynchronous sections (asyncDuration). Those variables are initialized by the programmer. The mutex lockSendsDone is also declared, initialized (locked), and destroyed with the other mutex in this process.

In the computing function, shown in Listing 7.12, the modifications consist of the insertion of the timer mechanism and the tests to differentiate the actions to be done in each mode. Some additional variables are also required to store the current operating mode in action during the execution (curMode),

the starting time of the current asynchronous section (asyncStart), and the
number of successive synchronous iterations done (nbSyncIter).

Listing 7.12. computing function in the final asynchronous scheme

```
// Variables declarations and initialization
...
OpMode curMode = SYNC; // Current operating mode (always begin in sync)
double asyncStart;      // Starting time of the current async section
int nbSyncIter = 0;     // Number of sync iterations done in async mode

// Computation loop
while(!Finished){
  // Determination of the dynamic operating mode
  if(curMode == ASYNC){
    // Entering synchronous mode when asyncDuration is reached
    if(MPI_Wtime() - asyncStart >= asyncDuration){
      // Waiting for the end of previous sends before starting sync mode
      omp_set_lock(&lockSendsDone);
      curMode = SYNC;                     // Entering synchronous mode
      stampData(dataToSend, SYNC);        // Mark data to send with sync flag
      nbSyncIter = 0;
    }
  }else{
    // In main async mode, going back to async mode when the max number
    // of sync iterations are done
    if(mainMode == ASYNC){
      nbSyncIter++; // Update of the number of sync iterations done
      if(nbSyncIter == 2){
        curMode = ASYNC;                  // Going back to async mode
        stampData(dataToSend, ASYNC);     // Mark data to send
        asyncStart = MPI_Wtime();         // Get the async starting time
      }
    }
  }

  // Sending of data dependencies
  if(curMode == SYNC || !SendsInProgress){
    ...
  }

  // Blocking data receptions in sync mode
  if(curMode == SYNC){
    omp_set_lock(&lockRec);
  }

  // Local computation
  // (init of residual, arrays swapping, and iteration computation)
  ...

  // Checking convergences (local & global) only in sync mode
  if(curMode == SYNC){
    // Local convergence checking (residual under threshold)
    ...
    // Blocking global exchange of local states of the nodes
    ...
    // Determination of global convergence (all nodes in local CV)
    //     Stopping the iterative process and sending end messages
    // or reinitialization of state information and iteration barrier
    ...
  }

  // Updating of the iteration number
  iter++;
}
```

In the sending function, the only modification is the replacement in line 11 of the assignment of variable SendsInProgress with the unlocking of lockSendsDone. Finally, in the reception function, the only modification is the insertion before line 21 of Listing 7.8 of the extraction of the stamp from the message and its counting among the receipts only if the stamp is SYNC.

The final step to get our complete scheme using GPU is to insert the GPU management in the computing thread. The first possibility, detailed in Listing 7.13, is to simply replace the CPU kernel (lines 42–44 in Listing 7.12) by a blocking GPU kernel call. This includes data transfers from the node RAM to the GPU RAM, the launching of the GPU kernel, the waiting for kernel completion, and the results transfers from GPU RAM to node RAM.

```
Listing 7.13. computing function in the final asynchronous scheme
// Variables declarations and initialization
...
dim3 Dg, Db; // CUDA kernel grids

// Computation loop
while(!Finished){
   // Determination of the dynamic operating mode, sendings of data
   // dependencies, and blocking data receptions in sync mode
   ...
   // Local GPU computation
   // Data transfers from node RAM to GPU
   CHECK_CUDA_SUCCESS(cudaMemcpyToSymbol(dataOnGPU, dataInRAM,
       inputsSize, 0, cudaMemcpyHostToDevice), "Data transfer");
   ... // There may be several data transfers: typically A and b in
       // linear problems of the form A.x = b
   // GPU grid definition
   Db.x = BLOCK_SIZE_X; // BLOCK_SIZE_# are kernel design dependent
   Db.y = BLOCK_SIZE_Y;
   Db.z = BLOCK_SIZE_Z;
   Dg.x = localSize/BLOCK_SIZE_X + (localSize%BLOCK_SIZE_X ? 1 : 0);
   Dg.y = localSize/BLOCK_SIZE_Y + (localSize%BLOCK_SIZE_Y ? 1 : 0);
   Dg.z = localSize/BLOCK_SIZE_Z + (localSize%BLOCK_SIZE_Z ? 1 : 0);
   // Use of shared memory (when possible)
   cudaFuncSetCacheConfig(gpuKernelName, cudaFuncCachePreferShared);
   // Kernel call
   gpuKernelName<<<Dg,Db>>>(... kernel parameters ...);
   // Waiting for kernel completion
   cudaDeviceSynchronize();
   // Results transfer from GPU to node RAM
   CHECK_CUDA_SUCCESS(cudaMemcpyFromSymbol(resultsInRam, resultsOnGPU,
       resultsSize, 0, cudaMemcpyDeviceToHost), "Results transfer");
   // Potential post-treatment of results on the CPU
   ...

   // Convergences checking
   ...
}
```

This scheme provides asynchronism through a cluster of GPUs as well as a complete overlap of communications with GPU computations (similar to the one described in Section 7.2). However, the autonomy of GPU devices according to their host can be further exploited in order to perform some computations on the CPU while the GPU kernel is running. The nature of computations that can be done by the CPU may vary depending on the application. For example, when processing data streams (pipelines), pre-processing of the next data item and/or post-processing of the previous result can be done on the CPU while the GPU is processing the current data item. In other cases, the CPU can perform *auxiliary* computations that are not absolutely required to obtain the result but that may accelerate the entire iterative process. Another possibility would be to distribute the main computations between the GPU and CPU. However, this usually leads to poor performance increases mainly due to data dependencies that often require additional transfers between CPU and GPU.

So, if we consider that the application enables such overlap of computations, its implementation is straightforward as it consists in inserting the additional CPU computations between lines 25 and 26 in Listing 7.13. Nevertheless, such a scheme is fully efficient only if the computation times on both sides are similar.

In some cases, especially with auxiliary computations, another interesting solution is to add a fourth CPU thread to perform them. This suppresses the duration constraint over those optional computations as they are performed in parallel with the main iterative process, without blocking it. Moreover, this scheme stays coherent with current architectures as most nodes include four CPU cores. The algorithmic scheme of such context of complete overlap of CPU/GPU computations and communications is described in Listings 7.14, 7.15, and 7.16, where we assume that auxiliary computations use intermediate results of the main computation process from any previous iteration. This may be different according to the application.

Listing 7.14. initialization of the main process of complete overlap with asynchronism

```
   // Variables declarations and initialization
   ...
   omp_lock_t lockAux;     // Informs main thread about new aux results
   omp_lock_t lockRes;     // Informs aux thread about new results
 5 omp_lock_t lockWrite;   // Controls exclusion of results access
   ... auxRes ... ;        // Results of auxiliary computations

   // Parameters reading, MPI initialization, and data initialization and
   // distribution
10 ...
   // OpenMP initialization
   ...
   omp_init_lock(&lockAux);
   omp_set_lock(&lockAux);//Unlocked when new aux results are available
15 omp_init_lock(&lockRes);
   omp_set_lock(&lockRes);  // Unlocked when new results are available
   omp_init_lock(&lockWrite);
   omp_unset_lock(&lockWrite); // Controls access to results from threads

20 #pragma omp parallel
   {
     switch(omp_get_thread_num()){
       case COMPUTATION :
       computations(... relevant parameters ...) ;
25     break;

       case AUX_COMPS :
```

```
         auxComps(... relevant parameters ...);
         break;
30
         case SENDINGS :
         sendings();
         break;

35       case RECEPTIONS :
         receptions();
         break;
       }
     }
40
     // Cleaning of OpenMP locks
     ...
     omp_test_lock(&lockAux);
     omp_unset_lock(&lockAux);
45   omp_destroy_lock(&lockAux);
     omp_test_lock(&lockRes);
     omp_unset_lock(&lockRes);
     omp_destroy_lock(&lockRes);
     omp_test_lock(&lockWrite);
50   omp_unset_lock(&lockWrite);
     omp_destroy_lock(&lockWrite);

     // MPI termination
     MPI_Finalize();
```

Listing 7.15. computing function in the final asynchronous scheme with CPU/GPU overlap

```
     // Variables declarations and initialization
     ...
     dim3 Dg, Db;  // CUDA kernel grids

5    // Computation loop
     while(!Finished){
       // Determination of the dynamic operating mode, sending of data
       // dependencies, and blocking data receptions in sync mode
       ...
10     // Local GPU computation
       // Data transfers from node RAM to GPU, GPU grid definition,
       // and init of shared memory
       CHECK_CUDA_SUCCESS(cudaMemcpyToSymbol(dataOnGPU, dataInRAM,
           inputsSize, 0, cudaMemcpyHostToDevice), "Data transfer");
       ...
15     // Kernel call
       gpuKernelName<<<Dg,Db>>>(... kernel parameters ...);
       // Potential pre-/post-treatments in pipeline-like computations
       ...
       // Waiting for kernel completion
20     cudaDeviceSynchronize();
       // Results transfer from GPU to node RAM
       omp_set_lock(&lockWrite); // Wait for write access to resultsInRam
       CHECK_CUDA_SUCCESS(cudaMemcpyFromSymbol(resultsInRam, resultsOnGPU,
           resultsSize, 0, cudaMemcpyDeviceToHost), "Results transfer");
       // Potential post-treatments in non pipeline computations
25     ...
       omp_unset_lock(&lockWrite); // Give back read access to aux thread
       omp_test_lock(&lockRes);
       omp_unset_lock(&lockRes);    // Informs aux thread of new results

30     // Auxiliary computations availability checking
       if(omp_test_lock(&lockAux)){
         // Use auxRes to update the iterative process
         ... // May induce additional GPU transfers
```

```
        }
35
    // Convergences checking
    if(curMode == SYNC){
        // Local convergence checking and global exchange of local states
        ...
40      // Determination of global convergence (all nodes in local CV)
        if(cvLocale == 1 && nbCVLocales == nbP-1){
            // Stopping the iterative process and sending end messages
            ...
            // Unlocking aux thread for termination
45          omp_test_lock(&lockRes);
            omp_unset_lock(&lockRes);
        }else{
            // Reinitialization of state information and iteration barrier
            ...
50      }
    }
}
```

Listing 7.16. auxiliary computing function in the final asynchronous scheme with CPU/GPU overlap

```
// Variables declarations and initialization
... auxInput ...  // Local array for input data

// Computation loop
5 while(!Finished){
    // Data copy from resultsInRam into auxInput
    omp_set_lock(&lockRes);     // Waiting for new results from main comps
    if(!Finished){
        omp_set_lock(&lockWrite); // Waiting for access to results
10      for(ind=0; ind<resultsSize; ++ind){
            auxInput[ind] = resultsInRam[ind];
        }
        omp_unset_lock(&lockWrite);//Give back write access to main thread
        // Auxiliary computations with possible interruption at the end
15      for(ind=0; ind<auxSize && !Finished; ++ind){
            // Computation of auxRes array according to auxInput
            ...
        }
        // Informs main thread that new aux results are available in auxData
20      omp_test_lock(&lockAux); // Ensures mutex is locked when unlocking
        omp_unset_lock(&lockAux);
    }
}
```

As can be seen in Listing 7.14, there are three additional mutex (`lockAux`, `lockRes`, and `lockWrite`) that are used to inform the main computation thread that new auxiliary results are available (lines 20–21 in Listing 7.16 and line 31 in Listing 7.15), to inform the auxiliary thread that new results from the main thread are available (lines 27–28 in Listing 7.15 and line 7 in Listing 7.16), and to perform exclusive accesses to the results from those two threads (lines 22 and 26 in Listing 7.15 and 9 and 13 in Listing 7.16). Also, an additional array (`auxRes`) is required to store the results of the auxiliary computations as well as a local array for the input of the auxiliary function (`auxInput`). That last function has the same general organization as the send/receive ones, that is, a global loop conditioned by the end of the global process. At each iteration in this function, the thread waits for the availability of new results produced by the main computation thread. This avoids performing the same computations several times with the same input data. Then, input data of auxiliary computations is copied with a mutual exclusion mechanism. Finally, auxiliary computations are performed. When they are completed, the associated mutex is unlocked to signal the availability of those auxiliary results to the main computing thread. The main thread regularly checks this availability at the end of its iterations and takes them into account whenever possible.

Finally, we obtain an algorithmic scheme allowing maximal overlap between CPU and GPU computations as well as communications. It is worth noticing that such scheme is also efficiently usable for systems without GPUs but with nodes having at least four cores. In such contexts, each thread in Listing 7.14 can be executed on distinct cores.

7.3.4 Experimental validation

As in Section 7.2, we validate the feasibility of our asynchronous scheme with some experiments performed with a representative example of scientific application. This three-dimensional version of the advection-diffusion-reaction process models the evolution of the concentrations of two chemical species in shallow waters. As this process is dynamic in time, the simulation is performed for a given number of consecutive time steps. This implies two nested loops in the iterative process, the outer one for the time steps and the inner one for solving the problem at each time. Full details about this PDE problem can be found in [21]. This two-stage iterative process implies a few adaptations of the general scheme presented above in order to include the outer iterations over the time steps, but the inner iterative process closely follows the same scheme.

We show two series of experiments performed with 16 nodes of the first cluster described in Section 7.2.5. The first one deals with the comparison of synchronous and asynchronous computations. The second one is related to the use of auxiliary computations. In the context of our PDE application, they consist of the update of the Jacobian of the system.

Synchronous and asynchronous computations

The first experiment allows us to check that the asynchronous behavior obtained with our scheme corresponds to the expected one according to its synchronous counterpart. So, we show in Figure 7.6 the computation times of our test application in both modes for different problem sizes. The size shown is the number of discrete spatial elements on each side of the cube representing the 3D volume. Moreover, for each of these elements, there are the concentrations of the two chemical species considered. So, for example, size 30 corresponds in fact to $30 \times 30 \times 30 \times 2$ values.

The results obtained show that the asynchronous version is significantly

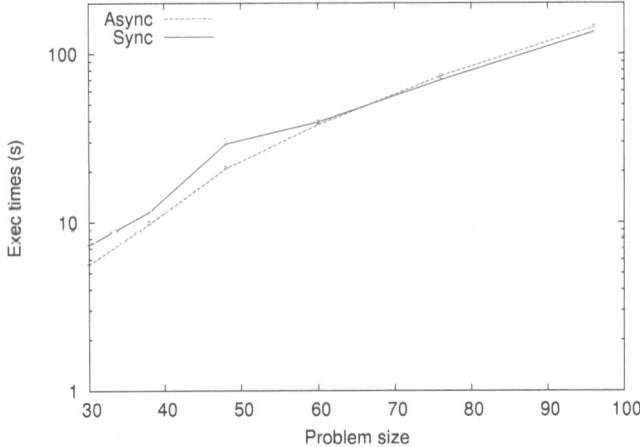

FIGURE 7.6. Computation times of the test application in synchronous and asynchronous modes.

faster than the synchronous one for smaller problem sizes, then it becomes similar or even a bit slower for larger problem sizes. A closer comparison of computation and communication times of each execution confirms that this behavior is consistent. The asynchronous version is interesting if communication time is similar or larger than computation time. In our example, this is the case up to a problem size between 50 and 60. Then, computations become longer than communications. Since asynchronous computations often require more iterations to converge, the gain obtained on the communication side becomes smaller than the overhead generated on the computation side, and the asynchronous version takes longer.

Overlap of auxiliary computations

In this experiment, we use only the asynchronous version of the application. In the context of our test application, we have an iterative PDE solver based on Netwon resolution. Such solvers are written under the form $x = T(x)$, $x \in \mathbb{R}^n$ where $T(x) = x - F'(x)^{-1}F(x)$ and F' is the Jacobian of the system. In such cases, it is necessary to compute the vector Δx in $F' \times \Delta x = -F$ to update x with Δx. There are two levels of iterations, the inner level to get a stabilized version of x, and the outer level to compute x at the successive time steps in the simulation process. In this context, classic algorithms either compute F' at only the first iteration of each time step or at some iterations but not all because the computation of F' is done in the main iterative process and it has a relatively high computing cost.

However, with the scheme presented above, it is possible to continuously compute new versions of F' in parallel with the main iterative process without penalizing it. Hence, F' is updated as often as possible and taken into account in the main computations when it is relevant. So, the Newton process should be accelerated a little bit.

We compare the performance obtained with overlapped Jacobian updatings and nonoverlapped ones for several problem sizes (see Figure 7.7).

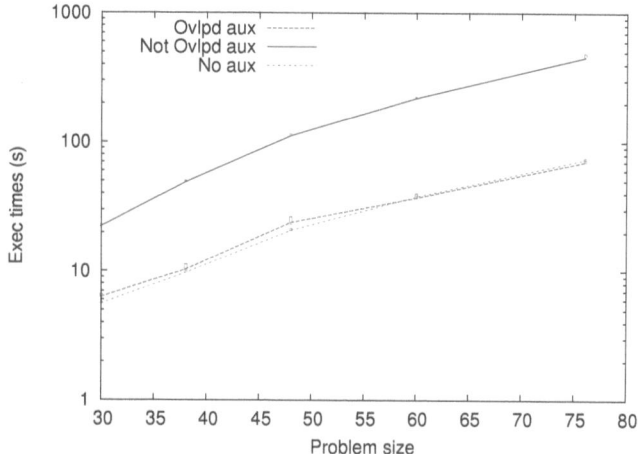

FIGURE 7.7. Computation times with or without overlap of Jacobian updatings in asynchronous mode.

The overlap is clearly efficient as the computation times with overlapping Jacobian updatings are much better than the ones without overlap. Moreover, the ratio between the two versions tends to increase with the problem size, which is as expected. Also, we have tested the application without auxiliary computations at all, that is, the Jacobian is computed only once at the beginning of each time step of the simulation. The results for this last version are quite similar to the overlapped auxiliary computations, and even better for small problem sizes. The fact that no significant gain can be seen on this range of problem sizes is due to the limited number of Jacobian updates taken into account in the main computation. This happens when the Jacobian update is as long as several iterations of the main process. So, the benefit is reduced in this particular case.

Those results show two things. First, auxiliary computations do not induce great overhead in the whole process. Second, for this particular application the choice of updating the Jacobian matrix as auxiliary computations does not speed up the iterative process. This does not question the parallel scheme in itself but merely points out the difficulty of identifying relevant auxiliary computations. Indeed, this identification depends on the considered application and requires a profound specialized analysis.

Another interesting choice could be the computation of load estimation for dynamic load balancing, especially in decentralized diffusion strategies where loads are transferred between neighboring nodes [8]. In such a case, the load evaluation and the comparison with other nodes can be done in parallel with the main computations without perturbing them.

7.4 Perspective: a unifying programming model

In the previous sections we have seen that controlling a distributed GPU application when using tools that are commonly available is quite a challenging task. To summarize, such an application has components that can be roughly classified as

CPU: CPU-bound computations, realized as procedures in the chosen programming language

CUDA$_{kern}$: GPU-bound computations, in our context realized as CUDA compute kernels

CUDA$_{trans}$: data transfer between CPU and GPU, realized with CUDA function calls

MPI: distributed data transfer routines, realized with MPI communication primitives

OpenMP: inter-thread control, realized with OpenMP synchronization tools such as mutexes

CUDA$_{sync}$: synchronization of the GPU, realized with CUDA functions

Among these, the last (CUDA$_{sync}$) is not strictly necessary on modern systems, but it is still recommended to obtain optimal performance. With or without that last step, such an application is highly complex: it is difficult to design or to maintain, and depends on a lot of different software components. The goal of this section is to present a new path of development that allows the replacement of the last three or four types of components that control the application (MPI, OpenMP, CUDA$_{sync}$, and eventually CUDA$_{trans}$) with a single tool: Ordered Read-Write Locks, ORWL (see [10, 13]). Besides the simplification of the algorithmic scheme that we have already mentioned, the ongoing implementation of ORWL allows the use of a feature of modern platforms that can improve the performance of CPU-bound computations: lock-free atomic operations to update shared data consistently. For these, ORWL relies on new interfaces that are available with the latest revision of the ISO standard for the C programming language (see [18]).

7.4.1 Resources

ORWL places all its concepts that concern data and control around a single abstraction: *resources*. An ORWL resource may correspond to a local or remote entity and is identified through a *location*, that is a unique identification through which it can be accessed from all different components of the same application. In fact, resources and locations (entities and their names, so to speak) are mostly identified by ORWL and these words will be used interchangeably.

Resources may be of very different kinds:

Data resources are entities that represents data and not specific memory buffers or locations. During the execution of an application they can be *mapped* repeatedly into the address space and in effect be represented at different addresses. Data resources can be made accessible uniformly in all parts of the application, provided that the locking protocol is observed (see below). Data resources can have different persistence:

> **RAM** data resources are typically temporary data that serve only during a single run of the application. They must be initialized at the beginning of their lifetime and the contents are lost at the end.
>
> **File** data resources are persistent and linked to a file in the file system of the platform.
>
> **Collective** data resources are data to which all tasks of an application contribute (see below). Examples for such resources are *broadcast*, *gather*, or *reduce* resources, e.g., to distribute initial data or to collect the result of a distributed computation.

Other types of data resources could be easily implemented with ORWL, e.g., web resources (through a ftp, http or whatever server address) or fixed hardware addresses.

Device resources represent hardware entities of the platform. ORWL can then be used to regulate the access to such device resources. In our context the most important such resource is the GPU, but we could easily use it to represent a CPU core, a camera, or another peripheral device.

Listing 7.17 shows an example of a declaration of four resources per task. Two (curBlock and nextBlock) are intended to represent the data in a block-cyclic parallel matrix multiplication (see Section 7.2.5), GPU represents a GPU device, and result will represent a collective "gather" resource among all the tasks.

```
Listing 7.17. declaration of ORWL resources for a block-cyclic matrix mul-
tiplication
#include "orwl.h"
...
ORWL_LOCATIONS_PER_TASK( curBlock , nextBlock , GPU, result );
ORWL_DATA_LOCATION( curBlock );
ORWL_DATA_LOCATION( nextBlock );
ORWL_DEVICE_LOCATION(GPU);
ORWL_GATHER_LOCATION( result );
```

7.4.2 Control

ORWL regulates access to all its resources; no "random access" to a resource is possible. It doesn't even have a user-visible data type for resources.

- All access is provided through *handles*. Similar to pointers or links, these only refer to a resource and help to manipulate it. Usually several handles to the same resource exist, even inside the same OS process or thread, or in the same application task.

- The access is locked with RW semantics, where R stands for concurrent Read access, and W for exclusive Write access. This feature replaces the control aspect of MPI communications, OpenMP inter-thread control, and $CUDA_{sync}$.

- This access is Ordered (or serialized) through a FIFO, *one FIFO per resource*. This helps to run the different tasks of an application in a controlled order and to always have all resources in a known state. This aspect largely replaces and extends the ordering of tasks that MPI typically achieves through the passing of messages.

- The access is transparently managed for remote or local resources. Communication, if necessary, is done asynchronously behind the scenes. This replaces the explicit handling of buffers and messages with MPI.

7.4.3 Example: block-cyclic matrix multiplication (MM)

Let us now have a look at how a block-cyclic matrix multiplication algorithm can be implemented with these concepts (Listing 7.18). Inside the loop there are mainly three different operations, the first two of which can be run concurrently, and the third must be done after the other two.

```
Listing 7.18. block-cyclic matrix multiplication, high level per task view
typedef double MBlock[N][N];
MBlock A;
MBlock B[k];
MBlock C[k];

<do some initialization>
```

```
for (size_t i = 0; i < k; ++i) {
  MBlock next;
  parallel-do {
    operation 1: <copy the matrix A of the left neighbor into next>;
    operation 2: {
      <copy the local matrix A to the GPU >;
      <on GPU perform C[i] = A * B[0] + ... + A * B[k-1]; >;
    }
  }
  operation 3: {
    <wait until the right neighbor has read our block A>;
    A = next;
  }
}
<collect the result matrix C consisting of all C blocks>
```

Listing 7.19 shows the local copy operation 3 from line 17 of Listing 7.18 as it could be realized with ORWL. It uses two resource handles `nextRead` and `curWrite` and marks nested *critical sections* for these handles. Inside the nested sections it obtains pointers to the resource data; the resource is *mapped* into the address space of the program, and then a standard call to `memcpy` achieves the operation itself. The operation is integrated in its own for-loop, such that it could run independently in an OS thread by its own.

Listing 7.19. an iterative local copy operation
```
for (size_t i = 0; i < k; ++i) {
  ORWL_SECTION(nextRead) {
    MBlock const* sBlock = orwl_read_map(nextRead);
    ORWL_SECTION(curWrite) {
      MBlock * tBlock = orwl_write_map(curWrite);
      memcpy(tBlock, sBlock, sizeof *tBlock);
    }
  }
}
```

Next, in Listing 7.20 we copy data from a remote task to a local task. Substantially the operation is the same, save for the different handles (`remRead` and `nextWrite`) that are used to represent the respective resources.

Listing 7.20. an iterative remote copy operation as part of a block cyclic matrix multiplication task
```
for (size_t i = 0; i < k; ++i) {
  ORWL_SECTION(remRead) {
    MBlock const* sBlock = orwl_read_map(remRead);
    ORWL_SECTION(nextWrite) {
      MBlock * tBlock = orwl_write_map(nextWrite);
      memcpy(tBlock, sBlock, sizeof *tBlock);
    }
  }
}
```

Now let us have a look into the operation that probably interests us the most, the interaction with the GPU in Listing 7.21. Again there is much structural resemblance to the copy operations from above, but we transfer the data to the GPU in the innermost block and then run the GPU MM kernel while we are still inside the critical section for the GPU.

Listing 7.21. an iterative GPU transfer and compute operation as part of a block cyclic matrix multiplication task

```
for (size_t i = 0; i < k; ++i) {
  ORWL_SECTION(GPUhandle) {
    ORWL_SECTION(curRead) {
      MBlock const* sBlock = orwl_read_map(curRead);
      transferToGPU(sBlock, i);
    }
    runMMonGPU(i);
  }
}
```

Now that we have seen how the actual procedural access to the resources is regulated, we will show how the association between handles and resources is specified. In our application of block-cyclic MM the curRead handle should correspond to the current matrix block of the corresponding task, whereas remRead should point to the current block of the neighboring task. Both read operations on these matrix blocks can be performed without creating conflicts, so we would like to express that fact in our resource specification. From a point of view of the resource "current block" of a particular task, this means that it can have two simultaneous readers, the task itself performing the transfer to the GPU, and the neighboring task transferring the data to its "next block."

Listing 7.22 first shows the local dynamic declarations of our application; it declares a block type for the matrix blocks, a result data for the collective resource, and the six handles that we have seen so far.

Listing 7.22. dynamic declaration of handles to represent the resources

```
/* A type for the matrix blocks */
typedef double MBlock[N][N];
/* Declaration to handle the collective resource */
ORWL_GATHER_DECLARE(MBlock, result);

/* Variables to handle data resources */
orwl_handle2 remRead    = ORWL_HANDLE2_INITIALIZER;
orwl_handle2 nextWrite  = ORWL_HANDLE2_INITIALIZER;
orwl_handle2 nextRead   = ORWL_HANDLE2_INITIALIZER;
orwl_handle2 curWrite   = ORWL_HANDLE2_INITIALIZER;
orwl_handle2 curRead    = ORWL_HANDLE2_INITIALIZER;

/* Variable to handle the device resources */
orwl_handle2 GPUhandle  = ORWL_HANDLE2_INITIALIZER;
```

With these declarations, we didn't yet tell ORWL much about the resources to which these handles refer, nor the type (read or write) or the priority (FIFO position) of the access. This is done in code Listing 7.23 that shows six insertions of handles into their respective FIFO locations. The handles GPUhandle and curRead are given first (lines 3 and 4), GPUhandle will be accessed exclusively (therefore the write) and, as said, curRead is used in shared access (so a read). Both are inserted in the FIFO of their respective resources with highest priority, specified by the 0s in the third function parameter. The resources to which they correspond are specified through calls to the macro ORWL_LOCATION, indicating the task (orwl_mytid is the ID of the current task) and the specific resource of that task, here GPU and curBlock.

Likewise, a second block of insertions concerns the handles remRead and nextWrite (lines 8 and 9). nextWrite reclaims an exclusive access and remRead a shared one. remRead corresponds to a resource of another task; the call to previous(orwl_mytid) is supposed to return the ID of the previous task in the cycle. Both accesses can be performed concurrently with the previous operation, so we insert them with the same priority 0 as previously.

Then, for the specification of the third operation related to handles nextRead and curWrite (lines 13 and 14), we need to use a different priority: the copy operation from data locations nextBlock to curBlock has to be performed after the other operations have terminated (so the priority 1).

As a final step, we then tell ORWL (line 16) that the specification of all accesses is complete and that it may schedule all these accesses in the respective FIFOs of the resources.

Listing 7.23. dynamic initialization of access mode and priorities
```
/* One operation with priority 0 (highest) consists    */
/* of copying from current to the GPU and running MM there. */
orwl_write_insert(&GPUhandle, ORWL_LOCATION(orwl_mytid, GPU), 0);
orwl_read_insert(&curRead, ORWL_LOCATION(orwl_mytid, curBlock), 0);

/* Another operation with priority 0 consists    */
/* of copying from remote to next                */
orwl_read_insert(&remRead, ORWL_LOCATION(previous(orwl_mytid),
    curBlock), 0);
orwl_write_insert(&nextWrite, ORWL_LOCATION(orwl_mytid, nextBlock), 0);

/* One operation with priority 1 consists    */
/* of copying from next to current           */
orwl_read_insert(&nextRead, ORWL_LOCATION(orwl_mytid, nextBlock), 1);
orwl_write_insert(&curWrite, ORWL_LOCATION(orwl_mytid, curBlock), 1);

orwl_schedule();
```

7.4.4 Tasks and operations

With this example we have now seen that ORWL distinguishes *tasks* and *operations*. An ORWL program is divided into tasks that can be seen as the algorithmic units that will concurrently access the resources that the program uses. A task for ORWL is characterized by

- a fixed set of resources that it manages, "*owns*," in our example the four resources that are declared in Listing 7.17.

- a larger set of resources that it accesses, in our example all resources that are used in Listing 7.23.

- a set of operations that act on these resources, in our example the three operations that are used in Listing 7.18, and that are elaborated in Listings 7.19, 7.20, and 7.21.

Each ORWL operation is characterized by

- one resource, usually one that is owned by the enclosing task, that it accesses exclusively. In our example, operation 1 has exclusive access to the next block, operation 2 has exclusive access the GPU resource, and operation 3 to the A block.

- several resources that are accessed concurrently with others.

In fact each ORWL operation can be viewed as a compute-and-update procedure of a particular resource with knowledge of another set of resources.

7.5 Conclusion

In this chapter, different methodologies that effectively take advantage of current cluster technologies with GPUs have been presented. Beyond the simple collaboration of several nodes that include GPUs, we have addressed parallel schemes to efficiently overlap communications with computations (including GPU transfers), and also computations on the CPU with computations on the GPU. Moreover, parallel schemes for synchronous as well as asynchronous iterative processes have been proposed. The proposed schemes have been validated experimentally and provide the expected behavior. Finally, as a prospect we have developed a programming tool that will, in middle or long term, provide all the required technical elements to implement the proposed schemes in a single tool, without requiring several external libraries.

We conclude that GPU computations are very well suited to achieve overlap with CPU computations and communications and they can be fully integrated in algorithmic schemes combining several levels of parallelism.

7.6 Glossary

AIAC: Asynchronous Iterations and Asynchronous Communications.

Asynchronous iterations: iterative process where each element is updated without waiting for the last updates of the other elements.

Auxiliary computations: optional computations performed in parallel to the main computations and used to complete them or speed them up.

BSP parallel scheme: bulk Synchronous Parallel, a parallel model that uses a repeated pattern (superstep) composed of computation, communication, barrier.

GPU stream: serialized data transfers and computations performed on a same piece of data.

Message loss/miss: can be said about a message that is either not sent or sent but not received (possible with unreliable communication protocols).

Message stamping: inclusion of a specific value in messages of the same tag to distinguish them (kind of secondary tag).

ORWL: Ordered Read-Write Locks, a programming tool proposing a unified programming model.

Page-locked data: data that are locked in cache memory to ensure fast accesses.

Residual: difference between results of consecutive iterations in an iterative process.

Streamed GPU sequence: GPU transfers and computations performed simultaneously via distinct GPU streams.

Bibliography

[1] Message passing interface. http://www.mpi-forum.org/docs.

[2] OpenMP multi-threaded programming API. http://www.openmp.org.

[3] J. M. Bahi, S. Contassot-Vivier, and R. Couturier. Asynchronism for iterative algorithms in a global computing environment. In *The 16th Annual International Symposium on High Performance Computing Systems and Applications (HPCS'2002)*, pages 90–97, Moncton, New-Brunswick, Canada, June 2002.

[4] J. M. Bahi, S. Contassot-Vivier, and R. Couturier. Evaluation of the asynchronous iterative algorithms in the context of distant heterogeneous clusters. *Parallel Computing*, 31(5):439–461, 2005.

[5] J. M. Bahi, S. Contassot-Vivier, and R. Couturier. Performance comparison of parallel programming environments for implementing AIAC algorithms. *Journal of Supercomputing. Special Issue on Performance Modelling and Evaluation of Parallel and Distributed Systems*, 35(3):227–244, 2006.

[6] J. M. Bahi, S. Contassot-Vivier, and R. Couturier. *Parallel Iterative Algorithms: From Sequential to Grid Computing*. Numerical Analysis & Scientific Computing Series. Chapman & Hall/CRC, 2007.

[7] J. M. Bahi, S. Contassot-Vivier, and R. Couturier. An efficient and robust decentralized algorithm for detecting the global convergence in asynchronous iterative algorithms. In *8th International Meeting on High Performance Computing for Computational Science, VECPAR'08*, pages 251–264, Toulouse, France, June 2008.

[8] J. M. Bahi, S. Contassot-Vivier, and A. Giersch. Load balancing in dynamic networks by bounded delays asynchronous diffusion. In J.M.L.M. Palma et al., editor, *VECPAR 2010*, volume 6449 of *LNCS*, pages 352–365. Springer, Heidelberg, 2011. DOI:~10.1007/978-3-642-19328-6\33.

[9] D. P. Bertsekas and J. N. Tsitsiklis. *Parallel and Distributed Computation*. Prentice Hall, Englewood Cliffs, New Jersey, 1999.

[10] P.-N. Clauss and J. Gustedt. Iterative computations with ordered read-write locks. *Journal of Parallel and Distributed Computing*, 70(5):496–504, 2010.

[11] S. Contassot-Vivier, T. Jost, and S. Vialle. Impact of asynchronism on GPU accelerated parallel iterative computations. In *PARA 2010 Conference: State of the Art in Scientific and Parallel Computing*, pages 43–53, Reykjavík, Iceland, June 2010.

[12] A. Frommer and D. B. Szyld. On asynchronous iterations. *J. Comput. and Appl. Math.*, 123:201–216, 2000.

[13] J. Gustedt and E. Jeanvoine. Relaxed synchronization with ordered read-write locks. In Michael Alexander et al., editors, *Euro-Par 2011: Parallel Processing Workshops*, volume 7155 of *LNCS*, pages 387–397, Bordeaux, France, May 2012. Springer.

[14] J. Gustedt, S. Vialle, and A. De Vivo. The parXXL environment: Scalable fine grained development for large coarse grained platforms. In Bo Kågström et al., editors, *PARA 06*, volume 4699, pages 1094–1104, Umeå, Sweden, 2007. Springer.

[15] T. Hoefler and A. Lumsdaine. Overlapping communication and computation with high level communication routines. In *IEEE International Symposium on Cluster Computing and the Grid*, pages 572–577, Lyon, France, 2008. IEEE Computer Society. http://doi.ieeecomputersociety.org/10.1109/CCGRID.2008.15.

[16] NVIDIA. *NVIDIA CUDA C Best Practices Guide 4.0*, May 2011.

[17] NVIDIA. *NVIDIA CUDA C Programming Guide 4.0*, June 2011.

[18] International standardization working group for the programming language C, editor. *Programming Languages – C*. Number 9899. ISO/IEC, cor. 1:2012 edition, 2011.

[19] L. G. Valiant. A bridging model for parallel computation. *Communications of the ACM*, 33(8):103–111, 1990.

[20] S. Vialle and S. Contassot-Vivier. *Patterns for parallel programming on GPUs*, chapter Optimization methodology for Parallel Programming of Homogeneous or Hybrid Clusters. Saxe-Coburg Publications, February 2013.

[21] S. Vialle, S. Contassot-Vivier, and T. Jost. *Handbook of Energy-Aware and Green Computing*, chapter Optimizing Computing and Energy Performances in Heterogeneous Clusters of CPUs and GPUs. Computer & Information Science Series. Chapman and Hall/CRC, Jan 2012.

Part IV

Optimization

Chapter 8

GPU-accelerated tree-based exact optimization methods

Imen Chakroun and Nouredine Melab

University of Lille 1, CNRS/LIFL/INRIA, France

8.1	Introduction	154
8.2	Branch-and-bound algorithm	156
8.3	Parallel branch-and-bound algorithms	157
	8.3.1 The parallel tree exploration model	158
	8.3.2 The parallel evaluation of bounds model	158
8.4	The flowshop scheduling problem	159
	8.4.1 Definition of the flowshop scheduling problem	159
	8.4.2 Lower bound for the flowshop scheduling problem	160
8.5	GPU-accelerated B&B based on the parallel tree exploration (GPU-PTE-BB)	161
8.6	GPU-accelerated B&B based on the parallel evaluation of bounds (GPU-PEB-BB)	162
8.7	Thread divergence	164
	8.7.1 The thread divergence issue	164
	8.7.2 Mechanisms for reducing branch divergence	165
8.8	Memory access optimization	168
	8.8.1 Complexity analysis of the memory usage of the lower bound	169
	8.8.2 Data placement pattern of the lower bound on GPU	170
8.9	Experiments	172
	8.9.1 Parameters settings	172
	8.9.2 Experimental protocol: computing the speedup	172
	8.9.3 Performance impact of GPU-based parallelism	174
	8.9.4 Thread divergence reduction	176
	8.9.5 Data access optimization	176
8.10	Conclusion and future work	178
	Bibliography	179

8.1 Introduction

In practice, a wide range of problems can be modeled as NP-hard combinatorial optimization problems (COPs). Those problems consist of choosing the best combination out of a large finite set of possible combinations and are known to be large in size and difficult to solve optimality. One of the most popular methods for solving exactly a COP (finding a solution having the optimal cost, is the Branch-and-Bound (B&B) algorithm. This algorithm is based on an implicit enumeration of all the feasible solutions of the tackled problem. Enumerating the solutions of a problem consists of building a dynamically generated search tree whose nodes are subsets of solutions of the considered problem. The construction of such a tree and its exploration is performed using four operators: branching, bounding, selection, and pruning. Due to the exponentially increasing number of potential solutions, the B&B algorithm explores only promising nodes of the search tree using an estimated optimal solution called "lower bound" of the associated subproblem.

Although this bounding mechanism allows the considerable reduction of the exploration time, often only small or moderatelysized instances of COPs can be practically solved. For this reason, over the last decades, parallel computing has been revealed as an attractive way to deal with larger instances of COPs. However, while many contributions have been proposed for parallel B&B methods using massively parallel processors [1], networks or clusters of workstations [15], and Shared Memory Multiprocessors (SMP) machines [3], very few contributions have been proposed for redesigning B&B algorithms on Graphical Processing Units (GPUs) [2]. For years, the use of GPU accelerators was limited to graphics and video applications. Driven by the demand for high-definition 3D graphics on personal computers, GPUs have evolved into a highly parallel, multithreaded and many-core environment. Their utilization has recently been extended to other application domains such as scientific computing [9].

In this chapter, we rethink the design and implementation of irregular tree-based algorithms such as the B&B algorithm on top of GPUs. During the execution of the B&B algorithm, the number of newly generated nodes and the number of not yet explored but promising nodes are variable and depend on the level of the tree being explored and on the best solution found so far. Therefore, due to such unstructured and unpredictable nature of its search tree, designing efficient B&B on top of GPUs is not straightforward. We investigate two different approaches for designing GPU-based B&B starting from the parallel models for B&B identified in [11]. The first one is based on the "parallel tree exploration" paradigm. This approach consists of exploring in parallel different subspaces of the tree. The second approach is based on the "parallel evaluation of bounds" approach. The two approaches have been applied to the permutation Flowshop Scheduling Problem (FSP; see Section 8.4) which is an NP-hard combinatorial optimization problem. The lower bound function used in this work for FSP is the one proposed in [8] for two machines and generalized in [10] to more than two machines.

When rethinking those two parallel models for GPU's architectures, our main focus was on the lower bound function. Indeed, preliminary experiments we carried out on some of Taillard's problem instances [16] show that computing the lower bounds takes on average between 98% and 99% of the total execution time of the B&B. The GPU-based lower bound's implementation raises mainly two challenges. On the one hand, having in mind that the execution model of GPUs is Single Instruction Multiple Data (SIMD), irregular computations (containing loops and conditional instructions) contained in the lower bound function may lead to a very challenging issue: the thread or branch divergence. This problem drops down the performance and arises when threads of a same warp (the smallest executable unit of parallelism on the GPU) execute different data-dependent instructions. On the other hand, the lower bound computation usually uses large and frequently accessed data structures. Since GPU is a many-core coprocessor device that provides a hierarchy of memories having different sizes and access latencies, the placement and sharing of these data sets become challenging.

The scope of this chapter is to design parallel B&B algorithms on GPU accelerators to allow highly efficient solving of permutation-based COPs. To do so, our contributions consist of: (1) rethinking two approaches for parallel B&B on top of GPUs, discussing the performances of each and identifying which best suits the GPU accelerators, (2) proposing a new approach for thread/branch divergence reduction through a thorough analysis of the different loops and conditional instructions of the bounding function, and (3) defining an optimal mapping of the data structures of the bounding function on the hierarchy of memories provided in the GPU device through a careful analysis of both the data structures (size and access frequency) and the GPU memories (size and access latency).

The chapter is organized into seven main sections. Section 8.2 presents the B&B algorithm. Section 8.3 introduces the different models used to parallelize B&B algorithms. Section 8.4 briefly describes the Flowshop Scheduling permutation Problem. In Section 8.5, we describe the GPU-accelerated B&B based on the parallel tree exploration. In Section 8.6, details about the second approach, the GPU-accelerated B&B based on the parallel evaluation of lower bounds, are given. In Section 8.7, the thread divergence issue related to the location of nodes in the B&B tree and to the control flow instructions within the bounding operator is described. In Section 8.8, the memory access optimization challenge is addressed and an overview of the GPU memory hierarchy and the used memory access pattern is given. In Section 8.9, we report experimental results showing the performances of each of two studied approaches compared to a sequential CPU-based execution of the B&B and demonstrating the efficiency of the proposed optimizations.

8.2 Branch-and-bound algorithm

Branch-and-bound algorithms are by far the most widely used methods for exactly solving large scale NP-hard combinatorial optimization problems. Indeed, they allow the finding of the optimal solution of a problem with proof of optimality.

The basic idea of the B&B algorithm consists in implicitly enumerating all the solutions of the original problem by only examining a subset of feasible solutions and eliminating the others when they are not likely to lead to a feasible or an optimal solution. Enumerating the solutions of a problem consists of building a dynamically generated search tree whose nodes are subsets of solutions of the considered problem. The construction of such tree and its exploration are performed using four operators: branching, bounding, selection and pruning.

The algorithm proceeds in several iterations during which the best solution found so far is progressively improved. During the exploration process, the search space is analyzed by a pool of unexplored nodes and the best solution found so far. The generated and not yet examined (pending) nodes are kept in a list initialized with the original problem. At each iteration of the algorithm, the following steps are performed:

- The *selection operator* chooses one node to process among the pending nodes according to a defined strategy. If the selection is based on the depth of the subproblem in the B&B tree, we speak about a depth-first exploration strategy. A selection based on the breadth of the subproblem is called a breadth-first exploration. A best-first selection strategy could also be used. It is based on the presumed capacity of the node to yield good solutions.

- The *branching operator* subdivides a solution space into two or more disjointed subspaces to be investigated in a subsequent iteration.

- The *bounding operator* computes a bound value of the optimal solution of each generated subproblem.

- Each subproblem having a greater bound than the upper-bound, i.e., the cost of the best solution found so far, is eliminated using the *pruning operator*.

Algorithm 6 gives the general template of the branch-and-bound method.

Algorithm 6: general template of the branch-and-bound algorithm

1 Create the initial problem;
2 Inset the initial problem into the tree;
3 Set the Upper_Bound to ∞;
4 Set the Best_Solution to $\emptyset$;
5 **while** *not_empty_tree()* **do**
6   Sub_Problem = Take_sub_problem();
7   **if** *Is_leaf (Sub_Problem)* **then**
8   Upper_Bound = Cost_Of(Sub_Problem);
9   Best_Solution = Sub_Problem;
10   **end**
11   **else**
12   Lower_Bound = compute_lower_bound(Sub_Problem);
13   **if** *Lower_Bound $\leq$ Upper_Bound* **then**
14   Branch(Sub_Problem);
15   Insert child subproblems into the tree;
16   **end**
17   **else**
18   Prune (Sub_Problem);
19   **end**
20   **end**
21 **end**

8.3 Parallel branch-and-bound algorithms

Thanks to the bounding operator, B&B allows the significant reduction of the computing time needed to explore the whole solution space. However, finding an optimal solution for large instances remains impractical using a sequential B&B. Therefore, parallel processing of these algorithms has been widely studied in the literature. In [11], a taxonomy of the various existing parallel paradigms used to parallelize the B&B algorithm is presented.

This taxonomy based on the classification proposed in [5] identified several models to accelerate the B&B search. The first model we consider in this chapter is called "parallel tree exploration model" and belongs to the "tree-based" strategies that aim to build and explore the B&B tree in parallel. The second model called "parallel evaluation of bounds model" (evaluation of bounds in parallel) belong to the parallelization approach called "node-based". This strategy aims to accelerate the execution of a particular operation at the node level.

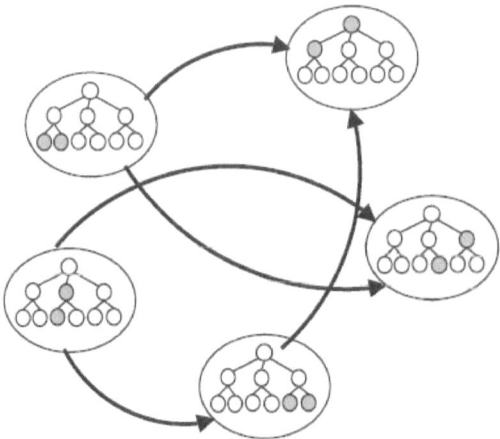

FIGURE 8.1. Illustration of the parallel tree exploration model.

8.3.1 The parallel tree exploration model

Tree-based strategies consist of building and/or exploring the solution tree in parallel by performing operations on several subproblems simultaneously. This coarse-grained type of parallelism affects the general structure of the B&B algorithm and makes it highly irregular.

The parallel tree exploration model, illustrated in Figure 8.1, consists of visiting in parallel different paths of the same tree. The search tree is explored in parallel by performing the branching, selection, bounding, and elimination operators on several subproblems simultaneously.

8.3.2 The parallel evaluation of bounds model

Node-based strategies introduce parallelism when performing the operations on a single problem. For instance, they consist of executing the bounding operation in parallel for each subproblem to accelerate the execution. This type of parallelism has no influence on the general structure of the B&B algorithm and is particular to the problem being solved.

The parallel evaluation of bounds model, as shown in Figure 8.2, allows the parallelization of the bounding of subproblems generated by the branching operator. This model is used in the case where the bounding operator is performed several times after the branching operator. Compared to the sequential B&B, the model does not change the order and the number of explored subproblems in the parallel B&B algorithm.

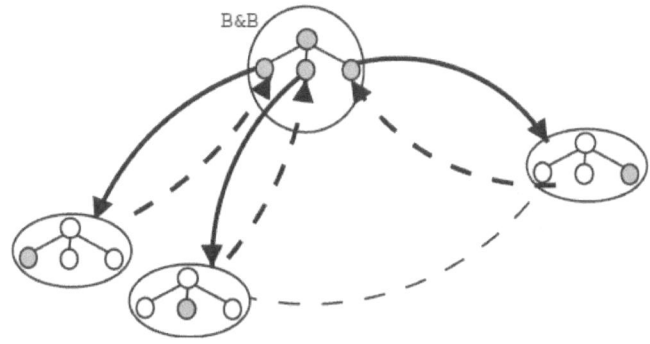

FIGURE 8.2. Illustration of the parallel evaluation of bounds model.

8.4 The flowshop scheduling problem

8.4.1 Definition of the flowshop scheduling problem

As a case study for our GPU-based branch-and-bound algorithm, we considered the NP-hard and well-known problem in the scheduling theory: the "Permutation Flow-shop Scheduling Problem" (FSP). In this work, the mono-objective case is considered. The FSP aims to find the optimal schedule of n jobs on m machines so that the overall completion time of all jobs, called *makespan*, is minimized.

Let us suppose the set of jobs is represented by $J = \{j_1, j_2, \ldots, j_n\}$ and the set of machines is represented by $M = \{m_1, m_2, \ldots, m_m\}$ organized in the line. Each job j_i is a sequence of operations $j_i = oi_1, oi_2, \ldots, oi_m$ where oi_m is the duration required for the job j_i on the machine m. A feasible solution of the flowshop permutation should satisfy these constraints:

- A machine cannot start processing a job if all the machines, which are located upstream, have not finished their treatment. Thus, the operation oi_j cannot be processed by the machine m_j if it is not completed on m_{j-1}.

- An operation cannot be interrupted, and the machines are critical resources, because a machine processes one job at a time.

- The sequence of jobs should be the same on every machine, e.g. if j_3 is treated in position 2 on the first machine, j_3 is also executed in position 2 on all machines.

Figure 8.3 illustrates a solution of a flow-shop problem instance defined by 6 jobs and 3 machines.

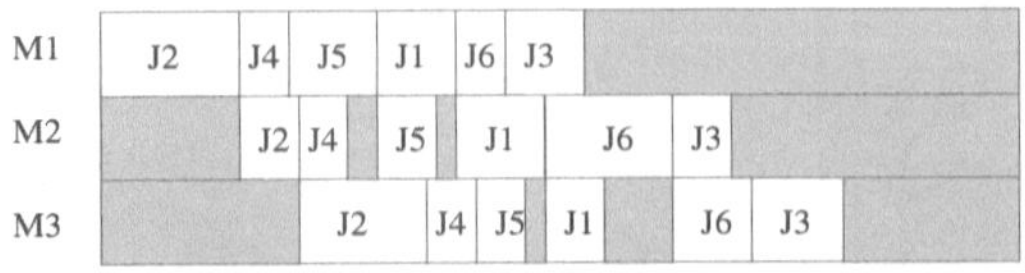

FIGURE 8.3. Flow-shop problem instance with 3 jobs and 6 machines.

8.4.2 Lower bound for the flowshop scheduling problem

The lower bounding technique provides a lower bound (LB) for each subproblem generated by the branching operator. The more the bound is accurate, the more it allows the elimination from the search tree that are not promising. Therefore, the efficiency of a B&B algorithm depends strongly on the quality of its lower bound function. In this chapter, we use the lower bound proposed by Lenstra et al. [10] for FSP, based on the Johnson's algorithm [8].

The Johnson's algorithm allows the optimal solution of FSP with two machines ($m = 2$) using the following transitive rule $\preceq$:

$$J_i \preceq J_j \Leftrightarrow \min(p_{i,1} \; ; \; p_{j,2}) \leq \min(p_{i,2} \; ; \; p_{j,1})$$

We recall that $p_{k,l}$ designates the processing time of the job J_k on the machine M_l. From the above rule, follows the Johnson's theorem:

Johnson's theorem *Given P and FSP with $m = 2$, if $J_i \preceq J_j$, there exists an optimal schedule for P in which job J_i precedes job J_j.*

According to Johnson's theorem, FSP with $m = 2$ is solved with a time complexity of $O(n.log n)$. The optimal solution is obtained by first sorting in increasing order the jobs having a processing time shorter on the first machine than on the second one, and second, sorting in decreasing order the jobs having a shorter processing time on the second machine.

In [7] and [13], the Johnson's rule extended by Jackson and Mitten with lags which further allowed Lenstra et al. to propose a lower bound for FSP with $m \geq 3$. A lag l_j designates the minimum duration between the starting time of the job J_j on the second machine and its finishing time on the first machine. Jackson and Mitten demonstrated that the optimal solution for FSP with $m = 2$ can be obtained using the following transitive rule $\preceq$:

$$J_i \preceq J_j \Leftrightarrow \min(p_{i,1} + l_i \; ; \; l_j + p_{j,2}) \leq \min(l_i + p_{i,2} \; ; \; p_{j,1} + l_j)$$

Based on this rule, Lenstra et al. [10] have proposed the following lower bound for a subproblem associated to a partial schedule where a set $\mathcal{J}$ of jobs have to be scheduled on m machines. $P^*_{Ja}(\mathcal{J}, M_k, M_l)$ represents the Jackson-Mitten optimal solution for the subproblem that consists in scheduling the set $\mathcal{J}$ of jobs on the two machines M_k and M_l. The term $r_{i,k} = \sum_{l<k} p_{i,l}$ designates the starting time of the job J_i on the machine M_k. The other term $q_{j,l} = \sum_{k>l} p_{j,k}$ refers to the latency between the finishing time of J_j on M_l and the finishing time of the schedule.

GPU-accelerated tree-based exact optimization methods

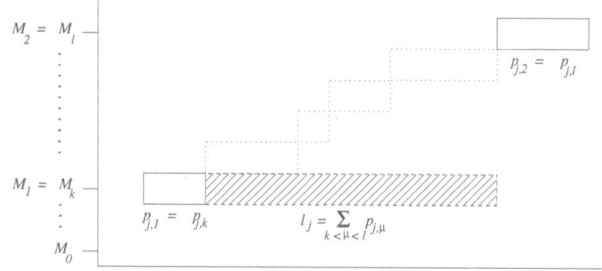

FIGURE 8.4. The lag l_j of a job J_j for a couple (k, l) of machines is the sum of the processing times of the job on all the machines between k and l.

$$LB(\jmath) = \max_{1 \leq k < l \leq m} \{P^*_{Ja}(\jmath, M_k, M_l) + \min_{(i,j) \in \jmath^2, i \neq j} (r_{i,k} + q_{j,l})\}$$

According to this LB expression, the lower bound for the scheduling of a subset $\jmath$ of jobs is calculated by applying the Johnson's rule with lags considering all the couples (k, l) for $1 \leq k, l \leq m$ and $k < l$. As illustrated in Figure 8.4, the lag l_j of a job J_j for a couple (k, l) of machines is the sum of the processing times of the job on all the machines between k and l.

8.5 GPU-accelerated B&B based on the parallel tree exploration (GPU-PTE-BB)

The first approach we investigate for designing B&B on GPUs consists of exploring in parallel the generated search tree. The idea is to divide the global search space into disjoint sub-spaces that are explored in parallel by the GPU threads. As explained in Section 8.2, during the execution of a B&B, the search space is described by a list of unexplored (pending) nodes and the best solution found so far. In the considered GPU-based scheme, a set of parent nodes is selected from this list according to their depth: deepest pending nodes are the first selected. The selected pool of nodes is off-loaded to the GPU where each thread builds its own local search tree by applying the *branching*, *bounding*, and *pruning* operators to the assigned node.

According to the CUDA threading model, each thread has a unique identifier used to determine its assigned role, which assigns specific input and output positions and selects work to perform. Therefore, each node (problem) from the pending list is mapped to a thread to ensure that each sub-space of the solution space is evaluated concurrently and is disjoint from others. Figure 8.5

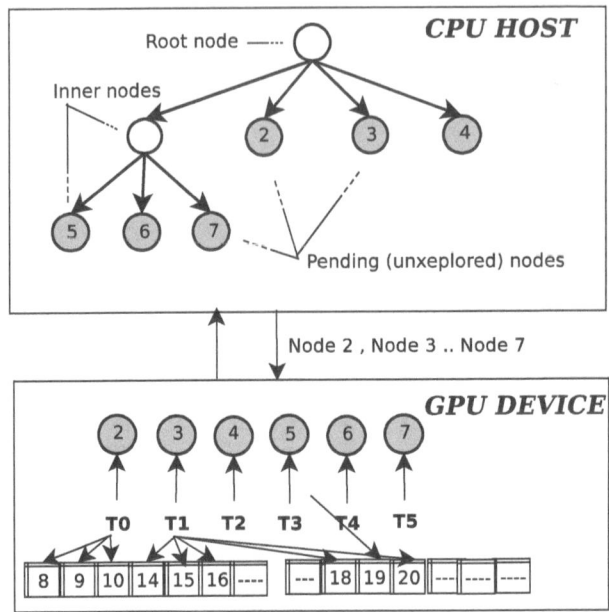

FIGURE 8.5. The overall architecture of the parallel tree exploration-based GPU-accelerated branch-and-bound algorithm.

illustrates the scheme of the parallel tree exploration-based GPU-accelerated B&B.

8.6 GPU-accelerated B&B based on the parallel evaluation of bounds (GPU-PEB-BB)

In the GPU-accelerated B&B based on the parallel evaluation of bounds, illustrated in Figure 8.6, the generation of the subproblems (elimination, selection and branching operations) to be solved is performed on CPU and the evaluation of their lower bounds (bounding operation) is executed on the GPU device. The pool of subproblems generated on CPU is off-loaded to the GPU device to be evaluated by a pool of threads partitioned into blocks. Each thread applies the lower bound function to one subproblem. Once the evaluation is completed, the lower bound values corresponding to the different subproblems are returned to the CPU to be used by the elimination operator to decide either to be pruned or to be decomposed. The process is iterated until the exploration is completed and the optimal solution is found.

In both approaches, GPU-PEB-BB and GPU-PTE-BB, the GPU-based

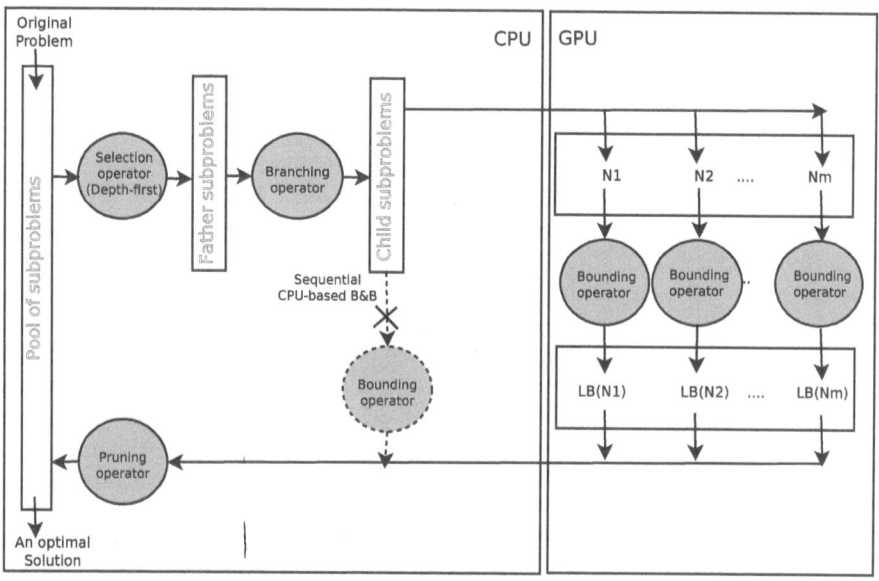

FIGURE 8.6. The overall architecture of the GPU-accelerated branch-and-bound algorithm based on the parallel evaluation of bounds.

lower bound's implementation raises mainly two challenges. The first one is related to the SIMD model of the GPU and to the implementation of the LB. Indeed, although typically every GPU thread will run the identical lower bound function, the body of the lower bound can contain conditions on thread identifiers and data. This implies that different instructions are executed in some threads. In SIMD architectures such as GPUs this behavior leads to the thread or branch divergence issue. This problem arises when threads of a same warp execute different data-dependent instructions. It might causes serious performance declining since computation occurs in parallel only when the same instructions are being performed. The second challenge consists of adjusting the pattern of accesses to the GPU device memory. Good placement of data over the different memory hierarchy allows programmers to further improve the throughput of many high-performance CUDA applications. For B&B applied to FSP, threads of the same block perform concurrent accesses to the six data structures of the problem when they execute the lower bound function. These data structures have different sizes and access frequencies and should be wisely placed on the different memories of the GPUs that also have different sizes and latencies.

In the following, we present how we dealt with the thread/branch divergence issue and map the different data structures on the memory hierarchy of the GPU device taking into account the characteristics of the data structures and those of the different GPU memories.

8.7 Thread divergence

8.7.1 The thread divergence issue

During the execution of an application on GPU, one or more thread block(s) are assigned to each GPU multiprocessor to execute. Those threads are partitioned into warps that get scheduled for execution. For each instruction of the flow, the multiprocessor selects a warp that is ready to be run. A warp executes one common instruction at a time, so full efficiency is realized when all threads of a warp agree on their execution path. In this chapter, the G80 model, in which a warp is a pool of 32 threads, is used. If threads of a warp diverge via a data-dependent conditional branch, the warp serially executes each branch path taken. Threads that are not on the taken path are disabled, and when all paths are complete, the threads converge back to the same execution path. This phenomenon is called thread/branch divergence and often causes serious performance degradations. Branch divergence occurs only within a warp; different warps execute independently regardless of whether they are executing common or disjointed code paths.

This section discusses thread divergence issues encountered when computing the bounds by GPU. The thread divergence occurs for two main reasons, namely, the locations of nodes in the search tree and the control flow instructions within the bounding operator.

Divergence related to the location of nodes

This divergence is related to the positions of the nodes in the B&B search tree. Below is given an example from the source code of the used LB showing that the execution flow depends on the position of the node in the search tree. In the following piece of code, three methods are used *is_leaf()*, *makespan()* and *lower_bound()*. *is_leaf()* tests if the node *_node* is a leaf or an internal node. If *_node* is a leaf, *makespan()* computes the cost of its makespan. Otherwise, *_node* is an internal node and *lower_bound()* computes the value of its lower bound.

```
if (_node.is_leaf())
    return _node.makespan();
else
    return _node.lower_bound();
```

Divergence related to the control flow instructions

Control flow refers to the order in which the instructions, statements, or function calls are executed in a program. This flow is determined by instructions such as if-then-else, for, while-do, switch-case. There are a dozen of such instructions in the implementation of our bounding operator. The source code examples given below show two scenarios in which this kind of instructions is used.

- Example 1:

```
if( pool[thread_idx].begin != 0 )
    time = TimeMachines[1] ;
else
    time = TimeArrival[1] ;
```

- Example 2:

```
for(int k = 0 ; k < pool[thread_idx].begin; k++)
    jobTime = jobEnd[k] ;
```

In these two examples, thread_idx is the index associated to the current thread. Let suppose that the code of Example 1 is executed by 32 threads, pool[thread_idx].begin is equal to 0 for the first thread, and pool[thread_idx].begin is not equal to 0 for the other 31 threads. When the first thread executes the statement time = TimeArrival[1];, all the other 31 threads remain idle. Therefore, the GPU cores on which these 31 threads are executed remain idle and cannot be used during the execution of the statement time = TimeArrival[1];.

The same scenario occurs during the execution of Example 2. Let us suppose that the instruction is executed by 32 threads, pool[thread_idx].begin is equal to 100 for the first thread, and pool[thread_idx].begin is equal to 0 for the other 31 threads. When the first thread executes the loop *for*, all the other 31 threads remain idle.

Existing techniques for handling branch divergence either demand hardware support [4] or require host-GPU interaction [17], which incurs overhead. Some other works such as [6] intervene at the code level. They expose a branch distribution method that aims to reduce the divergent portion of a branch by factoring out structurally similar code from the branch paths. In our work, we have also opted for software-based optimizations as in [6]. In fact, we figure out how to literally rewrite the branching instructions into basic ones in order to make thread execution paths uniform. We also demonstrate that we could ameliorate performances only by judiciously reordering data being assigned to each thread.

8.7.2 Mechanisms for reducing branch divergence

Thread-data reordering

At each iteration of our GPU-accelerated B&B approach, several thousands of subproblems are sent to the GPU. The GPU groups the received subproblems into several warps according to their reception order. The first 32 subproblems belong to the first warp, the following 32 subproblems belong to the second warp, etc. Therefore, thread-data reordering technique sorts subproblems before sending them to the GPU. These subproblems are sorted according to their position in the B&B tree. These sorts of subproblems allow warps containing more homogeneous subproblems and reduce the number of thread divergences.

Branch refactoring

As stated above, thread or branch divergence occurs when the kernel includes conditional instructions and loops that make the threads performing different control flows lead to their serial execution. In this chapter, we investigate the branch refactoring approach to deal with thread divergence. Branch refactoring consists of rewriting the conditional instructions so that threads of the same warp execute a uniform code avoiding their divergence. To do that, two major if scenarios are studied and some optimizations are proposed accordingly. These two scenarios correspond to the conditional instructions contained in the LB kernel code. In the first scenario, the conditional expression is a comparison of the content of a variable to 0. For instance, the following example extracted from the pseudocode of the lower bound LB illustrates such a scenario.

    if (pool[thread_idx].limit1 $\neq$ 0) tmp = MM[1];
    else tmp = RM[1] ;

The refactoring idea is to replace the conditional expression by two functions namely f and g as shown in Equation 8.1.

The behavior of f and g fits the cosine trigonometric function. These functions return values between 0 and 1. An integer variable is used to store the result of the cosine function. Its value is 0 or 1 since it is rounded to 0 if it is not equal to 1. In order to increase the performance the CUDA runtime math operations are used: sinf(x), expf(x), and so forth. Those functions are mapped directly to the hardware level [14]. They are faster but provide lower accuracy which does not matter in our case because the results are rounded to int.

$$if(x \neq 0) \ a = b[1]; \quad if(x \neq 0) \ a = b[1] + 0 \times c[1];$$
$$\Rightarrow$$
$$else \ a = c[1]; \quad else \ a = 0 \times b[1] + c[1];$$

$$\Rightarrow a = f(x) \times b[1] + g(x) \times c[1];$$

where

$$f(x) = \begin{cases} f(x) = 0 & if \quad x = 0 \\ 1 & else \end{cases}$$

and

$$g(x) = \begin{cases} g(x) = 1 & if \quad x = 0 \\ 0 & else \end{cases}$$

(8.1)

The throughput of sinf(x), cosf(x), expf(x) is one operation per clock cycle [14]. The refactoring result for the if pseudocode given above is the following:

```
int coeff = __cosf (pool[thread_idx].limit1);
tmp = (1 - coeff) × MM[1] + coeff × RM[1];
```

The second if scenario considered in our study compares two values between themselves as shown in Equation 8.2.

$$if(x > y)a = b[1]; \quad \Rightarrow if(x - y \geq 1)a = b[1];$$

$$\Rightarrow \quad if(x - y - 1 \geq 0) \quad a = b[1]; \quad (x, y) \in N$$

$$\Rightarrow \quad a = f(x, y) \times b[1] + g(x, y) \times a;$$

where:

$$f(x, y) = \begin{cases} 1 & if \quad x - y - 1 \geq 0 \\ 0 & if \quad x - y - 1 < 0 \end{cases}$$

and

$$g(x, y) = \begin{cases} 0 & if \quad x - y - 1 \geq 0 \\ 1 & if \quad x - y - 1 < 0 \end{cases}$$

(8.2)

For instance, the following example extracted from the pseudocode of the lower bound LB illustrates such a scenario.

```
if(RM[1]] > MIN ){ Best_idx = Current_idx; }
```

The same transformations as those applied for the first scenario are applied here using the exponential function. Recall that the exponential is a positive function which is equal to 1 when applied to 0. Thus, if x is greater than y then expf$(x - y - 1)$ returns a value between 0 and 1. If the result is rounded to an integer value 0 will be obtained. Now, if x is less than y then expf$(x - y - 1)$ returns a value greater than 1 and since the minimum between 1 and the exponential is get, the returned result would be 1. Such behavior satisfies exactly our prerequisites. The above if instruction pseudocode is now equivalent to

```
int coeff = min(1, _expf(RM[1] - MIN - 1));
Best_idx = coeff × Current_idx + ( 1 - coeff ) × Best_idx ;
```

8.8 Memory access optimization

Memory access optimizations are by far the most studied area for improving GPU-based application performances. Indeed, adjusting the pattern of accesses to the GPU device memory allows programmers to further improve the throughput of many high-performance CUDA applications. The goal of memory access optimizations is generally to use as much fast-access memory and as little slow-access memory as possible. This section discusses how best to set up data LB items on the various kinds of memory on the device.

CUDA-enabled devices use several memory spaces, which have different characteristics in term of sizes and access latencies. These memory spaces include global memory, local memory, shared memory, texture memory, and registers. Devices of compute capability 2.0 also have an L1/L2 cache hierarchy that is used to cache local and global memory accesses.

- At the thread-level, each thread has its own allocated registers and a private local memory. CUDA uses this local memory for thread-private variables that do not fit in the threads registers, as well as for stack frames and register spilling.

- At the thread block-level, each thread block has a shared memory visible to all its associated threads.

- At the grid-level, all threads have access to the same global memory. Texture and constant cached memories are two other memories accessible by all threads.

The data access optimization challenge is to find the best mapping of the data structures of the application at hand (different sizes and access frequencies) and the GPU hierarchy of memories (different sizes and access latencies). For instance, of these different memory spaces, global memory is the most plentiful but the one with the highest access latency. On the contrary, shared memory is smaller in size but has much higher bandwidth and lower latency than the global memory.

8.8.1 Complexity analysis of the memory usage of the lower bound

In this section, the characteristics of the data structures used by the lower bound function are studied in terms of sizes and access frequencies. For an efficient implementation of the LB, six data structures are required: the matrix PTM of the processing times of the jobs, the matrix of lags LM, the Johnson's matrix JM, the matrix RM of the earliest starting times of jobs, the matrix QM of their lowest latency times, and the matrix MM containing the couples of machines. The complexities of the different data structures are summarized in Table 8.1 where the columns represent, respectively, the name of the data structure, its size, and the number of times it is accessed.

In the LB expression, the computation of the term $P^*_{Ja}(J, M_k, M_l)$ requires the calculation of the lag of each remaining job to be scheduled on the couple (M_k, M_l) of machines using its processing times on these machines (Johnson's rule with lags). Such computation is repeated for each couple (M_k, M_l) of machines with $1 \leq k, l \leq m$ and $k < l$. To avoid the repetitive computation of the lags, they are computed once at the beginning of the algorithm and stored in the matrix LM. The dimension of LM is $n \times \frac{m \times (m-1)}{2}$, where n and m are respectively the number of jobs to be scheduled and m the number of machines. LM is accessed $n' \times \frac{m \times (m-1)}{2}$ times, n' being the number of remaining jobs to be scheduled in the subproblem for which the lower bound is being calculated. The processing times of all the jobs on all the machines are stored in the matrix PTM. This matrix has a dimension of $n \times m$ and is accessed $n' \times m \times (m-1)$ times.

In addition, in order to avoid relaunching the Johnson's algorithm for each couple of machines and each subset of jobs, the Johnson's algorithm is computed once to find the optimal solutions on the couples of machines. These optimal solutions are then stored in the Johnson's matrix JM. This matrix has the same dimension as LM and is accessed $n \times \frac{m \times (m-1)}{2}$ times during the computation of the lower bound. Finally, the MM matrix that contains all the couples of machines has a dimension and access frequency of $m \times (m-1)$.

To reduce the computation time cost of the term $\min_{(i,j) \in J^2, i \neq j} (r_{i,k} + q_{j,l})$ in the LB expression, two matrices are defined, namely RM and QM. They are used to store, respectively, the lowest starting and latency times of all the jobs on each machine. Their dimension is m and, are accessed $m \times (m-1)$ times and $\frac{m \times (m-1)}{2}$ times, respectively.

Matrix	Size	Number of accesses
PTM	$n \times m$	$n' \times m \times (m-1)$
LM	$n \times \frac{m \times (m-1)}{2}$	$n' \times \frac{m \times (m-1)}{2}$
JM	$n \times \frac{m \times (m-1)}{2}$	$n \times \frac{m \times (m-1)}{2}$
RM	m	$m \times (m-1)$
QM	m	$\frac{m \times (m-1)}{2}$
MM	$m \times (m-1)$	$m \times (m-1)$

TABLE 8.1. The different data structures of the LB algorithm and their associated complexities in memory size and numbers of accesses. The parameters n, m, and n' designate, respectively, the total number of jobs, the total number of machines and the number of remaining jobs to be scheduled for the subproblems the lower bound is being computed.

Prob. instance	JM	LM	PTM	RM, QM	MM
200 × 20	38,000 (38KB)	38,000 (76KB)	4,000 (4KB)	20 (0.04KB)	380 (0.76KB)
100 × 20	19,000 (19KB)	19,000 (38KB)	2,000 (2KB)	20 (0.04KB)	380 (0.76KB)
50 × 20	9,500 (9.5KB)	9,500 (19KB)	1,000 (1KB)	20 (0.04KB)	380 (0.76KB)
20 × 20	3,800 (3.8KB)	3,800 (7.6KB)	400 (0.4KB)	20 (0.04KB)	380 (0.76KB)

TABLE 8.2. The sizes of each data structure for the different experimented problem instances. The sizes are given in number of elements and in bytes (between parentheses).

8.8.2 Data placement pattern of the lower bound on GPU

This section discusses how best to map the six data structures identified above on the various kinds of memories of the GPU device.

The focus is put on the shared memory which is a key enabler for many high-performance CUDA applications. Indeed, because it is on-chip, shared memory has much higher bandwidth and lower latency than local and global memory. However, for large problem instances (large n and m) the data structures, especially JM and LM (see Table 8.2), do not fit in the shared memory of some GPU configurations.

In order to achieve further performances, we also take care of adequately using the global memory by judiciously configuring the L1 cache which greatly enables improved performance over direct access to global memory. Indeed, the GPU device we are using in our experiments is based on the NVIDIA Fermi architecture which introduced two new hierarchies of memories (L1/L2 cache) compared to older architectures.

Taking into consideration the sizes of each data structure presented in Table 8.2, our challenge is to find which data structure has to be mapped onto which memory and in some cases how to split the data structures onto different memories and efficiently manage their accesses. The sizes in bytes reported in Table 8.2 are computed knowing that in our implementation the elements of JM and PTM are unsigned chars (one byte) and that the elements of LM, RM, QM, and MM are unsigned short ints (2 bytes). It is important here to highlight that the types of the data of the used matrices impact the size of each matrix. For instance, a matrix of 100 integers has a size of 400 octets while the same matrix with 100 unsigned chars has a size of 100 octets. In order to minimize the size of each of the used matrices, we analyzed the ranges of their values and defined their data types accordingly. For instance, in PTM all the processing times have positive values varying between 0 and 100. Therefore, we defined PTM as a matrix of unsigned char having values in the range $[0, 255]$. Using the unsigned char type instead of the integer type allows us to reduce by 4 times the memory space occupied by PTM.

According to the Table 8.2 :

- The data structures RM, QM and MM are small sized-matrices. Therefore, their impact on the performances is not significant whatever is the memory to which they are off-loaded. In particular, preliminary experiments prove that putting them on the shared memory would allows a very poor performance improvement.

- The LM data structure is the double of the JM in memory size but with a much lower access frequency. It is thus better to map JM on the shared memory.

- The PTM has almost the same access frequency than JM but requires less memory space.

Consequently, the focus is put on the study of the performance impact of the placement of JM and PTM on the shared memory. Three placement scenarios of JM and PTM are experimented and studied: (1) Only PTM is stored in shared memory and all others are placed in global memory ; (2) Only JM is stored in shared memory and all others are placed on global memory ; (3) PTM and JM are stored together in shared memory and all others are placed on global memory.

Taking profit from the configurable storage space provided in the new Fermi-based devices, the 64 KB of local storage was split between the shared memory and the L1 cache according to the experimented scenario.

- For the scenario where the data structures are put on the shared memory, the 64 KB of available storage are split into 48 KB for shared memory and 16 KB for L1 cache.

- For the scenario where the data sets are put on global memory, we used 16 KB for shared memory and 48 KB for L1 cache.

8.9 Experiments

In the following, we present the experimental study we have performed with the aim of evaluating the performance impact of the GPU-accelerated bounding, the techniques for reducing the thread divergence, and the proposed approach for data placement on the GPU memories.

8.9.1 Parameters settings

In our experiments, we used the flow-shop instances defined by Taillard [16]. These standard instances are often used in the literature to evaluate the performance of methods that minimize the makespan. Optimal solutions of some of these instances are still not known. These instances are divided into groups of 10 instances. In each group, the 10 instances are defined by the same number of jobs and the same number of machines. The groups of 10 instances have different numbers of jobs, namely, 20, 50, 100, 200, and 500, and different numbers of machines, namely, 5, 10, and 20. For example, there are 10 instances with 200 jobs and 20 machines belonging to the same group of instances.

In this work, we used only the instances where the number of machines is equal to 20. Indeed, instances where the number of machines is equal to 5 or 10 are easy to solve. For these instances, the used bounding operator gives such good lower bounds that it is possible to solve them in a few minutes using a sequential B&B. Therefore, these instances do not require the use of a GPU.

Our approach has been implemented using C-CUDA 4.0. The experiments have been carried out using an Intel Xeon E5520 biprocessor coupled with a GPU device. The biprocessor is 64-bit, quad-core and has a clock speed of 2.27GHz. The GPU device is an NVIDIA Tesla C2050 with 448 CUDA cores (14 multiprocessors with 32 cores each), a clock speed of 1.15GHz, a 2.8GB global memory, a 49.15KB shared memory, and a warp size of 32.

8.9.2 Experimental protocol: computing the speedup

We need to compute the speedup of our approach to evaluate its performances. This speedup is obtained by comparing our GPU B&B version to a sequential B&B version deployed on one CPU core. However, all the instances used in our experiments are extremely hard to solve. Indeed, the resolution of each of these instances requires several months of computation on one CPU core. For example, the optimal solution of one of these instances defined by 50 jobs and 20 machines is obtained after 25 days of computation using an average of 328 CPU cores [12].

Using the approach defined in [12], it is possible to obtain a random list L

GPU-accelerated tree-based exact optimization methods 173

of subproblems such that the resolution of L lasts T minutes with a sequential B&B. So by initializing the pool of our sequential B&B with the subproblems of this list L, we are sure that the resolution of the sequential B&B will last $Tcpu$ minutes such as $Tcpu$ will be approximately equal to T. Therefore, it will be possible to initialize the pool of our GPU B&B with the same list L of subproblems in order to compute the speedup. Let us suppose that the resolution of the GPU B&B will last $Tgpu$ minutes. So the speedup of our GPU algorithm will be equal to $Tcpu/Tgpu$. With this experimental protocol, the subproblems explored by the GPU and CPU B&B versions will be exactly the same. So to find the speedup associated to an instance, we:

- compute, using the approach defined in [12], a list L of subproblems such as the resolution of L lasts T minutes with a sequential B&B;

- initialize the pool of our sequential B&B with the subproblems of this list L;

- solve the subproblems of this pool with our sequential B&B;

- get the sequential resolution time $Tcpu$ and the number of explored subproblems $Ncpu$;

- check that $Tcpu$ is approximately equal to T;

- initialize the pool of our GPU B&B with the subproblems of the list L;

- solve the subproblems of this pool with our GPU B&B;

- get the GPU resolution time $Tgpu$ and the number of explored subproblems $Ngpu$;

- check that $Ngpu$ is exactly equal to $Ncpu$;

- and finally compute the speedup associated to this instance by dividing $Tcpu$ by $Tgpu$ (i.e., $Tcpu/Tgpu$).

Table 8.3 gives, for each instance according to its number of jobs and its number of machines, the used resolution time with a sequential B&B. For example, the sequential resolution time of each instance defined with 20 jobs and 20 machines is approximately 10 minutes. Of course, the computation time of the lower bound of a subproblem defined with 20 jobs and 20 machines is on average greater than the computation time of the lower bound of a subproblem defined with 50 jobs and 20 machines. Therefore, as shown in this table, the sequential resolution time increases with the size of the instance in order to be sure that the number of subproblems explored is significant for all instances.

Instance (No. of jobs × No. of machines)	20×20	50×20	100×20	200×20
Sequential resolution time (minutes)	10	50	150	300

TABLE 8.3. The sequential resolution time of each instance according to its number of jobs and machines.

8.9.3 Performance impact of GPU-based parallelism

The objective of the experimental study presented in this section is to compared the performances of both proposed approaches for designing B&B on top of GPUs.

Table 8.4 and Table 8.5 report the speedups obtained with the GPU-PTE-BB and GPU-PEB-BB approaches, respectively, for different problem instances. The first part of both tables gives the size of the pool generated and evaluated on the GPU. The second part of the tables gives the average speedup for each group of instances and for each pool size. Each line corresponds to a group of 10 instances defined by the same number of jobs and the same number of machines.

The results obtained with the GPU-PTE-BB approach (see Table 8.4) show that exploring the tree search in parallel allows the speedup of the execution of the B&B compared to a CPU-based execution. Indeed, an acceleration factor up to 40.50 is obtained for the 20 × 20 problem instances using a pool of 262144 subproblems.

Pool size	4096	8192	16384	32768	65536	131072	262144
(N Jobs × N Machines)	Average speedup for each group of 10 instances						
200×20	1.12	2.89	3.57	4.23	6.442	8.32	13.4
100×20	1.33	1.88	3.45	6.45	12.38	20.40	28.76
50×20	2.70	3.80	6.82	13.04	23.53	30.94	37.66
20×20	6.43	11.43	20.14	27.78	30.12	35.74	40.50

TABLE 8.4. Speedups for different problem instances and pool sizes with the GPU-PTE-BB approach.

The results show also that the parallel efficiency decreases with the size of the problem instance. For a fixed number of machines (here 20 machines) and a fixed pool size, the obtained speedup declines accordingly with the number of jobs. For instance for a pool size of 262144, the acceleration factor obtained with 200 jobs is 13.4 while it is 40.50 for the instances with 20 jobs. This behavior is mainly due to the overhead induced by the transfer of the pool of resulting subproblems between the CPU and the GPU. For example, for the instances with 200 jobs the size of the pool to exchange between the CPU and the GPU is ten times bigger than the size of the pool for the instances with 20 jobs.

The results obtained with the GPU-PEB-BB approach (see Table 8.5) show that evaluating the bounds of a selected pool in parallel, allows significants speedup of the execution of the B&B. Indeed, an acceleration factor up to 71.69 is obtained for the 200 × 20 problem instances using a pool of 262144 subproblems. The results show also that the parallel efficiency grows with the size of the problem instance. For a fixed number of machines (here 20 machines) and a fixed pool size, the obtained speedup grows accordingly with the number of jobs. For instance for a pool size of 262144, the acceleration factor obtained with 200 jobs (71.69) is almost double obtained with 20 jobs (38.40).

Pool size	4096	8192	16384	32768	65536	131072	262144
(NJobs × NMachines)	Average speedup for each group of 10 instances						
200×20	42.83	56.23	57.68	61.21	66.75	68.30	**71.69**
100×20	42.59	56.18	57.53	60.95	65.52	65.70	**65.97**
50×20	42.57	**56.15**	55.69	55.49	55.39	55.27	55.14
20×20	38.74	**46.47**	45.37	41.92	39.55	38.90	38.40

TABLE 8.5. Speedups for different problem instances and pool sizes with the GPU-PEB-BB approach.

As far as the pool size tuning is considered, we could notice that this parameter depends strongly on the problem instance being solved. Indeed, while the best acceleration is obtained with a pool size of 8192 subproblems for the instances 50 × 20 and 20 × 20 (Tablerefch8:ParaGPU2 in bold), the best speedups are obtained with a pool size of 262144 subproblems with the instances 200 × 20 and 100 × 20 (Tablerefch8:ParaGPU2 in bold).

Compared to the parallel tree exploration-based GPU-accelerated B&B approach, the parallel evaluation of bounds approach is by far much more efficient wherever the instance is. For example, while the GPU-PEB-BB approach reaches speedup of ×71.69 for the instance with 200 jobs on 20 machines, a speedup of a ×13.4 is measured with the parallel tree exploration-based approach which corresponds to an acceleration of ×5.56. Moreover, to the contrary of the GPU-PEB-BB approach, in the GPU-PTE-BB the speedups decrease when the problem instance becomes higher. Remember here that while in the GPU-PEB-BB approach all threads evaluate only one node each whatever the permutation size is. In the GPU-PTE-BB, each thread branches all the children of its assigned parent node. Therefore, the bigger the size of the permutation, the bigger the amount of work performed by each thread is and the bigger the difference between the workload. Indeed, let us suppose that for the instance with 200 jobs, the thread 0 handles a node from the level 2 of the tree and the thread 100 handles a node from the level 170 of the tree. In this case, the thread 0 generates and evaluates 198 nodes while the thread 100 decomposes and bounds only 30 nodes. The problem in this example is that the kernel execution would last until the thread 0 finishes its work while the other threads might have completed their work and remains idle.

8.9.4 Thread divergence reduction

The objective of this section is to demonstrate that the thread divergence reduction mechanisms we propose have an impact on the performance of the GPU accelerated B&B and to evaluate how this impact is significant. In the following, the reported results are obtained with the GPU-accelerated B&B based on the parallel evaluation of bounds.

Pool size	4096	8192	16384	32768	65536	131072	262144
(N Jobs × N Machines)	Average speedup for each group of 10 instances							
200×20	46.63	60.88	63.80	67.51	73.47	75.94	**77.46**
100×20	45.35	58.49	60.15	62.75	66.49	66.64	**67.01**
50×20	44.39	**58.30**	57.72	57.68	57.37	57.01	56.42
20×20	41.71	**50.28**	49.19	45.90	42.03	41.80	41.65

TABLE 8.6. Speedups for different instances and pool sizes using thread divergence management.

Table 8.6 shows the experimental results obtained using the sorting process and the refactoring approach presented in Section 8.7. Results show that the proposed optimizations emphasize the GPU acceleration reported in Table 8.5 obtained without thread divergence reduction. For example, for the instances of 200 jobs over 20 machines and a pool size of 262144, the average reported speedup is 77.46 (Table 8.6 in bold) while the average acceleration factor obtained without thread divergence management for the same instances and the same pool size is 71.69 which corresponds to an improvement of 7.68%. Such considerable but not outstanding improvement is predictable, as claimed in [6], since the factorized part of the branches in the FSP lower bound is very small.

8.9.5 Data access optimization

The objective of the experimental study presented in this section is to find the best mapping of the six data structures of the LB kernel on the memories of the GPU device. In the following, the reported results are obtained with the GPU-accelerated B&B based on the parallel evaluation of bounds.

Table 8.7 reports the speedups obtained for the first experimental scenario where only the matrix PTM is put on the shared memory. Results show that the speedup grows on average with the growing of the pool size in the same way as in Table 8.6. For the largest problem instance and pool size, putting the PTM matrix on the shared memory improves the speedups up (14%) compared to those obtained when PTM is on global memory reaching an acceleration of ×90.51 for the problem instances 200 × 20 and a pool size of 262144 subproblems.

Table 8.8 reports the behavior of the speedup averaged on the different

Pool size	4096	8192	16384	32768	65536	131072	262144
(N Jobs × N Machines)	Average speedup for each group of 10 instances						
200×20	54.03	67.75	68.43	72.17	82.01	88.35	**90.51**
100×20	52.92	66.57	66.25	71.21	76.63	79.76	**83.01**
50×20	49.85	**65.68**	64.40	59.91	58.57	57.36	55.09
20×20	41.94	**60.10**	48.28	39.86	39.61	38.93	37.79

TABLE 8.7. Speedup for different FSP instances and pool sizes obtained with data access optimization. PTM is placed in shared memory and all others are placed in global memory.

Pool size	4096	8192	16384	32768	65536	131072	262144
(N Jobs × N Machines)	Average speedup for each group of 10 instances						
200×20	63.01	79.40	81.40	84.02	93.61	96.56	**97.83**
100×20	61.70	77.79	79.32	81.25	86.73	87.81	**88.69**
50×20	59.79	**75.32**	72.20	71.04	70.12	68.74	68.07
20×20	49.00	**60.25**	55.50	45.88	44.47	43.11	42.82

TABLE 8.8. Speedup for different FSP instances and pool sizes obtained with data access optimization. JM is placed in shared memory and all others are placed in global memory.

problem instances (sizes) as a function of the pool size for the scenario where the Johnson's matrix is put on the shared memory. Results show that putting the JM matrix on the shared memory improves the performances more than in the first scenario where PTM is put on the shared memory. Indeed, according to Table 8.1, matrix JM is accessed more frequently than matrix PTM. Putting JM matrix on the shared memory allows accelerations up to ×97.83 for the problem instances 200 × 20.

Table 8.9 reports the behavior of the average speedup for the different problem instances (sizes) with 20 machines for the data placement scenario where both PTM and JM are put on shared memory. According to the underlying tables, scenarios 3 (JM together with PTM in shared memory) is clearly better than the scenarios 1 and 2 (respectively PTM in shared memory and JM in shared memory) whatever the problem instance (size).

Pool size	4096	8192	16384	32768	65536	131072	262144
(NJobs × NMachines)	Average speedup for each group of 10 instances						
200×20	66.13	87.34	88.861	95.23	98.83	99.89	**100.48**
100×20	65.85	86.33	87.60	89.18	91.41	92.02	**92.39**
50×20	64.91	**81.50**	78.02	74.16	73.83	73.25	72.71
20×20	53.64	**61.47**	59.55	51.39	47.40	46.53	46.37

TABLE 8.9. Speedup for different FSP instances and pool sizes obtained with data access optimization. PTM and JM are placed together in shared memory and all others are placed in global memory.

By carefully analyzing each of the scenarios of data placement on the memory hierarchies of the GPU, the recommendation is to put in the shared memory the Johnson's and the processing time matrices (JM and PTM) if they fit in together. Otherwise, the whole or a part of the Johnson's matrix has to be given in priority in the shared memory. The other data structures are mapped to the global memory.

8.10 Conclusion and future work

In this chapter, we have revisited the design of parallel B&B algorithms on GPU accelerators to allow highly efficient solving of permutation-based COPs. To do so, our contributions consisted of: (1) rethinking two approaches for parallel B&B on top of GPUs, discussing the performances of each and identifying which best suits the GPU accelerators; (2) proposing a new approach for thread/branch divergence reduction through a thorough analysis of the different loops and conditional instructions of the bounding function; and (3) defining an optimal mapping of the data structures of the bounding function on the hierarchy of memories provided in the GPU device through a careful analysis of both the data structures (size and access frequency) and the GPU memories (size and access latency).

In the first parallel treeexploration-based B&B, a set of pending nodes is selected from this list according to their depth and off-loaded to the GPU where each thread builds its own local search tree by applying the branching, bounding, and pruning operators to the assigned node. In the GPU-accelerated B&B based on the parallel evaluation of bounds, the generation of the subproblems (branching, selection, and pruning operations) is performed on CPU and the evaluation of their lower bounds (bounding operation) is executed on the GPU device. Pools of subproblems are off-loaded from CPU to GPU to be evaluated by blocks of threads. After evaluation, the lower bounds are returned to the CPU.

In both considered approaches, our focus is on the GPU-based lower bound's implementation and the associated thread divergence and data placement challenges. The proposed mechanisms for reducing the thread divergence issue are based on a thorough analysis of the different loops and conditional instructions of the lower bound function. On the one hand, the sorting process aims to homogenize the data of the subproblems off-loaded to the GPU to minimize the number of threads that diverge on loop instructions. On the other hand, the technique of branch refactoring rewrite the conditional instructions into uniform instructions so that threads of the same warp execute the same code. The proposed data access optimization is based on a preliminary analysis of the lower bound function. Such analysis allowed us to identify six data structures for which we have proposed a complexity analysis in terms of memory size and access frequency. Due to the limited size of the shared memory the matrices do not fit in all together. According to the complexity study, the recommendation is to put the Johnson's and the processing time matrices (JM and PTM) in the shared memory if they fit in together. Otherwise, the whole or a part of the Johnson's matrix should be put in priority in the shared memory. The other data structures are mapped to the global memory. Such recommendation has been confirmed through extensive experiments using a recent C2050 Tesla GPU card.

The flowshop scheduling problem has been considered as a case study. The proposed approaches have been experimented using a Tesla C2050 GPU card on different classes of FSP instances. The experimental results show that the parallel evaluation of bounds is the parallelization paradigm that performs better on top of GPU accelerators. Compared to the parallel treeexploration model, accelerations up to ×5.56 are achieved.

Experiments show also that the proposed refactoring approach improves the parallel efficiency whatever the FSP instance and pool size. However, the improvement was not significant because the factorized part of the branches in the FSP lower bound is very small. The optimizations obtained with the proposed thread reduction mechanisms allowed us to achieve accelerations up to ×77.46 compared to a sequential B&B. The data access optimizations allow accelerations up to ×100 compared to a single CPU-based B&B.

In the near future, we plan to extend this work to a cluster of GPU-accelerated multicore processors. From the application point of view, the objective is to optimally solve challenging and unsolved flowshop instances as we did it for one 50×20 problem instance with grid computing [12]. Finally, we plan to investigate other lower bound functions to deal with other combinatorial optimization problems.

Bibliography

[1] R. Allen, L. Cinque, S. Tanimoto, L. Shapiro, and D. Yasuda. A parallel algorithm for graph matching and its maspar implementation. *IEEE Transactions on Parallel and Distributed Systems*, 8(5):490–501, 1997.

[2] T. Carneiro, A. E. Muritibab, M. Negreirosc, and G. A. Lima de Campos. A new parallel schema for branch-and-bound algorithms using GPGPU. In *23rd International Symposium on Computer Architecture and High Performance Computing (SBAC-PAD)*, pages 41–47, New York, USA, 2011.

[3] L. G. Casadoa, J. A. Martíneza, I. Garcíaa, and E. M. T. Hendrixb. Branch-and-bound interval global optimization on shared memory multiprocessors. *Optimization Methods and Software*, 23(5):689–701, 2008.

[4] W. Fung, I. Sham, G. Yuan, and T. Aamodt. Dynamic warp formation and scheduling for efficient GPU control flow. In *In MICRO '07: Proceedings of the 40th Annual IEEE/ACM International Symposium on Micro-architecture*, pages 407–420. Washington, DC, USA, 2007.

[5] B. Gendron and T. G. Crainic. Parallel branch and bound algorithms: Survey and synthesis. *Operations Research*, 42:1042–1066, 1994.

[6] T. Han and T. S. Abdelrahman. Reducing branch divergence in GPU programs. In *Proceedings of the Fourth Workshop on General Purpose Processing on Graphics Processing Units (GPGPU-4), ACM*, pages 1–8. New York, USA, 2011.

[7] J. R. Jackson. An extension of Johnson's results on job-lot scheduling. *Naval Research Logistis Quarterly*, 3(3):61–68, 1956.

[8] S. M. Johnson. Optimal two- and three-stage production schedules with setup times included. *Naval Research Logistis Quarterly*, 1:61–68, 1954.

[9] J. Kurzak, D. A. Bader, and J. Dongarra. *Scientific Computing with Multicore and Accelerators*. Chapman & Hall / CRC Press, 2010.

[10] J. K. Lenstra, B. J. Lageweg, and A. H. G. Rinnooy Kan. A general bounding scheme for the permutation flow-shop problem. *Operations Research*, 26(1):53–67, 1978.

[11] N. Melab. Contributions à la résolution de problèmes d'optimisation combinatoire sur grilles de calcul. LIFL, USTL, Novembre 2005. Habilitation to Direct Research.

[12] M. Mezmaz, N. Melab, and E.-G. Talbi. A grid-enabled branch and bound algorithm for solving challenging combinatorial optimization problems. In *Proceedings of 21th IEEE International Parallel and Distributed Processing Symposium (IPDPS)*, pages 1–9. Long Beach, California, March 2007.

[13] L. G. Mitten. Sequencing n jobs on two machines with arbitrary time lags. *Management Science*, 5(3):293–298, 1959.

[14] NVIDIA Corporation. NVIDIA CUDA C Programming Guide, version 4.0, 2011.

[15] M. J. Quinn. Analysis and implementation of branch-and-bound algorithms on a hypercube multicomputer. *IEEE Transactions on Computers*, 39(3):384–387, 1990.

[16] E. Taillard. Benchmarks for basic scheduling problems. *Journal of Operational Research*, 64:278–285, 1993.

[17] E. Z. Zhang, Y. Jiang, Z. Guo, and X. Shen. Streamlining GPU applications on the fly: Thread divergence elimination through runtime thread-data remapping. In *Proceedings of the 24th ACM International Conference on Supercomputing (ICS'10), ACM*, pages 115–126. New York, NY, USA, 2010.

Chapter 9

Parallel GPU-accelerated metaheuristics

Malika Mehdi and Ahcène Bendjoudi
CERIST Research Center, Algiers, Algeria

Lakhdar Loukil
University of Oran, Algeria

Nouredine Melab
University of Lille 1, CNRS/LIFL/INRIA, France

9.1	Introduction	184
9.2	Combinatorial optimization	184
9.3	Parallel models for metaheuristics	185
9.4	Challenges for the design of GPU-based metaheuristics	187
	9.4.1 Data placement on a hierarchical memory	188
	9.4.2 Threads synchronization	189
	9.4.3 Thread divergence	189
	9.4.4 Task distribution and CPU/GPU communication	190
9.5	State-of-the-art parallel metaheuristics on GPUs	190
	9.5.1 Implementing solution-based metaheuristics on GPUs	191
	9.5.2 Implementing population-based metaheuristics on GPUs	193
	9.5.3 Synthesis of the implementation strategies	198
9.6	Frameworks for metaheuristics on GPUs	201
	9.6.1 PUGACE: framework for implementing evolutionary computation on GPUs	201
	9.6.2 ParadisEO-MO-GPU	202
	9.6.3 libCUDAOptimize: an open source library of GPU-based metaheuristics	202
9.7	Case study: accelerating large neighborhood LS method on GPUs for solving the Q3AP	203
	9.7.1 The quadratic 3-dimensional assignment problem	204
	9.7.2 Iterated tabu search algorithm for the Q3AP	205
	9.7.3 Neighborhood functions for the Q3AP	206
	9.7.4 Design and implementation of a GPU-based iterated tabu search algorithm for the Q3AP	207

	9.7.5	Experimental results	208
9.8	Conclusion		210
	Bibliography		210

9.1 Introduction

This chapter presents GPU-based parallel metaheuristics, challenges, and issues related to the particularities of the GPU architecture and a synthesis on the different implementation strategies used in the literature. The implementation of parallel metaheuristics on GPUs is not straightforward. The traditional models used in CPUs must be rethought to meet the new requirements of GPU architectures. This chapter is organized as follows. Combinatorial optimization and resolution methods are introduced in Section 9.2. The main traditional parallel models used for metaheuristics are recalled in Section 9.3. Section 9.4 highlights the main challenges related to the GPU implementation of metaheuristics. A state-of-the-art of GPU-based parallel metaheuristics is summarized in Section 9.5. In Section 9.6, the main developed GPU-based frameworks for metaheuristics are described. Finally, a case study is presented in Section 9.7 and some concluding remarks are given in Section 9.8

9.2 Combinatorial optimization

Combinatorial optimization (CO) is a branch of applied and discrete mathematics. It consists in finding optimal configuration(s) among a finite set of possible configurations (or solutions) of a given combinatorial optimization problem (COP). The set of all possible solutions noted S is called solution space or search space. Each solution in S is defined by its real cost calculated by an objective function. COPs are generally defined as follows [5]:

A combinatorial problem $P = (S, f)$ can be defined by

- a set of decision variables X,

- an objective function f to optimize (minimize or maximize) over the set S,

- subject to constraints on the decision variables.

COPs are generally formulated as mixed integer programming problems (MIPS) and most of them are NP-hard [11]. According to the quality level of solutions and deadlines required for solving an optimization problem, two classes of optimization methods can be distinguished: *exact methods* and *approximate methods*. Exact methods allow one to reach optimal solution(s) of the handled optimization problem with a proof of its or their optimality. The known methods of this class are the *branch and bound technique, dynamic programming, constraint programming,* and *A* algorithm*. However, optimization problems, whether practical or academic, are often complex and NP-hard. Moreover, a large number of real-life optimization problems encountered in science, engineering, economics, and business are usually large-sized problems for which the size of the potential solution domain increases dramatically with the size of the problem instance. Such problems cannot be tackled using exact methods due to the excessive computation time needed by these methods to find optimal solution(s). In such a situation, approximate methods (or *metaheuristics*) offer an alternative approach to solve these problems. Indeed, these methods allow one to reach good quality solutions in reasonable computation time compared to exact methods but with no guarantee to find optimal or even bounded solutions. This is due to the nature of the search process adopted by these approaches which consists of performing a search in a subset of the whole search space.

Regarding the number of solutions considered at each iteration in the search process, two classes of metaheuristics can be distinguished [34]: *solution-based* and *population-based* metaheuristics. In the rest of this chapter, the term *s-metaheuristic* refers to solution-based metaheuristic and the term *p-metaheuristic* refers to population-based metaheuristic. In s-metaheuristics, the search process starts with a single solution (generally set at random) and iteratively improves it by exploring its neighborhood in the search space. The most known methods in this class are local search methods that include *simulated annealing* [17], *tabu search* [12], *iterated local search* [32], and *variable neighborhood search* [14].

Unlike s-metaheuristics, p-metaheuristics start with a population of solutions and implement an iterative process that evolves the current population towards a new population of better quality solutions. The process is repeated until a stopping criterion is satisfied. *Evolutionary algorithms, swarm optimization,* and *ant colonies* fall into this class.

9.3 Parallel models for metaheuristics

Optimization problems, whether real-life or academic, are more often NP-hard and CPU time and/or memory consuming. Metaheuristics allow the significant reduction of the computational time of the search process but remain time-consuming particularly when it comes dealing with large-sized problems.

The use of parallelism in the design of metaheuristics is a relevant approach that is widely adopted by the combinatorial optimization community for various reasons:

- One of the main goals of parallelism is to reduce the search time. This will allow the design of high performance optimization methods and the solving of large-sized optimization problems.

- Sequential processor architectures have reached their physical limit which prevents creating faster processors. The current trend of microprocessor manufacturers consists of placing multicores on a single chip. Nowadays, laptops and workstations are multicore processors. In addition, the evolution of network technologies and the proliferation of broadband networks have made possible the emergence of clusters of workstations (COWs), networks of workstations (NOWS), and large-scale networks of machines (grids) as platforms for high-performance computing.

From the granularity of parallelism point of view, three major parallel models for metaheuristics can be distinguished [34]: *algorithmic-level*, *iteration-level*, and *solution-level* as illustrated in Figure 9.1.

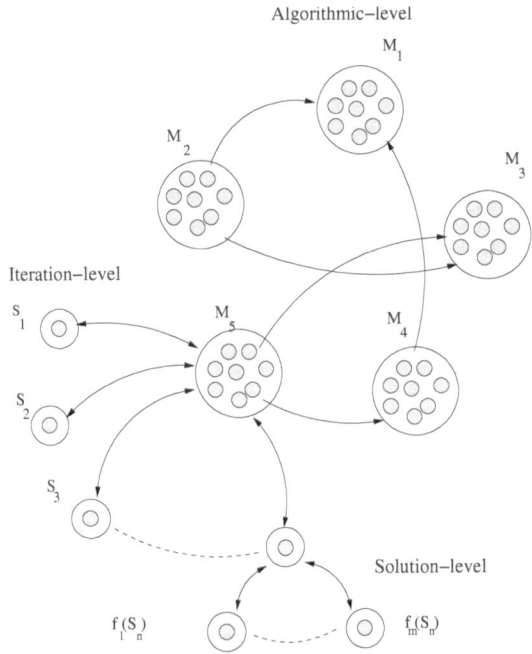

FIGURE 9.1. Parallel models for metaheuristics.

- In the algorithmic-level parallel model, several self-contained metaheuristics are launched in parallel. The parallel metaheuristics may start with identical or different solutions (s-metaheuristics case) or populations (p-metaheuristics case). Their parameter settings such as the size of tabu list for tabu search, transition probabilities for ant colonies, mutation and crossover probabilities for evolutionary algorithms may be the same or different. The parallel processes may evolve independently or in a cooperative manner. In cooperative parallel models, the algorithms exchange information related to the search during evolution in order to find better and more robust solutions.

- In the iteration-level parallel model, the focus is on the parallelization of each iteration of the metaheuristic. Indeed, metaheuristics are generally iterative search processes. Moreover, the most resource-consuming part of a metaheuristic is the evaluation of the generated solutions at each iteration. For s-metaheuristics (e.g., tabu search, simulated annealing, variable neighborhood search), the evaluation and generation of the neighborhood is the most time-consuming step of the algorithm particularly when it comes to dealing with large neighborhood sets. In this parallel model, the neighborhood is decomposed into partitions, and each partition is evaluated in a parallel way. For p-metaheuristics (e.g., evolutionary algorithms, ant colonies, swarm optimization), the iteration-level parallel model arises naturally since these metaheuristics deal with a population of independent solutions. In evolutionary algorithms, for instance, the iteration-level model consists of decomposing the whole population into several partitions where each partition is evaluated in parallel.

- In the solution-level parallel model, the focus is on the parallelization of the evaluation of a single solution. This model is useful when the objective function and/or the constraints are time and/or memory consuming. Unlike the two previous parallel models, the solution-level parallel model is problem-dependent.

9.4 Challenges for the design of GPU-based metaheuristics

Developing GPU-based parallel metaheuristics is not straightforward. The parallel models have to be rethought to meet the new requirements of the GPU architecture. Several major issues have to be taken into account both at design and implementation levels to develop efficient metaheuristics on GPU. These issues are mainly related to the size and latency of the GPU memories, thread synchronization and divergence, the distribution of tasks, and data transfer between the CPU and GPU [21].

9.4.1 Data placement on a hierarchical memory

During the execution of metaheuristics on GPU, the different threads may access multiple data structures from multiple memory spaces. These memories have different sizes and access latencies. Nevertheless, faster memories (registers, shared and constant memories) are generally very small in size, and the larger memories (global memory) are relatively slower. However, p-metaheuristics require the exploration of a large amount of individuals to diversify the search. Moreover, an efficient execution of s-metaheuristics requires exploring large neighborhoods. Thus, programmers have to take into account this point to efficiently place the different data structures of the metaheuristic on the different memories to benefit from both the faster memories and the larger ones. A deep study has to be performed on both the metaheuristic data structures (size and access frequency) and the GPU memories (size and access latency) to identify which data will be placed on which memory. Generally, the most accessed ones should be put on faster memories (registers, shared memory) and larger ones on the larger memories (global memory). Also, an efficient mapping between threads and the corresponding metaheuristic elements (one neighbor per thread, one individual per thread, single population per threads block, one ant per thread, etc.) must be defined to ensure a maximum occupancy of the GPU and to cover CPU/GPU communication and memory access times.

According to the used metaheuristic and to the handled problem, the data values may have different types and different ranges of their values. The data types should then be chosen carefully and the ranges of the data values should be analyzed to minimize the amount of occupied memory space.

In addition to the size and latency of GPU memory issues, the memory access pattern is another important issue to be dealt with to speedup GPU-based metaheuristics. Indeed, the different memories have been designed to achieve specific features that programmers must take into account to optimize their codes and then to benefit from these features. For instance, the global memory is optimized for coalesced accesses. The texture and the constant memory are read-only memories. The texture is optimized for uncoalesced accesses and the constant memory is optimized for simultaneously accesses of all threads to the same location [21]. Therefore, to improve the performance of the kernel execution, the programmers should put coalesced data on global memory, uncoalesced read-only data (e.g., metaheuristic instance data) on the texture, concurrent read-only data (e.g., data for fitness evaluation that all threads concurrently access) on the constant memory, and the most accessed data structures (e.g., population of individuals for a CUDA thread block) on the shared memory.

9.4.2 Threads synchronization

The thread synchronization issue is caused by both the GPU architecture and the synchronization requirements of the implemented method. Indeed, GPUs are based on a multicore architecture organized into several multiprocessors (Streaming Multiprocessor SM) supporting the SPMD model (Single Program Multiple Data). Each SM contains several cores executing the same instruction of different threads following the SIMD model (Single Instruction Multiple Data). These threads belong to a warp (a group of 32 threads) and handle different data elements. An efficient execution of applications on GPU is achieved when launching a large amount of threads (thousands of threads) [2]. However, the execution order of these thousands of threads is unknown by the programmer which makes the prediction of their execution order a challenging issue. Plus, the developer has to control explicitly the threads through the insertion of barrier synchronizations in the codes to avoid concurrent accesses to data structures and to meet some requirements related to data-dependent synchronizations.

9.4.3 Thread divergence

Thread divergence is another challenging issue in GPU-based metaheuristics [9, 10, 31]. Generally, metaheuristics contain irregular loops and conditional instructions when generating and evaluating the neighborhood (s-metaheuristics), and the population (p-metaheuristics) in the same block. In addition, the decision to apply a crossover or a mutation on an individual in a genetic algorithm and the exploration of different paths using an ant colony are random operations. Threads of the same warp have to execute instructions simultaneously leading to different branches whereas in an SIMD model the threads of a same warp execute the same instruction. Consequently, the different branches of a conditional instruction which is data-dependent lead to a serial execution of the different threads degrading the performance of the application in terms of execution time. The challenge here is then to revisit the traditional irregular metaheuristic codes to eliminate these divergences.

9.4.4 Task distribution and CPU/GPU communication

The performance of GPU-based metaheuristics in terms of execution time could be improved by choosing the most appropriate parallel model (algorithmic-level, instruction-level, solution-level). Moreover, an efficient decomposition of the metaheuristic and an efficient assignment of code portions between the CPU and GPU should be adopted. The objective is to take benefit from the GPU computing power without affecting the efficiency and the behavior of the metahcuristic and without losing performance in CPU/GPU communication and memory accesses. In order to decide which part of the metaheuristic will be executed on which component, one should perform a careful analysis on the serial code of the metaheuristic, identify the compute-intensive tasks (e.g., exploration of the neighborhood, evaluation of individuals), and then offload them to the GPU, while the remaining tasks still run on the CPU in a serial way.

The CPU/GPU communication is done through the global memory which is a slow memory making the memory transfer between the CPU and GPU time-consuming which can significantly degrade the performance of the application. Accesses to this memory should be optimized by minimizing the amount of transferred data between the two components in order to reduce the communication time and, therefore, the whole execution time of the metaheuristic.

9.5 State-of-the-art parallel metaheuristics on GPUs

After more than two decades of research by the combinatorial optimisation community devoted to developing adequate parallel metaheuristics for different types of parallel architectures (clusters, supercomputers and grids), the actual developement of General Perpose GPU (GPGPU) brings new challenges for parallel metaheuristics on SIMD architectures.

The first works on metaheuristic algorithms implemented on GPUs started on old graphics cards before the appearance of modern GPUs equipped with high-level programming interfaces such as CUDA and OpenCL. Among these pioneering works we cite the work of Wong et al. [38] dealing with the implementation of EAs on graphics processing cards and the work by Catala et al. in [8] where the ACO algorithm is implemented on old GPU architectures. Yu et al. [39] and Li et al. [18] proposed a full parallelization of genetic algorithms on old GPU architectures using shader libraries based on Direct3D and OpenGL.

Such architectures are based on preconfigured pipelined stages used to accelerate the transformation of 3D geometric primitives into pixels. Implementing a general-purpose algorithm on such preconfigured architectures is very hard and requires the tailoring of the algorithm's data and instructions to fit the pipelined stages of the GPU. Since then, GPU architectures have evolved to become programmable using high-level programming interfaces. In this section we will focus only on recent state-of-the-art works dealing with metaheuristics implementation on modern programmable GPUs. In this review two classes are considered: (1) s-metaheuristics on GPUs and (2) p-metaheuristics on GPUs. A comparative study is done of the main works and a classification of the different existing strategies is proposed in Section 9.5.3.

9.5.1 Implementing solution-based metaheuristics on GPUs

A very basic local search algorithm starts with an initial solution generated either at random or by the mean of a specific heuristic and is based on two elementary components: the generation of neighborhood structures using an elementary move function and a selection function that determines which solution in the current neighborhood will replace the actual search point. The search continues as long as there is improvement and stops when there is no better solution in the current neighborhood. The exploration (or evaluation) of the different moves of a given neighborhood is done independently for each move. Thus, the easiest way to accelerate a local search algorithm is to parallelize the evaluation of the neighborhood (instruction-level parallelism). This is by far the most used scheme in the literature for parallelizing local search algorithms on GPUs. Nevertheless, small neighborhoods may lead to nonoptimal occupation of the CUDA threads which may lead in turn to an overhead due to the communication and memory latencies. Therefore, large neighborhoods are necessary for efficient implementation of local searches on GPUs.

Luong et al. [20] proposed efficient mappings for large neighborhood structures on GPUs. In this work, three different neighborhoods are studied and mapped to the hierarchical GPU for binary problems. The three neighborhoods are based on the Hamming distance. The move operators used in the three neighborhoods consider Hamming distances of 1, 2, and 3 (this consists on flipping the binary value of one, two, or three positions at a time in the candidate binary vector). In [20], each thread is associated to a unique solution in the neighborhood. The addressed issue is how to efficiently map the different neighborhoods on the device memory, more explicitly, how to calculate the memory index of each solution associated to each CUDA thread's *id*. The three neighborhoods are implemented and experimented on the Permuted Perceptron Problem (PPP) using a tabu search algorithm (TS). Accelerations from 9.9× to 18.5× are obtained on different problem sizes.

In the same context, Deevacq et al. [10] proposed two parallelization strategies inspired by the multiwalk parallelization strategy, of a 3-opt iterated local search algorithm (ILS) over a CPU/GPU architecture. In the first strategy, each Local Search (LS) is associated to a unique CUDA thread and improves a unique solution by generating its neighborhood. The second strategy associates each solution to a CUDA block and the neighborhood exploration is parallelized on the block threads. In the first strategy, since several LS are executed on different solutions on each Multi Processor (MP), the data structures should be stored on the global memory, while the exploration of a single solution at a time in the second strategy allows the use of the shared memory to store the related data structures. The two strategies have been experimented on standard benchmarks of the Traveling Salesman Problem (TSP) with sizes varying from 100 to 3038 cities. The results indicate that increasing the number of solutions to be explored simultaneously improves the speedup in the two strategies, but at the same time it decreases final solution quality. The greater speedup factor reached by the second strategy is 6×.

The same strategy is used by Luong et al. in [21] to implement multistart parallel local search algorithms (a special case of the algorithmic-level parallel model where several homogeneous LS algorithms are used). The multistart model is combined with iteration-level parallelism: several LS algorithms are managed by the CPU and the neighborhood evaluation step of each algorithm is parallelized on the GPU (each GPU thread is associated with one neighbor and executes the same evaluation function kernel). The advantage of such a model is that it allows a high occupancy of the GPU threads. Nevertheless, memory management causes new issues due to the quantity of data to store and to communicate between CPU and GPU. A second proposition for implementing the same model on GPU consists of implementing the whole LS processes on GPU with each GPU thread being associated to a unique LS algorithm. This solves the communication issue encountered in the first model. In addition, a memory management strategy is proposed to improve the efficiency of the algorithmic-level model: texture memory is used to avoid memory latency due to uncoalesced memory accesses. The proposed approaches are implemented on the quadratic assignment problem (QAP) using CUDA. The acceleration rates obtained for the algorithmic-level with usage of texture memory rise from 7.8× to 12× (for different QAP benchmark sizes).

Janiak et al. [15] implemented two algorithms for TSP and the flow-shop scheduling problem (FSP). These algorithms are based on a multistart tabu search model. Both of the algorithms exploit multicore CPU and GPU. A full parallelization on GPU is adopted using shader libraries where each thread is mapped with one tabu search. However, even though their experiments report that the use of GPU speedups the serial execution almost 16×, the mapping of one thread with one tabu search requires a large number of local search algorithms to cover the memory access latency. The same mapping policy is adopted by Zhu et al. in [41] (one thread is associated to one local search) solving the quadratic assignment problem but using the CUDA toolkit instead of shader libraries.

Luong et al. [22] proposed a GPU-based implementation of hybrid metaheuristics on heterogeneous parallel architectures (multicore CPU coupled to one GPU). The challenge of using such a heterogeneous architecture is how to distribute tasks between the CPU cores and the GPU in such a way to have optimal performances. Among the three traditional parallel models (solution-level, iteration-level and algorithmic-level), the authors point out that the most convenient model for the considered heterogeneous architecture is a hybrid model combining iteration-level and algorithmic-level models. Several CPU threads execute several instances of the same S-metaheuristic in parallel while the GPU device is associated to one CPU core and used to accelerate the neighborhood calculation of several S-metaheuristics at the same time. In order to efficiently exploit the remaining CPU cores, a load-balancing heuristic is also proposed in order to decide on the number of additional S-metaheuristics to launch on the remaining CPU cores relative to the efficiency of the GPU calculations. The proposed approach is applied to the QAP using several instances of the Fast Ant Colony Algorithm (FANT) [33].

All the previously noted works exploit the same parallel models used on CPUs based on the task parallelism. A different implementation approach is proposed by Paul in [27] to implement a simulated annealing (SA) algorithm for the QAP on GPUs. Indeed, the author used a preinitialized matrix *delta* in which the incremental evaluation of simple swap moves are calculated and stored relative to the initial permutation p. For the GPU implementation, the authors used the parallel implementation of neighborhood exploration. The time-consuming tasks in the SA-matrix are identified by the authors as updating the matrix and passing through it to select the next accepted move. To initialize the delta matrix, several threads from different blocks explore different segments of the matrix (different moves) at the same time. In order to select the next accepted swap, several threads are also used. Starting from the last move, a group of threads explores different subsets of the delta matrix. The shared memory is used to preload all the necessary elements for a given group of threads responsible for the updating of the delta matrix. The main difference in this work compared to the previous works resides in the introduction of a data parallelism using the precalculated delta matrix. The use of this matrix allows the increase in the number of threads involved in the evaluation of a single move. Experimentations are done on standard QAP instances from the QAPLIB [6]. Speedups up to 10× are achieved by the GPU implementation compared to the same sequential implementation on CPU using SA-matrix.

9.5.2 Implementing population-based metaheuristics on GPUs

State-of-the-art works dealing with the implementation of p-metaheuristics on GPUs generally rely on parallel models and research efforts done for parallelizing different classes of p-metaheuristics over different types of parallel architectures: supercomputers, clusters, and computational grids. Three main classes of p-metaheuristics are considered in this section: evolutionary algorithms (EAs), ant colony optimization (ACO), and particle swarm optimization (PSO).

Evolutionary algorithms

Traditional parallel models for EAs are classified in three main classes: coarse-grain models using several subpopulations (islands), master-slave models used for the parallelization of CPU intensive steps (evaluation and transformation), and cellular models based on the use of one population disposed (generally) on a toroidal grid.

The three traditional models have been implemented on GPUs by several researchers for different optimization problems. The main chalenges to be raised when implementing the traditional models on GPUs concern (1) the saturation of the GPU in order to cover memory latency by calculations, and (2) efficent usage of the hierarchical GPU memories.

In [16], Kannan and Ganji present a CUDA implementation of the drug discovery application Autodock (molecular docking application). Autodock uses a genetic algorithm to find optimal docking positions of a ligand to a protein. The most time-consuming task in Autodock is the fitness function evaluation. The fitness function used for a docking problem consists of calculating the energy of the ligand-protein complex (sum of intermolecular energies). The authors explore two different approaches to evaluate the fitness function on GPU. In the first approach, each GPU thread calculates the energy function of a single individual. This approach requires the use of large-sized populations to saturate the GPU (thousands of individuals). In the second approach each individual is associated with one thread block. The evaluation of the energy function is performed by the threads of the same block. This allows the use of medium population sizes (hundreds of individuals) and the acceleration of a single fitness evaluation. Another great advantage of the per block approach resides in the use of shared memory instead of global memory to store all the information related to each individual. The obtained speedups range from $10\times$ to $47\times$ for population sizes ranging from 50 to 10000.

Maitre et al. [23] also exploited the master-slave parallelism of EAs on GPUs using EASEA. EASEA is a C-like metalanguage for easy development of EAs. The user writes a description of the problem-specific components (fitness function, problem representation, etc) in EASEA. The code is then compiled to obtain a ready-to-use evolutionary algorithm. The EASEA compiler uses genetic algorithm LIB and EO Libraries to produce C++ or JAVA written EA codes. In [23], the authors proposed an extension of EASEA to produce CUDA code from the EASEA files. This extension has been used to generate a master-slave parallel EA in which the sequential algorithm is maintained on CPU and the population is sent to GPU for evaluation. Two problems have been considered during the experiments: a benchmark mathematical function and a real problem (molecular structure prediction). In order to maximize the GPU occupation, very large populations are used (from 2000 to 20000). Even though transferring such large populations from the CPU to the GPU device memory at every generation is very costly, the authors report important speedups on the two problems on a GTX260 card: $105\times$ is reported for the benchmark function while for the real problem the reported speedup is 60×. This may be best explained by the complexity of the fitness functions. Nevertheless, there is no indication in the paper about the memory management of the populations on GPU.

The master-slave model is efficient when the fitness function is highly time intensive. Nevertheless, it requires the use of large-sized populations in order to saturate the GPU unless the per-block is used (one individual perblock) when the acceleration of the fitness function itself is possible. The use of many subpopulations of medium sizes is another way to obtain a maximum occupation of the GPU. This is coarse-grained parallelism (island model).

The coarse-grained model is used by Pospichal et al. in [30] to implement a parallel genetic algorithm on GPU. In this work the entire genetic algorithm is implemented on GPU. This choice is motivated by the overhead engendered by the CPU/GPU communication when only population evaluation is performed on GPU. Each population island is mapped with a CUDA thread block and each thread is responsible for a unique individual. Subpopulations are stored on shared memory of each block. Nevertheless, because interblock communications are not possible on the CUDA architecture, the islands evolve independently in each block, and migrations are performed asynchronously through the global memory. That is, after a given number of generations, selected individuals for migration from each island are copied to the GPU global memory part of the neighbor island and then to its shared memory to replace the worst individuals in the local population. The experiments are performed on three benchmark mathematical functions. During these experiments, the island sizes are varied from 2 to 256 individuals and island numbers from 1 to 1024. The maximum performance is achieved for high number of islands and increasing population sizes.

The same strategy is also adopted by Tsutsui and Fujimoto in [35] to implement a coarse-grained genetic algorithm on GPU to solve the QAP. Initially, several subpopulations are created on CPU and transferred to the global memory. The subpopulations are organized in the global memory into blocks of 8 individuals in such a way to allow coalesced memory access by the threads of the same thread block. Each sub-population is allocated to a single thread block in the GPU and transfered to the shared memory to start evolution. Population evaluation and transformation are done in parallel by the different threads of a given block. Migration is also done through the global memory. Experiments are performed on standard QAP benchmarks from the QAPLIB [6]. The GPU implementation reached speedups of 2.9× to 12.6× compared to a single core implementation of a coarse-grained genetic algorithm on a Intel i7 processor.

Nowotniak and Kucharski [26] proposed a GPU-based implementation of a Quantum Inspired Genetic Algorithm (QIGA). The parallel model used is a hierarchical model based on two levels: each thread in a block transforms a unique individual and a different population is assigned to each block. The algorithm is run entirely on GPU. A different instance of the QIGA is run on each block and the computations have been shared between 8 GPUs. This approach is very convenient to speed up the experimental process on metaheuristics that require several independent runs of the same algorithm in order to assess statistical efficiency. The populations are stored in the shared memory, the data matrix used for fitness evaluation is placed in read only constant memory, and finally seeds for random numbers generated on the GPU are stored in the global memory.

In coarse-grained parallelism, the use of the per-block approach to implement the islands (one subpopulation per thread block) is almost natural and it allows the use of shared memory to store the subpopulations. Fine-grained parallel models for EAs (cellular EAs) have been used by many authors to implement parallel EAs on GPUs. Indeed, the fine-grained parallelism of EAs fits perfectly to the SIMD architecture of the GPU.

Pinel et al. in [29] developed a highly parallel synchronous cellular genetic algorithm (CGA), called GraphCell, to solve the independent task scheduling problem on GPU architectures. In CGAs, the population is arranged into a two-dimensional toroidal grid where only neighboring solutions are allowed to interact with each other during the recombination step. In GraphCell, two recombination operators for CGA are especially designed to run efficiently on GPU. Indeed, instead of assigning a single thread to each solution of the population to perform the recombination, in GraphCell, a single thread is assigned to each task of each solution. Offsprings are created by independently modifying the assignment of some tasks in the current solution. Mainly, each thread chooses one neighboring solution in the grid as second parent using different selection strategies and assigns one task of the solution (first parent) to the machine for which the same task is assigned in the second parent. This way, the number of threads used for the recombination step is equal to population size × size of the solution (number of tasks). This leads to a high number of threads used to accelerate the recombination operators especially when dealing with large instances of the problem. In addition to the recombination operators, the rest of the CGA steps are also parallelized on GPU (fitness evaluation, mutation, and replacement).

A similar work is proposed by Vidal and Alba in [37] where a CGA using a toroidal grid is completely implemented on GPU. A direct mapping between the population and the GPU threads is done. At each step, several threads execute the same operations on each individual independently: initialization, computing the neighborhood, selection, crossover, mutation, and evaluation. A synchronization is done for all threads to perform the replacement stage and form the next generation. All the data of the algorithm is placed on the global memory. Several experiments have been performed on a set of standard benchmark functions with different grid sizes ranging from 32^2 to 512^2. The speedups reached by the GPU implementation against the CPU version range from 5× to 24× and increase as the size of the population increases. However, the CPU implementation runs faster than the GPU version for all the tested benchmarks when the size of the population is set to 32^2. When the size of the population is too small, there are not enough computations to cover the overhead created by the call of kernel functions, CPU/GPU communications, synchronization, and access to global memory. Finally, an interesting review on GPU parallel computation in bio-inspired algorithms is proposed by Arenas et al. in [4].

Ant colony optimization

Ant colony optimization (ACO) is another p-metaheuristic subject to parallelization on GPUs. State-of-the-art works on parallelizing ACO focus on accelerating the tour construction step performed by each ant by taking a task-based parallelism approach, with pheromone deposition on the CPU.

In [9], Cecilia et al. present a GPU-based implementation of ACO for TSP where the two steps (tour construction and pheromone update) are parallelized on the GPU. A data parallelism approach is used to enhance the performance of the tour construction step. The authors use two categories of artificial ants: queen ants associated with CUDA thread-blocks and worker ants associated with CUDA threads. A queen ant represents a simulated ant and worker ants collaborate with the queen ant to accelerate the decision about the next city to visit. The tour construction step of each queen ant is accelerated. Each worker ant maintains a history of the search in a tabu list containing a chronological ordering of the already visited cities. This memory is used to determine the feasible neighborhoods. After all queen ants have constructed their tours, the pheromone trails are updated. For pheromone update, several GPU techniques are also used to increase the data bandwidth of the application mainly by the use of precalculated matrices that are easily updated by several threads (one thread per matrix entry). The achieved speedups are $21\times$ for tour construction and $20\times$ for pheromone updates.

In another work, Tsutsui and Fujimoto [36] propose a hybrid algorithm combining ACO metaheuristic and Tabu Search (TS) implemented on GPU to solve the QAP. A solution of QAP is represented as a permutation of $1, 2, .., n$ with n being the size of the problem. The TS algorithm is based on the 2-opt neighborhood (swapping of two elements (i, j) in the permutation). The authors point out that the move cost of each neighbor depends on the couple (i, j). Two groups of moves are formed according to the move cost. In order to avoid thread divergence within the same warp, the neighborhood evaluation is parallelized in such a way to assign only moves of the same cost to each thread warp. This strategy is called MATA for Move-cost Adjusted Thread Assignment. Concerning the memory management, all the data of the ACO (population, pheromone matrix), QAP matrices, and tabu list are placed on the global memory of the GPU. Shared memory is used only for working data common to all threads in a given block. All the steps of the hybrid algorithm ACO-TS (ACO initialization, pheromone update, construct solutions, applying TS) are implemented as kernel functions on the GPU. The GPU/CPU communications are only used to transfer the best-so-far solution in order to verify termination conditions. The experiments performed on standard QAP benchmarks showed that the GPU implementation using MATA obtained a speedup of 19× compared to the CPU implementation, compared with a speedup of only 5× when MATA is not used.

Particle swarm optimization

In [40] Zhou and Tan propose a full GPU implementation of a standard PSO algorithm. All the data is stored in global memory (velocities, positions, swarm population, etc). Only working data is copied to shared memory by each thread. The four steps of the PSO have been parallelized on GPU: fitness evaluation of the swarm, update of local best and global best of each particle, and update of velocity and position of each particle. The same strategy is used to parallelize the first and last steps: the evaluation of fitness functions is performed in parallel for each dimension by several threads. It is the case for the update of position and velocities of each particle: one dimension at a time is updated for the whole swarm by several threads. Random numbers needed for updating the velocities and positions for the whole PSO processes are generated on CPU at the starting of the algorithm and transferred to the GPU global memory. For the steps 2 and 3 (update of local best and global best of each particle), the GPU threads are mapped to the N particles of the swarm one to one. Experiments done on four benchmark functions show speedups ranging from 3.7× to 9.0× for swarm sizes ranging from 400 to 2800.

9.5.3 Synthesis of the implementation strategies

After reviewing some works dealing with GPU-based implementation of metaheuristics, in this section we will try to come up with a classification of the different strategies used in the literature for the implementation of metaheuristics on GPUs. One may distinguish between the parallel models adopted in each metaheuristic (design level) and the way they are exploited on GPU architectures (implementation level). Indeed, even though the parallelization models used in most works for GPUs are derived from the traditional parallel models of each metaheuristic (on CPU), their implementation could take a different way and sometimes it may result in new parallel models customized for GPUs.

Traditional parallel models for metaheuristics are based on an intuitive task parallelism: the independent tasks of the algorithms are simply parallelized. For example, in the case of s-metaheuristics, the evaluation of large neighborhoods could be done in parallel since there is no synchronization at this step of the algorithm. That is the case of EAs when it comes to applying the evaluation step. Nevertheless, because of the particularity of the GPU architecture, some authors have used new implementation techniques to enhance the data parallelism in the sequential algorithms in order to increase the data throughput of the application.

From this observation we propose the following classification based on 2 levels: design level and implementation level as illustrated in Figure 9.2. The design level regroups the three classes of parallel models used in metaheuristics (solution-level, iteration-level, algorithmic-level) with examples for s-metaheuristics, EAs, ACO and PCO. This classification is principally built from the reviewed state-of-the-art works in the previous section. The implementation level refers to the way these parallel design models are implemented on GPU. This classification focuses only on the mapping strategies between the GPU threads and the parallelized tasks (neighborhood evaluation, solution construction, and so on). The different implementation strategies are explained in the following sections.

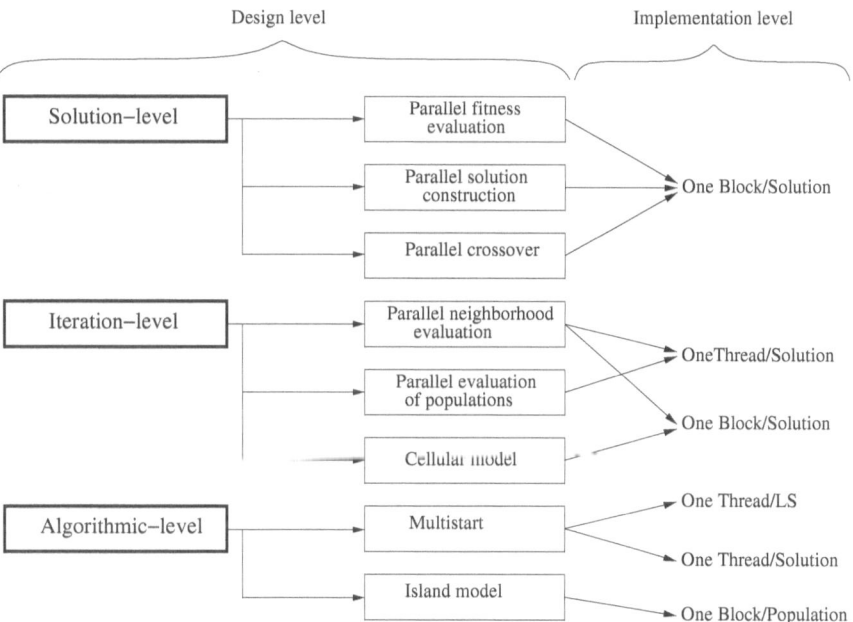

FIGURE 9.2. A two level classification of state-of-the-art GPU-based parallel metaheuristics.

GPU thread mapping for solution-level parallelism

Parallel models at solution level consist of parallelizing a time intensive atomic task of the algorithm. Generally, it consists of the fitness evaluation [16]. Nevertheless, crossover operators have also been parallelized by some authors [29]. These kinds of models are not always possible as they are problem-dependent. The GPU implementation of solution-level models uses the perblock mapping: each solution is associated to a block of threads. A second level of parallelism is used inside each block to parallelize the fitness evaluation of a single solution. This kind of mapping allows the use of shared memory to store the data structures of the solution and does not require the use of very large neighborhoods or populations.

GPU thread mapping for iteration-level parallelism

Iteration-level parallelism consists of parallelizing the tasks performed independently on different solutions. Different mapping strategies are adopted in the reviewed works to implement these models.

In Figure 9.2, the first example of iteration-level parallelism is the parallel evaluation of neighborhoods in s-metaheuristics. In most of the reviewed works, a per-thread mapping approach is used: each solution of the neighborhood is mapped to a unique thread in the GPU for evaluation [10, 20]. The same mapping is used for master-slave parallel EAs to accelerate the evaluation of large populations. This kind of mapping is only efficient for very large neighborhoods and very large populations (to saturate the GPU). Many authors have pointed out that the use of such large populations (or neighborhoods) may lead to an overhead due to the communication costs between the CPU and the GPU (if the sequential part of the algorithm is placed on CPU). In order to circumvent this issue, many authors have implemented the entire algorithm on GPU [30]. On the other hand, as several solutions may be mapped with the same thread block in the GPU, shared memory could not be used and all the data should be placed on global memory.

GPU thread mapping for algorithmic-level parallelism

Algorithmic-level parallelism consists of launching several self-contained algorithms in parallel. In the previously reviewed works two algorithmic-level models have been used: the multistart model and the island model (parallel EAs).

The implementation of the multistart model is based on two different mapping strategies [10, 21]: (1) each Local Search (LS) is associated to a unique thread and (2) each solution (from multiple neighborhoods) is associated to a unique thread. The first mapping strategy (one thread per LS) presents a big drawback: the number of LS to use should be very large to saturate the GPU and cover the memory access latency. On the other hand, the evaluation of the neighborhood of a single LS by one thread is time intensive. Furthermore, shared memory could not be used to store the huge data generated by the different neighborhoods. In the second mapping strategy, the LS algorithms are placed on CPU and the neighborhood evaluation of each LS is parallelized on GPU using per-thread mapping strategy (one thread per solution). This consists of a hierarchical parallel model combining algorithmic-level parallelism (multistart) with iteration-level parallelism (master-worker).

In the island model, the same mapping is used in all the reviewed works [23, 26, 35]: each subpopulation (island) is associated to one thread block in the GPU. A second level of parallelism is used inside the block to parallelize the evaluation step of the local population. Migrations are always performed through the global memory as interblock communications are impossible in CUDA. The first advantage of this hierarchical implementation is that it allows the occupation of a large number of threads even for medium population sizes. The second advantage consists of using shared memory to store subpopulations to benefit from the low latency of shared memory.

9.6 Frameworks for metaheuristics on GPUs

After the first pioneering works of metaheuristics on graphics processing units, the next challenge is to provide easy-to-use frameworks and libraries for rapid development of metaheuristics on GPUs. Although the works on this subject are not yet mature and do not cover the main metaheuristic algorithms, we will present the only three works to our knowledge, which propose open source frameworks for developing metaheuristics on GPUs.

The three frameworks reviewed in this section are PUGACE [31], ParadisEO-MO-GPU [24], and libCUDAOptimize [25]. PUGACE is a framework for implementing EAs on GPUs. ParadisEO-MO-GPU is an extension of the framework ParadisEO [7] for implementing s-metaheuristics on GPU. Finally, libCUDAOptimize is a library intended for the implementation of p-metaheuristics on GPU. The three frameworks are presented in more detail in the following.

9.6.1 PUGACE: framework for implementing evolutionary computation on GPUs

PUGACE is a generic framework for easy implementation of cellular evolutionary algorithms on GPUs implemented using C and CUDA. It is based on the frameworks MALLBA and JCell (a framework for cellular algorithms). The authors justified the choice of cellular evolutionary algorithm by the good feedback found in the literature concerning its efficient implementation on GPUs compared to other parallel models for EAs (island, master-slave). The main standard evolutionary operators are already implemented in PUGACE: different selection strategies, standard crossover, and mutation operators (PMX, swap, 2-exchange, etc.). Different problem encoding is also supported. The framework is organized as a set of modules in which the different functionalities are separated as much as possible in order to facilitate the extension of the framework. All the functions and procedures that execute on GPU are implemented in the same file kernel.cu.

The implementation strategy adopted on the GPU is as follows. Population initialization is done on the CPU side and the population is transferred to the

GPU. On the GPU side, each individual is associated to a unique CUDA thread. The function evaluation and mutation are done on the GPU while selection and replacement are maintained on the CPU. In order to avoid thread divergence appearing in the same CUDA thread block at the crossover step (because of the probability of application which may give different results from one thread to the other), the decision of whether to apply a crossover is taken at the block level and applied to all the individuals within the block. It is the decision on the choose of the cutting point for the crossover.

The framework is validated on standard benchmarks of the QAP. Speedups of 15.44× to 18.41× are achieved compared to a CPU implementation of a cEA using population sizes rising from 512 to 1024 and from 1024 to 2048.

9.6.2 ParadisEO-MO-GPU

Melab et al. [24] developed a reusable framework ParadisEO-MO-GPU for parallel local search metaheuristics (s-metaheuristics) on GPUs. It focuses on the iteration-level parallel model of s-metaheuristics which consists of exploring in parallel the neighborhood of a problem solution on GPU. The framework, implemented using C++ and CUDA, is an extension of the ParadisEO [7] framework previously developed by the same team for parallel and distributed metaheuristics on both dedicated parallel hardware platforms and computational grids. The objective of this framework is to facilitate the development of GPU-based metaheuristics providing a transparent use of GPUs to users who are unfamiliar with advanced features of all parallelization techniques and deployment on GPUs. The framework allows one to efficiently manage the hierarchical organization of the memories (latencies and sizes) of the GPU and its communication with the CPU as well as the minimizing of the user involvement in its management.

The framework is based on a master-worker model where the master is the CPU and the workers are threads executed by the processing cores of the GPU. The CPU executes the serial part of the metaheuristic and sends only the current solution to the GPU to minimize the transfer cost. The GPU, on its side, generates the neighboring of the received solution and evaluates them at each iteration. All the threads execute the same kernel and according to a static mapping table between the threads and the neighbors where each thread is associated with exactly one neighbor evaluation. After all the neighborhood is generated and evaluated, it is sent back to the CPU which selects the best solution (See Figure 9.3).

9.6.3 libCUDAOptimize: an open source library of GPU-based metaheuristics

LibCUDAOptimize [25] is a C++/CUDA open source library for the design and implementation of metaheuristics on GPUs. Until now, the metaheuristics supported by LibCUDAOptimize are: scatter search, differential evolution,

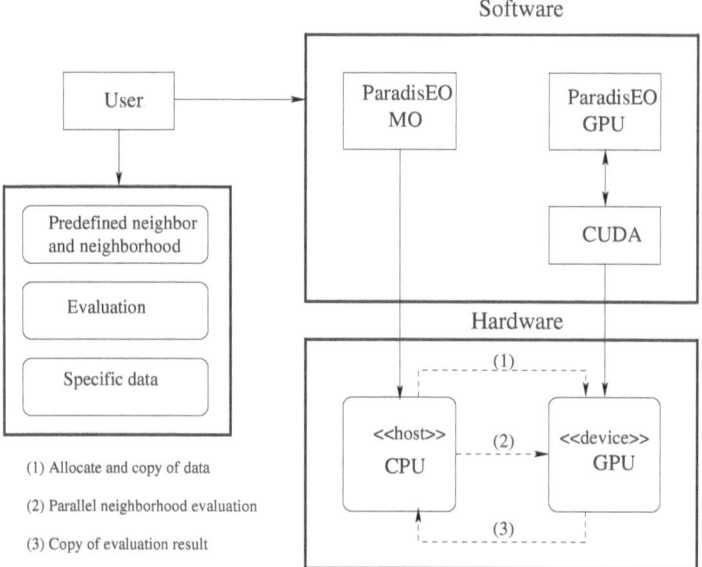

FIGURE 9.3. The skeleton of the ParadisEO-MO-GPU.

and particle swarm optimization. Nevertheless, the library is designed in such a way to allow further extensions for other metaheuristics and it is still in development phase by the authors. The parallelization strategy adopted on GPU is principally based on fitness evaluation. The sequential structure of the optimization algorithms is maintained on CPU.

9.7 Case study: accelerating large neighborhood LS method on GPUs for solving the Q3AP

In this case study, a large neighborhood GPU-based local search method for solving the Quadratic 3-dimensional Assignment Problem (Q3AP) will be presented. The local search method is an Iterated Local Search (ILS) [32] using an embedded TS algorithm. The ILS principle consists of executing iteratively the embedded local search, each iteration which starts from a disrupted local optima reached by the previous local search process. The disruption heuristic is a performance parameter of an ILS algorithm and should be judiciously defined. A template of an ILS algorithm is given by the Algorithm 7.

Algorithm 7: iterated local search template

1 s_0=GenerateInitSol();
2 s^*=LocalSearch(s_0);
3 **repeat**
4   s'=Perturbate($s^*, history$);
5     $s^{*'}$=LocalSearch(s');
6   s^*=AcceptationCriteria($s^*, s^{*'}, history$);
7 **until** *a maximum number of iterations is reached*;

9.7.1 The quadratic 3-dimensional assignment problem

The Q3AP is an extension of the well-known NP-hard QAP. The latter is one of the most studied problems by the combinatorial optimization community due to its wide range of practical applications (site planning, schedule problem, computer-aided design, etc.) and to its computational challenges since it is considered as one of the most computationally difficult optimization problems.

The Q3AP was first introduced by William P. Pierskalla in 1967 [28] and, unlike the QAP, the Q3AP is a less studied problem. Indeed, the Q3AP was revisited only this past year and has recently been used to model some advanced assignment problems such as the symbol-mapping problem encountered in wireless communication systems and described in [13]. The Q3AP optimization problem can be mathematically expressed as follows:

$$min \left\{ \sum_{i=0}^{n-1}\sum_{j=0}^{n-1}\sum_{l=0}^{n-1}\sum_{k=0}^{n-1}\sum_{s=0}^{n-1}\sum_{r=0}^{n-1} c_{ijlksr} x_{ijl} x_{ksr} + \right.$$

$$\left. \sum_{i=0}^{n-1}\sum_{j=0}^{n-1}\sum_{l=0}^{n-1} b_{ijl} x_{ijl} \right\} \quad (9.1)$$

where

$$X = (x_{ijl}) \in I \cap J \cap L, \quad (9.2)$$
$$x_{ijl} \in \{0,1\}, \quad i,j,l = 0,1,...,n-1 \quad (9.3)$$

$I, J,$ and L sets are defined as follows:

$$I = \{X = (x_{ijl}) : \sum_{j=0}^{n-1}\sum_{l=0}^{n-1} x_{ijl} = 1, \quad i = 0,1,...,n-1\};$$

$$J = \{X = (x_{ijl}) : \sum_{i=0}^{n-1}\sum_{l=0}^{n-1} x_{ijl} = 1, \quad j = 0,1,...,n-1\};$$

$$L = \{X = (x_{ijl}) : \sum_{i=0}^{n-1}\sum_{j=0}^{n-1} x_{ijl} = 1, \quad l = 0, 1, ..., n-1\}$$

Other equivalent formulations of the Q3AP can be found in the literature. A particularly useful one is the *permutation-based formulation* wherein the Q3AP can be expressed as follows:

$$min \left\{ f(\pi_1, \pi_2) = \sum_{i=0}^{n-1}\sum_{j=0}^{n-1} c_{i\pi_1(i)\pi_2(i)j\pi_1(j)\pi_2(j)} + \sum_{i=0}^{n-1} b_{i\pi_1(i)\pi_2(i)} \right\} \quad (9.4)$$

where π_1 and π_2 are permutations over the set $\{0, 1, \ldots, n-1\}$. According to this formulation, minimizing the Q3AP consists of finding a double permutation (π_1, π_2) which minimizes the objective function (9.4).

The Q3AP is proven to be an NP-hard problem since it is an extension of the quadratic assignment problem and of the axial 3-dimensional assignment problem which are both NP-hard problems. It is particularly computationally difficult since the number of feasible solutions of an instance of size n is $n! \times n!$.

9.7.2 Iterated tabu search algorithm for the Q3AP

To tackle large-sized instances of the Q3AP and speed up the search process, a parallel ILS algorithm has been designed. The local search embedded in the ILS is a TS. A TS procedure [12] starts from an initial feasible solution and tries, at each step, to move to a neighboring solution that minimizes the fitness (for a minimization case). If no such move exists, the neighbor solution that less degrades the fitness is chosen as a next move. This enables the TS process to escape local optima. However, this strategy may generate cycles, i.e., previous moves can be selected again. To avoid cycles, the TS manages a short-term memory that contains the moves that have been recently performed. A TS template is given by Algorithm 8.

Algorithm 8: tabu search template

1 GenerateInitSol(s_0);
2 TabuList=ϕ;
3 $t = 0$;
4 **repeat**
5     $m(t)$ = SelectBestMove($s(t)$);
6     $s(t+1)$ = ApplyMove($m(t), s(t)$);
7   TabuList = TabuList $\bigcup \{m(t)\}$;
8     $t = t + 1$;
9 **until** *a maximum number of iterations is reached*;

9.7.3 Neighborhood functions for the Q3AP

The neighborhood function is a crucial parameter in any local search algorithm. Indeed, if the neighborhood function is not adequate to the problem and/or does not consider the targeted computing framework, any local search algorithm will fail to reach good quality solutions of the search space.

Regarding the Q3AP, many neighborhood structures can be considered. A basic neighborhood was proposed and investigated in different works of the literature [13, 19] and consists of the set of all solutions (double-permutations) generated from the current one by an exchange of two positions in either the first (π_1) or the second (π_2) permutation. This neighborhood that we denote by N_b can be formalized as follows:

$$
\begin{aligned}
N_b(\pi_1, \pi_2) = \{ \quad & (\pi'_1, \pi'_2) : \pi'_1[k] = \pi_1[l], \pi'_1[l] = \pi_1[k] \\
& 0 \le k \ne l < n; \\
& \pi'_1[i] = \pi_1[i], \quad 0 \le i \ne k, l < n; \\
& \pi'_2[j] = \pi_2[j], \quad 0 \le j < n \\
\} & \\
\cup & \\
\{ \quad & (\pi'_1, \pi'_2) : \pi'_2[k] = \pi_2[l], \pi'_2[l] = \pi_2[k] \\
& 0 \le k \ne l < n; \\
& \pi'_2[i] = \pi_2[i], \quad 0 \le i < n; \\
& \pi'_1[j] = \pi_1[j], \quad 0 \le j \ne k, l < n \\
\} &
\end{aligned}
\quad (9.5)
$$

where (π_1, π_2) is the current solution and n is the size of the Q3AP instance. The size $|N_b|$ of such neighborhood is equal to

$$|N_b| = n \times (n-1) \quad (9.6)$$

In our GPU-based implementations, a large-sized neighborhood structure has been used for experimentation. In fact, theoretical and experimental studies have shown that the use of large neighborhood structures may improve the effectiveness of LS algorithms [3]. However, for such a neighborhood, the generation/evaluation step of an LS becomes a time-consuming task and may dramatically increase the computational time of the LS process. This justifies the use of intensive data-parallelism provided by GPUs where all neighboring solutions may be concurrently evaluated.

The considered large-sized neighborhood consists of swapping two positions in both permutations π_1 and π_2. This neighborhood structure, $N(\pi_1, \pi_2)$, can be formalized as follows:

$$N(\pi_1, \pi_2) = \{ \quad (\pi_1', \pi_2') : \pi_1'[k] = \pi_1[l], \pi_1'[l] = \pi_1[k],$$
$$\pi_2'[r] = \pi_2[s], \pi_2'[s] = \pi_2[r]$$
$$0 \leq k \neq l < n, 0 \leq r \neq s < n;$$
$$\pi_1'[i] = \pi_1[i], \quad 0 \leq i \neq k, l < n;$$
$$\pi_2'[j] = \pi_2'[j], \quad 0 \leq j \neq r, s < n$$
$$\}$$

(9.7)

So, for a given Q3AP instance of size n, the size $|N|$ of the advanced neighborhood set can be expressed by the following formula:

$$|N| = \left(\frac{n \times (n-1)}{2} \right)^2 \qquad (9.8)$$

9.7.4 Design and implementation of a GPU-based iterated tabu search algorithm for the Q3AP

The use of GPU devices to speed up the search process of local search methods is not a straightforward task. It requires one to consider, at the same time, the characteristics and underlying issues of the GPU architecture and the parallel models of LS methods. The main challenges that must be faced when designing a local search algorithm are the efficient distribution of the search process between the CPU and the GPU minimizing the data transfer between them, the hierarchical memory management and the capacity constraints of GPU memories, and the thread synchronization. All these issues must be regarded when designing parallel LS models to allow solving of large scale optimization problems on GPU architectures.

To go back to our problem (i.e., Q3AP), we propose in Algorithm 9 an iterated tabu search on GPU (GPU-ITS). The parallel model is in agreement with the iteration-level parallel model of LS methods presented in Section 9.3 (Fig. 9.1). This algorithm can be seen as a cooperative model between the CPU and the GPU. Indeed, the GPU is used as a coprocessor in a synchronous manner. The resource-consuming part, i.e., the generation and evaluation kernel, is calculated by the GPU and the rest of the LS process is handled by the CPU. First of all, at initialization stage, memory allocations on GPU are made; the input matrices (distance and flow matrices) and the initial candidate solution of the Q3AP must be allocated (lines 4 and 5). Since GPUs require massive computations with predictable memory accesses, a structure has to be allocated for storing all the neighborhood fitnesses at different addresses (line 6). Second, the matrices and the initial candidate solution have to be copied on the GPU (lines 7 and 8). We notice that the input matrices are read-only structures and do not change during all the execution of the LS algorithm. Therefore, their associated memory is copied only once during all the execution. Third, comes the parallel iteration-level, in which each neighboring solution is generated, evaluated, and copied into the neighborhood fitnesses structure (from lines 10 to 14). Fourth, since the order in which candidate neighbors are evaluated is undefined, the neighborhood fitnesses structure has to be copied to the host CPU (line 15). Then the selection strategy is applied to this structure (line 16): the exploration of the neighborhood fitnesses structure is done by the CPU. Finally, after a new candidate has been selected, this information is copied to the GPU (line 18). The process is repeated until a given number of iterations has been reached.

Algorithm 9: template of an iterated tabu search on GPU for solving the Q3AP

1 Choose an initial solution;
2 Evaluate the solution;
3 Initialize the tabu list;
4 Allocate the Q3AP input data on GPU device memory;
5 Allocate a solution on GPU device memory;
6 Allocate a neighborhood fitnesses structure on GPU device memory;
7 Copy the Q3AP input data on GPU device memory;
8 Copy the solution on GPU device memory;
9 $t = 0$;
10 **repeat**
11   **for** *each generated neighbor on GPU* **do**
12   Incremental evaluation of the candidate solution;
13   Insert the resulting fitness into the neighborhood fitnesses structure;
14   **end**
15   Copy the neighborhood fitnesses structure on CPU host memory;
16   Select the best admissible neighboring solution;
17   Update the tabu list;
18   Copy the chosen solution on GPU device memory;
19 **until** *a maximum number of iterations is reached*;

9.7.5 Experimental results

In this section, some experimental results related to the approach presented in Section 9.7.4 are reported. We recall that the approach is a GPU-based iterated tabu search (GPU-ITS) method consisting in an iterated local search (ILS) embedding a tabu search (TS) and where the generation/evaluation step of the TS process is executed on GPU. The ILS is used to improve the quality of successive local optima provided by TS methods. This is achieved by perturbing the local optima reached by the current TS process and reconsidering it as initial solution of the following TS process. Regarding our algorithm, the applied perturbation is a random number μ of swaps in either the first or the second permutation where $\mu \in [2 : n]$ (n is the instance size).

Experiments have been carried out on a node of the Chirloute cluster of the Lille site. This is one of the 10 sites that currently make up Grid5000 [1],

Instance	Optimal /BKV[1]	Average value	Maximal value	Hits	CPU time	GPU time	Speed-up	ITS iter.
Nug12-d	580	615.4 41.7	744	35%	87.7 40.9	2.5 0.9	34.7×	16 6.3
Nug13-d	1912	1985.4 51.0	2100	20%	209.2 1.3	3.3 1.0	63.5×	17 5.6
Nug15-d	2230	2418.1 135.3	2580	30%	305.5 164.5	5.2 1.3	58.8×	17 5.0
Nug18-d	17836	18110.2 157.8	18506	10%	1375.9 123.5	12.8 2.6	107.4×	19 4.2
Nug22-d	42476	43282.1 529.6	44140	25%	4506.5 341.1	32.7 6.6	137.8×	18 4.0

TABLE 9.1. Results of the GPU-based iterated tabu search for different Q3AP instances.

the French experimental computational grid. A Chirloute cluster node consists of an Intel Xeon E5620 CPU and a NVIDIA Tesla Fermi M2050 (448 cores) GPU type. The number of ILS iterations and the number of TS iterations were set respectively to 20 and 500. The tabu list size has been initalized to $\frac{m}{4}$, m being the size of the neighborhood set.

Table 9.1 reports the obtained results for the GPU-ITS using our large-sized neighborhood structure. The method has been tested on 5 Q3AP instances derived from the QAP Nugent instances in QAPLIB [6]. The average time measurement for 20 executions is reported in seconds and acceleration factors compared to a standalone CPU are also considered. The algorithm is stopped when a maximum number of iterations has been reached or when the optimal (or best known) value has been discovered. Average and max values of the evaluation function of the parallel GPU version have been measured. The number of successful tries (hits) and the average number of ILS iterations to converge to the optimal/best known value are also represented. The associated standard deviation for each average measurement is shown in small type characters.

Regarding the execution time, the generation and evaluation of the neighborhood in parallel on GPU provides an efficient way to speedup the search process in comparison with a single CPU. In fact, for the smaller instance Nug12-d, the GPU version starts to be faster than the CPU one (acceleration factor of 34.7 ×). As long as the problem size increases, the speedup grows significantly (up to 137.8× for the Nug22-d instance). This means that the use of GPU provides an efficient way to deal with large neighborhoods. Indeed, the proposed neighborhoods were unpractical in terms of single CPU computational resources for large Q3AP instances such as Nug22-d (estimated to around 2 hours per run). So, implementing this algorithm on GPU has al-

---
[1] Best known value.

lowed the exploitation of parallelism in such neighborhood and improved the robustness/quality of provided solutions.

9.8 Conclusion

This chapter has presented state-of-the-art GPU-based parallel metaheuristics and a case study on implementing large neighborhood local search methods on GPUs for solving large benchmarks of the quadratic three-dimensional assignment problem (Q3AP).

Implementing parallel metaheuristics on GPU architectures poses new issues and challenges such as memory management; finding efficient mapping strategies between tasks to parallelize; and the GPU threads, thread divergence, and synchronization. Actually, most of metaheuristics have been implemented on GPU using different implementation strategies. In this chapter, a two-level classification of the reviewed works has been proposed: design level and implementation level. Design level regroups traditional parallel models used for metaheuristics while implementation level refers to the way those models are mapped to the GPU architecture. This classification focuses mainly on the mapping between the metaheuristic tasks to parallelize and the GPU threads. Indeed, the choice of a given mapping strategy strongly influences the other challenges (memory usage, communication, thread divergence).

Bibliography

[1] Grid'5000 French nation-wide grid. https://www.grid5000.fr.

[2] NVIDIA CUDA C Programming Best Practices Guide. http://docs.nvidia.com/cuda/cuda-c-best-practices-guide/.

[3] R. K. Ahuja, K. C. Jha, J. B. Orlin, and D. Sharma. Very large-scale neighborhood search for the quadratic assignment problem. *INFORMS Journal on Computing*, 19(4):646–657, October 2007.

[4] M. G. Arenas, A. M. Mora, G. Romero, and P. A. Castillo. GPU computation in bioinspired algorithms: A review. In *Proceedings of the 11th International Conference on Artificial Neural Networks Conference on Advances in Computational Intelligence, Volume Part I*, IWANN'11, pages 433–440, 2011.

[5] C. Blum and A. Roli. Metaheuristics in combinatorial optimization: Overview and conceptual comparison. *ACM Comput. Surv.*, 35:268–308, September 2003.

[6] R. E. Burkard, E. Cela, S. Karisch, and F. Rendl. QAPLIB - A quadratic assignment problem library.

[7] S. Cahon, N. Melab, and E.-G. Talbi. ParadisEO: A framework for the reusable design of parallel and distributed metaheuristics. *Journal of Heuristics*, 10(3):357–380, 2004.

[8] A. Catala, J. Jaen, and J. A. Modioli. Strategies for accelerating ant colony optimization algorithms on graphical processing units. In *IEEE Congress on Evolutionary Computation, 2007. CEC 2007.*, pages 492–500. IEEE, 2007.

[9] J. M. Cecilia, J. M. García, A. Nisbet, M. Amos, and M. Ujaldón. Enhancing data parallelism for ant colony optimization on GPUs. *Journal of Parallel and Distributed Computing*, 73(1):42 – 51, 2013.

[10] A. Delévacq, P. Delisle, M. Gravel, and M. Krajecki. Parallel ant colony optimization on graphics processing units. *Journal of Parallel and Distributed Computing*, 73(1):52 – 61, 2013.

[11] M. R. Garey and D. S. Johnson. *Computers and Intractability: A Guide to the Theory of NP-Completeness*. W. H. Freeman & Co., New York, NY, USA, 1990.

[12] F. Glover. Tabu search. *ORSA Journal on Computing*, 1(3):190 – 206, 1989.

[13] P. M. Hahn, B. J. Kim, T. Stützle, S. Kanthak, W. L. Hightower, Z. Ding, H. Samra, and M. Guignard. The quadratic three-dimensional assignment problem: Exact and approximate solution methods. *European Journal of Operational Research*, 184:416–428, 2008.

[14] P. Hansen and N. Mladenovic. Variable neighborhood search. *Computers and Operations Research*, 24(11):1097 – 1100, 1997.

[15] A. Janiak, W. Janiak, and M. Lichtenstein. Tabu search on GPU. *Journal of Universal Computer Science*, 14(14):2416–2427, 2008.

[16] S. Kannan and R. Ganji. Porting AutoDock to CUDA. In *IEEE Congress on Evolutionary Computation*, CEC2010, 2010.

[17] S. Kirkpatrick, C. D. Gellat, and M. P. Vecchi. Optimization by simulated annealing. *Science*, 220:671 – 680, 1983.

[18] J. M. Li, X. J. Wang, R. S. He, and Z. X. Chi. An efficient fine-grained parallel genetic algorithm based on GPU-accelerated. In *Proceedings of the 2007 IFIP International Conference on Network and Parallel Computing Workshops*, NPC '07, pages 855–862, Washington, DC, USA, 2007. IEEE Computer Society.

[19] L. Loukil, M. Mehdi, N. Melab, E.-G. Talbi, and P. Bouvry. Parallel hybrid genetic algorithms for solving Q3AP on computational grid. *Int. J. Found. Comput. Sci.*, 23(2):483–500, 2012.

[20] T. V. Luong, N. Melab, and E-G. Talbi. Large neighborhood local search optimization on graphics processing units. In *Workshop on Large-Scale Parallel Processing (LSPP) in Conjunction with the International Parallel & Distributed Processing Symposium (IPDPS)*, Atlanta, USA, 2010.

[21] T. V. Luong, N. Melab, and E.-G. Talbi. GPU-based multi-start local search algorithms. In *Learning and Intelligent Optimization*, volume 6683 of *Lecture Notes in Computer Science*, pages 321–335. 2011.

[22] T. V. Luong, E. Taillard, N. Melab, and E-G. Talbi. Parallelization strategies for hybrid metaheuristics using a single GPU and multi-core resources. In *Parallel Problem Solving from Nature - PPSN XII*, volume 7492 of *Lecture Notes in Computer Science*, pages 368–377. Springer Berlin Heidelberg, 2012.

[23] O. Maitre, L. Baumes, N. Lachiche, A. Corma, and P. Collet. Coarse grain parallelization of evolutionary algorithms on GPGPU cards with EASEA. In *Proceedings of the 11th Annual Conference on Genetic and Evolutionary Computation*, GECCO '09, pages 1403–1410, 2009.

[24] N. Melab, T. V. Luong, K. Boufaras, and E.-G. Talbi. Towards ParadisEO-MO-GPU: a framework for GPU-based local search metaheuristics. In *Advances in Computational Intelligence*, volume 6691 of *Lecture Notes in Computer Science*, pages 401–408. 2011.

[25] Y. S. G. Nashed, R. Ugolotti, P. Mesejo, and S. Cagnoni. libCudaOptimize: An open source library of GPU-based metaheuristics. In *Proceedings of the Fourteenth International Conference on Genetic and Evolutionary Computation Conference Companion*, GECCO Companion '12, pages 117–124, 2012.

[26] R. Nowotniak and J. Kucharski. GPU-based massively parallel implementation of metaheuristic algorithms. *Automatyka*, 15(3):595–611, 2011.

[27] G. Paul. A GPU implementation of the simulated annealing heuristic for the quadratic assignment problem. *CoRR*, abs/1208.2675, 2012.

[28] W. P. Pierskalla. The multi-dimensional assignment problem. Technical Memorandum No. 93, Operations Research Department, CASE Institute of Technology, September 1967.

[29] F. Pinel, B. Dorronsoro, and P. Bouvry. Solving very large instances of the scheduling of independent tasks problem on the GPU. *Journal of Parallel and Distributed Computing*, 73(1):101 – 110, 2012.

[30] P. Pospichal, J. Jaros, and J. Schwarz. Parallel genetic algorithm on the CUDA architecture. In *Applications of Evolutionary Computation*, volume 6024 of *Lecture Notes in Computer Science*, pages 442–451. Springer Berlin Heidelberg, 2010.

[31] N. Soca, J. L. Blengio, M. Pedemonte, and P. Ezzatti. PUGACE, a cellular evolutionary algorithm framework on GPUs . In *IEEE Congress on Evolutionary Computation (CEC)*, 2010.

[32] T. Stutzle. Iterated local search for the quadratic assignment problem. *European Journal of Operational Research*, 174(3):1519–1539, November 2006.

[33] E. D. Taillard. FANT: Fast ant system. Technical report, 1998.

[34] E.-G. Talbi. *Metaheuristics: From Design to Implementation*. Wiley, 2009.

[35] S. Tsutsui and N. Fujimoto. Solving quadratic assignment problems by genetic algorithms with GPU computation: A Case Study. In *Proceedings of the 11th Annual Conference Companion on Genetic and Evolutionary Computation Conference: Late Breaking Papers*, GECCO '09, pages 2523–2530, 2009.

[36] S. Tsutsui and N. Fujimoto. ACO with tabu search on a GPU for solving QAPs using move-cost adjusted thread assignment. In *Proceedings of the 13th Annual Conference on Genetic and Evolutionary Computation*, GECCO '11, pages 1547–1554, 2011.

[37] P. Vidal and E. Alba. Cellular genetic algorithm on graphic processing units. In *Nature Inspired Cooperative Strategies for Optimization (NICSO 2010)*, volume 284 of *Studies in Computational Intelligence*, pages 223–232. 2010.

[38] T. T. Wong and M. L. Wong. Parallel evolutionary algorithms on consumer-level graphics processing unit. In *Parallel Evolutionary Computations*, volume 22 of *Studies in Computational Intelligence*, pages 133–155. 2006.

[39] Q. Yu, C. Chen, and Z. Pan. Parallel genetic algorithms on programmable graphics hardware. In *Advances in Natural Computation*, volume 3612 of *Lecture Notes in Computer Science*, pages 1051–1059. 2005.

[40] Y. Zhou and Y. Tan. GPU-based parallel particle swarm optimization. In *Proceedings of the Eleventh Conference on Congress on Evolutionary Computation*, CEC'09, pages 1493–1500, Piscataway, NJ, USA, 2009. IEEE Press.

[41] W. Zhu, J. Curry, and A. Marquez. SIMD tabu search for the quadratic assignment problem with graphics hardware acceleration. *International Journal of Production Research*, 48(4):1035–1047, 2010.

Chapter 10

Linear programming on a GPU: a case study

Xavier Meyer and Bastien Chopard
Department of Computer Science, University of Geneva, Switzerland

Paul Albuquerque
*Institute for Informatics and Telecommunications, HEPIA,
University of Applied Sciences of Western Switzerland – Geneva, Switzerland*

10.1	Introduction	216
10.2	Simplex algorithm	217
	10.2.1 Linear programming model	217
	10.2.2 Standard simplex algorithm	217
	10.2.3 Revised simplex method	219
	10.2.4 Heuristics and improvements	221
10.3	Branch-and-bound algorithm	224
	10.3.1 Integer linear programming	224
	10.3.2 Branch-and-bound tree	224
	10.3.3 Branching strategy	225
	10.3.4 Node selection strategy	226
	10.3.5 Cutting-plane methods	227
10.4	CUDA considerations	228
	10.4.1 Parallel reduction	228
	10.4.2 Kernel optimization	228
10.5	Implementations	229
	10.5.1 Standard simplex	229
	10.5.2 Revised simplex	232
	10.5.3 Branch-and-bound	236
10.6	Performance model	237
	10.6.1 Levels of parallelism	238
	10.6.2 Amount of work done by a thread	238
	10.6.3 Global performance model	239
	10.6.4 A kernel example: *steepest-edge*	239
	10.6.5 Standard simplex GPU implementation model	240
10.7	Measurements and analysis	241
	10.7.1 Performance model validation	241
	10.7.2 Performance comparison of implementations	241

| 10.8 | Conclusion and perspectives | 245 |
| | Bibliography | 246 |

10.1 Introduction

The simplex method [4] is a well-known optimization algorithm for solving linear programming (LP) models in the field of operations research. It is part of software often employed by businesses for finding solutions to problems such as airline scheduling problems. The original standard simplex method was proposed by Dantzig in 1947. A more efficient method, named the revised simplex, was later developed. Nowadays its sequential implementation can be found in almost all commercial LP solvers. But the always increasing complexity and size of LP problems from the industry, drives the demand for more computational power. In this context, parallelization is the natural idea to investigate [17]. Already in 1996, Thomadakis and Liu [23] implemented the standard method on a massive parallel computer and obtained an increase in performances when solving dense or large problems.

A few years back, in order to meet the demand for processing power, graphics card vendors made their graphical processing units (GPU) available for general-purpose computing. Since then GPUs have gained a lot of popularity as they offer an opportunity to accelerate algorithms having an architecture well-adapted to the GPU model. The simplex method falls into this category. Indeed, GPUs exhibit a massive parallel architecture optimized for matrix processing. To our knowledge, there are only a few simplex implementations on GPU. Bieling et al. [10] presented encouraging results while solving small to mid-sized LP problems with the revised simplex. However, the complexity of their algorithm seems to be rather close to the one of the standard simplex with similar heuristics. Following the steps of this first work, an implementation of the revised simplex [8] showed interesting results on dense and square matrices. Furthermore, an implementation of the interior point method [18] outperformed its CPU equivalent on mid-sized problems.

The Branch-and-Bound (B&B) algorithm is extensively used for solving integer linear programming (ILP) problems. This tree-based exploration method subdivides the feasible region of the relaxed LP model into successively smaller subsets to obtain bounds on the objective value. Each corresponding submodel can be solved with the simplex method, and the bounds computed determine whether further branching is required. Hence, a natural parallelization of the B&B method is to let the CPU manage the B&B tree and dispatch the relaxed submodels to an LP solver on a GPU. We refer the reader to Chapter 8, and the references therein, for a more complete introduction to parallel B&B algorithms.

In this chapter, we present a GPU implementation of the standard and revised simplex methods, based on the CUDA technology of NVIDIA. We also derive a performance model and establish its accurateness. Let us mention that there are many technicalities involved in CUDA programming, in particular regarding the management of tasks and memories on the GPU. Thus, fine tuning is indispensable to avoid a breakdown on performance.

The chapter is organized as follows. First, we begin with a description of the standard and revised simplex methods and introduce the heuristics used in our implementations. This is followed by a presentation of the B&B algorithm. The next section points out CUDA aspects which are important for our implementations. In the following one, we focus on the GPU implementations of the simplex method as well as the B&B algorithm. In the sixth section, we describe a performance model for the standard simplex implementation. In the seventh section, a performance comparison between our implementations is made on real-life problems and an analysis is given. Finally, we summarize the results obtained and consider new perspectives.

10.2 Simplex algorithm

10.2.1 Linear programming model

An LP model in its canonical form can be expressed as the following optimization problem:

$$\begin{aligned} \text{maximize} \quad & z = \mathbf{c}^\mathbf{T}\mathbf{x} \\ \text{subject to} \quad & \mathbf{A}\mathbf{x} \leq \mathbf{b} \\ & \mathbf{x} \geq \mathbf{0} \end{aligned} \quad (10.1)$$

where $\mathbf{x} = (x_j), \mathbf{c} = (c_j) \in \mathbb{R}^n, \mathbf{b} = (b_i) \in \mathbb{R}^m$, and $\mathbf{A} = (a_{ij})$ is the $m \times n$ constraints matrix. The objective function $z = \mathbf{c}^\mathbf{T}\mathbf{x}$ is the inner product of the cost vector $\mathbf{c}$ and the unknown variables $\mathbf{x}$. An element $\mathbf{x}$ is called a solution which is said to be feasible if it satisfies the m linear constraints imposed by $\mathbf{A}$ and the bound $\mathbf{b}$. The feasible region $\{\mathbf{x} \in \mathbb{R}^n \mid \mathbf{A}\mathbf{x} \leq \mathbf{b}, \mathbf{x} \geq \mathbf{0}\}$ is a convex polytope. An optimal solution to the LP problem will therefore reside on a vertex of this polytope.

10.2.2 Standard simplex algorithm

The simplex method [4] is an algorithm for solving LP models. It proceeds by iteratively visiting vertices on the boundary of the feasible region. This amounts to performing algebraic manipulations on the system of linear equations.

We begin by reformulating the model. So-called *slack variables* x_{n+i} are added to the canonical form in order to replace inequalities by equalities in equation (10.1):

$$x_{n+i} = b_i - \sum_{j=1}^{n} a_{ij}x_j \quad (i = 1, 2, ..., m) \quad (10.2)$$

The resulting problem is called the *augmented form* in which the variables are divided into two disjoint index sets, $\mathcal{B}$ and $\mathcal{N}$, which correspond to the basic and nonbasic variables. Basic variables, which form the basis of the problem, are on the left-hand side of equation (10.2), while nonbasic variables, which form the core of the equations, appear on the right-hand side. We can thus consider the following LP form:

$$\begin{aligned}\text{maximize} \quad & z = \mathbf{c}^T\mathbf{x} \\ \text{subject to} \quad & \mathbf{Ax} = \mathbf{b} \\ & \mathbf{x} \geq \mathbf{0}\end{aligned} \quad (10.3)$$

where $\mathbf{x} \in \mathbb{R}^{n+m}$ and $\mathbf{A}$ is now the $m \times (n+m)$ matrix obtained by concatenating the constraints matrix with the $m \times m$ identity matrix $\mathbf{I_m}$. The cost vector has been padded with zeros so that $\mathbf{c} \in \mathbb{R}^{n+m}$.

The basic and nonbasic variables can be separated given the expression $\mathbf{A}_\mathcal{N}\mathbf{x}_\mathcal{N} + \mathbf{x}_\mathcal{B} = \mathbf{b}$. Similarly, $z = \mathbf{c}^T\mathbf{x} = \mathbf{c}_\mathcal{N}^T\mathbf{x}_\mathcal{N} + \mathbf{c}_\mathcal{B}^T\mathbf{x}_\mathcal{B}$ with $\mathbf{c}_\mathcal{B} = \mathbf{0}$. By definition, a basic solution is obtained by assigning null values to all nonbasic variables ($\mathbf{x}_\mathcal{N} = \mathbf{0}$). Hence, $\mathbf{x} = \mathbf{x}_\mathcal{B} = \mathbf{b}$ is a basic solution.

The simplex algorithm searches for the optimal solution through an iterative process. For the sake of simplicity, we will assume here that $\mathbf{b} > \mathbf{0}$, that is, the origin belongs to the feasible region. Otherwise a preliminary treatment is required to generate a feasible initial solution (see Section 10.2.4). A typical iteration then consists of three operations (summarized in Algorithm 10).

1. **Choosing the entering variable.** The entering variable is a nonbasic variable whose increase will lead to an increase in the value of the objective function z. This variable must be selected with care so as to yield a substantial leap towards the optimal solution. The standard way of making this choice is to select the variable x_e with the largest positive coefficient $c_e = \max\{c_j > 0 \,|\, j \in \mathcal{N}\}$ in the objective function z. However, other strategies, such as choosing the positive coefficient with the smallest index, also prove to be useful.

2. **Choosing the leaving variable.** This variable is the basic variable which first violates its constraint as the entering variable x_e increases. The choice of the leaving variable must guarantee that the solution remains feasible. More precisely, setting all nonbasic variables except x_e to zero, x_e is bounded by

$$t = \min\left\{\frac{b_i}{a_{ie}} \,\middle|\, a_{ie} > 0, i = 1, \ldots, m\right\}$$

If $t = +\infty$, the LP problem is unbounded.

Otherwise, $t = \dfrac{b_\ell}{a_{\ell e}}$ for some $\ell \in \mathcal{B}$, and x_ℓ is the leaving variable; the element $a_{\ell e}$ is called the pivot.

3. **Pivoting.** Once both variables are defined, the pivoting operation switches these variables from one set to the other: the entering variable enters the basis $\mathcal{B}$, taking the place of the leaving variable which now belongs to $\mathcal{N}$, namely, $\mathcal{B} \leftarrow (\mathcal{B} \setminus \{\ell\}) \cup \{e\}$ and $\mathcal{N} \leftarrow (\mathcal{N} \setminus \{e\}) \cup \{\ell\}$. Correspondingly, the columns with index ℓ and e are exchanged between $\mathbf{I_m}$ and $\mathbf{A}_\mathcal{N}$, and similarly for $\mathbf{c}_\mathcal{B} = \mathbf{0}$ and $\mathbf{c}_\mathcal{N}$. We then update the constraints matrix $\mathbf{A}$, the bound $\mathbf{b}$ and the cost $\mathbf{c}$ using Gaussian elimination. More precisely, denoting $\tilde{\mathbf{I}}_\mathbf{m}$, $\tilde{\mathbf{A}}_\mathcal{N}$, $\tilde{\mathbf{c}}_\mathcal{B}$, and $\tilde{\mathbf{c}}_\mathcal{N}$ as the resulting elements after the exchange, Gaussian elimination then transforms the tableau

$$\begin{array}{|c|c||c|} \hline \tilde{\mathbf{A}}_\mathcal{N} & \tilde{\mathbf{I}}_\mathbf{m} & \mathbf{b} \\ \hline \tilde{\mathbf{c}}_\mathcal{N}^T & \tilde{\mathbf{c}}_\mathcal{B}^T & z \\ \hline \end{array}$$

into a tableau with updated values for $\mathbf{A}_\mathcal{N}$, $\mathbf{c}_\mathcal{N}$, and $\mathbf{b}$

$$\begin{array}{|c|c||c|} \hline \mathbf{A}_\mathcal{N} & \mathbf{I_m} & \mathbf{b} \\ \hline \mathbf{c}_\mathcal{N}^T & \mathbf{c}_\mathcal{B}^T & z - tc_e \\ \hline \end{array}$$

The latter is obtained by first dividing the ℓ-th row by $a_{\ell e}$; the resulting row multiplied by a_{ie} ($i \neq \ell$) is then subtracted to the ith row; the same operation is performed using c_e and the last row. Hence, $\mathbf{c}_\mathcal{B} = \mathbf{0}$. These operations amount to jumping from the current vertex to an adjacent vertex with objective value $z = tc_e$.

The algorithm ends when no more entering variable can be found, that is, when $\mathbf{c}_\mathcal{N} \leq \mathbf{0}$.

10.2.3 Revised simplex method

The operation that takes the most time in the standard method is the pivoting operation, and more specifically, the update of the constraints matrix $\mathbf{A}$. The revised method tries to avoid this costly operation by updating only a smaller part of this matrix.

The revised simplex method uses the same operations as in the standard method to choose the entering and leaving variables. However, since the constraints matrix $\mathbf{A}$ need not be fully updated, some additional reformulation is required.

At any stage of the algorithm, basic and nonbasic variables can be separated according to $\mathbf{Ax} = \mathbf{A}_\mathcal{N}\mathbf{x}_\mathcal{N} + \mathbf{A}_\mathcal{B}\mathbf{x}_\mathcal{B} = \mathbf{b}$. We can then transform the

Algorithm 10: standard simplex algorithm

1 //1. Find entering variable;
2 **if** $c_\mathcal{N} \leq 0$ **then**
3 | **return** Optimal;
4 **end**
5 Choose an index $e \in \mathcal{N}$ such that $c_e > 0$;
6 //2. Find leaving variable;
7 **if** $(A_\mathcal{N})_e \leq 0$ **then**
8 | **return** Unbounded
9 **end**
10 Let $\ell \in \mathcal{B}$ be the index such that ;
11 $t := \dfrac{b_\ell}{a_{\ell e}} = \min\left\{ \dfrac{b_i}{a_{ie}} \,\middle|\, a_{ie} > 0,\ i = 1, \ldots, m \right\}$;
12 //3. Pivoting, update ;
13 $\mathcal{B} := (\mathcal{B} \setminus \{\ell\}) \cup \{e\},\ \mathcal{N} := (\mathcal{N} \setminus \{e\}) \cup \{\ell\}$;
14 Compute $z_{best} := z_{best} + tc_e$;
15 Exchange $(\mathbf{I_m})_\ell$ and $(\mathbf{A}_\mathcal{N})_e$, $(\mathbf{c}_\mathcal{B})_\ell$ and $(\mathbf{c}_\mathcal{N})_e$;
16 Update $\mathbf{A}_\mathcal{N}$, $\mathbf{c}_\mathcal{N}$ and $\mathbf{b}$;
17 Go to 1.;

system $\mathbf{Ax} = \mathbf{b}$ into $\mathbf{A}_\mathcal{B} \mathbf{x}_\mathcal{B} = \mathbf{b} - \mathbf{A}_\mathcal{N} \mathbf{x}_\mathcal{N}$. Let us denote $\mathbf{B} = \mathbf{A}_\mathcal{B}, \mathbf{N} = \mathbf{A}_\mathcal{N}$. Since $\mathbf{B}$ is invertible, we can write

$$\mathbf{x}_\mathcal{B} = \mathbf{B}^{-1}\mathbf{b} - \mathbf{B}^{-1}\mathbf{N}\mathbf{x}_\mathcal{N}$$
$$z = \mathbf{c}_\mathcal{B}^T \mathbf{B}^{-1}\mathbf{b} + (\mathbf{c}_\mathcal{N}^T - \mathbf{c}_\mathcal{B}^T \mathbf{B}^{-1}\mathbf{N})\mathbf{x}_\mathcal{N}$$

The vector $\mathbf{c}_\mathcal{N}^T - \mathbf{c}_\mathcal{B}^T \mathbf{B}^{-1}\mathbf{N}$ is called the reduced cost vector.

The choice of the leaving variable can be rewritten. Setting all nonbasic variables except the entering variable x_e to zero, x_e is then bounded by

$$t = \min\left\{ \dfrac{(\mathbf{B}^{-1}\mathbf{b})_i}{(\mathbf{B}^{-1}\mathbf{N})_{ie}} \,\middle|\, (\mathbf{B}^{-1}\mathbf{N})_{ie} > 0,\ i = 1, \ldots, m \right\}$$

If $t = +\infty$, the LP problem is unbounded. Otherwise, $t = \dfrac{(\mathbf{B}^{-1}\mathbf{b})_\ell}{(\mathbf{B}^{-1}\mathbf{N})_{\ell e}}$ for some $\ell \in \mathcal{B}$, and x_ℓ is the leaving variable.

Recall that the pivoting operation begins by exchanging columns with index ℓ and e between $\mathbf{B}$ and $\mathbf{N}$ and similarly for $\mathbf{c}_\mathcal{B}$ and $\mathbf{c}_\mathcal{N}$.

To emphasize the difference between standard and revised simplex, we first express the updating for the standard simplex. This amounts to

$$[\mathbf{B}\ \ \mathbf{N}\ \ \mathbf{b}] \leftarrow [\mathbf{I_m}\ \ \mathbf{B}^{-1}\mathbf{N}\ \ \mathbf{B}^{-1}\mathbf{b}]$$
$$[\mathbf{c}_\mathcal{B}^T\ \ \mathbf{c}_\mathcal{N}^T] \leftarrow [\mathbf{0}\ \ \mathbf{c}_\mathcal{N}^T - \mathbf{c}_\mathcal{B}^T \mathbf{B}^{-1}\mathbf{N}]$$

which moves the current vertex to an adjacent vertex $\mathbf{x} = \mathbf{B}^{-1}\mathbf{b}$ with objective value $z = \mathbf{c}_{\mathcal{B}}^T \mathbf{B}^{-1}\mathbf{b}$ (the end condition remains $\mathbf{c}_{\mathcal{N}} \leq \mathbf{0}$). Many computations performed in this update phase can in fact be avoided.

The main point of the revised simplex method is that the only values which really need to be computed at each step are $\mathbf{B}^{-1}$, $\mathbf{B}^{-1}\mathbf{b}$, $\mathbf{B}^{-1}\mathbf{N}_e$, $\mathbf{c}_{\mathcal{B}}^T \mathbf{B}^{-1}$, and $\mathbf{c}_{\mathcal{B}}^T \mathbf{B}^{-1}\mathbf{b}$. However, matrix inversion is a time-consuming operation (of cubic order for Gaussian elimination). Fortunately, there are efficient ways of computing an update for $\mathbf{B}^{-1}$. One way is to take advantage of the sparsity of the LP problem by using the so-called LU decomposition[1] for sparse matrices [3]. This decomposition may be updated at a small cost at each iteration [22]. Another way is to use the *product form of the inverse* [6], which we describe hereafter.

Let $\mathbf{b}_1, \ldots, \mathbf{b}_m$ be the columns of $\mathbf{B}$, $\mathbf{v} \in \mathbb{R}^m$, and $\mathbf{a} = \mathbf{B}\mathbf{v} = \sum_{i=1}^m v_i \mathbf{b}_i$. Denote $\mathbf{B}_\mathbf{a} = (\mathbf{b}_1, \ldots, \mathbf{b}_{p-1}, \mathbf{a}, \mathbf{b}_{p+1}, \ldots, \mathbf{b}_m)$ for a given $1 \leq p \leq m$ such that $v_p \neq 0$. We want to compute $\mathbf{B}_\mathbf{a}^{-1}$. We first write

$$\mathbf{b}_p = \frac{1}{v_p}\mathbf{a} + \sum_{i \neq p} \frac{-v_i}{v_p}\mathbf{b}_i$$

Define

$$\boldsymbol{\eta} = \left(\frac{-v_1}{v_p}, \ldots, \frac{-v_{p-1}}{v_p}, \frac{1}{v_p}, \frac{-v_{p+1}}{v_p}, \ldots, \frac{-v_m}{v_p}\right)^T$$

and

$$\mathbf{E} = (\boldsymbol{\varepsilon}_1, \ldots, \boldsymbol{\varepsilon}_{p-1}, \boldsymbol{\eta}, \boldsymbol{\varepsilon}_{p+1}, \ldots, \boldsymbol{\varepsilon}_m)$$

where $\boldsymbol{\varepsilon}_j = (0, \ldots, 0, 1, 0, \ldots, 0)^T$ is the jth element of the canonical basis of $\mathbb{R}^m$. Then $\mathbf{B}_\mathbf{a} \mathbf{E} = \mathbf{B}$, so $\mathbf{B}_\mathbf{a}^{-1} = \mathbf{E}\, \mathbf{B}^{-1}$.

We apply these preliminary considerations to the simplex algorithm with $\mathbf{a} = \mathbf{N}_e, \mathbf{v} = (\mathbf{B}^{-1}\mathbf{N})_e$ (recall that x_e is the entering variable). If initially $\mathbf{B}$ is the identity matrix $\mathbf{I}_\mathbf{m}$, at the kth iteration of the algorithm, the inverse matrix is given by $\mathbf{B}^{-1} = \mathbf{E}_k \mathbf{E}_{k-1} \cdots \mathbf{E}_2 \mathbf{E}_1$, where $\mathbf{E}_i$ is the matrix constructed at the ith iteration.

10.2.4 Heuristics and improvements

Specific heuristics or methods are needed to improve the performance and stability of the simplex algorithm. We will explain below how we find an initial feasible solution and how we choose the entering and leaving variables.

Finding an initial feasible solution

Our implementations use the *two-phase simplex* [4]. The first phase aims at finding a feasible solution required by the second phase to solve the original problem. If the origin $\mathbf{x} = \mathbf{0}$ is not a feasible solution, we proceed to find

---

[1] The LU decomposition is a linear algebra decomposition which allows to write a matrix as a product of a lower and an upper triangular matrix.

such a solution by solving a so-called *auxiliary problem* with the simplex algorithm. This can be achieved by adding a nonnegative artificial variable to each constraint equation corresponding to a basic variable which violates its nonnegativity condition, before minimizing the sum of these artificial variables. The algorithm will thus try to drive all artificial variables towards zero. If it succeeds, then a basic feasible solution is available as an initial solution for the second phase in which it attempts to find an optimal solution to the original problem. Otherwise, the problem is infeasible.

To avoid having to introduce these additional auxiliary variables, we use an alternate version of this procedure. For basic variables with index in $\mathcal{I} = \{j \in \mathcal{B} \mid x_j < 0\}$, we temporarily relax the nonnegativity condition in order to have a feasible problem and then apply the simplex algorithm to minimize $w = -\sum_{j \in \mathcal{I}} x_j$. However, we update $\mathcal{I}$ at each iteration and modify accordingly the objective function, whose role is to drive these infeasible basic variables towards their original bound. If at some stage $\mathcal{I} = \emptyset$, we end up with an initial feasible solution. Otherwise, the algorithm terminates with $\mathcal{I} \neq \emptyset$ and $w > 0$, which indicates that the original problem is infeasible. The alteration introduced above involves more computations during the first phase, but it offers the advantage of preserving the problem size and making good use of the GPU processing power.

Choice of the entering variable

The number of iterations required to solve a problem depends on the method used to select the entering variable. The one described in Section 10.2.2 chooses the most promising variable x_e in terms of cost. While being inexpensive to compute, this method can lead to an important number of iterations before the best solution is found.

There exist various heuristics to select this variable x_e. One of the most commonly used is the *steepest-edge* method [15]. To improve the speed at which the best solution is found, this method takes into account the coefficients of $\mathbf{B}^{-1}\mathbf{N}$. This can be explained from the geometrical point of view. The constraints $\mathbf{Ax} = \mathbf{b}$ form the hull of a convex polytope. The simplex algorithm moves from one vertex (i.e., a solution) to another while trying to improve the objective function. The steepest-edge method searches for the edge along which the rate of improvement of the objective function is the best. The entering variable x_e is then determined by

$$e = \arg\max\left\{\frac{c_j}{\sqrt{\gamma_j}} \;\middle|\; c_j > 0, j \in \mathcal{N}\right\}$$

with $\gamma_j = \|\mathbf{B}^{-1}\mathbf{N}_j\|^2$.

This method is quite costly to compute but it reduces significantly the number of iterations required to solve a problem. This heuristic can be directly applied to the standard simplex since the full tableau is updated at each iteration. However, it defeats the purpose of the revised simplex algorithm since the aim is precisely to avoid updating the whole constraints matrix at each iteration. Taking this into account, the steepest-edge coefficients γ are updated based only on their current value. For the sake of clarity, the hat notation is used to differentiate the next iteration value of a variable from its current value: for example if γ denotes the steepest-edge coefficients at the current iteration, then $\hat{\gamma}$ are the updated coefficients.

Given the column of the entering variable $\mathbf{d} = (\mathbf{B}^{-1}\mathbf{N})_e$, we may process afresh the steepest-edge coefficient of the entering variable as $\gamma_e = \|\mathbf{d}\|^2$. The updated steepest-edge coefficients are then given by

$$\hat{\gamma}_j = \max\left\{\gamma_j - 2\hat{\alpha}_j \beta_j + \gamma_e \hat{\alpha}_j^2, 1 + \hat{\alpha}_j^2\right\} \quad \text{for } j \neq e$$

$$\hat{\gamma}_e = \gamma_e / d_e^2$$

with $\hat{\boldsymbol{\alpha}} = \mathbf{N}^{\mathbf{T}}((\hat{\mathbf{B}}^{-1})^{\mathbf{T}})_\ell$ and $\boldsymbol{\beta} = \mathbf{N}^{\mathbf{T}}(\mathbf{B}^{-1})^{\mathbf{T}}\mathbf{d}$

Choice of the leaving variable

The stability and robustness of the algorithm depend considerably on the choice of the leaving variable. With respect to this, the *expand* method [13] proves to be very useful in the sense that it helps to avoid cycles and reduces the risks of encountering numerical instabilities. This method consists of two steps of complexity $\mathcal{O}(m)$. In the first step, a small perturbation is applied to the bounds of the variables to prevent stalling of the objective value, thus avoiding cycles. These perturbed bounds are then used to determine the greatest gain on the entering variable imposed by the most constraining basic variable. The second phase uses the original bounds to define the basic variable offering the gain closest to the one of the first phase. This variable will then be selected for leaving the basis.

10.3 Branch-and-bound algorithm

10.3.1 Integer linear programming

In some problems, variables are integer-valued. For example, in a vehicle routing problem, one cannot assign one third of a vehicle to a specific route. ILP problems restrict LP problems by imposing an integrality condition on the variables. While this change may seem to have little impact on the model, the aftermaths on the resolution method are quite important.

From a geometrical perspective, the simplex algorithm is a method for finding an optimum in a convex polytope defined by an LP problem. However, the additional integrality condition results in the loss of this convexity property. The resolution method must then be altered in order to find a solution to the ILP problem. The idea is to explore the solution space by a divide-and-conquer approach in which the simplex algorithm is used to find a local optimum in subspaces.

10.3.2 Branch-and-bound tree

The solution space is explored by the B&B algorithm [1]. The strategy used is conceptually close to a tree traversal.

At first, the ILP problem is considered as an LP problem by relaxing the integrality condition before applying an LP solver to it. This initial relaxed problem represents the root of the tree about to be built. From the obtained solution $\boldsymbol{\xi}$, if ξ_j is not integral for some j, then x_j can be chosen as a branching variable. The current problem will then be divided, if possible, into two subproblems: one having $x_j \leq \lfloor \xi_j \rfloor$ and the other $x_j \geq \lceil \xi_j \rceil$. Each of these LP subproblems represents a child node waiting to be solved.

This is repeated for each subproblem in such a way that all variables are led towards integral values. At some point a subproblem will either be infeasible or lead to a feasible ILP solution (leaf nodes). The algorithm ends when the tree has been fully visited, returning the best feasible ILP solution.

During the exploration, lower and upper bounds on the objective value are computed. The upper bound is represented by the best objective value encountered in a child node. This nonadmissible solution hints towards what the objective value of the ILP problem could be. The lower bound is the best ILP solution yet found, in other words the objective value of the most promising leaf node. The bounds may be used to prune subtrees when further branching cannot improve the best current solution.

In the B&B algorithm, the main two operations that impact the convergence of the bounds towards the optimal solution are the branching strategy and the node selection strategy.

Example

Figure 10.1 illustrates the use of the B&B algorithm for solving an ILP. The ILP problem is the following:

$$\text{Maximize} \quad z = x_1 + 4x_2$$
$$\text{Subject to} \quad 5x_1 + 8x_2 \leq 40$$
$$-2x_1 + 3x_2 \leq 9$$
$$x_1, x_2 \geq 0 \quad \text{integer-valued}$$

The nodes are solved with the simplex method in the order written on the figure. At the 3rd step of the B&B, the first feasible solution is encountered (light grey leaf). The optimal solution is encountered at the 7th step (dark grey leaf). However, this is assessed only after solving the last two nodes (8th and 9th) whose objective value z is inferior to the best one yet found.

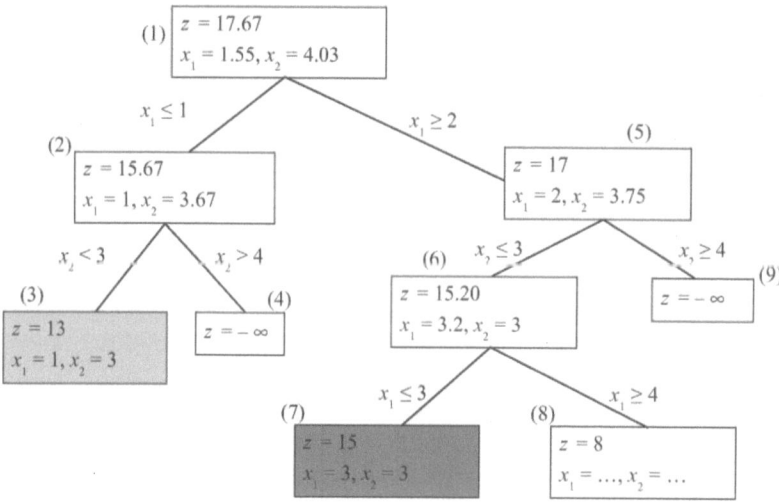

FIGURE 10.1. Solving an ILP problem using a branch-and-bound algorithm.

10.3.3 Branching strategy

The branching strategy defines the method used to select the variable on which branching will occur. The objective value of the child node depends greatly on the choice of this variable. Branching on a variable may lead to a drop on the upper bound and thus speed up the exploration, while branching on other variables could leave this bound unchanged.

Several branching strategies exist [19]. Let us briefly comment on some of them in terms of improving the objective function and processing cost.

- *Smallest-index strategy* is a greedy approach that always selects the variable with the smallest index as branching variable. This method is simple, has a cheap processing cost, but does not try to select the best variable.

- *Strong branching strategy* is an exhaustive strategy. The branching variable selected is the one among all the potential variables that leads to the best solution. This means that for each potential branching variable, its potential child nodes must be solved. This method is easy to implement, but its computational cost makes it inefficient.

- *Reliability branching strategy* is a strategy which maintains a pseudocost [12] for each potential branching variable. The pseudocost of a variable is based on the result obtained when branching on it at previous steps. Since at the start, pseudocosts are unreliable, a limited strong branching approach is used until pseudocosts are deemed reliable. This method is more complex than the two others and requires fine tuning. It offers, however, the best trade-off between improvement and computional cost.

10.3.4 Node selection strategy

The node selection strategy defines the methodology used to explore the tree. While the usual depth-first search and breadth-first search are considered and used, some remarks about the tree exploration must be made. First let us mention a few facts:

1. Solutions obtained from child nodes cannot be better than the one of their parent node.

2. An LP solver is able to quickly find a solution if the subproblem (child node) is only slightly different from its parent.

Given the above statements, it is of interest to quickly find a feasible solution. Indeed, this allows the pruning of all pending nodes which do not improve the solution found. However, the quality of the latter solution impacts the amount of nodes pruned. It takes more time to produce a good solution because one must search for the best nodes in the tree. Consequently, a trade-off must be made between the quality and the time required to find a solution.

Two types of strategies [21] can then be considered:

- *Depth-first search* aims at always selecting one of the child nodes until a leaf (infeasible subproblem or feasible solution) is reached. This strategy is characterized by fast solving, and it quickly finds a feasible solution. It mostly improves the lower bound.

- *Best-first search* aims at always selecting one of the most promising nodes in the tree. This strategy is characterized by slower solving but guarantees that the first feasible solution found is the optimal. It mostly improves the upper bound.

A problem occurs with the best-first search strategy: there might be numerous nodes having solutions of the same quality, thus making the choice of a node difficult. To avoid this problem, the *best-estimate search* strategy [14] differentiates the best nodes by estimating their potential cost with the help of a pseudocost (see previous section).

The most promising variant is a hybrid strategy in which the base strategy is the best-estimate search with the subtrees of the best-estimated nodes being then subject to a limited depth-first search, more commonly called *plunging*. This method launches a fast search for a feasible solution in the most promising subtrees, thus improving the upper and lower bounds at the same time.

10.3.5 Cutting-plane methods

Cutting-planes [25] (also simply *cuts*) are new constraints whose role is to cut off parts of the search space. They may be generated from the first LP solution (*cut-and-branch*) or periodically during the B&B (*branch-and-cut*). On the one hand cutting-planes may considerably reduce the size of the solution space, but on the other hand they increase the problem size. Moreover, generating cutting-planes is costly since it requires a thorough analysis of the current state of the problem.

Various types or families of cutting-planes exist. Those that are most applied are the *Gomory cutting-planes* and the *complemented mixed integer rounding inequalities* (c-MIR). Other kinds of cutting-planes target specific families of problems, for example, the *0-1 knapsack cutting-planes* or *flow cover cuts*.

The cutting-planes generated must be carefully selected in order to avoid a huge increase in the problem size. They are selected according to three criteria: their efficiency, their orthogonality with respect to other cutting-planes, and their parallelism with respect to the objective function. Cutting-planes having the most impact on the problem are then selected, while the others are dropped.

10.4 CUDA considerations

The most expensive operations in the simplex algorithm are linear algebra functions. The NVIDIA CUDA Basic Linear Algebra Subroutines[2] (CUBLAS) library is a GPU-accelerated version of the complete standard BLAS library. It can be used for dense matrix-vector multiplications (cublasDgemv) and dense vector sums (cublasDaxpy). The CUSPARSE library is used for the sparse matrix-vector multiplication (cusparseDcsrmv).

However, complex operations are required to implement the simplex and must be coded from scratch. For example, reduction operations, such as *argmax* or *argmin*, are fundamental for selecting variables. In the following section, we first quickly describe the CUDA reduction operation, before making some global remarks on kernel optimization.

10.4.1 Parallel reduction

A parallel reduction operation is performed in an efficient manner inside a GPU block as shown in Figure 10.2. Shared memory is used for a fast and reliable way to communicate between threads. However, at the grid level, reduction cannot be easily implemented due to the lack of direct communication between blocks. The usual way of dealing with this type of limitation is to apply the reduction in two separate steps. The first one involves a GPU kernel reducing the data over multiple blocks, the local result of each block being stored on completion. The second step finishes the reduction on a single block or on the CPU. An optimized way of doing the reduction can be found in the examples[3] provided by NVIDIA.

10.4.2 Kernel optimization

Optimizing a kernel is a difficult task. The most important point is to determine whether the performances are limited by the bandwidth or by the instruction throughput. Depending on the case and the specificities of the problem, various strategies may be applied. This part requires a good understanding of the underlying architecture and its limitations. The *CUDA Programming Guide* offers some insight into this subject, as do the interesting articles and presentations by Vassily Volkov [24]. The CUDA profiler is the best way to monitor the performances of a kernel. This tool proposes multiple performance markers giving indications about potential bottlenecks.

---

[2] The libraries CUBLAS and CUSPARSE are available at https://developer.nvidia.com/cublas and https://developer.nvidia.com/cusparse.

[3] Available at http://docs.nvidia.com/cuda/cuda-samples/index.html#advanced

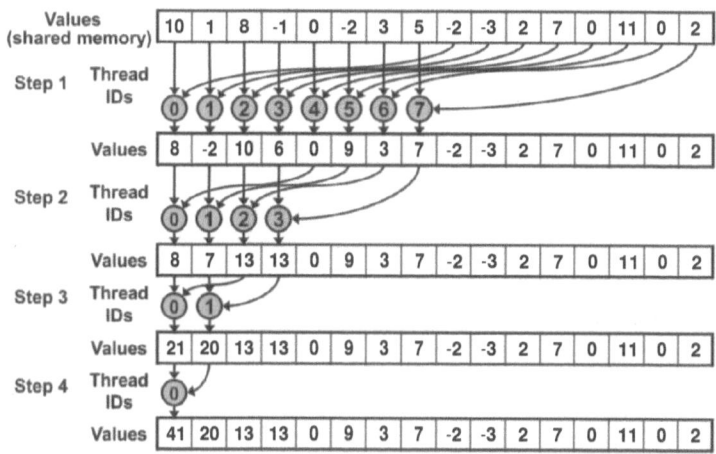

FIGURE 10.2. Example of a parallel reduction at block level. (Courtesy NVIDIA).

10.5 Implementations

In this section, the implementations of the simplex algorithm are studied. The emphasis is put on the main algorithms and kernels having a major impact on performance. The *expand* method used for choosing the leaving variable will not be detailed for that reason. Then the B&B implementation is quickly explained focusing on the interaction between the B&B algorithm and the simplex solver.

10.5.1 Standard simplex

The implementation of the standard simplex algorithm (see Section 10.2.2) is rather straightforward. The main difference with Algorithm 10 is that the basis matrix is not stored in memory since it is equal to the identity matrix most of the time. Instead a proper data structure keeps track of the basic variables and their values.

The first step of Algorithm 11 is the selection of the entering variable using the steepest-edge heuristic. This step is discussed thoroughly in the next section. Then the leaving variable index ℓ and the potential gain on the entering variable t are determined using the *expand* method. Finally, the pivoting is done. The nonbasic matrix $\mathbf{A}_\mathcal{N}$ and the objective function coefficients $\mathbf{c}$ are updated using the CUBLAS library (respectively, with cublasDger and cublasDaxpy). This requires the entering variable column $\mathbf{d}$ and the leaving variable row $\mathbf{r}$ to be copied and slightly altered. The new value of the variables

Algorithm 11: standard simplex algorithm

1 // Find entering variable x_e;
2 $\gamma_j \leftarrow \|(\mathbf{A}_\mathcal{N})_j\|^2$;
3 $e \leftarrow argmax\ (\mathbf{c}/\sqrt{\gamma})$;
4 **if** $e < 0$ **then**
5 $\quad$ **return** optima_found
6 **end**
7 // Find leaving variable x_ℓ;
8 $\ell, t \leftarrow expand((\mathbf{A}_\mathcal{N})_e)$;
9 **if** $\ell < 0$ **then**
10 $\quad$ **return** unbounded
11 **end**
12 // Pivoting;
13 $\mathbf{d} \leftarrow (\mathbf{A}_\mathcal{N})_e$, $d_e \leftarrow (A_\mathcal{N})_{\ell e} - 1$;
14 $\mathbf{r} \leftarrow \mathbf{N}^T \ell$;
15 $x_e \leftarrow x_e + t$;
16 $\mathbf{x}_\mathcal{B} \leftarrow \mathbf{x}_\mathcal{B} - t(\mathbf{A}_\mathcal{N})_e$;
17 $\mathbf{A}_\mathcal{N} \leftarrow \mathbf{A}_\mathcal{N} - \mathbf{d}^T \mathbf{r}/(A_\mathcal{N})_{\ell e}$ $\quad$ // cublasDger;
18 $\mathbf{c} \leftarrow \mathbf{c} - c_\ell \mathbf{r}/(A_\mathcal{N})_{\ell e}$ $\quad$ // cublasDaxpy;
19 $swap(x_e, x_\ell)$;

is computed and, since we use a special structure instead of the basis matrix, the entering and leaving variables are swapped.

Choice of the entering variable

Let us discuss two different approaches for the selection of the entering variable. The first one relies on the CUBLAS library. The idea is to split this step into several small operations, starting with the computation of the norms one by one with the cublasDnrm2 function. The score of each variable is then obtained by dividing the cost vector **c** by the norms previously computed. The entering variable is finally selected by taking the variable with the biggest steepest-edge coefficient using cublasDamax.

While being easy to implement, such an approach would lead to poor performances. The main problem is a misuse of data parallelism. Each column can be processed independently, and thus at the same time. Moreover, slicing this step into small operations requires that each temporary result be stored in global memory. This creates a huge amount of slow data transfers between kernels and global memory.

Listing 10.1. a kernel for the choice of the entering variable

```
extern __shared__ volatile double sData[];
__global__ void
selectInVar(int m, int n, double *c, double *AN, uint pitchAN,
            uint *resIdx, double *resVal) {
```

```
uint i, maxIdx = -1, bid = blockIdx.x;
double val, locSum, xScore, maxScore = 0.0;
while (bid < n) { // Processing multiple columns
    i = threadIdx.x;
    locSum = 0.0;
    if (isPotentialEnteringVar(bid)) { // Do the local processing
        while (i < m) { // Each thread processes multiple elements
            val = AN[i+bid*pitchAN];
            locSum += val*val;
            i += blockDim.x;
        }
        // Reduce the value using shared memory
        reduceSum(locSum);
        if (tid == 0){ // Is this the best variable encountered ?
            // on tid=0 locSum equals the steepest edge coeffcient
            xScore = cVal*rsqrt(locSum);
            if (fabs(maxScore) < fabs(xScore)) {
                maxIdx = bid;
                maxScore = xScore;
            }
        }
        __syncthreads();
    }
    bid += gridDim.x;
}
// Write the result into global memory
if (tid == 0) {
    resIdx[blockIdx.x] = maxIdx;
    resVal[blockIdx.x] = maxScore;
}
}
```

To avoid this, the whole step could be implemented as a unique kernel, as shown in the simplified Listing 10.1. Each block evaluates multiple potential entering variables. The threads of a block process part of the norm of a column. Then a reduction (see Section 10.4.1) is done inside the block to form the full norm. The thread having the final value can then compute the score of the current variable and determine if it is the best one encountered in this block. Once a block has evaluated its variables, the most promising one is stored in global memory. The final reduction step is finally done on the CPU.

The dimension of the blocks and grid have to be chosen wisely as a function of the problem size. The block size is directly correlated to the size of a column while the grid size is a trade-off between giving enough work to the scheduler in order to hide the latency and returning as few potential entering variables as possible for the final reduction step.

CPU-GPU interactions

The bulk of the data representing the problem is stored on the GPU. Only variables required for decision-making operations are updated on the CPU. The communications arising from the aforementioned scheme are illustrated in Figure 10.3. The amount of data exchanged at each iteration is independent of the problem size ensuring that this implementation scales well as the problem size increases.

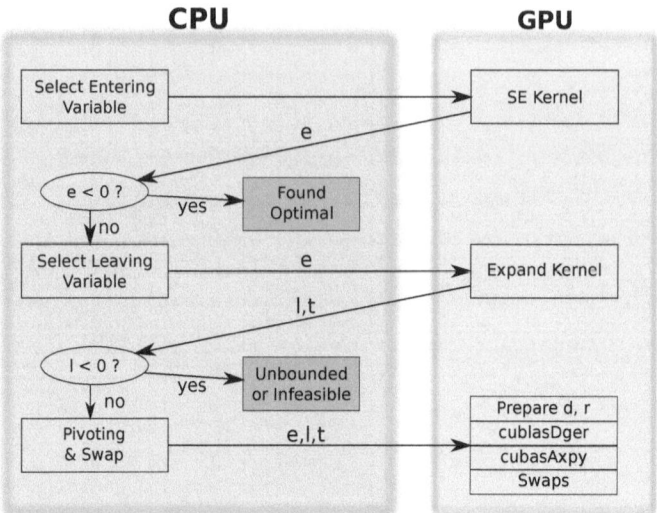

FIGURE 10.3. Communications between CPU and GPU.

10.5.2 Revised simplex

The main steps found in the previous implementation are again present in the revised simplex. The difference comes from the fact that a full update of the matrix $\mathbf{N}$ is not necessary at every iteration. As explained in Section 10.2.3, the basis matrix $\mathbf{B}$ is updated so that the required values, such as the entering variable column, can be processed from the now *constant* matrix $\mathbf{N}$. Due to these changes, the steepest-edge method is adapted. The resulting implementation requires more operations as described in Algorithm 12.

Let us compare the complexity, in terms of level 2 and level 3 BLAS operations, of both implementations. The standard one has mainly two costly steps: the selection of the entering variable and the update of the matrix $\mathbf{N}$. These are level 2 BLAS operations or the equivalent, and thus the approximate complexity is $\mathcal{O}(2mn)$.

The new implementation proposed has three operations `cublasDgemv` where the matrix $\mathbf{N}$ is involved, and three others with the matrix $\mathbf{B}^{-1}$. The complexities of these operations are respectively $\mathcal{O}(mn)$ and $\mathcal{O}(m^2)$. The update of the matrix $\mathbf{B}$ is a level 3 BLAS operation costing $\mathcal{O}(m^3)$. The approximate complexity is then $\mathcal{O}(3mn + 3m^2 + m^3)$.

It appears clearly that, in the current state, the revised simplex implementation is less efficient than the standard one. However this method can be improved and might well be better than the standard one based on two characteristics of LP problems: their high sparsity and the fact that usually $m \ll n$. In the next sections, we will give details about these improvements.

Algorithm 12: revised simplex algorithm

1 // Find entering variable x_e, $e \in \mathcal{B}$;
2 $\tau \leftarrow (\mathbf{B}^{-1})^{\mathbf{T}}\mathbf{c}_{\mathcal{B}}$ // cublasDgemv;
3 $\boldsymbol{v} \leftarrow \mathbf{c}_{\mathcal{N}} - \mathbf{N}^{\mathbf{T}}\boldsymbol{\tau}$ // cublasDgemv;
4 $e \leftarrow argmax\left(\boldsymbol{v}/\sqrt{\gamma}\right)$;
5 **if** $e < 0$ **then**
6   **return** optima_found
7 **end**
8 // Find leaving variable x_ℓ, $\ell \in \mathcal{N}$;
9 $\mathbf{d} \leftarrow \mathbf{B}^{-1}\mathbf{N}_e$ // cublasDgemv;
10 $\ell, t \leftarrow expand(\mathbf{d})$;
11 **if** $\ell < 0$ **then**
12   **return** unbounded
13 **end**
14 // Pivoting, basis update;
15 $\boldsymbol{\kappa} \leftarrow (\mathbf{B}^{-1})^{\mathbf{T}}\mathbf{d}$ // cublasDgemv;
16 $\boldsymbol{\beta} \leftarrow \mathbf{N}^{\mathbf{T}}\boldsymbol{\kappa}$ // cublasDgemv;
17 $\mathbf{B}^{-1} \leftarrow \mathbf{E}\,\mathbf{B}^{-1}$;
18 $x_e \leftarrow x_e + t$;
19 $\mathbf{x}_{\mathcal{B}} \leftarrow \mathbf{x}_{\mathcal{B}} - t\,\mathbf{d}$;
20 $swap(x_\ell, x_e)$;
21 $\hat{\boldsymbol{\alpha}} \leftarrow \mathbf{N}^{\mathbf{T}}\left((\mathbf{B}^{-1})^{\mathbf{T}}\right)_\ell$ // cublasDgemv;
22 $\gamma_e \leftarrow \|\mathbf{d}\|^2$;
23 $\hat{\gamma}_j \leftarrow \max\left\{\gamma_j - 2\hat{\alpha}_j\beta_j + \gamma_e\hat{\alpha}_j{}^2, 1 + \hat{\alpha}_j{}^2\right\}$, $\forall j \neq e$, $\hat{\gamma}_e \leftarrow \gamma_e/d_e^2$;

Choice of the entering variable

Once again, this part of the algorithm can be substantially improved. First, algorithmic optimizations must be considered. Since row $\boldsymbol{\alpha}$ of the leaving variable is processed to update the steepest-edge coefficients, the cost vector $\boldsymbol{v}$ can be updated directly without using the basis matrix $\mathbf{B}$. This is done as follow

$$v_j = \bar{v}_j - \bar{v}_e \alpha_j, \; \forall j \neq e, \qquad v_e = -\frac{\bar{v}_e}{\alpha_e}$$

where $\bar{v}_j$ denotes the value of v_j at the previous iteration. It is important to note that all these updated values $(\boldsymbol{v}, \boldsymbol{\gamma})$ must be processed afresh once in a while to reduce numerical errors.

Including this change, the operations required for the selection of the entering variable e are detailed in Algorithm 13. The values related to the entering variable at the previous iteration $\bar{e} = q$ are used. The reduced costs are updated, followed by the steepest-edge coefficients. With these values the entering variable is determined.

Algorithm 13: choice of the entering variable
---
1   $q \leftarrow \bar{e}$;
2   $\bar{v}_q \leftarrow c_q - \mathbf{c}_B^T \bar{\mathbf{d}}$;
3   $\bar{\gamma}_q \leftarrow \|\bar{\mathbf{d}}\|$;
4   // *Update of the reduced costs*;
5   $v_j \leftarrow \bar{v}_j - \alpha_j \bar{v}_q, \; \forall j \neq q$;
6   $v_q \leftarrow \dfrac{\bar{v}_q}{\bar{d}_q^2}$;
7   // *Update of the steepest edge coefficients*;
8   $\gamma_j \leftarrow \max\left\{\bar{\gamma}_j - 2\alpha_j \bar{\beta}_j + \bar{\gamma}_q \alpha_j^2, 1 + \alpha_j^2\right\}, \; \forall j \neq q$;
9   $\gamma_q \leftarrow \bar{\gamma}_q / \bar{d}_q^2$;
10 // *Selection of the entering variable*;
11 $e \leftarrow argmax\left(\boldsymbol{v}/\sqrt{\boldsymbol{\gamma}}\right)$;

Coupling these operations into a single kernel is quite beneficial. This leads $\boldsymbol{v}$ and $\boldsymbol{\gamma}$ to be loaded only once from global memory. Their updated values are stored while the entering variable e is selected. With these changes, the global complexity of this step is reduced from $\mathcal{O}(mn + m^2 + n)$ to $\mathcal{O}(n)$. Moreover the remaining processing is done optimally by reusing data and by overlapping global memory access and computations.

Basis update

The basis update $\mathbf{B}^{-1} \leftarrow \mathbf{E}\,\mathbf{B}^{-1}$ is a matrix-matrix multiplication. However, due to the special form of the matrix $\mathbf{E}$ (see Section 10.2.3), the complexity of this operation can be reduced from $\mathcal{O}(m^3)$ to $\mathcal{O}(m^2)$ [10]. The matrix

E is merely the identity matrix having the ℓth column replaced by the vector $\boldsymbol{\eta}$. The update of the matrix $\mathbf{B}^{-1}$ can be rewritten as

$$\hat{B}^{-1}_{ij} = B^{-1}_{ij}\left(1 - \frac{d_i}{d_\ell}\right), \quad \forall i \neq \ell, \qquad \hat{B}^{-1}_{\ell j} = \frac{B^{-1}_{\ell j}}{d_\ell}$$

As shown in Listing 10.2, each block of the kernel processes a single column while each thread may compute multiple elements of a column. This organization ensures that global memory accesses are coalescent since the matrix $\mathbf{B}$ is stored column-wise.

Listing 10.2. basis update

```
extern __shared__ volatile double sdata[];
__global__ void
updateBasisKernel(int m, uint l, double d_l, double *B,
                  uint pitch_B, double *d) {
    uint bId = blockIdx.x, tId = threadIdx.x;
    uint colStart = bId*pitch_B;
    double Bij, d_i, B2ij;

    // First thread loads Blj so it can be
    // broadcast via shared memory to each thread
    if (tId == 0)
        sdata[0] = B[colStart+l] / d_l;
    __syncthreads();

    // Each thread proccesses multiple elements
    while (tId < m) {
        // Load di and Bij
        d_i = d[tId];
        Bij = B[colStart+tId];
        // Update Bij
        B2ij = sdata[0];
        if (tId != q) {
            B2ij *= -d_i;
            B2ij += Bij;
        }
        __syncthreads();
        B[colStart+tId] = B2ij;

        tId += blockDim.x;
    }
}
```

Sparse matrix operations

The complexity of the implementation is now $\mathcal{O}(2mn+3m^2)$ which is close to the one of the standard simplex implementation. The operations where the matrix $\mathbf{N}$ is involved remain the more expensive (considering $m \ll n$). However, this matrix is *constant* in the revised simplex allowing the use of sparse matrix-vector multiplication from the CUSPARSE library. On sparse problems, this leads to important gains in performance. The sparse storage of the matrix $\mathbf{N}$ reduces significantly the memory space used by the algorithm. All these improvements come at a small cost: columns are no longer directly available in their dense format and must be decompressed to their dense representation when needed.

The complexity of the revised simplex implementation is thus finally $\mathcal{O}(2\theta + 3m^2)$ where θ is a function of m, n, and the sparsity of the problem. We shall see in Section 10.7.2, what kind of performances we may obtain on various problems.

10.5.3 Branch-and-bound

The B&B algorithm is operated from the CPU. The simplex implementations, also referred to as LP solver, are used to solve the nodes of the B&B tree. Algorithm 14 contains the main operations discussed in Section 10.3. It starts by solving the root node. The branching strategy is then used to select the next variable to branch on. From thereon, until no node remains to be solved, the node selection strategy is used to select the next one to be processed.

The critical factor for coupling the LP solver and the B&B is the amount of communications exchanged between them. Since CPU-GPU communications are expensive, the informations required to solve a new node must be minimized. Upon solving a new node, the full transfer of the problem must be avoided. This will be the focus of this section.

Algorithm 14: branch-and-bound

1 Solution ← Solve(InitialProblem);
2 BranchVar ← SelectNextVariable(Solution);
3 NodeList ← AddNewNodes(BranchVar);
4 **while** *NodeList* $\neq \emptyset$ **do**
5 $\quad$ Node ← SelectNextNode(NodeList);
6 $\quad$ Solution ← Solve(Node);
7 $\quad$ **if** *exists(Solution)* **then**
8 $\quad\quad$ BranchVar ← SelectNextVariable(Solution);
9 $\quad\quad$ NodeList ← AddNewNodes(BranchVar);
10 $\quad$ **end**
11 **end**

Algorithmic choices

The first critical choice is to select an efficient tree traversal strategy that also takes advantage of the locality of the problem. At first glance, the best-first search would be a poor choice, since from one node to the next the problem could change substantially. However, when combined with the plunging heuristic, this same strategy becomes really interesting. Indeed, the plunging phase can be completely decoupled from the B&B. The CPU, acting as a decision maker, chooses a promising node and spawns a task that will independently explore the subtree of this node. Once done, this task will report its findings to the decision maker. Moreover, when plunging, the next node to be solved differs by only the new bound on the branching variable. Communications are then minimized and the LP solver is usually able to quickly find the new solution since the problem has been only slightly altered.

Warmstart

There are two cases where starting from a fresh problem is required or beneficial. The first one is imposed by the numeric inaccuracy appearing after several iterations of the LP solver. The second is upon the request of a new subtree. To avoid the extensive communication costs of a full restart, the GPU keeps in memory an intermediate stable state of the problem, a *warmstart*. This state could, for example, be the one found after solving the root node of the tree.

Multi-GPU exploration

Having the CPU act as a decision maker and the GPU as an explorer, allows for the possibility of using multiple GPUs to explore the tree. The global knowledge is maintained by the CPU task. The CPU assigns to the GPUs the task of exploring subtrees of promising nodes. Since each plunging is done locally, no communications are required between the GPUs. Moreover, the amount of nodes processed during a plunging can be used to tune the load of the CPU task.

10.6 Performance model

Performance models are useful to predict the behaviour of implementations as a function of various parameters. Since the standard simplex algorithm is the easiest to understand, we will focus in this section on its behaviour as a function of the problem size.

CUDA kernels require a different kind of modeling than usually encountered in parallelism. The key idea is to capture in the model the decomposition into threads, warps, and blocks. One must also pay a particular attention to global memory accesses and to how the pipelines reduce the associated latency.

In order to model our implementation, we follow the approach given by K. Kothapalli et al. [11] (see also [9]). First, we examine the different levels of parallelism on a GPU. Then, we determine the amount of work done by a single task. By combining both analyses, we establish the kernel models. The final model then consists of the sum of each kernel.

This model has also been used to model a multi-GPUs version of the standard simplex method [2].

10.6.1 Levels of parallelism

A kernel can be decomposed into an initialization phase followed by a processing phase. During the initialization phase the kernel context is set up. Since this operation does not take a lot of time compared to the processing phase, it is needless to incorporate it into the model. The processing phase is more complex and its execution time depends directly on the amount of work it must perform. We shall now focus on this phase and on the various levels of parallelism on a GPU.

At the first level, the total work is broken down into components, the blocks. They are then dispatched on the available multiprocessors on the GPU. The execution time of a kernel depends on the number of blocks N_B per SM (streaming multiprocessor) and on the number N_{SM} of SM per GPU.

At a lower level, the work of a block is shared among the various cores of its dedicated SM. This is done by organizing the threads of the block into groups, the warps. A warp is a unit of threads that can be executed in parallel on an SM. The execution time of a block is then linked to the number N_W of warps per block, the number N_T of threads per warp, and the number N_P of cores per SM.

The third and lowest level of parallelism is a pipeline. This pipeline enables a pseudoparallel execution of the tasks forming a warp. The gain produced by this pipeline is expressed by its depth D.

10.6.2 Amount of work done by a thread

In the previous section, we defined the different levels of parallelism down to the smallest, namely the thread level. We must now estimate how much work is done by a single thread of a kernel in terms of cycles. It depends on the number and the type of instructions. In CUDA, each arithmetic instruction requires a different number of cycles. For example, a single precision add costs 4 cycles while a single precision division costs 36 cycles.

Moreover, since global memory access instructions are nonblocking operations, they must be counted separately from arithmetic instructions. The number of cycles involved in a global memory access amounts to a 4 cycle instruction (read/write) followed by a nonblocking latency of 400–600 cycles. To correctly estimate the work due to such accesses, one needs to sum only the latency that is not hidden. Two consecutive read instructions executed by a thread would cost twice the 4 cycles, but only once the latency due to its nonblocking behaviour. Once the amount of cycles involved in these memory accesses has been determined, it is then necessary to take into account the latency hidden by the scheduler (warp swapping).

The total work C_T done by a thread can be defined either as the sum or as the maximum of the memory access cycles and the arithmetic instructions cycles. Summing both types of cycles means we consider that latency cannot be used to hide arithmetic instructions. The maximum variant represents the opposite situation where arithmetic instructions and memory accesses are concurrent. Then only the biggest of the two represents the total work of a thread. This could occur for example when the latency is entirely hidden by the pipeline.

It is not trivial to define which scenario is appropriate for each kernel. There are many factors involved: the dependency on the instructions, the behaviour of the scheduler, the quantity of data to process, and so on.

If latency cannot be used to hide arithmetic instructions, the number of cycles C_T done by a thread can be defined as the sum of the global memory access and the arithmetic instructions. Otherwise, one must consider the maximum instead of the sum.

10.6.3 Global performance model

We now turn to the global performance model for a kernel. We must find out how many threads are run by a core of an SM. Recall that a kernel decomposes into blocks of threads, with each SM running N_B blocks, each block having N_W warps consisting of N_T threads. Also recall that an SM has N_P cores which execute batches of threads in a pipeline of depth D. Thus, the total work C_{core} done by a core is given by

$$C_{core} = N_B \cdot N_W \cdot N_T \cdot C_T \cdot \frac{1}{N_P \cdot D} \qquad (10.4)$$

which represents the total work done by an SM divided by the number of cores per SM and by the depth of the pipeline. Finally, the execution time of a kernel is obtained by dividing C_{core} by the core frequency.

10.6.4 A kernel example: *steepest-edge*

The selection of the entering variable is done by the *steepest-edge* method as described in Section 10.5.1.

The processing of a column is done in a single block. Each thread of a block has to compute $N_{el} = \frac{m}{N_W \cdot N_T}$ coefficients of the column. This first computation consists of N_{el} multiplications and additions. The resulting partial sum of squared variables must then be reduced on a single thread of the block. This requires $N_{red} = \log_2(N_W \cdot N_T)$ additions. Since the shared memory is used optimally, there are no added communications. Finally, the coefficient c_j must be divided by the result of the reduction.

Each block of the kernel computes $N_{col} = \frac{n}{N_B \cdot N_{SM}}$ columns, where N_{SM} is the number of SM per GPU. After processing a column, a block keeps only the minimum of its previously computed steepest-edge coefficients. Thus, the number of arithmetic instruction cycles for a given thread is given by

$$C_{Arithm} = N_{col} \cdot (N_{el} \cdot (C_{add} + C_{mult}) + N_{red} \cdot C_{add} + C_{div} + C_{cmp}) \qquad (10.5)$$

where C_{ins} denotes the number of cycles to execute instruction $ins \in \{add, div, mult, cmp\}$.

Each thread has to load N_{el} variables to compute its partial sum of squared variables. The thread computing the division also loads the coefficient c_j. This must be done for the N_{col} columns with which a block has to deal. We must also take into account that the scheduler hides some latency by swapping the warps, so the total latency $C_{latency}$ must be divided by the number of warps N_W. Thus, the number of cycles relative to memory accesses is given by

$$C_{Accesses} = \frac{N_{col} \cdot (N_{el} + 1) \cdot C_{latency}}{N_W} \qquad (10.6)$$

At the end of the execution of this kernel, each block stores in global memory its local minimum. The CPU will then have to retrieve the $N_B \cdot N_{SM}$ local minimums and reduce them. It is then profitable to minimize the number N_B of blocks per SM. With a maximum of two blocks per SM, the cost of this final operation done by the CPU can be neglected when m and n are big.

It now remains to either maximize or sum C_{Arithm} and $C_{Accesses}$ to obtain C_T. The result of Equation (10.4) divided by the core frequency yields the time $T_{KSteepestEdge}$ of the steepest-edge kernel.

10.6.5 Standard simplex GPU implementation model

As seen in Section 10.5.1, the standard simplex implementation requires only a few communications between the CPU and the GPU. Since all of these communications are constant and small, they will be neglected in the model. For the sake of simplicity, we will consider the second phase of the *two-phase simplex* where we apply iteratively the three main operations: selecting the entering variable, choosing the leaving variable, and pivoting. Each of these operations is computed as a kernel. The time of an iteration $T_{Kernels}$ then amounts to the sum of all three kernel times:

$$T_{Kernels} = T_{KSteepestEdge} + T_{KExpand} + T_{KPivoting} \qquad (10.7)$$

The times $T_{KExpand}$ and $T_{KPivoting}$ for the expand and pivoting kernels are obtained in a similar way as for the steepest-edge kernel described in the previous section.

With the estimated time per iteration $T_{Kernels}$, we can express the total time T_{prob} required for solving a problem as

$$T_{prob} = T_{init} + r \cdot T_{Kernels} \qquad (10.8)$$

where r is the number of iterations. Note that research by Dantzig [5] asserts that r is proportional to $m \log_2(n)$.

10.7 Measurements and analysis

Two sets of measurements are presented here. The first one is a validation of the performance model presented in Section 10.6. The second is a comparison of the performances of the various implementations of the simplex method detailed in Section 10.5.

As a preliminary, we ensured that our implementations were functional. For that matter, we used a set of problems available on the NETLIB Repository [7]. This dataset usually serves as a benchmark for LP solvers. It consists of a vast variety of real and specific problems for testing the stability and robustness of an implementation. For example, some of these represent real-life models of large refineries, industrial production/allocation, or fleet assignments problems. Our implementations are able to solve almost all of these problems.

10.7.1 Performance model validation

In order to check that our performance model for the standard simplex implementation is correct, a large range of problems of varying size is needed. As none of the existing datasets offered the desired diversity of problems, we used a problem generator. It is then possible to create problems of given size and density. Since usual LP problems have more variables than equations, our generated problems have a ratio of 5 variables for 1 equation ($n = 5 \cdot m$).

The test environment is composed of a CPU server (2 Intel Xeon X5570, 2.93GHz, with 24GB DDR3 RAM) and a GPU computing system (NVIDIA Tesla S1070) with the 3.2 CUDA Toolkit. This system connects 4 GPUs to the server. Each GPU has 4GB GDDR3 graphics memory and 30 streaming multiprocessors, each holding 8 cores (1.4GHz).

We validated our performance model by showing that it accurately fits our measurements. The correlation between the measurements and the model is above 0.999 (see Figure 10.4).

10.7.2 Performance comparison of implementations

Four different implementations are compared in this section. We will refer to each of them according to the terminology introduced below.

- Standard simplex method improved ($\mathcal{O}(2mn)$): *Standard*

- Revised simplex method using basis kernel

    - without improvements ($\mathcal{O}(3mn + 4m^2)$): *Revised-base*
    - optimized ($\mathcal{O}(2mn + 3m^2)$): *Revised-opti*

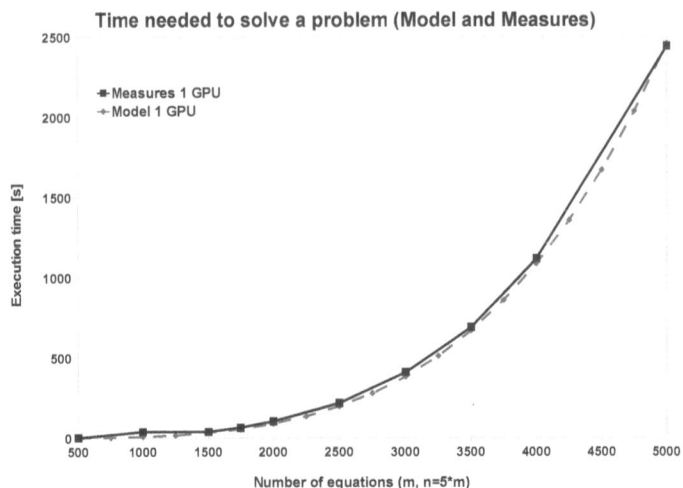

FIGURE 10.4. Performance model and measurements comparison.

- optimized with sparse matrix-vector multiplication ($\mathcal{O}(2\theta + 3m^2)$): *Revised-sparse*

These implementations all use the *steepest-edge* and *expand* methods for the choice of, respectively, the entering and the leaving variables.

We used problems from the NETLIB Repository to illustrate the improvements discussed in Section 10.5. The results expected from the first three methods are quite clear. The *Standard* method should be the best, followed by the *Revised-opti*, and then the *Revised-base*. However, the performance of the *Revised-sparse* implementation remains unclear since the value of θ is unknown and depends on the density and size of the constraints matrix. This is the main question we shall try to answer with our experiments.

The test environnement is composed of a CPU server (2 Intel Xeon X5650, 2.67GHz, with 96GB DDR3 RAM) and a GPU computing system (NVIDIA Tesla M2090) with the 4.2 CUDA Toolkit. This system connects 4 GPUs to the server. Each GPU has 6GB GDDR5 graphics memory and 512 cores (1.3GHz).

Let us begin by discussing the performances on problems solved in less than one second. The name, size, number of nonzero elements (NNZ), and columns to rows ratio (C-to-R) of each problem are reported in Table 10.1. The performances shown in Figure 10.5 corroborate our previous observations. On these problems the *Revised-sparse* method doesn't outperform the *Standard* one. This can be explained by two factors: the added communications (kernel calls) for the revised method, and the small size and density of the problems.

Problem Name	Rows	Cols	NNZ	C-to-R
etamacro.mps	401	688	2489	1.7
fffff800.mps	525	854	6235	1.6
finnis.mps	498	614	2714	1.2
gfrd-pnc.mps	617	1092	3467	1.8
grow15.mps	301	645	5665	2.1
grow22.mps	441	946	8318	2.1
scagr25.mps	472	500	2029	1.1

TABLE 10.1. NETLIB problems solved in less than 1 second.

It is likely that sparse operations on a GPU require larger amounts of data to become more efficient than their dense counterparts.

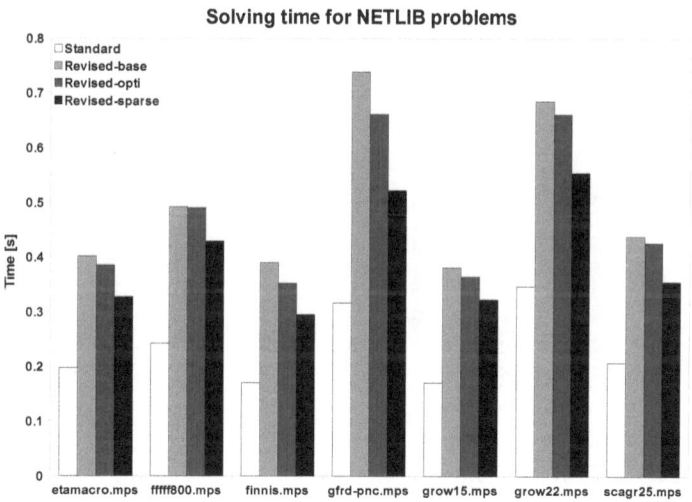

FIGURE 10.5. Time required to solve problems of Table 10.1.

The problems shown in Table 10.2 are solved in less than 4 seconds. As we can see in Figure 10.6, the expected trend for the *Revised-base* and the *Revised-opti* methods is now confirmed. Let us presently focus on the *Standard* and *Revised-sparse* methods. Some of the problems, in particular czprob.mps and nesm.mps, are solved in a comparable amount of time. The performance gain of the *Revised-sparse* is related to the C-to-R ratio of these problems, displaying, respectively, a 3.8 and a 4.4 ratio.

Finally, the biggest problems, and slowest to solve, are given in Table 10.3. A new tendency can be observed in Figure 10.7. The *Revised-sparse* method is the fastest on most of the problems. The performances are still close between

Problem Name	Rows	Cols	NNZ	C-to-R
25fv47.mps	822	1571	11127	1.9
bnl1.mps	644	1175	6129	1.8
cycle.mps	1904	2857	21322	1.5
czprob.mps	930	3523	14173	3.8
ganges.mps	1310	1681	7021	1.2
nesm.mps	663	2923	13988	4.4
maros.mps	847	1443	10006	1.7
perold.mps	626	1376	6026	1.0

TABLE 10.2. NETLIB problems solved in the range of 1 to 4 seconds.

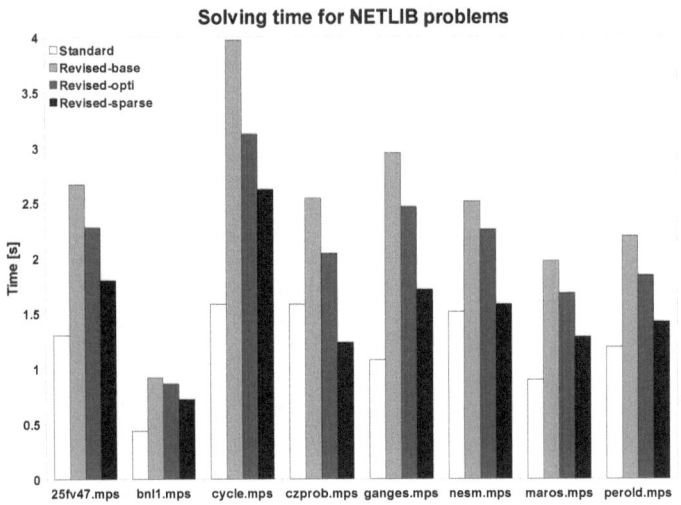

FIGURE 10.6. Time required to solve problems of Table 10.2.

the best two methods on problems having a C-to-R ratio close to 2 such as bnl2.mps, pilot.mps, or greenbeb.mps. However, when this ratio is greater, the *Revised-sparse* can be nearly twice as fast as the standard method. This is noticeable on 80bau3b.mps (4.5) and fit2p.mps (4.3). Although the C-to-R ratio of d6cube.mps (14.9) exceeds the ones previously mentioned, the *Revised-sparse* method doesn't show an impressive performance, probably due to the small amount of rows and the density of this problem, which doesn't fully benefit from the lower complexity of sparse operations.

Problem Name	Rows	Cols	NNZ	C-to-R
80bau3b.mps	2263	9799	29063	4.3
bnl2.mps	2325	3489	16124	1.5
d2q06c.mps	2172	5167	35674	2.4
d6cube.mps	416	6184	43888	14.9
fit2p.mps	3001	13525	60784	4.5
greenbeb.mps	2393	5405	31499	2.3
maros-r7.mps	3137	9408	151120	3.0
pilot.mps	1442	3652	43220	2.5
pilot87.mps	2031	4883	73804	2.4

TABLE 10.3. NETLIB problems solved in more than 5 seconds.

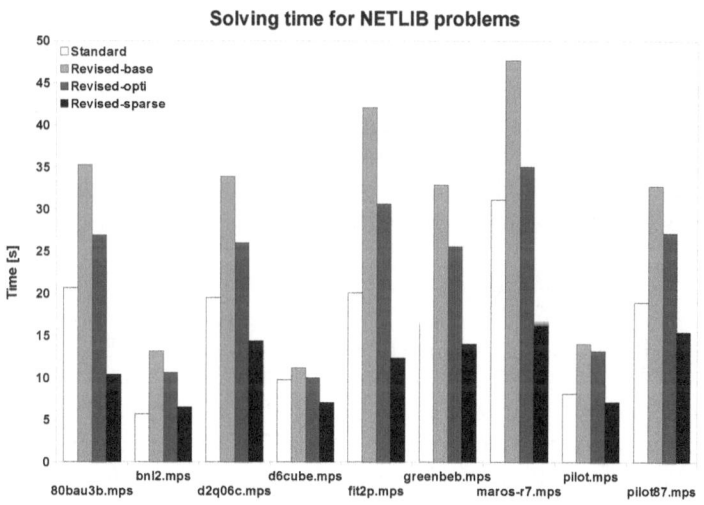

FIGURE 10.7. Time required to solve problems of Table 10.3.

10.8 Conclusion and perspectives

In this chapter, we have tried to present the knowledge and understanding necessary in our view to produce an efficient integer linear programming solver on a GPU. We proposed various solutions to implement the standard and revised simplex. We have discussed the main concepts behind a branch-and-bound algorithm and pointed out some major concerns when it is coupled with a GPU solver. The results obtained with the standard simplex implementation allowed us to validate a detailed performance model we had proposed. Finally, we used problems from the NETLIB library to compare the performances of our various implementations. The revised simplex implementation with sparse matrix operations showed the best performances on time-consuming problems, while the standard simplex one was more competitive on easier problems. However, sequential open-source solvers such as CLP of the COIN-OR project still outperform such GPU implementations.

We shall now discuss some remaining potential improvements. A first step towards performance would be to consider the dual revised simplex [20]. While being similar to the methods treated in this chapter, it has the capacity to greatly reduce the time needed to solve problems. Yet, the main improvement would be seen when tackling larger problems than the ones considered here. Indeed, problems having hundreds of thousands of variables would technically be solvable on GPU devices with 2GB to 4GB of global memory. Moreover, such amounts of data would fully benefit from the potential of these devices. However, solving problems of this size raises an issue not addressed here: numerical instabilities. This phenomenon is due to the limited precision of mathematical operations on computing devices. Let us illustrate this problem on the inverse $\mathbf{B}^{-1}$ of the basis matrix $\mathbf{B}$. At each iteration of the revised simplex, $\mathbf{B}^{-1}$ is updated from its previous values. Performing this update adds a small error to the result. Unfortunately, these errors accumulate at each iteration, bringing at some point the $\mathbf{B}^{-1}$ matrix into a degenerate state. To avoid this situation, the matrix $\mathbf{B}^{-1}$ must be recomputed afresh once in a while by inverting the matrix $\mathbf{B}$.

Instead of directly processing the inverse of the matrix $\mathbf{B}$, it is more common to use some specific arithmetical treatment. Most simplex implementations use the so-called LU decomposition of $\mathbf{B}$ as a product of a lower and an upper triangular matrix [3]. Since the matrix $\mathbf{B}$ is sparse, the sparse version of this decomposition can be used and the corresponding update performed for $\mathbf{B}^{-1}$ at each iteration [16]. The sparse LU decomposition on CPU has been recently the subject of a large amount of research, yielding many improvements. Once its GPU counterpart is available, this might result in improved and competitive simplex implementations on GPU.

Bibliography

[1] T. Achterberg. *Constraint integer programming*. PhD thesis, TU Berlin, 2007.

[2] X. Meyer, P. Albuquerque, and B. Chopard. A multi-GPU imple-

mentation and performance model for the standard simplex method. http://spc.unige.ch/lib/exe/fetch.php?media=pub:meyer_balcor2011.pdf. Presented at the 1st Int'l Symp. and 10th Balkan Conf. on Operational Research, BalcOR, Thessaloniki, Greece, Sept. 22-24, 2011.

[3] R. H. Bartels and G. H. Golub. The simplex method of linear programming using LU decomposition. *Commun. ACM*, 12(5):266–268, May 1969.

[4] V. Chvatal. *Linear Programming*. W.H. Freeman, New-York, 1983.

[5] G. B. Dantzig. Expected number of steps of the simplex method for a linear program with a convexity constraint. Technical report, Stanford University, 1980.

[6] G. B. Dantzig and W. Orchard-Hays. *The Product Form for the Inverse in the Simplex Method*. Defense Technical Information Center, Santa Monica, California, 1953.

[7] J. Dongarra and E. Grosse. The NETLIB Repository at UTK and ORNL. http://www.netlib.org.

[8] M. E. Lalami, D. El-Baz, and V. Boyer. Multi-GPU implementation of the simplex algorithm. *IEEE International Conference on High Performance Computing and Communications*, pages 179–186, 2011.

[9] D. G. Spampinato, A. C. Elster, and T. Natvig. Modeling multi-GPU systems in parallel computing: From multicores and GPU's to petascale. *Advances in Parallel Computing*, 19:562–569, 2010.

[10] J. Bieling et al. An efficient GPU implementation of the revised simplex method. In *Proc. of the 24th Int'l Parallel and Distributed Processing Symp.* IEEE, 2010.

[11] K. Kothapalli et al. A performance prediction model for the CUDA GPGPU platform. Technical report, Int'l Inst. of Information Technology, Hyderabad, 2009.

[12] M. Benichou et al. Experiments in mixed-integer linear programming. *Mathematical Programming*, 1(1):76–94, 1971.

[13] P. E. Gill et al. A pratical anti-cycling procedure for linearly constrained optimization. *Mathematical Programming*, 45:437–474, 1989.

[14] J. J. H. Forrest, J. P. H. Hirst, and J. A. Tomlin. Practical solution of large mixed integer programming problems with umpire. *Management Science*, 20(5):736–773, 1974.

[15] D. Goldfarb and J. K. Reid. A practicable steepest-edge simplex algorithm. *Mathematical Programming*, 12:361–371, 1977.

[16] R. K. Brayton, F. G. Gustavson, and R. A. Willoughby. Some results on sparse matrices. *Mathematics of Computation*, 24(112):937–954, 1970.

[17] J. A. J. Hall. Towards a practical parallelisation of the simplex method, 2007.

[18] J. H. Jung and D. P. O'Leary. Implementing an interior point method for linear programs on a CPU-GPU system. *Electronic Transactions on Numerical Analysis*, 28:174–189, 2008.

[19] T. Achterberg, T. Koch and A. Martin. Branching rules revisited. *Oper. Res. Lett.*, 33(1):42–54, January 2005.

[20] C. E. Lemke. The dual method of solving the linear programming problem. *Naval Research Logistics Quarterly*, 1(1):36–47, 1954.

[21] J. T. Linderoth and M. W. P. Savelsbergh. A computational study of search strategies for mixed integer programming. *INFORMS Journal on Computing*, 11:173–187, 1997.

[22] L. M. Suhl and U. H. Suhl. A fast LU update for linear programming. *Annals of Operations Research*, 43:33–47, 1993.

[23] M. E. Thomadakis and J.-C. Liu. An efficient steepest-edge simplex algorithm for SIMD computers, 1996.

[24] V. Volkov. Better performance at lower occupancy. http://www.eecs.berkeley.edu/~volkov. Presented at the GPU Technology Conference 2010 (GTC 2010), San José, California, USA.

[25] K. Wolter. *Implementation of cutting plane separators for mixed integer programs*. PhD thesis, TU Berlin, 2006.

Part V

Numerical applications

Chapter 11

Fast hydrodynamics on heterogeneous many-core hardware

Allan P. Engsig-Karup, Stefan L. Glimberg, Allan S. Nielsen, and Ole Lindberg

Technical University of Denmark

11.1	On hardware trends and challenges in scientific applications ...	253
11.2	On modeling paradigms for highly nonlinear and dispersive water waves	255
11.3	Governing equations	256
	11.3.1 Boundary conditions	260
11.4	The numerical model	260
	11.4.1 Strategies for efficient solution of the Laplace problem .	261
	11.4.2 Finite difference approximations	262
	11.4.3 Time integration	263
	11.4.4 Wave generation and absorption	266
	11.4.5 Preconditioned Defect Correction (PDC) method	267
	11.4.6 Distributed computations via domain decomposition ...	269
	11.4.7 Assembling the wave model from library components ..	271
11.5	Properties of the numerical model	276
	11.5.1 Dispersion and kinematic properties	276
11.6	Numerical experiments	277
	11.6.1 On precision requirements in engineering applications ..	277
	11.6.2 Acceleration via parallelism in time using parareal	282
	11.6.3 Towards real-time interactive ship simulation	286
11.7	Conclusion and future work	288
11.8	Acknowledgments	289
	Bibliography	290

In this chapter, we use our library for heterogeneous and massively parallel GPU implementations. The library is written in Compute Unified Device Architecture (CUDA) C/C++ and a fully nonlinear and dispersive free surface water wave model [18] is implemented. We describe how flexible-order finite difference (stencil) approximations to the partial differential equations of the model can be prototyped using library components provided in an in-house library. In this library hardware-specific implementation details are hidden via

251

FIGURE 11.1. Snapshot of steady state wave field generated by a Series 60 ship hull.

template-based components, as described in Chapter 6. We provide details of the modeling basis and important unique numerical properties which have been made tunable to create a powerful and robust tool that can be tailored for specific purposes in engineering analysis. The model is based on unified potential flow theory and can be applied in scientific applications related to maritime engineering. It can be applied for cost-efficient estimation of broad banded wave propagation, transformation of irregular multidirectional waves over variable depth, kinematics and structural wave loads over large areas and scales.

A main motivation of this work is to deliver exceptional performance to minimize calculation times, using modern parallel hardware technologies in combination with a proper choice of discretization methods and data-local algorithms with optimal complexity. This enable work and memory requirements to grow (scale) linearly with problem size on a suitable hardware system. For the wave model this is achieved by explicit time integration and iterative solution of a large nonsymmetric and sparse linear σ-transformed Laplace problem. For the latter, we use an iterative Preconditioned Defect Correction (PDC) method, accelerated using a geometric multigrid preconditioning strategy. We use modern programming paradigms in the form of Message Passing Interface (MPI) and CUDA for development of a novel massively parallel wave modeling tool, executable on modern heterogeneous many-core hardware.

One purpose of the developed numerical model is to ultimately perform hydrodynamic calculations in the time domain for practical analysis and simulation, e.g., to enable computationally intensive interactive real-time simulations. Realistic interactive simulations are, with present technology and available computational resources, a tremendous challenge in this setting. Yet, our aim is to take a first step in this direction and compute first-order accurate hydrodynamics for near-realistic simulations of unsteady ship-wave dynamics in a large ship simulator, used for training purposes in seakeeping operations.

For this type of application, a mandatory ingredient for real-time and interactive simulation is a truly high-performance parallel implementation to ensure data processing in time for interactive visualization and responses. Details of the model properties, implementation, and promising novel combinations of techniques and algorithms for acceleration of performance are presented. Numerical experiments and benchmarks are provided to demonstrate the accuracy and efficiency of the model across recent generations of many-core CUDA-enabled hardware.

11.1 On hardware trends and challenges in scientific applications

During the last two decades we have seen how computer graphics hardware has been developed from fixed pipeline processors with no level of programmability, to flexible high-performance hardware platforms, suitable for general purpose scientific computations other than computer graphics. This trend has contributed to a disruptive breakthrough in high-performance computing on mass-produced commodity hardware and fuelled new opportunities for computational science and engineering for a broad range of scientific as well as modern business applications. This emphasizes the increasingly important role of computers in simulation of real world dynamics [27]. In recent years, the CUDA programming model, based on the standard C/C++ programming language and introduced by NVIDIA Corporation worldwide, has become popular as a proprietary and widely used standard in high performance communities. It is, by design and supported functionality, easy and sufficient to be deployed for wide improvement of existing and new applications across science and engineering fields, that can benefit from the the use of heterogeneous hardware.

We should be careful about speculating about the future and extrapolating from current trends. The TOP 500 list[1] of supercomputers shows that there are some general noticeable hardware trends and gives indication of what to expect in the near future. First, since 2005 we have seen how power constraints and resulting heat dissipation problems forced chip producers to increase the number of cores rather than clock frequency. Multicore processors have become the new standard and many-core processors are becoming available as a standard in commodity hardware, from personal laptops to supercomputing clusters.

This trend suggests that there will be less fast low-latency memory available per core in the future, favoring data-locality in algorithms. In addition, we have also seen how communication speed to computation speed ratio de-

---

[1] http://www.top500.org.

creases, making it increasingly difficult to supply data to hungry floating point units. In addition, there will likely be increasing amounts of data to store as a result of increasing processing capabilities. The rapidly increasing floating point performance following Moore's law for transistor production has resulted in a significant memory gap which leaves most scientific applications based on partial differential equations (PDEs) bandwidth bound rather than compute bound. This trend is driven by pure commercial needs and not the needs of high-performance computing. Roads to better performance include standardization of software infrastructure, rethinking algorithms to better exploit memory hierarchies optimally to boost strong scaling properties, increasing locality in algorithms, and introducing as much concurrency and work as possible to both utilize and exploit the many cores. Also, software that can utilize many cores should be fault-tolerant to maximize time to solution for application users. We should also expect to see multiple layers of parallelism that will have to be exploited and possibly autotuned to optimally utilize available hardware resources. This introduces new challenges in compilers, requires programming experts with hardware knowledge, and introduces new trends in software developments to leverage productivity and utilize available computational resources in more optimal ways. We have observed a fundamental paradigm shift of underlying hardware design towards much more heterogeneity and parallelism.

A key problem is that improvements in performance require porting legacy codes[2] to new hardware, and possibly changing algorithms which have been developed for the conventional single core processors decades ago. Without this, it may be impossible to utilize and scale algorithmic work optimally to achieve high performance on modern and emerging hardware. This problem is currently addressed with rapid progress by researchers and industry by development of new optimized libraries that can utilize such new hardware. While we have seen significant improvements in such efforts, and today see much more rapid development of applications, there are still relatively few scientific applications running entirely on heterogeneous hardware.

In this work, we explore some of these trends by developing, by bottom-up-design, a water-wave model which can be utilized in maritime engineering and with the intended use on affordable office desktops as well as on more expensive modern compute clusters for engineering analysis purposes.

---

[2]In the worst case, a legacy code is an undocumented serial code developed a long time ago by a developer no longer around.

11.2 On modeling paradigms for highly nonlinear and dispersive water waves

We see the development of new or improved hardware technologies as a key driver for exploring new and revisiting existing approaches that can contribute to next-generation modeling techniques.

For instance, the dominant wave modeling paradigm today for numerical simulation in coastal engineering tools is the use of Boussinesq-type formulations for approximate solution of unified potential flow equations over varying bathymetry [39]. The use of Boussinesq-type models in engineering tools was pioneered in 1978 by Abott et. al. [1, 2] based on the original Boussinesq equations due to Peregrine [41] for calculations of waves in a harbor area. New formulations for highly nonlinear and dispersive water waves, useful for description of wave propagation in the important application range from deep to shallow areas, have been the subject of intense research for more than two decades. Such higher order formulations can be derived by first introducing an infinite Mclaurin series solution to the Laplace equation as described in [3]. This technique was later generalized to arbitrary expansion levels [37]. By analytical truncation of such series solutions, a polynomial variation in the vertical is assumed and provides the basis for efficient higher order Boussinesq-type formulations [6, 38] for fully nonlinear and dispersive water waves. It is attractive, since it is then possible to eliminate the vertical coordinate in the analytical formulation of the Laplace problem. The resulting approximate model contains higher order derivatives to describe dispersion and these require careful treatment in numerical models. Thus, this truncation procedure inherently limits the practical application range; however, it can be significantly improved via Padé approximations together with the introduction of free parameters for extending the finite application range by mathematical optimization to enhance accuracy in dispersion, kinematics, and shoaling characteristics.

Main challenges of Boussinesq-type models are accurate and large-scale simulation of waves propagating towards near-shore from deep to shallow waters through surf zones, while accounting for high-order dispersive and nonlinear effects [8]. Within the last two decades, much research has focused on extending the application range through improved formulations in terms of dispersion, shoaling, kinematic and nonlinear properties. The ultimate high-order Boussinesq-type model due to [38] was at the time considered a breakthrough in this direction, and since then promising new formulations have been proposed. For example, the methodology behind Boussinesq-type formulations can be extended via a multilayer approach [9, 35, 36] that makes it possible to achieve a similar range of application and levels of accuracy, but without higher derivatives in the formulation that can cause numerical difficulties.

Boussinesq-type formulations for free surface waves are conventionally evaluated against the unified potential flow theory to evaluate limits to application range and accuracy limits. The use of unified potential theory as a basis for numerical models has traditionally been perceived as too expensive [32] to solve in comparison with the typically more efficient Boussinesq-type models. This may be true in a strict comparison between the models, especially with respect to applications towards the more restricted shallow regions. However, this is in spite of the fact that a numerical unified potential flow model can be used for a larger range of practical scientific applications. A unified potential flow model has at most second-order derivatives in the formulation. In a numerical setting it has good opportunities for balancing accuracy and work effort by appropriate tuning of discrete parameters. This comes without a need for changing the underlying wave model to extend application range towards deep waters. Thus, the main problem related to the practical use of a unified model in maritime applications is arguably an issue of numerical efficiency.

To address this issue, we have recently proposed a new approach in a proof-of-concept that combines modern many-core hardware with appropriate numerical and parallel strategies to facilitate efficient, accurate, and scalable solution of water wave problems [18]. The use of potential theory for unsteady water wave computations can be traced at least back to 1975 [25], and the fully nonlinear potential equations have been solved using various numerical methods since then, e.g., see reviews [12,32,50,52]. In the context of the finite-difference method, an efficient and scalable second-order geometric multigrid approach was first proposed by Li and Fleming in 1997 [31]. Since then, the numerical strategy has been significantly improved in several works [7, 15] that have led to more efficient and robust discretization techniques, with the objective of developing a general purpose strategy, that can be used for a broad range of practical maritime applications. Recently, a comparison with a High-Order Spectral (HOS) model [13] was also reported to assess accuracy and relative differences in efficiency on single-core hardware against a superior spectral modeling basis for a numerical wave tank setup in a structured domain with a flat sea bed.

11.3 Governing equations

We describe how, by physical principles via mathematical procedures and assumptions, it is possible to formulate a fully nonlinear and dispersive water wave model, describing incompressible, irrotational, and inviscid fluid flow above an uneven seabed.

Conservation of mass for an infinitely small control volume can be stated

as
$$\frac{\partial \rho}{\partial t} + \nabla \cdot (\rho \mathbf{u}) = 0, \tag{11.1}$$

where ρ is fluid density, $\mathbf{u} = (u, v, w)$ is the velocity field vector, and $\nabla = (\partial/\partial x, \partial/\partial y, \partial/\partial z)$ is a Cartesian gradient operator. If we assume that fluid density is constant, i.e., that the fluid is incompressible, the mass continuity equation simplifies to

$$\nabla \cdot \mathbf{u} = 0, \tag{11.2}$$

which shows that the divergence of the velocity field is zero everywhere.

Conservation of momentum for an infinitely small control volume is expressed by the famous Navier-Stokes equations

$$\rho \frac{D\mathbf{u}}{Dt} = -\nabla p + \mu \nabla^2 \mathbf{u} + \mathbf{F}, \tag{11.3}$$

where p is pressure and $\mathbf{F}$ is the net force vector acting on the fluid volume, assumed to be of the form $\mathbf{F} = \rho \mathbf{g}$ with $\mathbf{g} = (0, 0, -g_z)$ accounting for gravitational effects in the vertical direction. This implies that surface tension effects on the free surface are neglected. Exact solutions to the Navier-Stokes equations are in general difficult to obtain, and this motivates the use of numerical methods for direct simulation of fluid flow and if necessary analytical simplifications to account only for physics of interest.

The material derivative for a comoving coordinate system used in Lagrangian formulations

$$\frac{D}{Dt} \equiv \frac{\partial}{\partial t} + (\mathbf{u} \cdot \nabla) \tag{11.4}$$

is defined as the sum of a time derivative and a convective term measured in a static (Eularian) coordinate system and accounts for the time rate of change following the motion. Thus, for the velocity vector $\mathbf{u}$, the total acceleration is defined as

$$\frac{D\mathbf{u}}{Dt} \equiv \frac{\partial \mathbf{u}}{\partial t} + \frac{1}{2}\nabla \mathbf{u}^2 + \boldsymbol{\omega} \times \mathbf{u}, \tag{11.5}$$

where the curl of the velocity field $\boldsymbol{\omega} \equiv \nabla \times \mathbf{u}$ is referred to as the vorticity vector field accounting for rotation of fluid particles. If we assume that the flow is irrotational

$$\nabla \times \mathbf{u} = 0 \tag{11.6}$$

and make use of the following relationship known from vector calculus

$$\frac{1}{2}\nabla(\mathbf{u} \cdot \mathbf{u}) = (\mathbf{u} \cdot \nabla)\mathbf{u} + \mathbf{u} \times (\nabla \times \mathbf{u}), \tag{11.7}$$

we can rewrite (11.5) as

$$\frac{D\mathbf{u}}{Dt} = \frac{\partial \mathbf{u}}{\partial t} + \frac{1}{2}\nabla(\mathbf{u} \cdot \mathbf{u}). \tag{11.8}$$

Since the Navier-Stokes equations (11.3) can be derived by application of Newton's second law to an infinitely small fluid volume, we can establish the following. Changes in momentum in an infinitely small control volume of a fluid is simply the sum of forces due to pressure gradients, dissipative forces, gravitational forces, and possibly other forces acting inside the fluid volume.

To accurately simulate propagation of long gravity waves and high Reynolds number flows in the context of maritime applications it is for many applications acceptable to assume that viscous forces are small in comparison with inertial forces. In this case, it is reasonable to assume that the fluid is inviscid and we can neglect the viscous terms in (11.3). Then we obtain the set of Euler equations

$$\frac{D\mathbf{u}}{Dt} = -\frac{1}{\rho}\nabla p + \frac{1}{\rho}\mathbf{F}, \tag{11.9}$$

which does not account for any losses in energy via dissipative physical mechanisms.

If we introduce a scalar velocity potential function ϕ that relates to the velocity components in the following way

$$\mathbf{u} \equiv \nabla \phi = \left(\frac{\partial \phi}{\partial x}, \frac{\partial \phi}{\partial y}, \frac{\partial \phi}{\partial z} \right), \tag{11.10}$$

the number of unknowns can be lowered and we find that the scalar velocity potential function due to (11.2) must satisfy the Laplace equation

$$\nabla^2 \phi = 0. \tag{11.11}$$

Solutions to this equation are completely determined by the boundary conditions. Thus, it is possible, given appropriate boundary conditions, to determine the scalar velocity potential ϕ in all of the domain by solving the resulting Laplace problem. When the scalar velocity potential function is known, detailed information of the kinematics can immediately be obtained.

With the definition of the vector field (11.10), we can collect the momentum equations (11.8) and (11.9) and express the momentum equation as

$$\rho \left(\frac{\partial \nabla \phi}{\partial t} + \nabla \left(\frac{1}{2} \nabla \phi \cdot \nabla \phi \right) \right) = -\nabla p - \nabla(\rho g z), \tag{11.12}$$

having assumed that the net force can be decomposed into pressure and gravity forces only. This set of equations can be rewritten as

$$\nabla \left[\rho \frac{\partial \phi}{\partial t} + \rho \frac{1}{2} \nabla \phi \cdot \nabla \phi + p + \rho g z \right] = 0, \tag{11.13}$$

and by integration in space we arrive at the unsteady Bernoulli's equation

$$\rho\frac{\partial \phi}{\partial t} + \rho\frac{1}{2}\nabla\phi \cdot \nabla\phi + p + \rho g z = G(t), \qquad (11.14)$$

where $G(t)$ is an arbitrary function of the integration that can be assumed to be zero as it defines only a reference value for the unphysical scalar velocity potential function. Bernoulli's equation is typically used as a dynamic condition for the free fluid surface, expressing that the fluid pressure at the free surface is equal to the pressure in the air above the free surface.

In the following, we assume that the displacement of the free surface $z = \eta(\mathbf{x}, t)$ is described in a Cartesian coordinate system with the xy-plane located at the still water level and the positive z-axis pointing upwards. It is typical to assume a constant atmospheric pressure at the free surface by defining $p = 0$ as reference. This leaves us with a dynamic boundary condition for the free surface velocity potential function stated as

$$\frac{\partial \phi}{\partial t} = -\frac{1}{2}\nabla\phi \cdot \nabla\phi - gz, \quad z = \eta. \qquad (11.15)$$

At the free surface, we can determine a kinematic free surface condition by determining the rate of change of the streamline for the surface as

$$\frac{\partial \eta}{\partial t} = -\frac{\partial \phi}{\partial x}\frac{\partial \eta}{\partial x} - \frac{\partial \phi}{\partial y}\frac{\partial \eta}{\partial y} + \frac{\partial \phi}{\partial z}, \quad z = \eta. \qquad (11.16)$$

Spatial and temporal differentiations of the free surface variables are related by the chain rule

$$\nabla\tilde{\phi} = (\nabla\phi)_{z=\eta} + \left(\frac{\partial \phi}{\partial z}\right)_{z=\eta}\nabla\eta, \qquad (11.17)$$

$$\frac{\partial \tilde{\phi}}{\partial t} = \left(\frac{\partial \phi}{\partial t}\right)_{z=\eta} + \frac{\partial \eta}{\partial t}\left(\frac{\partial \phi}{\partial z}\right)_{z=\eta}, \qquad (11.18)$$

and can be used to transform the free surface problem to variables defined solely at the free surface as

$$\frac{\partial}{\partial t}\eta = -\nabla\eta \cdot \nabla\tilde{\phi} + \tilde{w}(1 + \nabla\eta \cdot \nabla\eta), \qquad (11.19a)$$

$$\frac{\partial}{\partial t}\tilde{\phi} = -g\eta - \frac{1}{2}\left(\nabla\tilde{\phi} \cdot \nabla\tilde{\phi} - \tilde{w}^2(1 + \nabla\eta \cdot \nabla\eta)\right), \qquad (11.19b)$$

with the ∇-operator from here conveniently re-defined as a horizontal gradient operator $\nabla = (\partial_x, \partial_y)$ and tilde's used for free surface variables. To solve the set of unsteady free surface equations (11.19), we need a closure between the horizontal and vertical free surface velocity variables. This can be established by solving the Laplace equation (11.11) in the interior domain together with suitable boundary conditions.

A kinematic bottom condition can be derived by assuming that the fluid particles follow a streamline along the solid sea bed. Consider the rate of change of such a streamline at still-water depth $z = -h(\mathbf{x}, t)$ and we find

$$\frac{\partial z}{\partial t} = -\frac{\partial h}{\partial x}\frac{\partial \phi}{\partial x} - \frac{\partial h}{\partial y}\frac{\partial \phi}{\partial y} - \frac{\partial h}{\partial t}, \quad z = -h(\mathbf{x}, t). \quad (11.20)$$

We assume that the sea bed is static allowing us to neglect the last term. Thus, specifying $\tilde{\phi}$ as a Dirichlet condition at the free surface together with a kinematic bottom boundary condition at the sea bed defines a Laplace problem

$$\phi = \tilde{\phi}, \quad z = \eta(\mathbf{x}, t), \quad (11.21\text{a})$$
$$\nabla^2 \phi + \partial_{zz}\phi = 0, \quad -h \leq z < \eta(\mathbf{x}, t), \quad (11.21\text{b})$$
$$\partial_z \phi + \nabla h \cdot \nabla \phi = 0, \quad z = -h, \quad (11.21\text{c})$$

where we have used $\partial_t z \equiv \partial_z \phi$ to rewrite the first term of the kinematic bottom condition (11.20). The moving free surface makes the spatial fluid domain Ω vary in time. The main challenges in solving these equations numerically are dealing with the time-dependent fluid domain and nonlinearity of the equations.

11.3.1 Boundary conditions

We consider three types of boundaries, namely, fully reflective boundaries, incident wave boundaries, and absorbing boundaries. The fully reflective boundaries are handled through numerical approximations of the boundary conditions for solid walls and bottom surfaces stating that the velocity in the normal direction is zero

$$\mathbf{n} \cdot \nabla \phi = 0, \quad \mathbf{x} \in \partial\Omega, \quad (11.22)$$

where $\mathbf{n} = (n_x, n_y)$ is a two-dimensional normal vector pointing outwards from the solid surface. A complementary condition for the free surface elevation variable is

$$\mathbf{n} \cdot \nabla \eta = 0. \quad (11.23)$$

Incident wave and absorbing boundary conditions are imposed via an embedded penalty forcing technique as described in Section 11.4.4.

11.4 The numerical model

The unified potential flow model is attractive as a basis due to the underlying analytical properties. From a numerical point of view, an efficient and scalable discretization strategy should be based on using a data-local method, e.g., a flexible-order finite difference method for discretely approximating the governing equations and imposing boundary conditions via fictitious ghost points techniques as described in [7, 15]. Such an approach has several attractive features from a scientific computing perspective. For example, finite difference methods are among the simplest and most efficient methods due to the use of structured grids and data structures. This results in low implementation and computational complexity which maps efficiently to modern computer architectures. Formal accuracy and tunable numerics are achieved by employing flexible-order finite difference (local stencil) approximations.

We present scalability and performance tests based on the same two test environments outlined in Chapter 6, Section 6.1.1, plus a fourth test environment based on the most recent hardware generation:

Test environment 4. Desktop with dual-socket Sandy Bridge Intel Xeon E5-2670 (2.60 GHz) processors, 64GB RAM, 2x NVIDIA Tesla K20 GPUs.

Performance results can be used to predict actual runtimes as described in [18], e.g., for estimation of whether a real-time constraint for a given problem size can be met.

11.4.1 Strategies for efficient solution of the Laplace problem

As explained in Section 11.2, for the formulation of potential flow problems there are two widely used paradigms for solving the Laplace problem efficiently. The most widely used approach is the Boussinesq-type where essentially the three-dimensional formulation is reduced to a two-dimensional formulation. The main argument for this type of model reduction procedure is the resulting efficiency in the numerical models. The price paid is typically high-order derivatives in the approximate formulation and is justified by the efficient solution of an approximate Laplace problem.

A second approach is to transform the partial differential equation mathematically to provide a basis for an efficient direct solution of the discrete Laplace problem for the entire volume. This strategy is based on a paradigm where approximations are done by discrete approximations rather than analytical manipulations of the equation. At a first look, this approach introduces more complexity in the formulation, e.g., by the introduction of mixed derivatives, however, essentially does not limit the application range beyond the numerical approximations and properties hereof. Using this second approach, it is standard to introduce a σ-transformation in the vertical coordinate

$$\sigma \equiv \frac{z + h(\boldsymbol{x})}{d(\boldsymbol{x}, t)}, \quad 0 \leq \sigma \leq 1, \tag{11.24}$$

where $d(\boldsymbol{x}, t) = \eta(\boldsymbol{x}, t) + h(\boldsymbol{x})$ is the height of the water column above the bottom. This enables a transformation of the problem to a time-constant computational domain at the expense of time-varying coefficients.

The fluid domain is mapped to a time-invariant computational domain in which the Laplace problem (11.21) is expressed as

$$\Phi = \tilde{\phi}, \quad \sigma = 1, \tag{11.25a}$$

$$\nabla^2 \Phi + \nabla^2 \sigma (\partial_\sigma \Phi) + 2\nabla \sigma \cdot \nabla (\partial_\sigma \Phi) +$$
$$(\nabla \sigma \cdot \nabla \sigma + (\partial_z \sigma)^2) \partial_{\sigma\sigma} \Phi = 0, \quad 0 \leq \sigma < 1, \tag{11.25b}$$
$$\mathbf{n} \cdot (\nabla, \partial_z \sigma \partial_\sigma) \Phi = 0, \quad (\mathbf{x}, \sigma) \in \partial\Omega, \tag{11.25c}$$

where the scalar velocity function $\Phi(\mathbf{x}, \sigma, t) = \phi(\mathbf{x}, z, t)$ contains all information about the flow kinematics in the entire fluid volume. The spatial derivatives of the coordinate σ appearing in (11.25) are expressed as

$$\nabla \sigma = \tfrac{1-\sigma}{d} \nabla h - \tfrac{\sigma}{d} \nabla \eta, \tag{11.26a}$$

$$\nabla^2 \sigma = \tfrac{1-\sigma}{d}\left(\nabla^2 h - \tfrac{\nabla h \cdot \nabla h}{d}\right) - \tfrac{\sigma}{d}\left(\nabla^2 \eta - \tfrac{\nabla \eta \cdot \nabla \eta}{d}\right)$$
$$- \tfrac{1-2\sigma}{d^2} \nabla h \cdot \nabla \eta - \tfrac{\nabla \sigma}{d} \cdot (\nabla h + \nabla \eta), \tag{11.26b}$$

$$\partial_z \sigma = \tfrac{1}{d}. \tag{11.26c}$$

All of these coefficients can be computed explicitly from the known two-dimensional free surface and bottom positions at given instances of time.

The velocity field can be determined from a known Φ using the relation

$$(\mathbf{u}, w) = (\nabla, \partial_z \sigma \partial_\sigma) \Phi. \tag{11.27}$$

The flow can be computed from the scalar velocity potential and used for estimating nonhydrostatic pressure and resulting wave loads. An exact expression for local pressure as a function of the vertical coordinate can be found by vertical integration of the vertical momentum equation to be of the form

$$p(z) = \rho g(\eta - z) + \int_z^\eta \partial_t w \, dz + \frac{1}{2}(\tilde{u}^2 - u(z)^2 + \tilde{v}^2 - v(z)^2 + \tilde{w}^2 - w(z)^2). \tag{11.28}$$

The integral term can be estimated numerically using an accurate quadrature rule. Estimation of structural forces is determined by integration of pressure

$$\mathbf{F} = -\iint_S p \mathbf{n} \, dS, \tag{11.29}$$

where S is a structural surface.

11.4.2 Finite difference approximations

The numerical scheme is implemented as a flexible-order finite difference collocation scheme where all finite difference approximations of derivatives are constructed from one-dimensional approximations in a standard way, each having the maximum possible accuracy. In explicit numerical schemes, finite difference approximations can be implemented using a matrix-free technique to exploit that only a few different stencils are in fact needed. This can significantly reduce memory requirements of the implemented model by exploiting that the same small set of stencils can be reused. See Chapter 6 for more details about matrix-free stencil operations supported in our in-house library for heterogeneous and massively parallel computing using GPUs.

11.4.3 Time integration

For users of scientific applications, robustness is of paramount importance for the solution of time-dependent PDEs. This makes stability considerations relevant in the context of both explicit and iterative numerical methods often considered most suitable for massively parallel applications. In the following, we address aspects of explicit time integration schemes which are associated with a stability requirement on time steps.

A method of lines approach is used for the discretization of the wave model. The spatial discretization yields a system of ordinary differential equations which can be expressed as a semidiscrete system. We use the classical fourth-order Explicit Runge-Kutta Method (ERK4). This algorithm is suitable for massive parallel computations via a data-parallel implementation where the spatial discretization terms are processed. As a means to introduce more concurrency into the time integration, we consider the Parareal algorithm as described in Section 11.6.2. For explicit time-integration schemes a Courant-Levy-Friedrichs (CFL) condition defines a necessary restriction for temporal stability of the form

$$\Delta t \leq \frac{C}{\max_n |\lambda_n(\mathcal{J}_h)|}, \qquad (11.30)$$

with $C \in \mathbb{R}_+$ a CFL constant typically of size $\mathcal{O}(1)$ dependent on chosen scheme, and $\mathcal{J}_h \in \mathbb{R}^{2m \times 2m}$, where $m = N_x N_y$ is a discrete Jacobian matrix obtained by local linearization in time of (11.19). For ERK4, $C = 2\sqrt{2}$ if all eigenvalues are purely imaginary.

To gain insight into necessary conditions for stability, we employ a linear stability analysis based on the semi-discrete linear system

$$\frac{d}{dt}\begin{bmatrix} \eta \\ \tilde{\phi} \end{bmatrix} = \mathcal{J} \begin{bmatrix} \eta \\ \tilde{\phi} \end{bmatrix}, \quad \mathcal{J} = \begin{bmatrix} 0 & \mathcal{J}_{12} \\ -g & 0 \end{bmatrix}, \qquad (11.31)$$

where $\mathcal{J}_{12} = k \tanh(kh)$ for the continuous problem which results in purely imaginary eigenvalues of the Jacobian $\lambda(\mathcal{J}) = \pm i\sqrt{gk \tanh(kh)}$ that grow unbounded in the limit $kh \to \infty$ corresponding to deep water relative to wave length. The waves are described in terms of wave number $k = 2\pi/L$ where L is the wave length.

For any numerical method, a numerical discretization of the same system of equations should mimic the continuous eigenspectrum in well-resolved parts of the spectrum. For a given discretization method, the stability of the discrete problem (11.31) is governed by the eigenvalues of the discrete Neumann-Dirichlet operator $\mathcal{J}_{12,h} \in \mathbb{R}^{m \times m}$, which connects the vertical velocity at the free surface with that of the velocity potential at the free surface. It is possible to partition the discrete σ-transformed Laplace problem and the differential in the z-direction in terms of equations for the interior (i) and those at the free surface boundary (b) such that

$$\mathcal{A}\phi = \begin{bmatrix} \mathcal{A}_{bi} & \mathcal{A}_{bb} \\ \mathcal{A}_{ii} & \mathcal{A}_{ib} \end{bmatrix} \begin{bmatrix} \tilde{\phi} \\ \phi_i \end{bmatrix}, \quad \mathcal{D}\phi = \begin{bmatrix} \mathcal{D}_{bi} & \mathcal{D}_{bb} \\ \mathcal{D}_{ii} & \mathcal{D}_{ib} \end{bmatrix} \begin{bmatrix} \tilde{\phi} \\ \phi_i \end{bmatrix}. \quad (11.32)$$

From these equations we find for the discrete block Jacobian operator

$$\left.\frac{\partial \phi}{\partial z}\right|_{z=\eta} = \mathcal{J}_{12,h}\tilde{\phi}, \quad \mathcal{J}_{12,h} = \mathcal{D}_{bb} - \mathcal{D}_{bi}\mathcal{A}_{ii}^{-1}\mathcal{A}_{ib}. \quad (11.33)$$

As shown in [44] the eigenvalues of $\mathcal{J}_{12,h}$ are related to the eigenvalues of the discrete Jacobian $\mathcal{J}_h$ through the following relationship

$$\lambda(\mathcal{J}_h) = \pm i\sqrt{\lambda(\mathcal{J}_{12,h})g}. \quad (11.34)$$

and are all imaginary confirming the hyperbolic (energy-conserving) nature of the potential flow formulation. Thus, for a given discretization of the linearized equations, it is possible to compute the eigenvalues of the discrete block operator to determine the eigenspectrum of the full operator. A discrete analysis of the eigenvalues is given in [15, Section 4.1], but it is not clearly pointed out that in fact the discrete eigenspectrum is compact (bounded) for a fixed polynomial order in the vertical, i.e., that for a constant depth h

$$\max|\lambda(\mathcal{J}_h)| = \lim_{kh \to \infty} |\lambda(\mathcal{J}_h)| \leq C(N_z)\sqrt{\frac{g}{h}}. \quad (11.35)$$

Similar results were reported for the first time in the context of high-order Boussinesq-type equations in [14, 16] and recently it has been shown [19] that widely used implicitly-implicit Boussinesq-type equations can be reformulated to have bounded eigenspectra using high-order discretisation methods. This is an important practical property of the discrete scheme as it is favorable to numerical stability. It implies that the linear model is not severely limited by the spatial resolution in the horizontal for a specific choice of the number of collocation nodes (N_z) in the vertical. This suggests that the model is quite robust due to insensitivity in the choice of time step, with the implication that local grid adaptivity can be used for improving spatial accuracy. Interestingly, for the unified potential flow model we find that this also holds for nonlinear simulations. Large time steps can be chosen when using dense grids and high-order numerics without severely degrading overall numerical stability and efficiency. This is confirmed in numerical experiments and demonstrated in Figure 11.2. However, for very steep nonlinear waves and very densely clustered nonuniform grids, stability is found to be compromised without filtering. A proper filtering strategy, e.g., based on a super collocation technique [28], can be used to remedy stability problems without destroying accuracy.

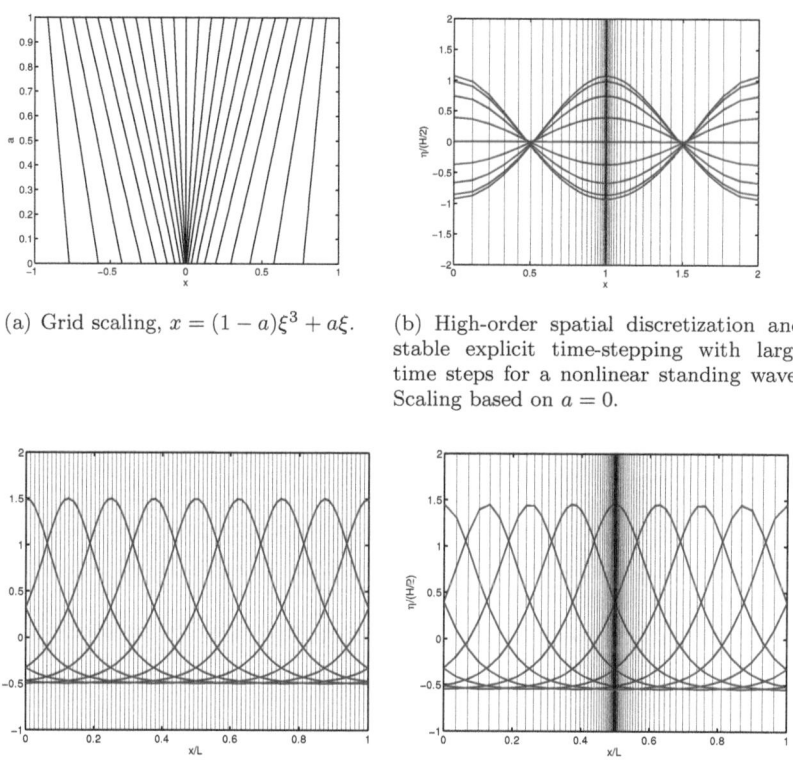

(a) Grid scaling, $x = (1-a)\xi^3 + a\xi$.

(b) High-order spatial discretization and stable explicit time-stepping with large time steps for a nonlinear standing wave. Scaling based on $a = 0$.

(c) Uniform grid ($a = 1$).

(d) Clustered grid ($a = 0.05$).

FIGURE 11.2. Numerical experiments to assess stability properties of numerical wave model. In three cases, computed snapshots are taken of the wave elevation over one wave period of time. In (a) the grid distribution of nodes in a one-parameter mapping for the grid is illustrated. Results from changes in wave elevation are illustrated for (b) a mildly nonlinear standing wave on a highly clustered grid, (c) a regular stream function wave of medium steepness in shallow water $(kh, H/L) = (0.5, 0.0292)$ on a uniform grid ($N_x = 80$), and (d) a nonuniform grid with a minimal grid spacing 20 times smaller(!). In every case the step size remains fixed at $\Delta t = T/160$ s corresponding to a Courant number $C_r = c\frac{\Delta t}{\Delta x} = 0.5$ for the uniform grid. A sixth order scheme and explicit EKR4 time-stepping are used in each test case.

11.4.4 Wave generation and absorption

To simulate waves using a numerical model, a general purpose technique for both generating and absorbing waves inside the finite numerical domain is needed. It is preferable that the technique is suitable for easy integration in a software library component setup. One such technique is the line relaxation method attributed to [29]. This is a simple ad hoc technique sufficiently accurate for engineering purposes. It modifies the computed solution every time step during simulation by a postprocessing step where the relaxed solution becomes

$$g^*(x_i,t) = \Gamma(x_i)g(x_i,t) + (1-\Gamma(x_i))g_e(x_i,t), \quad x_i \in \Omega_\Gamma. \quad (11.36)$$

Here $g(x,t)$ is one of the free surface variables $\tilde{\phi}, \eta$ at an instant in time, and $0 \leq \Gamma(x) \leq 1$ is a single-valued function within the relaxation region $x_i \in \Omega_\Gamma$. The first term acts as a sponge layer which is responsible for effectively dissipating energy inside a specified relaxation zone Ω_Γ. The terms containing $g_e(x,t)$, where g_e is an analytical solution (e.g., such as the stream function wave theory [10]), act as source terms in the relaxation zone. This makes it possible to generate arbitrary waves accurately in the computational domain in accordance with an analytical representation of incident waves.

We can interpret (11.36) as a discrete update of the solution at an isolated spatial point inside a relaxation zone. We introduce the notations $g^n = g(x_i,t_n)$ and $g^{*,n+1} = g^*(x_i,t_{n+1})$ and assume that $t_{n+1} = t_n + \tau$ is an instant in time. Then we can rewrite (11.36) to motivate an analytical modification to time-dependent equations that can provide similar modification (forcing) in simulation.

Subtracting g^n in (11.36) and dividing by a pseudo time step size τ, we obtain the equivalent form

$$\frac{g^{*,n+1} - g^n}{\tau} = \frac{(1-\Gamma)}{\tau}(g_e^n - g^n). \quad (11.37)$$

The first term is similar to a first-order accurate Forward Euler approximation of a rate of change term. This motivates an *embedded penalty forcing technique* based on adding a correction term of the form

$$\partial_t g = \mathcal{N}(g) + \frac{1-\Gamma(x)}{\tau}(g_e(t,x) - g(t,x)), \quad \mathbf{x} \in \Omega_\Gamma, \quad (11.38)$$

where $\mathcal{N}$ represents a general nonlinear operator for the right-hand side. The immediate advantage is that a time-stepping scheme can easily be interchanged in a model implementation. The added term is a source term resulting in forcing inside relaxation zones when $g_e(t,x) \neq g(t,x)$ and $\Gamma(x) \neq 1$ and otherwise has no effect. The strength of the forcing is influenced by the arbitrary parameter $\tau \in \mathbb{R}_+$ which can be defined to match the time scale of the dynamics. We have found that $\tau \approx \Delta t$ works well; however, it is possible that a more optimal choice exist. Note that a too small τ may degrade the numerical stability of the model.

A simple validation of the zones is shown in Figure 11.3 where waves are generated at the left wall and propagate to the right wall, where reflection occurs leading to formation of standing waves due to the resulting interaction with the incident waves inside the numerical wave tank.

The following relaxation functions proposed in [14] guarantee continuity across the interface of the relaxation zone and computational domain and are used in simulations for sponge layers and wave generation zones, respectively

$$\Gamma_s(\hat{x}) = 1 - (1-\hat{x})^p, \quad \Gamma_g(\hat{x}) = -2\hat{x}^3 + 3\hat{x}^2, \quad \hat{x} \in [0,1] \tag{11.39}$$

where $\hat{x} \equiv x/x_L$ is coordinate normalised by the length x_L of the zone in question.

The profiles can be reversed by a change of coordinate, i.e., $\Gamma(1-\hat{x})$, and scaled to interval sizes of interest. The first function satisfies the condition that any derivative at the left boundary vanishes at $\hat{x} = 1$. The first derivatives of the second function vanish at both ends. The relaxation zones are positioned appropriately where waves are to be both/either generated and/or absorbed. A practical rule of thumb is that a relaxation used for absorption has a spatial length of at least two wave lengths. For absorption zones, we find that this technique is more efficient in velocity formulation of the free surface equations often used in Boussinesq-type formulations, e.g., see [14,16,17] in comparison with scalar potential formulations (11.19). However, similar performance can be achieved by merely increasing the length of relaxation zones in such regions. Demonstrations of the technique are soon in Figure 11.3 where vertical dashed lines indicate interfaces between relaxation zones and the computational region. Incident waves propagate from left to right in both examples.

11.4.5 Preconditioned Defect Correction (PDC) method

For the solution of sparse linear systems

$$\mathcal{A}\boldsymbol{\Phi} = \mathbf{b}, \quad \mathcal{A} \in \mathbb{R}^{n \times n}, \quad \mathbf{b} \in \mathbb{R}^n, \tag{11.40}$$

it is attractive to use iterative methods for large system sizes $n = N_x N_y N_z$ and for parallel implementations. Acceleration of suitable iterative methods can be done, e.g., by solving a left-preconditioned system of the form

$$\mathcal{M}^{-1}(\mathcal{A}\boldsymbol{\Phi} = \mathbf{b}), \quad \mathcal{M} \in \mathbb{R}^{n \times n}, \tag{11.41}$$

where $\mathcal{M}$ is a preconditioner with the property that $\mathcal{M}^{-1} \approx \mathcal{A}^{-1}$ can be computed at low cost.

The bottleneck problem in a unified potential flow model is the solution of a discrete σ-transformed Laplace problem stated in the compact forms

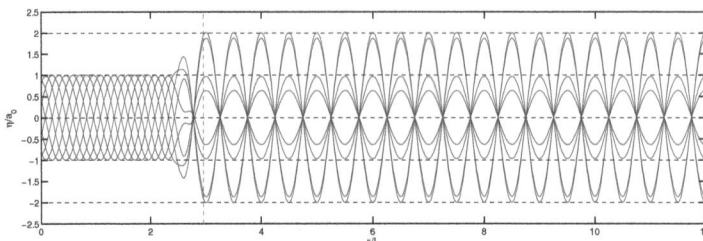

(a) Wave generation, reflection and absorption of small-amplitude waves.

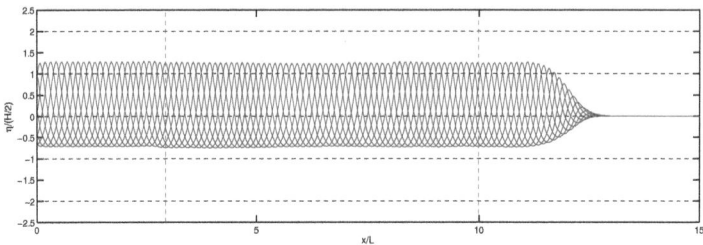

(b) Wave generation and absorption of steep finite-amplitude waves.

FIGURE 11.3. Snapshots at intervals $T/8$ over one wave period in time of computed (a) small-amplitude $(kh, kH) = (0.63, 0.005)$ and (b) finite-amplitude $(kh, kH) = (1, 0.41)$ stream function waves elevations having reached a steady state after transient startup. Combined wave generation and absorption zones occur in the left relaxation zone of both (a) and (b). In (b) an absorption zone is positioned next to the right boundary and causes minor visible reflections.

(11.40) or (11.41). It is attractive to find an efficient iterative strategy where convergence is understood via a convergence theory, has modest storage requirements, has minimal global communication requirements (in the form of global inner products), and is suitable for flexible-order discretizations. The class of geometric Multigrid methods fulfills these requirements and has been shown to be among the most efficient iterative strategies for a wide class of problems [49]. In particular, the time required to solve a system of linear equations to a given accuracy level can be made to scale proportional to the number of unknowns.

There are several known approaches to multigrid methods [45] for high-order discretizations. Among these, Defect Correction Methods (DCMs) [4,5,24,46,49] have been employed successfully, e.g., in computational fluid dynamics [30], in numerical simulations since the early 1970s. The fundamental idea of DCMs is to combine the good stability properties of low-order discretizations with high-order accuracy discretizations for explicit residual evaluations. These iterative methods impose low storage requirements, have scalable work effort under suitable choices of preconditioning strategies, and may be accelerated using a multigrid method based on low-order discretizations while still achieving high-order accuracy. Furthermore, it has been shown, that the rate of convergence of DCM combined with standard multigrid methods can achieve rates of convergence corresponding to the most efficient multigrid methods [4].

Therefore, for the efficient and scalable solution of the unified potential flow model, we have recently [18] proposed a Preconditioned Defect Correction (PDC) method for efficient iterative low-storage solution of high-order accurate discretization of the σ-transformed Laplace problem (11.25). The proposed strategy can be seen as a generalization of the multigrid strategy proposed by [31]. The PDC method enables significant improvement of overall efficiency and accuracy with the preconditioning based on a second-order linearized version of the full coefficient matrix $\mathcal{A}$ as described in [15].

Starting from some initial guess $\Phi^{[0]} \in \mathbb{R}^n$, the PDC method for solving (11.41) can be stated compactly as a three-step recurrence for $k = 1, 2, \ldots$

$$\Phi^{[k]} = \Phi^{[k-1]} + \delta^{[k-1]}, \quad \delta^{[k-1]} = \mathcal{M}^{-1}\mathbf{r}^{[k-1]}, \quad \mathbf{r}^{[k-1]} = \mathbf{b} - \mathcal{A}\Phi^{[k-1]}, \quad (11.42)$$

where $\Phi^{[k]}, \delta^{[k]}, \mathbf{r}^{[k]} \in \mathbb{R}^n$ are the approximate solution, the defect (preconditioned residual), and the residual of (11.41) at the kth iteration, respectively. The algorithm can be speeded up by using mixed-precision calculations on modern many-core hardware as demonstrated in [23].

11.4.6 Distributed computations via domain decomposition

Numerical modeling of large ocean areas to account for nonlinear wave-wave interactions and wave-structure interactions requires large degrees of spatial resolution, significant computational resources, and parallel computations to be practical. The recent generations of programmable GPUs are heavily optimized for on-chip bandwidth performance but not capacity. This implies that for the solution of large-scale PDE problems, distributed computing on multiple GPU devices is required due to limited capacity in the global memory space of current GPUs. Via a combination of MPI and CUDA we have recently demonstrated how both small and large systems can be solved efficiently by heterogeneous computations using a data domain decomposition technique in parallel [22]. The idea is to distribute the computational tasks to multiple GPUs, to enable reduced computational times and increased problem sizes. A homogenous partitioning of the data is used to ensure that the load balance across the GPUs is uniform. Data distribution and message passing introduce a data transfer bottleneck in the form of the Peripheral Component Interconnect express (PCIe) link and network interconnection. Thus, if the computational intensity of the local problem is not large enough to enable sufficient latency hiding of this bottleneck, the whole application is likely to be (severely) limited by the PCIe link or network bandwidth performance rather than the high on-chip bandwidth of the individual GPUs.

The ratio between necessary data transfers and computational work for the proposed numerical model for free surface water waves is high enough to expect reasonable latency hiding. The data domain decomposition method consists of a logically structured division of the computational domain into multiple subdomains. Each of these subdomains is connected via fictitious ghost layers at the artificial boundaries of width corresponding to the half-width of the finite difference stencils employed. This results in a favorable volume-to-boundary ratio as the problem size increases, and diminishing communication overhead for message passing. Information between subdomains is exchanged through ghost layers at every step of the iterative PDC method, in connection with the matrix-vector evaluation for the σ-transformed Laplace problem, and before relaxation steps in the multigrid method. A single global synchronization point occurs at most once each iteration, if convergence is monitored, where a global reduction step (inner product) between all processor nodes takes place. The main advantage of this decomposition strategy is that the decomposition into multiple subdomains is straightforward. However, it comes with the cost of extra data transfers to update the set of fictitious ghost layers.

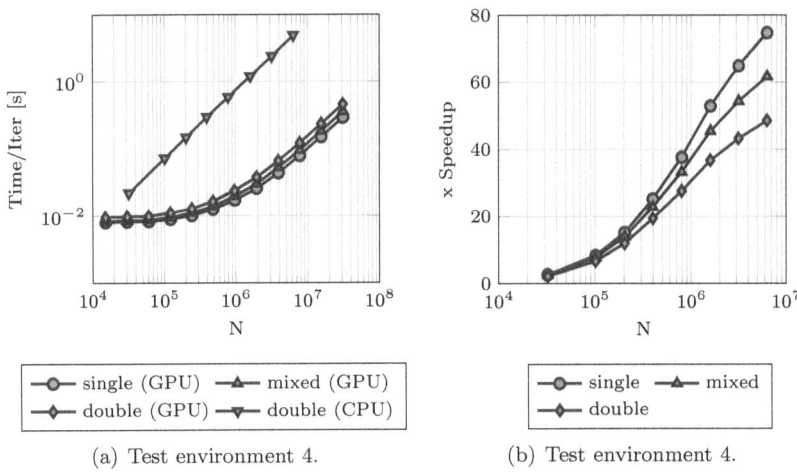

(a) Test environment 4.   (b) Test environment 4.

FIGURE 11.4. Performance timings per PDC iteration as a function of increasing problem size N, for single, mixed, and double precision arithmetics. Three-dimensional nonlinear waves, using sixth order finite difference approximations, preconditioned with one multigrid V-cycle and with one pre- and post- red-black Gauss-Seidel smoothing operation. Speedup compared to fastest known serial implementation. Using test environment 4, CPU timings represent starting points for our investigations and have been obtained using the Fortran 90 code. These references results are based on a single-core (nonparallel) run on a Intel Core i7, 2.80GHz processor.

The parallel domain decomposition solver has been validated against the sequential solvers with respect to algorithmic efficiency to establish that the code produces correct results. An analysis of the numerical efficiency has also been carried out on different GPU systems to identify comparative behaviors as both the problems sizes and number of compute nodes vary. For example, performance scalings on Test environment 1 and Test environment 3 are presented in Figure 11.5. The figure confirms that there is only a limited benefit from using multiple GPUs for small problem sizes, since the computational intensity is simply too low to efficiently hide the latency of message passing. A substantial speedup is achieved compared to the single GPU version, while being able to solve even larger systems. With the linear scaling of memory requirements and improved computational speed, the methodology based on multiple GPUs makes it possible to simulate water waves in very large numerical wave tanks with improved performance.

Future work involves extending the domain decomposition method to include support for more general curvilinear boundary-fitted meshes for representing the free surface plane and to include bottom-mounted structures which are typically encountered in near-costal areas. This will enable opportunities to resolve wave-structure interactions such as those encountered in large harbor regions and isolated human-made structures in offshore regions, e.g., legs of oil rigs or offshore windmill foundations.

11.4.7 Assembling the wave model from library components

It is described in Chapter 6 how we have developed a heterogeneous library has for fast prototyping of PDE solvers, utilizing the massively parallel architecture of CUDA-enabled GPUs. The objective is to provide a set of generic components within a single framework, such that software developers can assemble application-specific solvers efficiently at a high abstraction level, requiring a minimum of CUDA specific kernel implementations and parameter tuning.

The CUDA-based numerical wave model has been developed based on all the numerical techniques described in preceding sections. These techniques are a part of the library implemented as generic components, which makes them useful for the numerical solutions of PDE systems. The components of the numeral model as described in Section 11.4 include an ERK4 time integrator, flexible-order finite difference approximations on regular grids, and an iterative multigrid PDC solver for the σ-transformed Laplace equation (11.25). Application developers either either can use these components directly from the library or are free to combine their own implementations with library components, if they need alternative strategies that are not present in the library.

For the unified potential flow model the user will need to provide implementations of the following components: the right-hand side operator for the semidiscrete free surface variables (11.19), the matrix-vector operator for the discretized σ-transformed Laplace equation (11.25), a smoother for the multigrid relaxation step, and the potential flow solver itself, that reads initial data

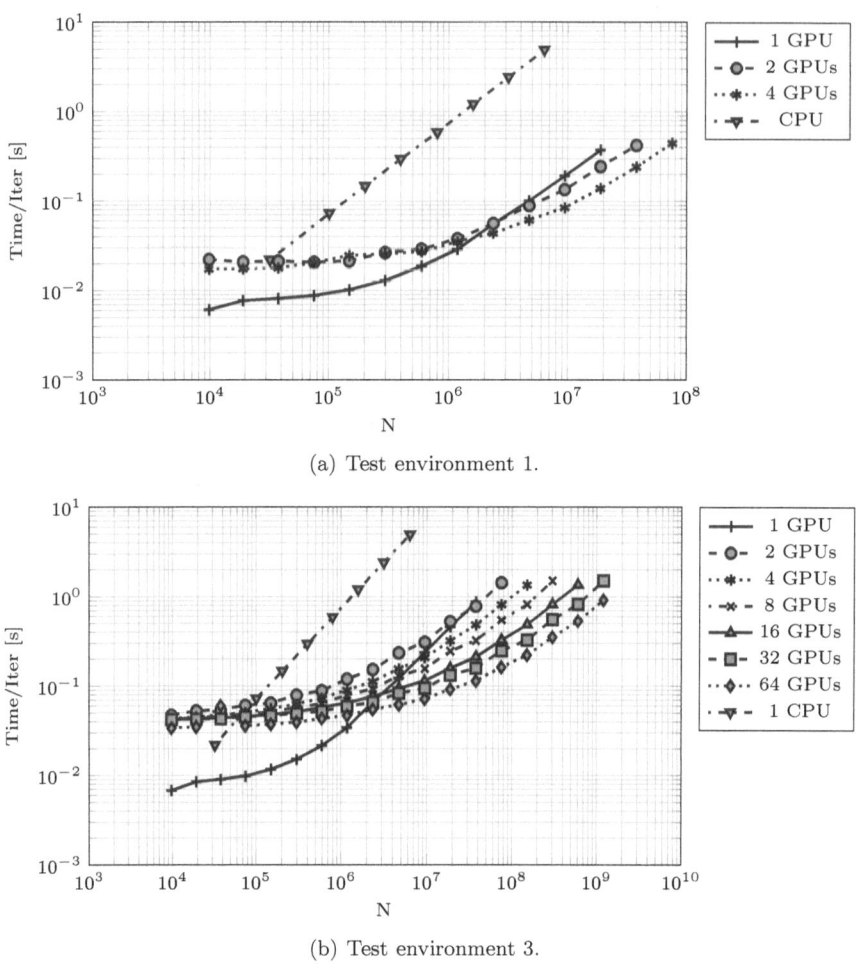

FIGURE 11.5. Domain decomposition performance on multi-GPU systems. In single precision, performance timings per PDC iteration are a function of increasing problem sizes. Same setup as in Figure 11.4.

and advances the solution in time. In order to make the library as generic as possible, all components are template-based, which makes it possible to assemble the PDE solver by combining type definitions in the preamble of the application. An excerpt of the potential flow assembling is given in Listing 11.1.

Fast hydrodynamics on heterogeneous many-core hardware

Listing 11.1. generic assembling of the potential flow solver for fully nonlinear free surface water waves

```
// Basics
typedef double value_type;
typedef gpulab::grid<value_type> vector_type;
typedef free_surface::laplace_sigma_stencil_3d<vector_type>
    matrix_type;

// Multigrid setup
typedef gpulab::solvers::multigrid_types<
    vector_type
    , matrix_type
    , free_surface::gs_rb_low_order_3d
    , gpulab::solvers::grid_handler_3d_boundary> multigrid_types;
typedef gpulab::solvers::multigrid<multigrid_types>
    preconditioner_type;

// Defect Correction setup
typedef gpulab::solvers::defect_correction_types<
    vector_type
    , matrix_type
    , monitor_type
    , preconditioner_type> dc_types;
typedef gpulab::solvers::defect_correction<dc_types>
    laplace_solver_type;

// Potential flow solver setup
typedef free_surface::potential_flow_solver_types<
    vector_type
    , laplace_solver_type
    , gpulab::integration::ERK4> potential_flow_types;
typedef free_surface::potential_flow_solver_3d<potential_flow_types>
    potential_flow_solver_type;
```

Hereafter, the potential flow solver is aware of all component types that should be used to solve the entire PDE system, and it will be easy for developers to exchange parts at later times. The laplace_sigma_stencil_3d class implements both the matrix-vector and right-hand side operator. The flexible-order finite difference kernel for the matrix-free matrix-vector product for the two-dimensional Laplace problem is presented in Listing 11.2. Library macros and reusable kernel routines are used throughout the implementations to enhance developer productivity and hide hardware specific details. This kernel can be used both for matrix-vector products for the original system and for the preconditioning.

Listing 11.2. CUDA kernel implementation for the two-dimensional finite difference approximation to the transformed Laplace equation

```
template <typename value_type, typename size_type>
__global__ void laplace_sigma_transformed(
          value_type* out           , value_type const* p
        , value_type const* eta     , value_type const* etax
        , value_type const* etaxx   , value_type const* h
        , value_type const* hx      , value_type const* hxx
        , value_type dx             , value_type ds
        , size_type Nx              , size_type Ns
        , size_type xstart          , size_type alpha
        , value_type const* stencil1, value_type const* stencil2)
{
    size_type j = IDX1Dx;
    size_type i = IDX1Dy;
```

```
            size_type rank = alpha*2+1; // Total stencil size

            // Get shared memory pointers
            value_type* ss1 = device::shared_memory<value_type>::get_pointer();
            value_type* ss2 = ss1 + rank*rank;

            // Load stencils into ss1 and ss2
            gpulab::device::load_shared_memory(ss1,stencil1,rank*rank);
            gpulab::device::load_shared_memory(ss2,stencil2,rank*rank);

            // Only internal points
            if(j>=xstart && j<Nx-xstart && i>0 && i<Ns-1)
            {
                size_type offset_i = i < alpha ? 2*alpha-i : i >= Ns-alpha ? Ns
                    -1-i : alpha;
                size_type row_i    = offset_i*rank;
            // Always centered stencils in x-dir
                size_type offset_j = alpha;
                size_type row_j    = alpha*rank;

                value_type dhdx    = hx[j];
                value_type dhdxx   = hxx[j];
                value_type detadx  = etax[j];
                value_type detadxx = etaxx[j];

                // Calculate the sigma transformation variables
                value_type sig   = kernel::sigma(i,ds);
                value_type d     = h[j]+eta[j];
                value_type dsdz  = kernel::dsdz(d);
                value_type dsdx  = kernel::dsdx(d,sig,detadx,dhdx);
                value_type dsdxx = kernel::dsdxx(d,sig,detadx,detadxx,dhdx,dhdxx
                    ,dsdx);

                // dp/dxx
                value_type dpdxx = values<value_type>::zero();
                for(size_type s = 0; s<rank; ++s)
                {
                    dpdxx += ss2[row_j+s]*p[i*Nx+j-alpha+s];
                }
                dpdxx /= (dx*dx);

                // Calculate dp/dss and dp/ds
                value_type dpdss = values<value_type>::zero();
                value_type dpds  = values<value_type>::zero();
                value_type ps;
                for(size_type s = 0; s<rank; ++s)
                {
                    ps = p[(i+offset_i-s)*Nx+j];
                    dpds  += ss1[row_i+s]*ps;
                    dpdss += ss2[row_i+s]*ps;
                }
                dpds  /= ds;
                dpdss /= (ds*ds);

                // Calculate dp/dxds
                value_type dpdxds = values<value_type>::zero();
                for(size_type ss = 0; ss<rank; ++ss)
                {
                    value_type dpdx = values<value_type>::zero();
                    for(size_type sx = 0; sx<rank; ++sx)
                    {
                        dpdx += ss1[row_j+sx]*p[(i+offset_i-ss)*Nx+j-offset_j+sx];
                    }
                    dpdx /= dx;
                    dpdxds += ss1[row_i+ss]*dpdx;
                }
                dpdxds /= ds;
```

```
                // Calculate total
                out[i*Nx+j] = dpdxx + dsdxx*dpds + 2.0*dsdx*dpdxds + (dsdx*dsdx
                    + dsdz*dsdz)*dpdss;
            }
        }
```

In a similar template-based approach, the kernel for the right-hand side operator of the two-dimensional problem is implemented and given in Listing 20.3. The kernel computes the right-hand side updates for both surface variables, η and $\tilde{\phi}$, and applies an embedded penalty forcing, cf. (11.38), for all nodes within generation or absorption zones. The penalty forcing functions are computed based on linear or nonlinear wave theory in a separate device function.

Listing 11.3. CUDA kernel implementation for the 2D right-hand side

```
    template <typename value_type, typename size_type>
    __global__ void rhs(value_type const* p      , value_type const* p_surf
                      , value_type const* eta   , value_type* dp_surf_dt
                      , value_type* deta_dt     , value_type const* etax
5                     , value_type const* h     , value_type dx
                      , value_type ds           , size_type Nx
                      , size_type xstart        , size_type xend
                      , unsigned int alpha      , value_type const* stencil
                      , value_type t            , value_type dt
10                    , bool generate           , bool absorb)
    {
        size_type j    = IDX1D;
        size_type rank = alpha*2+1; // Total stencil size

15      // Load shared memory for stencil into ss
        value_type* ss = gpulab::device::shared_memory<value_type>::
            get_pointer();
        gpulab::device::load_shared_memory(ss,stencil,rank*rank);

        // Only interior points
20      if(j>=xstart && j<Nx-xstart)
        {
            value_type g     = values<value_type>::gravity();
            value_type detax = etax[j];

25          // Calculate the sigma transformation variables
            value_type d    = h[j]+eta[j];
            value_type dsdz = kernel::dsdz(d);

            // Calculate dp/ds on surface
30          value_type dpds = values<value_type>::zero();
            for(size_type s=0; s<rank; ++s)
            {
                dpds += p[j+s*Nx]*ss[rank*rank-1-s];
            }
35          dpds /= ds;
            value_type w    = dsdz * dpds;

            // Calculate dp/dx on surface
            value_type dpdx = values<value_type>::zero();
40          for(size_type s=0; s<rank; ++s)
            {
                dpdx += p_surf[j-alpha+s]*ss[rank*alpha+s];
            }
            dpdx /= dx;
45
            // Update right-hand side function
```

```
              dp_surf_dt[j]  = - g*eta[j] - 0.5*(dpdx*dpdx - w*w*(1.0 + (detax*
                  detax)));
              deta_dt[j]     = - detax*dpdx + w*(1.0 + (detax*detax));

50            if(generate || absorb)
              {
                  // Relaxation terms
                  value_type reta  = values<value_type>::zero();
                  value_type rp    = values<value_type>::zero();
55                free_surface::kernel::rhs_relax(reta, rp, eta[j], p_surf[j
                      ], (((int)j)-(int)xstart)*dx, (xend-xstart-1)*dx, t+dt
                      , dt, false, generate, absorb);

                  deta_dt[j]     = deta_dt[j]    + reta;
                  dp_surf_dt[j]  = dp_surf_dt[j] + rp;
              }
60        }
      }
```

11.5 Properties of the numerical model

We now consider different properties of the numerical model in order to shed light on unique features and limits of the model with respect to maritime engineering applications. The presented results extend and complement earlier studies [7,15] for the same model. In particular, we seek to highlight that the properties are tunable to the practical applications of interest through proper choice of discretization parameters, and we therefore also provide details of numerical properties.

11.5.1 Dispersion and kinematic properties

The dispersion and kinematic properties of the unified model are determined by the tunable discretization parameters and should in general be chosen for specific problems. For assessment of errors, we introduce the metrics proposed in [38]

$$E_c(N_x, N_z, kh, h/L) = \frac{c^2}{c_s^2}, \tag{11.43a}$$

$$E_m(N_x, N_z, kh, h/L) = \frac{1}{h}\int_{-h}^{\eta}\left(\frac{\phi(z)-\phi_s(z)}{\phi_s(z)}\right)^2 dz, \tag{11.43b}$$

where m is one of the scalar functions ϕ, u, w describing kinematics; c is the numerical phase celerity of regular waves; and $c_s = \sqrt{g\tanh(kh)/k}$ is the exact phase celerity according to linear Stokes Theory [47]. Measurements of the errors are taken in the vertical below the crest of a wave which is well resolved in the horizontal direction.

The accuracy of the dispersion and kinematic properties are found to be

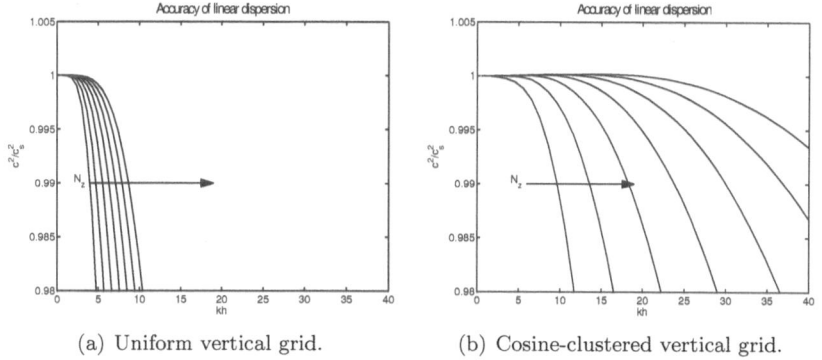

(a) Uniform vertical grid.   (b) Cosine-clustered vertical grid.

FIGURE 11.6. The accuracy in phase celerity c determined by (11.43a) for small-amplitude (linear) wave. $N_z \in [6, 12]$. Sixth order scheme.

excellent with small differences between the accuracy of computed profiles for linear and nonlinear waves. Figure 11.6 shows curves from a discrete analysis of how the accuracy can be improved by increasing the number of nodes in the vertical for (a) uniform grids and (b) cosine-clustered grids. Similarly, Figure 11.7 shows how the accuracy of the kinematic properties of the model can be controlled by choosing an appropriate number of cosine clustered vertical collocation points.

11.6 Numerical experiments

The numerical model detailed has been subject to careful verification and validation utilizing a range of standard benchmarks, cf. [15, 18]. Here we exclusively focus on properties and performance of the numerical wave model. We provide several new results that highlight possibilities for acceleration of the wave model via simple and readily applicable techniques that work well on massively parallel hardware. Finally, we describe how we can extend the implementation of the wave model into a novel GPU implementation of a linear ship-wave model for fast hydrodynamics calculations.

11.6.1 On precision requirements in engineering applications

Practical engineering applications are widely used for analysis purposes and give support to decision making in engineering design. For engineering purposes the turn-around time for producing analysis results is of crucial importance as it affects cost-benefit of work efforts. The key interest is often just

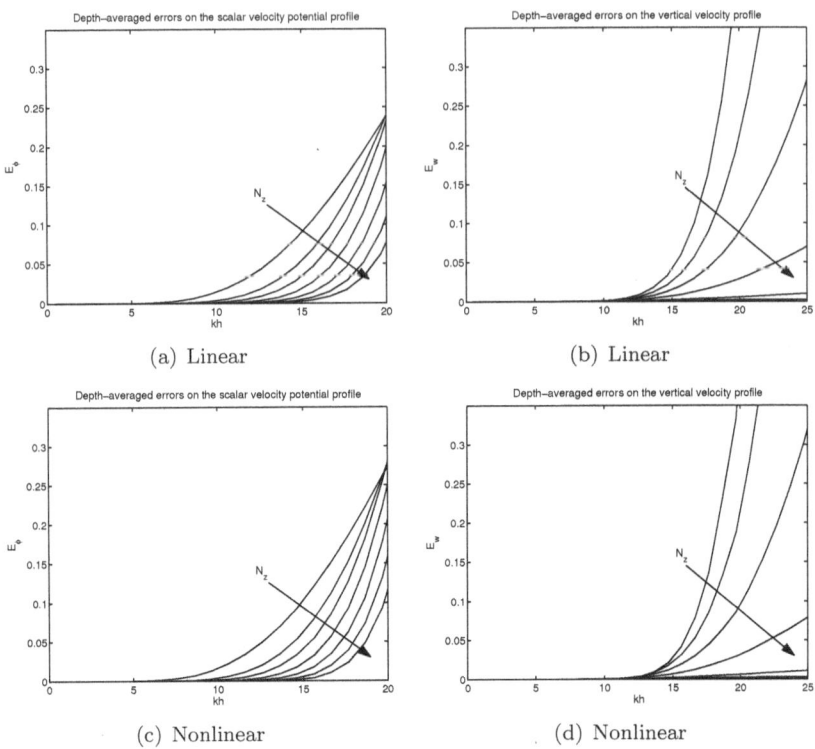

FIGURE 11.7. Assessment of kinematic error is presented in terms of the depth-averaged error determined by (11.43b) for, (a) scalar velocity potential, (b) vertical velocity for a small-amplitude (linear) wave, (c) scalar velocity potential, and (d) vertical velocity for a finite-amplitude (nonlinear) wave with wave height $H/L = 90\%(H/L)_{\max}$. $N_z \in [6, 12]$. Sixth order scheme. Clustered vertical grid.

"engineering accuracy" in computed end results which suggests that we can do with less precision in calculations. One may ask: *what are the precision requirements for engineering applications?*

In a recent study [18, 23], it was shown that the PDC method when executed on GPUs can be utilized to efficiently solve water-wave problems. This was done by trading accuracy for speed in parts of the PDC algorithm, e.g., by using either single, or mixed-precision computations. Without preconditioning the PDC method reduces to a classical iterative refinement technique, which is known to be fault tolerant [26].

Previously reported performance results for the wave model can be taken a step further. We seek to demonstrate how single-precision computations can be used for engineering analysis without significantly affecting accuracy in final computational results. At the same time improvements in computational speed can be almost a factor of two for large problems as a direct result of reduced data transfer, cf. Figure 11.4. Therefore, in pursue of high performance, it is of interest to exploit the reduced data transfers associated with replacing double-precision with single-precision floating point calculations. In a well-organized code this step can be taken with minimal programming effort.

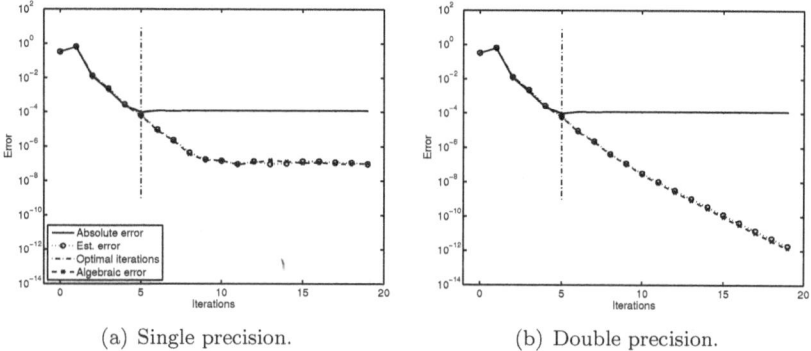

(a) Single precision. (b) Double precision.

FIGURE 11.8. Comparison between convergence histories for single- and double-precision computations using a PDC method for the solution of the transformed Laplace problem. Estimated errors are qualitatively very close to the algebraic errors. Very steep nonlinear stream function wave in intermediate water $(kh, H/L) = (1, 0.0903)$. Discretization based on $(N_x, N_z) = (15, 9)$ with sixth order stencils.

Most scientific applications [20] use double-precision calculations to minimize accumulation of round-off errors, to employ higher precision for ill-conditioned problems, or to stabilize critical sections in the code that require higher precision. Round-off errors tend to accumulate more slowly when higher precision is used, thereby avoiding significant losses of accuracy due to round-off errors. Paradoxically, for many computational tasks, the need for high precision connected with the above-mentioned restrictions does not apply at all, or only applies to a portion of the tasks. In addition, on modern hardware there can be relative differences in peak floating-point performance of up to about one order of magnitude in favor of single-precision over double-precision math calculations depending on choice of hardware architecture. However, for bandwidth-bound applications, the key performance metric is not floating-point performance, but rather bandwidth performance. In both cases, data transfer can be effectively halved by switching to single-precision storage, in which case bandwidth performance increases and at the same time makes it possible to feed floating-point units with data at effectively twice the speed. If maximizing performance is an ultimate goal, such considerations suggests that it can be possible to compute faster by using single- over double-precision arithmetics if accuracy requirements can be fulfilled.

As a simple illustrative numerical experiment, we can consider the iterative solution of (11.41) using the PDC method using single- and double-precision math, respectively. As a simple test case, we consider the solution of periodic stream function waves in two spatial dimensions. Computed convergence histories are presented in Figure 11.8 where it is clear that the main difference is the attainable accuracy level achievable before stagnation. In both cases, the attainable accuracy, defined in terms of the absolute error for the exact stream function solution to the governing equations, is associated with accuracy of approximately 10^{-4} (solid line). In single-precision math, the algebraic and estimated errors measured in the two-norm can reach the level of machine precision. These results suggest that single-precision math is sufficient for calculating accurate solutions at the chosen spatial resolution. We find that the iterative solution of the σ-transformed Laplace problem by the PDC method does not immediately lead to significant accumulation of round-off errors and further investigations are warranted for unsteady computations.

Elaborating on this example, we examine the propagation of a regular stream function wave in time. We consider the errors in wave elevation as a function of time with and without a filtering strategy for single-precision calculations in comparison with double-precision calculations. With an objective to exert control on accumulation of round-off errors that appear as high-frequency noise, the idea is to employ an inexpensive stencil-based filtering strategy, for example, a central filter in one spatial dimension

$$\mathcal{F}u(x_i) = \sum_{n=-\alpha}^{\alpha} c_n u(x_{i+n}), \qquad (11.44)$$

where $c_n \in \mathbb{R}$ are the stencil coefficients and $\alpha \in \mathbb{Z}_+$ is the stencil half-width. An active filter can for example be based on employing a Savitzky-Golay smoothening filter [42], e.g., the mild 7-point SG(6,10) filter, and applying it after every 10th time step to each of the collocation nodes defining the free surface variables. The same procedure can be used for stabilization of nonlinear simulations to remove high-frequency "saw-tooth" instabilities as shown in [15]. This filtering technique can also remove high-frequency noise resulting from round-off errors in computations that would otherwise potentially pollute the computational results and in the worst case leave them useless. The effect of this type of filtering on the numerical efficiency of the model is insignificant.

Results from numerical experiments are presented in Figure 11.9, and most of the errors can be attributed to phase errors resulting from difference in exact versus numerical phase speed. In numerical experiments, we find that while results computed in double-precision are not significantly affected by accumulation of round-off errors, the single-precision results are. In Figures 11.9 (a) and (b), a direct solver based on sparse Gaussian elimination within

MATLAB[3] is used to solve the linear system at every stage and a comparison is made between single- and unfiltered double-precision calculations. It is shown in Figure 11.9 a) that without a filter, the single-precision calculations result in "blow-up" after which the solver fails just before 50 wave periods of calculation time. However, in Figure 11.9 (b) it is demonstrated that invoking a smoothening filter, cf. (11.44), stabilizes the accumulation of round-off errors and the calculations continue and achieve reduced accuracy compared to the computed double-precision results. Thus, it is confirmed that such a filter can be used to control and suppress high-frequency oscillations that results from accumulation of round-off errors. In contrast, replacing the direct solver with an iterative PDC method using the GPU-accelerated wave model appears to be much more attractive upon inspection of Figures 11.9 (c) and (d). The single-precision results are found to be stable with and *without* the filter-based strategy for this problem. The calculations show that single-precision math leads to slightly faster error accumulation for this choice of resolution, however, with only small differences in error level during long time integration. This highlights that fault-tolerance of the iterative PDC method contributes to securing robustness of the calculations.

Last, we demonstrate using a classical benchmark for propagation of non-linear waves over a semicircular shoal that single-precision math is likely to be sufficient for achieving engineering accuracy. The benchmark is based on Whalin's experiment [51] which is often used in validation of dispersive water wave models for coastal engineering applications, e.g., see previous work [15]. Experimental results exist for incident waves with wave periods $T = 1, 2, 3$ s and wave heights $H = 0.0390, 0.0150, 0.0136$ m. All three test cases have been discretized with a computational grid of size $(257 \times 41 \times 7)$ to resolve the physical dimensions of $L_x = 35$ m, $L_y = 6.096$ m. The still water depth decreases in the direction of the incident waves as a semicircular shoal from 0.4572 m to 0.1524 m with an illustration of a snapshot of the free surface given in Figure 11.10(a). The time step Δt is computed based on a constant Courant number of $Cr = c\Delta x/\Delta t = 0.8$, where c is the incident wave speed and Δx is the grid spacing. Waves are generated in the generation zone $0 \leq x/L \leq 1.5$, where L is the wave length of incident waves, and absorbed again in the zone $35 - 2L \leq x \leq 35$ m.

A harmonic analysis of the wave spectrum at the shoal center line is computed and plotted in Figure 11.10 for comparison with the analogous results obtained from the experiments data. The three harmonic amplitudes are computed via a Fast Fourier Transform (FFT) method using the last three wave periods up to $t = 50$ s. There is a satisfactory agreement between the computed and experimental results and no noticeable loss in accuracy resulting from the use of single-precision math.

---

[3]http://www.mathworks.com.

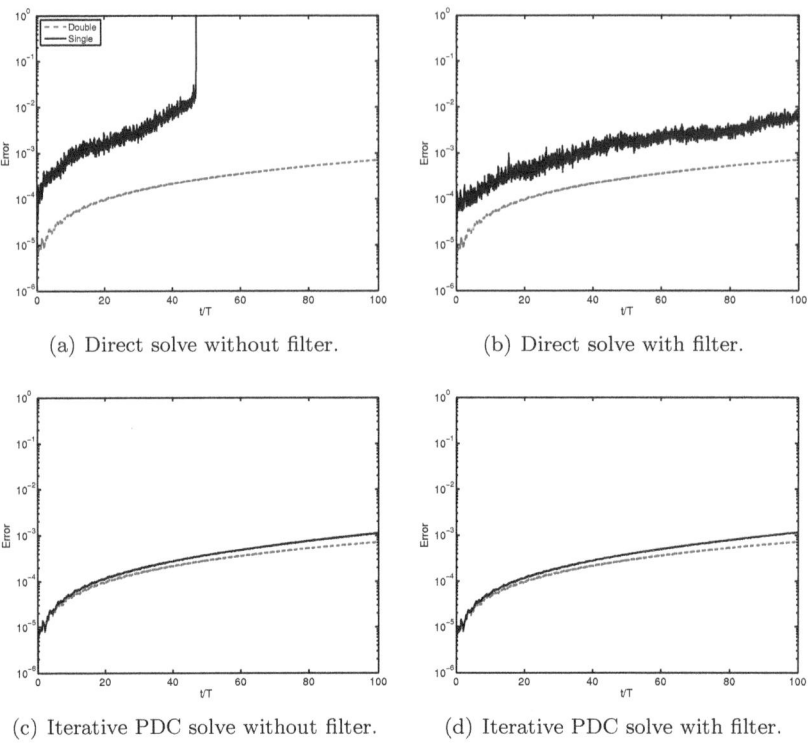

FIGURE 11.9. Comparison of accuracy as a function of time for double-precision calculations vs. single-precision with and without filtering. The double-precision result is unfiltered in each comparison and shows to be less sensitive to round-off errors. Medium steep nonlinear stream function wave in intermediate water $(kh, H/L) = (1, 0.0502)$. Discretization is based on $(N_x, N_z) = (30, 6)$, a courant number of $C_r = 0.5$, and sixth order stencils.

11.6.2 Acceleration via parallelism in time using parareal

With modern many-core architectures, performance is no longer intrinsic and free on new generations of hardware. Added performance with new hardware now comes with the requirement of sufficient parallelism in the application to be accelerated. Methods, tricks, and techniques for extracting parallelism in scientific applications are thus becoming increasingly relevant to enable added numerical accuracy as well as minimization of time to solution in the pursuit of faster and better analysis for engineering applications.

The parareal algorithm has been introduced as a component in our in-house GPU library as described in Section 6.4.2. The parareal library component makes it possible to easily investigate potential opportunities for further acceleration of the water wave model on a heterogeneous system and to assess

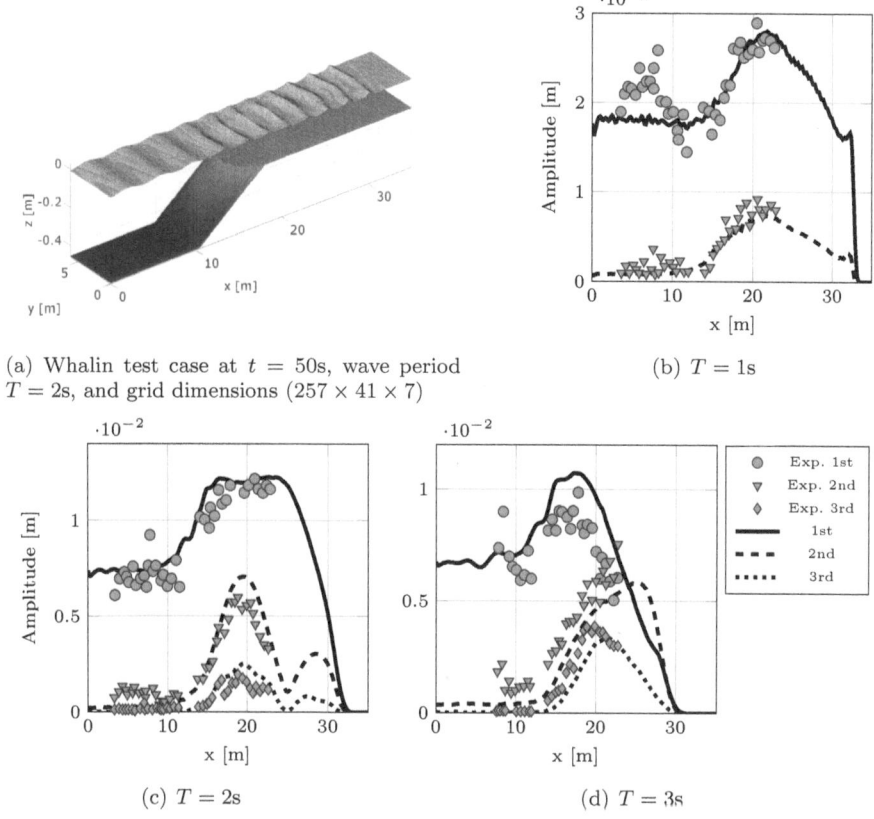

(a) Whalin test case at $t = 50\,\text{s}$, wave period $T = 2\,\text{s}$, and grid dimensions $(257 \times 41 \times 7)$

(b) $T = 1\,\text{s}$

(c) $T = 2\,\text{s}$

(d) $T = 3\,\text{s}$

FIGURE 11.10. Harmonic analysis for the experiment of Whalin for $T = 1, 2, 3\,\text{s}$. Measured experimental and computed results (single-precision) are in good agreement. Test environment 1.

practical feasibility of this algorithmic strategy for various wave types. We omit a detailed review of the parareal algorithm and refer to details given in Section 6.4.2 together with recent reviews [21, 34, 40].

In Section 6.4.2 it is assumed that communication costs can be neglected and a simple model for the algorithmic work complexity is derived. It is found that there are four key discretization parameters for parareal that need to be balanced appropriately in order to achieve high parallel efficiency: the number of coarse-grained time intervals N, the number of iterations K, the ratio between the computational cost of the coarse to the fine propagator $\mathcal{C}_\mathcal{G}/\mathcal{C}_\mathcal{F}$, and the ratio between fine and coarse time step sizes $\delta t/\delta T$.

Ideally, the ratio $\mathcal{C}_\mathcal{G}/\mathcal{C}_\mathcal{F}$ is small and convergence happens in $k = 1$ iteration. This is seldom the case though, as it requires the coarse propagator to achieve accuracy close to that of the fine propagator while at the same time being substantially cheaper computationally, these two objectives obviously being conflicting. Obtaining the highest possible speed-up is a matter of trade-off, typically, the more GPUs used, the faster the coarse propagator should be. The performance of parareal is problem- and discretization-dependent and as such one would suspect that different wave parameters influence the suitability of the method. This was investigated in [40] and indeed the performance does change with wave parameters. Typically the method works better for deep water waves with low- to medium-wave amplitudes.

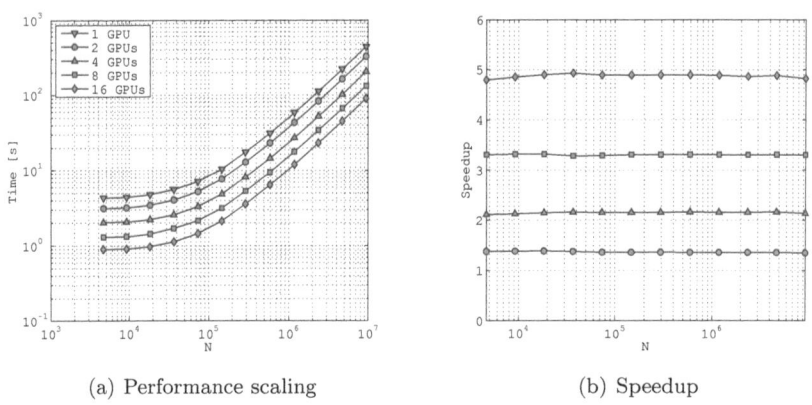

(a) Performance scaling (b) Speedup

FIGURE 11.11. (a) Parareal absolute timings for an increasingly number of water waves traveling one wave length; each wave resolution is (33×9). (b) Parareal speedup for two to sixteen compute nodes compared to the purely sequential single GPU solver. Is is noticeable how insensitive the parareal scheme is to the size of the problem solved. Test environment 3.

We have performed a scalability study for parareal using 2D nonlinear stream function waves based on a discretization with $(N_x, N_z) = (33, 9)$ collocation points, cf. Figure 11.11. The study shows that moderate speedup is possible for this hyperbolic system. Using four GPU nodes, a speedup of slightly more than two was achieved while using sixteen GPU nodes resulted in a speedup of slightly less than five. As noticed in Figure 11.11, parallel efficiency decreases quite fast when using more GPUs. This limitation is due to the usages of a fairly slow and accurate coarse propagator and is linked to a known difficulty with parareal applied to hyperbolic systems. For hyperbolic systems, instabilities tend to arise when using a very inaccurate coarse propagator. This prevents using a large number of time subdomains, as this by Amdahl's law also requires a very fast coarse propagator. The numbers are still impressive though, considering that the speedup due to parareal comes as additional speedup to an already efficient and fast code. Performance results for the Whalin test case are also shown in Figure 11.12. There is a natural limitation to how much we can increase R (the ratio between the complexity of the fine and coarse propagators), because of stability issues with the coarse propagator. In this test case we simulate from $t = [0, 1]$s, using up to 32 GPUs. For low R and only two GPUs, there is no speedup gain, but for the configuration with eight or more GPUs and $R \geq 6$, we are able to get more than 2 times speedup. Though these hyperbolic systems are not optimal for performance tuning using the parareal method, results still confirm that reasonable speedups are in fact possible on heterogeneous systems.

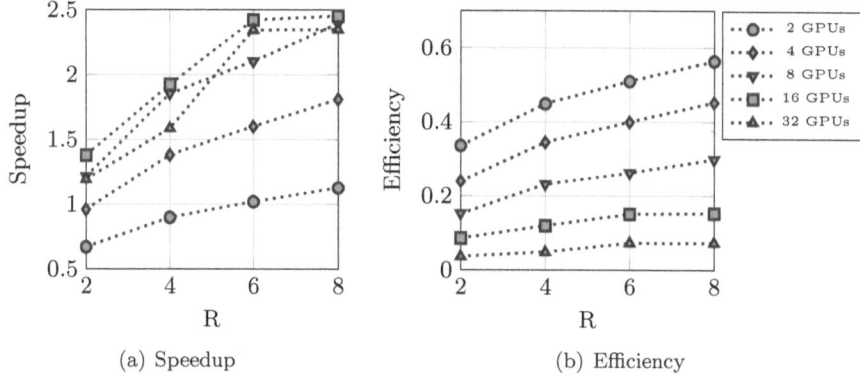

(a) Speedup (b) Efficiency

FIGURE 11.12. Parallel time integration using the parareal method. R is the ratio between the complexity of the fine and coarse propagators. Test environment 3.

The parareal method is observed to be the most viable approach at speeding up small-scale problems due to the reduced communication and overhead involved. For sufficiently large problems, where sufficient work is available to hide the latency in data communication, we find the spatial domain decomposition method to be more favorable as it does not involve the addition of computational work and thereby allows for ideal speedup, something usually out of reach for the parareal algorithm. An important thing to note here is that it is technically possible to extend the work and wrap the parareal method around the domain decomposition method, thereby obtaining a multiplication of the combined speedup of both methods. This is of great interest in the sense that for any problem size, increasing the number of spatial subdomains will eventually degrade speedup due to the latency in communication of boundaries. By exploiting the latency robustness of parareal in conjunction with domain decomposition parallelism, it may be possible to go large scale for problems that would otherwise be too small to exploit a large number of GPUs. These investigations are subject to future work.

Finally, we remark that the parareal algorithm is also a fault-tolerant algorithm. This property follows from the iterative nature of the algorithm and implies that a process can be lost during computations and regenerated with-

11.6.3 Towards real-time interactive ship simulation

out restarting the computations. This can be exploited to minimize total run time in case of such failures.

A fast GPU-accelerated ship hydrodynamic model is developed for real-time interactive ship simulation by modification of the unified potential flow model presented in Section 11.3. The target scientific application is an interactive full mission marine simulator, where multiple ships controlled by naval officers can navigate in a near-realistic virtual marine environment. Full mission simulators are used for education and training of naval officers in critical maneuvering operations and for evaluation of ship and marine infrastructure designs. To predict the motion of ships, a hydrodynamics model is required for prediction of forces by (11.29) which is affected by the kinematic properties of the model, cf. Section 11.5.1. The state-of-the-art for such a hydrodynamic model in today's realtime ship simulators is based on fast interpolation and proper scaling of experimental model data. The amount of experimental model data is limited with respect to hull forms and configurations, requiring the need for extrapolation that compromises the accuracy.

The objective of current and ongoing work is aimed at removing these limitations by replacing the existing hydrodynamic model and instead calculating at full-scale the flow field, wave field, ship-structure, and ship-ship interaction forces in real-time using massive parallel computation technology. The potential flow model (OceanWave3D) presented in Section 11.3 is suitable as the modeling basis for this purpose since it is robust, accurate, efficient, and scalable to arbitrarily large domains. Furthermore, it can accurately account for dispersive waves in the range from shallow to deep waters in marine settings where the sea bed may be uneven.

The inclusion of ships in the wave model requires an approximate representation of such ships in the model. These ship approximations have to be chosen carefully with consideration to the computational performance of the numerical model to enable interactive real-time computing on today's modern hardware. For a first simple proof-of-concept we develop a linear wave and ship model to be used as the model basis. This implies that wave heights and ship draft are assumed to be of small amplitude corresponding and derived by a linearization technique around the mean sea level $z = 0$ m.

The physical domain for the computation is bounded from below by the seabed and from above by the free surface of the sea or the hull of the ship. If the ship is navigating in open water, the ship's physical spatial domain is unbounded in the horizontal direction and in confined waters it is bounded by harbor structures, etc. The representation of the physical domain surrounding the ship is done by finite truncation in the horizontal directions. The resulting time-varying finite physical domain $\Omega(t)$ is a box fixed to follow the ship motion, with the ship in the middle of the top face and with Cartesian coordinate axes aligned with the horizontal components of the forward and sideward ship directions and the upward is the opposite of the direction of gravitational acceleration.

A linear model can be formulated in terms of kinematic and dynamic boundary conditions at the mean sea level (11.19) together with a Laplace problem subject to a variable depth kinematic boundary condition (11:21). The effects of ship hulls can be accounted for by splitting the scalar velocity potential function into steady ϕ_0 and unsteady ϕ_1 potentials such that $\phi = \phi_0 + \phi_1$ and with a quasi-static approximation of the pressure acting on the ship hull as suggested in [33]. This leads to a relatively simple ship model that enables a flexible and computationally efficient approximation of the ship geometry. The steady potential ϕ_0 is calculated using a double-body approximation [43] of the ship and the unsteady potential ϕ_1 is calculated by a linear free surface flow model. The resulting double-body flat-ship problem becomes

$$\nabla^2 \phi_0 = 0, \quad -h \leq z \leq 0 \tag{11.45a}$$

$$\frac{\partial \phi_0}{\partial n} = 0, \quad z = -h, \tag{11.45b}$$

$$\frac{\partial \phi_0}{\partial z} = -U \frac{\partial \eta_0}{\partial x}, \quad z = 0, \tag{11.45c}$$

where U is the velocity of the ship. The unsteady linear water problem is used to calculate the unsteady wave evolution around and away from the ship

$$\nabla^2 \phi_1 = 0, \quad -h \leq z \leq 0, \tag{11.46a}$$

$$\frac{\partial \phi_1}{\partial n} = 0, \quad z = -h, \tag{11.46b}$$

$$\frac{\partial \phi_1}{\partial t} + (u_0 - U)\frac{\partial \phi_1}{\partial x} + v_0 \frac{\partial \phi_1}{\partial y} = -\left(\frac{1}{2}\mathbf{u}_0 \cdot \mathbf{u}_0 + g\eta_0 + \frac{p_{ship}}{\rho}\right), \quad z = 0, \tag{11.46c}$$

$$\frac{\partial \eta_1}{\partial t} + (u_0 - U)\frac{\partial \eta_1}{\partial x} + v_0 \frac{\partial \eta_1}{\partial y} = \frac{\partial \phi_1}{\partial z}, \quad z = 0, \tag{11.46d}$$

where the pressure on the ship hull p_{ship} is calculated explicitly based on a quasi-static approximation which is determined by assuming $\partial_t \phi_1 \approx 0$ and rewriting (11.46c). In general, a ship hull is a complex surface in three-dimensional space, but its draft can be approximated by a single-valued function of the horizontal coordinates $\eta_0 = \eta_0(x, y)$, and the no-flux condition on the ship hull is approximated by a flat-ship approximation. Radiation boundary conditions are approximated by a Sommerfelt absorbing boundary condition [11] on the vertical sides of the physical domain to let waves escape the domain.

The modified numerical model can still be based on flexible-order finite difference method as discussed in Section 11.4. The computational bottleneck problem is the efficient solution of the Laplace problem twice which can be done efficiently by the GPU-accelerated iterative PDC method as explained in section 11.4.5. A snapshot of the steady state wave field is provided in the introduction to this chapter. Computed resistance curves for a Series 60 hull moving at forward speed corresponding to Froude number $F_n = 0.316$ knots in calm water are compared to experimental data [48] in Figure 11.13 (a). The computed Kelvin wave system is shown in Figure 11.13 (b). The computed results compare well with experiments at moderate ship Froude numbers $F_n = U/\sqrt{gh}$ in the range 0.1–0.25 as expected for a linear model. The real-time constraint required to fulfill the interactive and visualization requirements can currently be met with the GPU-accelerated hydrodynamics model for problem sizes of approximately 10^6 for ship Froude numbers in the range 0.1–0.3. The modeling and real-time aspects will be addressed in more detail in ongoing work.

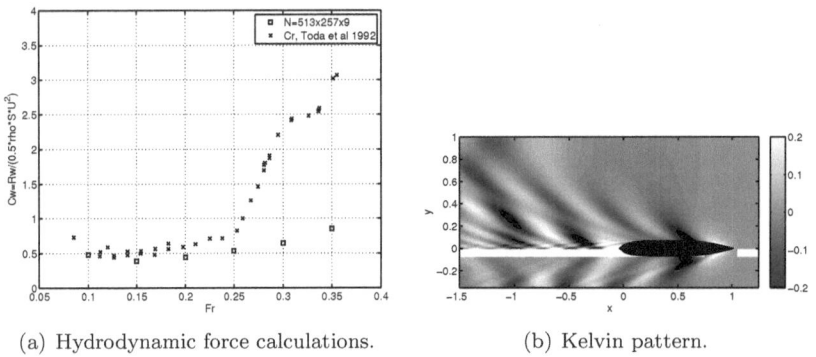

(a) Hydrodynamic force calculations.   (b) Kelvin pattern.

FIGURE 11.13. Computed results. Comparison with experiments for hydrodynamics force calculations confirming engineering accuracy for low Froude numbers.

11.7 Conclusion and future work

We have presented implementation details together with several novel results on development of a new massively parallel and scalable tool for simulation of nonlinear free surface water waves on heterogeneous hardware. The tool is based on the unified potential flow model referred to as OceanWave3D [15] which provides the basis for efficient and scalable simulation of water waves over uneven bottoms on arbitrary domain sizes. We have demonstrated in a few examples how we can accelerate performance by using single-precision math without compromising accuracy. We have shown that performance can be accelerated by introducing concurrency in the time integration using the parareal algorithm and for the first time in a heterogeneous setup based on the use of multiple GPUs. Interestingly, we find that parallel computations using parareal may be more efficient than using conventional data-parallel distributed computations in a multi-GPU setup for moderate problem sizes. We have measured absolute performance and scalability using several of the most recent generations of NVIDIA GPUs to detail the efficiency of the current code. This is useful to predict time to results as explained in [18] and may be compared against other wave models in fair comparisons.

Work in progress focuses on extending the governing equations to account for lack of physics such as wave runup and wave breaking. Also, we plan to extend the domain decomposition method to unstructured grids of blocks that can be boundary-fitted to more general bottom-mounted structures to be able to address wave-structure problems, cf. [16,17]. For example, this will provide the basis for simulations of wave transformations in large harbor areas or predict wave climates in near-coastal areas.

We anticipate that a tool based on the proposed parallel solution strategies will be useful for further advancement in fast and robust analysis techniques and large-scale simulation of free surface wave simulation (e.g., for use as an efficient far-field solver at large scales) and be a basis for next-generation wave models. We also expect that the tool can be useful for hybrid-solution strategies with local flow features possibly resolved by other models and for advancing state-of-the-art in fast physics-based wave-body simulations, e.g., ship-wave interactions in ship simulation where real-time constraints are imposed due to visualization. These subjects will be part of ongoing work addressing application aspects.

11.8 Acknowledgments

This work was supported by grant no. 09-070032 from the Danish Research Council for Technology and Production Sciences. A special thank goes to Professor Jan S. Hesthaven for supporting parts of this work. Scalability and performance tests was done in the GPUlab at DTU Informatics, Technical University of Denmark and using the GPU-cluster at Center for Computing and Visualization, Brown University, USA. NVIDIA Corporation is acknowledged for generous hardware donations to facilities of the GPUlab.

Bibliography

[1] M. B. Abott, A. D. McCowan, and I. R. Warren. Accuracy of short-wave numerical models. *ASCE Journal of Hydraulic Engineering*, 110(10):1287–1301, 1984.

[2] M. B. Abott, H. M. Petersens, and O. Skovgaard. On the numerical modelling of short waves in shallow water. *Journal of Hydraulic Research*, 16(3):173–203, 1978.

[3] Y. Agnon, P. A. Madsen, and Schäffer. A new approach to high-order Boussinesq models. *J. Fluid Mech.*, 399:319–333, 1999.

[4] W. Auzinger. Defect corrections for multigrid solutions of the dirichlet problem in general domains. *Mathematics of Computation*, 48(178):471–484, 1987.

[5] W. Auzinger and H. J. Stetter. Defect Corrections and Multigrid Iterations. In *Multigrid Methods*, volume 960, pages 327–351. Springer-Verlag, New York., 1982.

[6] H. B. Bingham, P. A. Madsen, and D. R. Fuhrman. Velocity potential formulations of highly accurate Boussinesq-type models. *Coastal Engineering*, 56(4):467–478, 2009.

[7] H. B. Bingham and H. Zhang. On the accuracy of finite-difference solutions for nonlinear water waves. *J. Engineering Math.*, 58:211–228, 2007.

[8] L. Cavaleri, J.-H. G. M. Alves, F. Ardhuin, A. Babanin, M. Banner, K. Belibassakis, M. Benoit, M. Donelan, J. Groeneweg, T.H.C. Herbers, P. Hwang, P. A. E. M. Janssen, T. Janssen, I. V. Lavrenov, R. Magne, J. Monbaliu, M. Onorato, V. Polnikov, D. Resio, W. E. Rogers, A. Sheremet, J. McKee Smith, H. L. Tolman, G. van Vledder, J. Wolf, and I. Young. Wave modelling–the state of the art. *Progress in Oceanography*, 75(4):603–674, 2007.

[9] F. Chazel, M. Benoit, A. Ern, and S. Piperno. A double-layer Boussinesq-type model for highly nonlinear and dispersive waves. *Proc. Roy. Soc. London Ser. A*, 465(2108):2319–2346, 2010.

[10] R. G. Dean. Stream function representation of nonlinear ocean waves. *J. Geophys. Res.*, 70:4561–4572, 1965.

[11] K. Dgaygui and P. Joly. Absorbing boundary conditions for linear gravity waves. *SIAM Journal of Applied Mathematics*, 54:93–131, 1994.

[12] F. Dias and T. J. Bridges. The numerical computation of freely propagating time-dependent irrotational water waves. *Fluid Dynam. Res.*, 38(12):803–830, 2006.

[13] G. Ducrozet, H. B. Bingham, A.P. Engsig-Karup, F. Bonnefoy, and P. Ferrant. A comparative study of two fast nonlinear free-surface water wave models. *International Journal for Numerical Methods in Fluids*, 69(11):1818–1834, 2012.

[14] A. P. Engsig-Karup. *Unstructured Nodal DG-FEM solution of high-order Boussinesq-type equations*. PhD thesis, Department of Mechanical Engineering, Technical University of Denmark, 2006.

[15] A. P. Engsig-Karup, H. B. Bingham, and O. Lindberg. An efficient flexible-order model for 3D nonlinear water waves. *J. Comput. Phys.*, 228:2100–2118, 2009.

[16] A. P. Engsig-Karup, J. S. Hesthaven, H. B. Bingham, and P. Madsen. Nodal DG-FEM solutions of high-order Boussinesq-type equations. *J. Engineering Math.*, 56:351–370, 2006.

[17] A. P. Engsig-Karup, J. S. Hesthaven, H. B. Bingham, and T. Warburton. DG-FEM solution for nonlinear wave-structure interaction using Boussinesq-type equations. *Coastal Engineering*, 55:197–208, 2008.

[18] A. P. Engsig-Karup, M. G. Madsen, and S. L. Glimberg. A massively parallel GPU-accelerated model for analysis of fully nonlinear free surface waves. *Int. J. Num. Meth. Fluids*, 2011.

[19] C. Eskilsson and A. P. Engsig-Karup. On devising Boussinesq-type models with bounded eigenspectra: One horizontal dimension. *Submitted to Journal of Computational Physics*, 2013.

[20] M. Feldman. Less is More: Exploiting Single Precision Math in HPC. http://archive.hpcwire.com/hpc/692906.html, 2006.

[21] M. Gander and S. Vandewalle. Analysis of the parareal time-parallel time-integration method. *SIAM Journal of Scientific Computing*, 29(2):556–578, 2007.

[22] S. L. Glimberg and A. P. Engsig-Karup. On a Multi-GPU Implementation of a Free Surface Water Wave Model for Large-scale Simulations. *Submitted to a Special Issue of the Journal Parallel Computing*, 7th Special Issue devoted to PMAA, 2012.

[23] S. L. Glimberg, A. P. Engsig-Karup, and M. G. Madsen. A Fast GPU-accelerated Mixed-precision Strategy for Fully Nonlinear Water Wave Computations. In A. Cangiani, R. L. Davidchack, E. Georgoulis, A.N.

Gorban, J. Levesley, and M. V. Tretyakov, editors, *Proceedings of ENU-MATH 2011, the 9th European Conference on Numerical Mathematics and Advanced Applications, Leicester, September*. Springer, 2011.

[24] W. Hackbusch. On multigrid iterations with defect correction. in: Hackbusch, w.; trottenberg, u. (eds): Lecture notes in math. In *Multigrid Methods*, volume 960, pages 461–473. Springer-Verlag, New York., 1982.

[25] H. J. Haussling and R. T. Van Eseltine. Finite-difference methods for transient potential flows with free surfaces. In J. W. et al., editor, *Proceedings of the Int. Conf. Num. Ship. Hydrodyn.*, pages 295–313. Berkely: Univ. Extension Publ., 1975.

[26] N. J. Higham. *Accuracy and Stability of Numerical Algorithms*. Society for Industrial and Applied Mathematics, Philadelphia, PA, USA, 2nd edition, 2002.

[27] D. E. Keyes. Exaflop/s: The why and the how. *Comptes Rendus Mecanique*, 339(23):70–77, 2011.

[28] R.M. Kirby and G.E. Karniadakis. De-aliasing on non-uniform grids: algorithms and applications, 2003.

[29] J. Larsen and H. Dancy. Open boundaries in short wave simulations - a new approach. *Coastal Engineering*, 7:285–297, 1983.

[30] W. Layton, H. K. Lee, and J. Peterson. A defect-correction method for the incompressible Navier-Stokes equations. *Appl. Math. Comput.*, 129(1):1–19, 2002.

[31] B. Li and C. A. Fleming. A three dimensional multigrid model for fully nonlinear water waves. *Coastal Engineering*, 30:235–258, 1997.

[32] P. Lin. *Numerical Modeling of Water Waves*. Taylor & Francis, 2008.

[33] O. Lindberg, H. B. Bingham, A. P. Engsig-Karup, and P. Madsen. Towards real time simulation of ship-ship interaction. In *Proceedings of The 27th International Workshop on Water Waves and Floating Bodies (IWWWFB)*, Kongens Lyngby, Denmark, 2012.

[34] J.-L. Lions, Y. Maday, and G. Turinici. Résolution d'edp par un schéma en temps pararéel. *C.R. Acad Sci. Paris Sér. I Math*, 332:661–668, 2001.

[35] P. Lynett and P. L.-F. Liu. Linear analysis of the multi-layer model. *Coastal Engineering*, 51:439–454, 2004.

[36] P. Lynett and P. L.-F. Liu. A two-layer approach to wave modelling. *Proc. Roy. Soc. London Ser. A*, 460:2637–2669, 2004.

[37] P. A. Madsen, H. B. Bingham, and H. Liu. A new Boussinesq method for fully nonlinear waves from shallow to deep water. *J. Fluid Mech.*, 462:1–30, 2002.

[38] P. A. Madsen, H. B. Bingham, and H. A. Schäffer. Boussinesq-type formulations for fully nonlinear and extremely dispersive water waves: derivation and analysis. *Proc. R. Soc. Lond. A*, 459:1075–1104, 2003.

[39] P. A. Madsen and H. A. Schäffer. Higher order Boussinesq-type equations for surface gravity waves–derivation and analysis. *Advances in Coastal and Ocean Engineering*, 356:3123–3181, 1998.

[40] A. S. Nielsen. Feasibility study of the parareal algorithm. Master thesis, Technical University of Denmark, Department of Informatics and Mathematical Modeling, 2012.

[41] D. H. Peregrine. Long waves on a beach. *Journal of Fluid Mechanics*, 27(4):815–827, 1967.

[42] W. H. Press and S. A. Teukolsky. Savitzky-Golay smoothening filters. *Computers in Physics*, 4:669–672, 1990.

[43] H. C. Raven. Validation of an approach to analyse and understand ship wave making. *Journal of Maritime Science and Technology*, 15:331–344, 2010.

[44] I. Robertson and S. Sherwin. Free-surface flow simulation using hp/spectral elements. *J. Comput. Phys.*, 155:26–53, 1999.

[45] S. Schaffer. Higher order multigrid methods. *Math. Comp.*, 43(167):89–115, 1984.

[46] H. J. Stetter. The defect correction principle and discretization methods. *Numer. Math.*, 29:425–443, 1978.

[47] I. A. Svendsen and I. G. Jonsson. *Hydrodynamics of Coastal Regions*. Technical University of Denmark, Kongens Lyngby, 2001.

[48] Y. Toda, F. Stern, and J. Longo. Mean-flow measurement in the boundary-layer and wave and wave field of a series 60-cb 0.6 ship model. froude numbers 0.16 and 0.316. *Journal of Ship Research*, 36:360–377, 1992.

[49] U. Trottenberg, C. W. Oosterlee, A. Schuller, contributions by A. Brandt, P. Oswald, and K. Stuben. *Multigrid*. Academic Press, San Diego, CA, 2001.

[50] W. Tsai and D. K. P. Yue. Computation of nonlinear free-surface flows. In *Annual Review of Fluid Mechanics, Vol. 28*, pages 249–278. Annual Reviews, Palo Alto, CA, 1996.

[51] R. W. Whalin, United States Army. Corps of Engineers, and Waterways Experiment Station (U.S.). *The Limit of Applicability of Linear Wave Refraction Theory in a Convergence Zone*. Research report. Waterways Experiment Station, 1971.

[52] R. W. Yeung. Numerical methods in free-surface flows. In *Annual Review of Fluid Mechanics, Vol. 14*, pages 395–442. Annual Reviews, Palo Alto, Calif., 1982.

Chapter 12

Parallel monotone spline interpolation and approximation on GPUs

Gleb Beliakov and Shaowu Liu
School of Information Technology, Deakin University, Burwood, Australia

12.1	Introduction	295
12.2	Monotone splines	298
	12.2.1 Monotone quadratic splines	298
	12.2.2 Monotone Hermite splines	302
12.3	Smoothing noisy data via parallel isotone regression	303
12.4	Conclusion	308
	Bibliography	308

12.1 Introduction

Monotonicity preserving interpolation and approximation have received substantial attention in the last thirty years because of their numerous applications in computer aided-design, statistics, and machine learning [9,10,19]. Constrained splines are particularly popular because of their flexibility in modeling different geometrical shapes, sound theoretical properties, and availability of numerically stable algorithms [9,10,26]. In this work we examine parallelization and adaptation for GPUs of a few algorithms of monotone spline interpolation and data smoothing, which arose in the context of estimating probability distributions.

Estimating Cumulative Probability distribution Functions (CDF) from data is quite common in data analysis. In our particular case we faced this problem in the context of partitioning univariate data with the purpose of efficient sorting. It was necessary to partition large data sets into chunks of approximately equal size, so that these chunks could be sorted independently and subsequently concatenated. In order to do that, empirical CDF of the data was used to find the quantiles, which served to partition the data. CDF was estimated from the data based on a number of pairs $(x_i, y_i), i = 1, \ldots, n$, where y_i was the proportion of data no larger than x_i. As data could come from a variety of distributions, a distribution-free nonparametric fitting procedure was required to interpolate the above pairs. Needless to say the whole process was aimed at GPU, and hence the use of CPU for invoking serial algorithms had to be minimized.

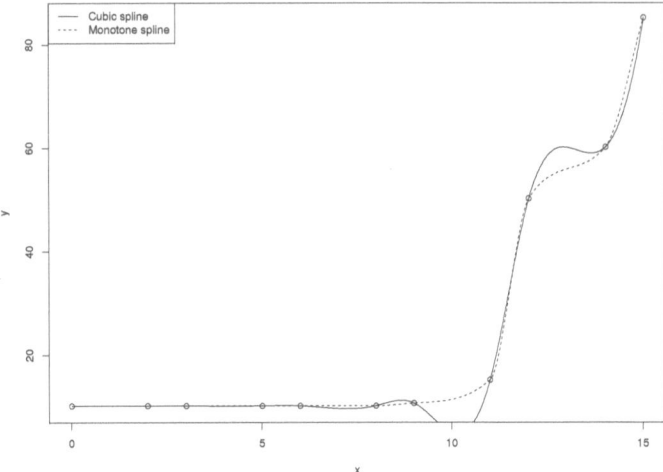

FIGURE 12.1. Cubic spline (solid) and monotone quadratic spline (dashed) interpolating monotone data from [14]. Cubic spline fails to preserve monotonicity of the data.

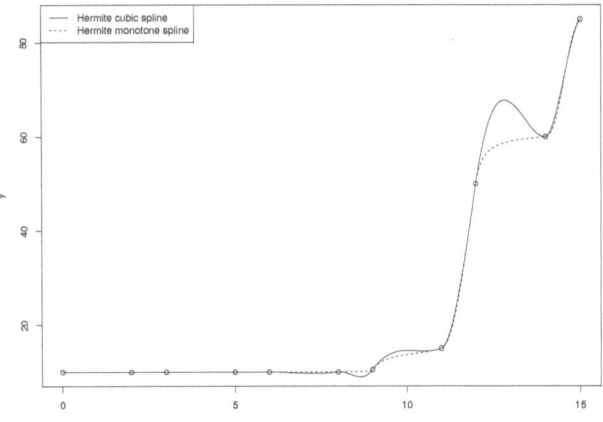

FIGURE 12.2. Hermite cubic spline (solid) and Hermite rational spline interpolating monotone data from [14] with nonnegative prescribed slopes. Despite nonnegative slopes, the Hermite cubic spline is not monotone.

The above mentioned application is one of many examples (e.g., mass spectrography [18] and global warming data [2]) where univariate data needs to be fitted by monotonicity preserving interpolants. Of course, CDF is a monotone increasing function, whose inverse, called quantile function, can be used to calculate the quantiles. Spline interpolation would be the most suitable nonparametric method to fit the CDF, except that polynomial splines do not preserve monotonicity of the data, as illustrated on Figure 12.1.

The failure of splines to preserve monotonicity has prompted fundamental research in this area since the 1960s. One of the first methods to remedy this problem was splines in tension by Schweikert [28], where a tension parameter controlled the shape of exponential splines [29]. Later on several monotonicity preserving polynomial spline algorithms were proposed [3, 4, 21, 22, 27]. These algorithms typically rely on introducing additional spline knots between the abscissae of the data. Algorithmic developments are active to this day; see, for example, [1, 19].

When in addition to the pairs (x_i, y_i) the slopes of the function are available, i.e., the data comes in triples (x_i, y_i, p_i), the interpolation problem is called Hermite, and the Hermite splines are used. However, even when the sequence y_i is increasing and the slopes p_i are nonnegative, cubic Hermite splines may still fail to be monotone, as illustrated in Figure 12.2. Thus, monotone Hermite splines are needed [14].

Another issue with monotone approximation is noisy data. In this case inaccuracies in the data make the input sequence y_i itself nonmonotone; and hence monotone spline interpolation algorithms will fail. Monotone spline smoothing algorithms are available, e.g., [4, 11]. Such algorithms are based on solving a quadratic (or another convex) programming problem numerically, and have not yet been adapted to parallel processing.

In this work we examine several monotone spline fitting algorithms, and select the ones that we believe are most suitable for parallelization on GPUs. We pay attention to numerical efficiency in terms of numerical calculations and memory access pattern, and favor one-pass algorithms. We also look at smoothing noisy data and developed a parallel version of the Minimum Lower Sets (MLS) algorithm for the isotonic regression problem [7, 23].

The rest of the chapter is organized as follows. Section 12.2 discusses monotone spline interpolation methods and presents two parallel algorithms. Section 12.3 deals with the smoothing problem. It presents the isotonic regression problem and discusses the Pool Adjacent Violators (PAV) and MLS algorithms. Combined with monotone spline interpolation, the parallel MLS method makes it possible to build a monotone spline approximation to noisy data entirely on GPU. Section 12.4 concludes.

12.2 Monotone splines

Splines are piecewise continuous functions very popular in numerical approximation and computer-aided design [9, 10]. An example of a spline is the broken line interpolation. Typically, polynomial splines are used, and the first (and often second) derivatives of the polynomial pieces are required to match at the knots. The knots of the splines are usually the abscissae of the input data, although this condition is not always required (e.g., splines with free knots [6, 10, 17]).

Polynomial splines are often represented in the B-spline basis, in which case their coefficients are computed from the input data by solving a banded system of linear equations [9, 10, 20]. Tridiagonal systems arise in cubic spline interpolation, while pentadiagonal systems arise in cubic spline smoothing [20]. Splines possess important extremal properties [16, 20], in particular splines of degree $2m - 1$ are the most "smooth" functions that interpolate (or approximate, in the least squares sense) the data. The smoothness term is Tihkonov regularization functional, the L_2 norm of the mth derivative of the interpolant [20].

When the data are known to come from a monotone function, the interpolant needs to be monotone as well. Even if the sequence of data ordinates $y_i, i = 1, \ldots, n$ is nondecreasing, cubic (and higher degree) interpolating splines are not necessarily monotone; an example is shown in Figure 12.1. To deal with the problem of extraneous inflection points, Schweikert [28] proposed splines in tension, which are piecewise exponential functions. Splines in tension have been further explored in [24, 25, 29] and many subsequent works.

12.2.1 Monotone quadratic splines

For polynomial splines, monotone or otherwise constrained splines were developed in [3, 4, 21, 22, 27]. Two monotone quadratic spline algorithms were published in the early 1980s [21, 27]. Both algorithms are based on introducing additional interpolation knots under certain conditions, to facilitate preservation of monotonicity of the data. McAllister and Roulier's algorithm [21] introduces at most two extra knots between two neighbouring data, while Schumaker's algorithm [27] introduces only one extra knot. In addition, Schumaker's algorithm is one pass, which is particularly suited for parallelization, as no system of equations needs to be solved. While parallel tridiagonal linear systems solvers have been developed for GPUs [13], the obvious advantage of a one-pass algorithm is the speed. Because of that, we chose Schumaker's algorithm for GPU parallelization.

Let us formally describe Schumaker's algorithm, with Butland's slopes [8]. The spline is a piecewise quadratic polynomial in the form

$$s(x) = \alpha_i + \beta_i(x - t_i) + \gamma_i(x - t_i)^2, \quad x \in [t_i, t_{i+1}], i = 1, \ldots, T,$$

where the knot sequence t_i is obtained from the data x_i by inserting at most one additional knot per subinterval. Let $\delta_i = (y_{i+1} - y_i)/(x_{i+1} - x_i), i = 1, \ldots, n - 1$. We define Butland's slopes for $i = 2, \ldots, n - 1$ as

$$d_i = \begin{cases} \frac{2\delta_{i-1}\delta_i}{\delta_{i-1}+\delta_i}, & \text{if } \delta_{i-1}\delta_i > 0, \\ 0 & \text{otherwise.} \end{cases}$$

The first and the last Butland's slopes are

$$d_1 = \begin{cases} 2\delta_1 - d_2, & \text{if } \delta_1(2\delta_1 - d_2) > 0, \\ 0 & \text{otherwise,} \end{cases}$$

$$d_n = \begin{cases} 2\delta_{n-1} - d_{n-1}, & \text{if } \delta_{n-1}(2\delta_{n-1} - d_{n-1}) > 0, \\ 0 & \text{otherwise.} \end{cases}$$

When $d_i + d_{i+1} = 2\delta_i$, then a single quadratic polynomial interpolates the data on $[x_i, x_{i+1}]$ and $t_i = x_i$, $\alpha_i = y_i$, $\beta_i = d_i$, and $\gamma_i = \frac{d_{i+1}-d_i}{2(x_{i+1}-x_i)}$. Otherwise an additional knot t_i is required, and

$$\alpha_i = y_i, \beta_i = d_i, \gamma_i = \frac{(\bar{d}_i - d_i)}{2(t_i - x_i)}, x \in [x_i, t_i],$$

$$\bar{\alpha}_i = y_i + d_i(t_i - x_i) + \frac{(\bar{d}_i - d_i)}{2(t_i - x_i)}, \bar{\beta}_i = \bar{d}_i, \bar{\gamma}_i = \frac{(d_{i+1} - \bar{d}_i)}{2(x_{i+1} - t_i)}, x \in [t_i, x_{i+1}],$$

where

$$\bar{d}_i = (2\delta_i - d_{i+1}) + \frac{(d_{i+1} - d_i)(t_i - x_i)}{(x_{i+1} - x_i)}.$$

The position of the additional knot is selected as follows

$$t_i = \begin{cases} x_{i+1} + \frac{(d_i - \delta_i)(x_{i+1} - x_i)}{(d_{i+1} - d_i)}, & \text{if } (d_{i+1} - \delta_i)(d_i - \delta_i) < 0, \\ \frac{1}{2}(x_{i+1} + x_i) & \text{otherwise.} \end{cases}$$

It is almost straightforward to parallelize this scheme for GPUs, by processing each subinterval $[x_i, x_{i+1}]$ independently in a separate thread. However, it is not known in advance whether an extra knot t_i needs to be inserted, and therefore calculation of the position of the knot in the output sequence of knots t_i is problematic for parallel implementation (for a sequential algorithm no such issue arises). To avoid serialization, we decided to insert an additional knot in every interval $[x_i, x_{i+1}]$, but set $t_i = x_i$ when the extra knot is not actually needed. This way we know in advance the position of the output knots and the length of this sequence is $2(n-1)$, and therefore all calculations can now be performed independently. The price we pay is that some of the spline knots can coincide. However, this does not affect spline evaluation, as one of the coinciding knots is simply disregarded, and the spline coefficients are replicated (so for a double knot $t_i = t_{i+1}$, we have $\alpha_i = \alpha_{i+1}$, $\beta_i = \beta_{i+1}$, $\gamma_i = \gamma_{i+1}$). Our implementation is presented in Listings 12.1-12.2.

Listing 12.1. calculation of monotone spline knots and coefficients.

```
template<typename Tx, typename Ty>
__global__ void CalculateBeta(Tx *u, Ty *v, double *b, int N)
{
    int tid = threadIdx.x + blockIdx.x * blockDim.x;
    while(tid<=(N-2)) {
        b[tid]=(v[tid+1]-v[tid])/fmax(1e-20,double(u[tid+1]-u[tid]));
        tid += blockDim.x * gridDim.x;
    }
    __syncthreads();
}

__global__ void CalculateDGeneral( double *b, double *c, int N)
{
    int tid = threadIdx.x + blockIdx.x * blockDim.x;
    while(tid<=(N-2)) {
        if((b[tid-1]*b[tid])<=0) c[tid]=0;
        else c[tid]=(2*b[tid-1]*b[tid])/(b[tid-1]+b[tid]);
        tid += blockDim.x * gridDim.x;
    }
    __syncthreads();
}

__global__ void CalculateD( double *b, double *c, int N )
{
    if((b[0]*(2*b[0]-c[1]))<=0)    c[0]=0;
    else  c[0]=2*b[0] - c[1];
    if((b[N-2]*(2*b[N-2]-c[N-2]))<=0) c[N-1]=0;
    else  c[N-1]=2*b[N-2] - c[N-2];
    __syncthreads();
}

template<typename Tx, typename Ty>
int BuildMonotonSpline(Tx *d_X, Ty *d_Y, int N, double *t, double *
    alpha, double *beta, double *gamma)
{
    int T = (N-1)*2+1; // length of the output array
    double *b, *c; // temp variables
    cudaMalloc( (void**)&b, 1*N*sizeof(double) );
    cudaMalloc( (void**)&c, 2*N*sizeof(double) );
    int threads=256;
    int blocks = (N-1)/threads + 1;
    CalculateBeta<<<blocks,threads>>>(d_X,d_Y,b,N);
    CalculateDGeneral<<<blocks,threads>>>(b,c,N);
    CalculateD<<<1,1>>>(b,c,NN);    // calculate d_1 and d_N
    CalculateCoefficientsKnots<<<blocks,threads>>>(d_X,d_Y,b,c,h,alpha,
        beta,gamma,N);
    cudaFree(b); cudaFree(c);
    return T;
}
```

At the spline evaluation stage we need to compute $s(z_k)$ for a sequence of query values $z_k, k = 1, \ldots, K$. For each z_k we locate the interval $[t_i, t_{i+1}]$ containing z_k, using the bisection algorithm presented in Listing 12.3, and then apply the appropriate coefficients of the quadratic function. This is also done in parallel. The bisection algorithm could be implemented using texture memory (to cache the array z), but this is not shown in Listing 12.3.

Listing 12.2. implementation of the kernel for calculating spline knots and coefficients; function fmax is used to avoid division by zero for data with coinciding abscissae.

```
template<typename Tx, typename Ty>
__global__ void CalculateCoefficientsKnots( Tx *u, Ty *v, double *b,
    double *c, double *t, double *alpha, double *beta, double *gamma,
    int N )
{
    int tid = threadIdx.x + blockIdx.x * blockDim.x;
    int s = tid*2;
    while(tid<=(N-2))
    {
        // decide whether an additional knot is necessary
        if(fabs(c[tid]+c[tid+1]- 2*b[tid])<=0.1e-5) // tolerance
        {
            //no additional knot
            h[s]=h[s+1]=u[tid];
            alpha[s]=alpha[s+1]=v[tid];
            beta[s]=beta[s+1]=c[tid];
            gamma[s]=gamma[s+1]=(c[tid+1]-c[tid])/(2*(fmax(1e-10,u[tid+1]-u[
                tid])));
        }
        else {
            //adding a knot
            h[s]=u[tid];
            //determine the position of the knot
            if((c[tid+1] - b[tid])*(c[tid] - b[tid])<0)
                h[s+1]=u[tid+1] + (c[tid] - b[tid])*(fmax(1e-10,u[tid+1]-u[tid
                    ]))/fmax(1e-10,(c[tid+1] - c[tid]));
            else
                h[s+1]=0.5*(u[tid+1] + u[tid]);
            //calculate coefficients
            double dtemp = (2*b[tid] - c[tid+1])+((c[tid+1] - c[tid])*
                (h[s+1] - u[tid]))/ fmax(1e-10,(u[tid+1] - u[tid]));
            alpha[s]=v[tid];    beta[s]=c[tid];
            gamma[s]=(dtemp - c[tid])/(2*fmax(1e-10,(h[s+1] - u[tid])));
            alpha[s+1]=v[tid] + c[tid]*(h[s+1] - u[tid]) +
                (dtemp - c[tid])*(h[s+1] - u[tid])/2;
            gamma[s+1]=(c[tid+1]-dtemp)/(2*fmax(1e-10,(u[tid+1]-h[s+1])));
            beta[s+1]=dtemp;
        }
        tid += blockDim.x * gridDim.x;   s = tid*2;
    }
    __syncthreads();
    // Select a single thread  to perform the last operation
    if((threadIdx.x  ) == 0)  {
        s = (N-1) * 2;    h[s]=u[N-1];
    }
    __syncthreads();
}
```

Listing 12.3. implementation of the spline evaluation algorithm for GPU.

```
template<typename T>
__device__ void Bisection_device(T z, T* t, int mi,int ma,int* l)
{
    int i; ma--;
    while(1) {
        i=(mi+ma)/2;
        if(z >= t[i]) mi=i+1;
                else ma=i;
        if(mi>=ma) break;
    }
    *l = mi-1;
```

```
}
// Kernel to evaluate monotone spline for a sequence of query points
// residing in the array z of size m

template<typename Tx, typename Ty>
__global__ void d_MonSplineValue(Tx* z, int K, double* t, double *
    alpha, double * beta, double * gamma, int T, Ty *value)
{
    int tid = threadIdx.x + blockIdx.x * blockDim.x;
    int mi=0, ma=T, i=0;
    Ty r;
    while(tid<K)
    {
        Bisection_device(z[tid], t, mi, ma, &i);
        r= z[tid]-t[i];
        r= alpha[i] + r*(beta[i] + gamma[i]*r);
        value[tid]=r;
        tid += blockDim.x * gridDim.x;
    }
    __syncthreads();
}

template<typename Tx, typename Ty>
void MonotoneSplineValue(Tx *z, int K, double* t, double * alpha,
    double * beta, double * gamma, int T, Ty* result)
{
    int blocks,threads=256;
    blocks=(K-1)/threads+1;
    d_MonSplineValue<<<blocks,threads>>>(z,K,t,alpha,beta,gamma,T,result
        );
}
```

12.2.2 Monotone Hermite splines

In this section, in addition to the points (x_i, y_i) we have the slopes p_i, and hence, we consider monotone Hermite interpolation. In our motivating application of CDF estimation, the values p_i are easily obtained together with y_i, and their use may help to build a more accurate interpolant. Of course, for monotone nondecreasing functions we must have $p_i \geq 0$. However, this does not guarantee that the spline interpolant is monotone, as can be seen in Figure 12.2. Fritsch and Carlson [12] showed that nonnegative p_i is not a sufficient condition to guarantee monotonicity, and designed a process for modification of derivatives, so that the necessary and sufficient conditions for monotonicity of a piecewise cubic are met. Hence, the values p_i are not matched exactly. In contrast, Gregory and Delbourgo [14] designed piecewise rational quadratic spline, for which the nonnegativity of p_i is both a necessary and sufficient condition.

The rational quadratic spline in [14] is constructed as

$$s(x) = \begin{cases} \frac{P_i(\theta)}{Q_i(\theta)}, & \text{if } \Delta_i \neq 0, \\ y_i & \text{otherwise,} \end{cases}$$

where

$$\theta = (x - x_i)/(x_{i+1} - x_i), \quad \Delta_i = (y_{i+1} - y_i)/(x_{i+1} - x_i),$$

and
$$P_i(\theta)/Q_i(\theta) = y_i + \frac{(y_{i+1} - y_i)[\Delta_i\theta^2 + p_i\theta(1-\theta)]}{\Delta_i + (p_{i+1} + p_i - 2\Delta_i)\theta(1-\theta)}.$$

This rational spline is continuously differentiable and interpolates both the values y_i and the derivatives p_i. Its derivative is given by

$$s'(x) = \Delta_i^2[p_{i+1}\theta^2 + 2\Delta_i\theta(1-\theta) + p_i(1-\theta)^2]/Q_i(\theta)^2,$$

with
$$Q_i(\theta) = \Delta_i + (p_{i+1} + p_i - 2\Delta_i)\theta(1-\theta),$$

provided $\Delta_i \neq 0$ ($s'(x) = 0$ otherwise), and this expression is nonnegative.

It is clear that Gregory and Delbourgo's Hermite interpolant [14] is trivially parallel, and the parameters $h_i = x_{i+1} - x_i$ and Δ_i are easily computed in a simple kernel. Evaluation of the spline and its derivative is accomplished by locating the interval containing the query point x using bisection, as in Listing 12.3, and applying the above-mentioned formulas.

12.3 Smoothing noisy data via parallel isotone regression

Inaccuracies in the data are common in practice and need to be accounted for during the spline approximation process. Smoothing polynomial splines were presented in [20], where the data were fitted in the least squares sense while also minimizing the L_2 norm of the mth derivative of the spline. Monotone smoothing splines have been dealt with in several works, in particular we mention [4, 11]. The presented algorithms rely on solving quadratic programming problems. Monotone approximating splines with fixed knots distinct form the data have been presented in [5], where an instance of a quadrating programming problem is solved as well.

Another approach consists of monotonizing the data, so that the sequence y_i becomes monotone. This approach is known as isotone regression [7, 23]. It is different from monotone spline smoothing, as the regularization term controlling the L_2 norm of the mth derivative is not taken into account. Usually the data is monotonized by minimizing the squared differences to the inputs. It becomes a quadratic programming problem, usually solved by active sets methods [7]. A popular PAV algorithm (PAVA) is one method that provides efficient numerical solution.

PAVA consists of the following steps. The sequence y_i is processed from the start. If violation of monotonicity $y_i > y_{i+1}$ is found, both values y_i and y_{i+1} are replaced with their average y'_i, and both values form a block. Since the new value y'_i is smaller than y_i, monotonicity may become violated with respect to y_{i-1}. If this is the case, the $i-1$st, ith, and $i+1$st data are merged into a block and their values are replaced with their average. We continue to back-average as needed to get monotonicity.

Various serial implementations of the PAVA exist. It is noted [18] that in PAVA, which is based on the ideas from convex analysis, a decomposition theorem holds, namely, performing PAVA separately on two contiguous subsets of data and then performing PAVA on the result produces isotonic regression on the whole data set. Thus, isotonic regression is parallelizable, and the divide-and-conquer approach, decomposing the original problem into two smaller subproblems, can be implemented on multiple processors. However, to our knowledge, no parallel PAVA for many-core systems such as GPUs exist.

Another approach to isotonic regression is called the MLS algorithm [7,23]. It provides the same solution as the PAVA, but works differently. For each datum (or block), MLS selects the largest contiguous block of subsequent data with the smallest average. If this average is smaller than that of the preceding block, the blocks are merged, and the data in the block are replaced with their average. MLS is also an active set method [7], but its complexity is $O(n^2)$ as opposed to $O(n)$ of the PAVA, and of another active set algorithm proposed in [7] by the name of Algorithm A.

In terms of GPU parallelization, neither PAVA nor Algorithm A appears to be suitable, as the techniques that achieve $O(n)$ complexity are inheritably serial. In this work we focus on parallelizing MLS. First, we precompute the values

$$z_i = \sum_{j=i}^{n} y_i$$

and $z_{n+1} = 0$ using the parallel partial sum algorithm (scan algorithm from Thrust [15] library). From these values we can compute the averages of the blocks of data with the indices $\{i, i+1, \ldots, j\}$

$$P_{ij} = \frac{1}{j-i+1} \sum_{k=i}^{j} y_k = \frac{1}{j-i+1}(z_i - z_{j+1}). \tag{12.1}$$

As per MLS algorithm, for each fixed i from 1 to n, we compute the smallest P_{ij} starting from $j = i+1$ and fix the index j^*. If $y_i > P_{ij^*}$, we replace the values $y_i, \ldots, y_{j^*}$ with their average P_{ij^*}; otherwise we keep the value y_i. In case of replacement, we advance i to position $j^* + 1$. We check the condition $y_i > P_{i,j^*}$ to form a block, which is equivalent to $y_i > P_{i+1,j^*}$ as $P_{ij} = \frac{1}{j-i+1}((j-i)P_{i+1,j} + y_i)$, from which we deduce that both inequalities hold simultaneously.

Now the presented algorithm can be parallelized for GPUs: each datum y_i is treated in its own thread. Calculation of the smallest P_{ij} is performed serially within the ith thread, or in parallel by starting children threads. Replacing the values $y_i, \ldots, y_{j^*}$ with P_{ij^*} leads to potential clashes, as several threads can perform this operation on the same elements y_k. This can be circumvented by using max operation, i.e., $y_k \leftarrow \max(y_k, P_{ij})$. Note that thread

i replaces the value y_k, $k \geq i$ if $P_{ij} < y_i$. Now, if two threads i_1 and i_2 need to replace y_k, and $i_1 < i_2$, we must have $P_{i_1 j_1} \geq P_{i_2 j_2}$, as formalized in the following.

Proposition 1 *If partial averages P_{ij} are defined by (12.1) and $i_1 < i_2$, $j_1, j_2 \geq i_1, i_2$, where j_1, j_2 denote the minimizers of $P_{i_1 j}$ over $j \geq i_1$ (respectively $P_{i_2 j}$ over $j \geq i_2$), then $P_{i_1 j_1} \geq P_{i_2 j_2}$.*

Proof 1 *Because the average satisfies*

$$\frac{1}{k}\sum_{i=1}^{k} y_i \geq \frac{1}{n-k}\sum_{i=k+1}^{n} y_i \Rightarrow \frac{1}{k}\sum_{i=1}^{k} y_i \geq \frac{1}{n}\sum_{i=1}^{n} y_i \geq \frac{1}{n-k}\sum_{i=k+1}^{n} y_i,$$

we must have $P_{i_1 i_2 - 1} \geq P_{i_1 j_1} \geq P_{i_2 j_1}$. At the same time we have $P_{i_2 j_1} \geq P_{i_2 j_2}$, which implies $P_{i_1 j_1} \geq P_{i_2 j_2}$.

The order in which the steps are performed is not guaranteed in parallel computations. By the proposition above, $P_{i_2 j_2} \leq P_{i_1 j_1} < y_{i_1}$ whenever the value y_{i_1} needs replacement by the average of its block, which leads to overriding all $y_k, i_1 \leq k \leq j_1$ with $P_{i_1 j_1}$, which is no smaller than $P_{i_2 j_2}$. Thus, in the serial algorithm y_k may only be replaced with a larger value as the algorithm progresses. Therefore, the max operation in the parallel algorithm ensures that y_k is replaced with the same value as in the serial algorithm, regardless of the order of the steps.

We present the source code of the parallel MLS in Listing 12.4. Here we reduce the number of writes to the global memory by using an indexing array keys_d to encode blocks and subsequently performing a scan operation with the maximum operator and indexed by keys_d, so that maximum is taken within each block.

Listing 12.4. fragments of implementation of a parallel version of the MLS algorithm using Thrust library.

```
template<typename Tx>
__device__ Tx Aver(Tx z, int i, int j, Tx *z)
    { return (z-z[j+1])/(j-i+1); }

template<typename Tx>
__global__ void monotonizekernel(Tx *y, Tx *z, Tx *u, int *key, int n)
{
    int i = threadIdx.x + blockIdx.x * blockDim.x;
    if(i<n) {
        int smallestJ = i;
        Tx curP, smallestP, curz=z[i];
        smallestP=Aver(curz,i,i,z);
        for(int j = i+1; j < n; j++) {
            curP=Aver(curz,i,j,z);
            if(smallestP>curP) {
                smallestJ = j;
                smallestP = curP;
            }
        }
        curP=y[i];
        if(curP > smallestP)
```

```
                t=smallestP ;
           else
                  smallestJ=i ;
            key[i]=smallestJ ;
            u[i]=t ;
      }
}

template< typename Tx >
void MonotonizeData(Tx *y, int n, Tx *u)
{
      thrust::less_equal<int> binary_pred ;
      thrust::maximum<Tx>       binary_op2 ;
      thrust::device_vector<Tx> z_d(n+1) ;
      thrust::device_vector<int> keys_d(n) ;
      thrust::device_ptr<Tx> y_d(y) , u_d(u) ;
      thrust::fill(u_d, u_d+n, -1e100) ;
      thrust::fill(keys_d.begin(), keys_d.end(), 0) ;

      thrust::reverse_iterator< typename thrust::device_vector<Tx>::
          iterator>  y_reverse_b(y_d+n), y_reverse_end(y_d), z_reverse_b
          (z_d.end()) ;

      thrust::inclusive_scan(y_reverse_b, y_reverse_end, z_reverse_b+1) ;

      monotonizekernel<<<grid, block>>>(y, thrust::raw_pointer_cast(&z_d
          [0]), u, thrust::raw_pointer_cast(&keys_d[0]), n ) ;

      thrust::sort(keys_d.begin(), keys_d.end()) ;
      thrust::inclusive_scan_by_key(keys_d.begin(), keys_d.end(), u_d,
          u_d, binary_pred , binary_op2) ;
}
```

As we mentioned, the complexity of the MLS algorithm is $O(n^2)$, due to the fact that for each datum, the smallest average P_{ij} of the blocks of subsequent data is needed. Thus, each thread needs to perform $O(n)$ comparisons (the averages themselves are precomputed in $O(n)$ operations using the partial sum algorithm). It is interesting to compare the runtime of the PAVA algorithm on CPU and parallel MLS on GPU to establish for which n parallel MLS is preferable. We performed such experiments on Tesla 2050 device connected to a four-core Intel i7 CPU with 4 GB RAM clocked at 2.8 GHz, running Linux (Fedora 16).

First we compared the serial versions of PAV and MLS algorithms. For this we used two packages in R environment, `stats` and `fdrtool`. The package `stats` offers function `isoreg`, which implements the MLS algorithm in C language, whereas package `fdrtool` offers PAVA, also implemented in C. Overheads of R environment can be neglected, as the input data are simply passed to C code, so we can compare the running time of both algorithms head to head. We generated input data of varying lengths n from 10^4 to 5×10^7 randomly, using $y_i = f(x_i) + \varepsilon_i$, where f is a monotone test function and ε is random noise. We also tried completely ordered isotone data, and antitone data, to check the performance for adversary inputs. Subsequently, we measured the runtime of our parallel version of MLS algorithm on Tesla 2050 GPU. The results are presented in Table 12.1.

As expected, the runtimes of both methods differed significantly, as shown in Table 12.1, and clearly linear PAVA was superior to serial MLS algorithm.

Data	PAVA	MLS	GPU MLS
monotone increasing f			
$n = 5 \times 10^4$	0.01	5	0.092
$n = 10^5$	0.03	40	0.35
$n = 5 \times 10^5$	0.4	1001	8.6
$n = 10^6$	0.8	5000	38
$n = 2 \times 10^6$	1.6	–	152
$n = 10 \times 10^6$	2	–	3500
$n = 20 \times 10^6$	4.5	–	–
$n = 50 \times 10^6$	12	–	–
constant or decreasing f			
$n = 10^6$	0.2	0.1	38
$n = 10 \times 10^6$	1.9	1.9	3500
$n = 20 \times 10^6$	3.5	4.0	–
$n = 50 \times 10^6$	11	11	–

TABLE 12.1. The average CPU time (sec) of the serial PAVA, MLS, and parallel MLS algorithms.

Even though for some special cases, e.g., test function $f = const$, both serial methods gave the same running time; this can be explained by the fact that large blocks of data allowed MLS to skip the majority of tests. This did not happen in the parallel version of MLS, where for each datum the smallest value of P_{ij^*} was computed (in parallel), so the average CPU times were the same for all data.

From the results in Table 12.1 we conclude that serial PAVA is superior to MLS for $n > 10^4$. While it is possible to transfer data from GPU to CPU and run PAVA there, it is warranted only for sufficiently large data $n \geq 5 \times 10^5$, for otherwise the data transfer overheads will dominate CPU time. For smaller n, isotone regression is best performed on GPU.

We also see that the use of GPU accelerated MLS by a factor of at least 100, except for antitone data. The cost of serial MLS is prohibitive for $n > 10^6$.

We should mention that not all isotone regression problems allow a PAV-like algorithm linear in time. When the data may contain large outliers, monotonizing the data is better done not in the least squares sense, but using other cost functionals, such as by minimizing the sum of absolute deviations [30] or using M-estimators [2], which are less sensitive to outliers. It is interesting than in all such cases the solution to an isotone regression problem can be found by solving maximin problem

$$u_i = \max_{k \leq i} \min_{l \geq i} \hat{y}(k, l),$$

with $\hat{y}(k, l)$ being the unrestricted maximum likelihood estimator of $y_k \ldots, y_l$. For the quadratic cost function $\hat{y}(k, l)$ corresponds to the mean of these data (as in PAV and MLS algorithms), for the absolute deviations $\hat{y}(k, l)$ corresponds to the median, and for other cost functions it corresponds to an M-estimator of location. The MLS algorithm can be applied to such isotone regression problems with very little modification. However, we are unaware of other algorithms for solving the modified problem that linear in time. Our parallel MLS algorithm will be valuable in such cases.

12.4 Conclusion

We presented three GPU-based parallel algorithms for approximating monotone data: monotone quadratic spline, monotone Hermite rational spline, and minimum lower sets algorithm for monotonizing noisy data. These tools are valuable in a number of applications that involve large data sets modeled by monotone nonlinear functions. The source code of the package monospline is available from www.deakin.edu.au/~gleb/monospline.html

Bibliography

[1] M. Abbas, A. A. Majid, M. N. H. Awang, and J. M. Ali. Monotonicity preserving interpolation using rational spline. In *International MultiConference of Engineers and Computer Scientists (IMECS '11)*, volume 1, page 278–282, Hong Kong, 2011.

[2] E. Alvarez and V. Yohai. M-estimators for isotonic regression. *J. Stat. Planning and Inference*, 142:2351–2368, 2012.

[3] L-E. Andersson and T. Elfving. An algorithm for constrained interpolation. *SIAM J. Sci. Stat. Comput.*, 8:1012–1025, 1987.

[4] L.-E. Andersson and T. Elfving. Interpolation and approximation by monotone cubic splines. *Journal of Approximation Theory*, 66(3):302–333, 1991.

[5] G. Beliakov. Shape preserving approximation using least squares splines. *Approximation Theory and Applications*, 16:80–98, 2000.

[6] G. Beliakov. Least squares splines with free knots: global optimization approach. *Applied Mathematics and Computation*, 149:783–798, 2004.

[7] M. Best and N. Chakravarti. Active set algorithms for isotonic regression; a unifying framework. *Mathematical Programming*, 47:425–439, 1990.

[8] J. Butland. A method of interpolating reasonably-shaped curves through any data. In R.J. Lansdown, editor, *Computer Graphics '80*, pages 409–422, Middlesex, U.K., 1980. Online Publications Ltd.

[9] C. De Boor. *A Practical Guide to Splines*. Revised, Springer, New York, 2001.

[10] P. Dierckx. *Curve and Surface Fitting with Splines*. Clarendon Press, Oxford, 1995.

[11] T. Elfving and L.-E. Andersson. An algorithm for computing constrained smoothing spline functions. *Numer. Math.*, 52:583–595, 1989.

[12] F. N. Fritsch and R. E. Carlson. Monotone piecewise cubic interpolation. *SIAM J. Numer. Anal.*, 17:238–246, 1980.

[13] D. Göddeke and R. Strzodka. Cyclic reduction tridiagonal solvers on GPUs applied to mixed precision multigrid. *IEEE Transactions on Parallel and Distributed Systems*, 22:22–32, 2011.

[14] J. A. Gregory and R. Delbourgo. Piecewise rational quadratic interpolation to monotonic data. *IMA Journal of Numerical Analysis*, 2:123–130, 1982.

[15] J. Hoberock and N. Bell. Thrust, 2012. http://thrust.github.com/ last accessed July 31.

[16] J. C. Holladay. A smoothest curve approximation. *Mathematical Tables and Other Aids to Computation*, 11(60):233–243, 1957.

[17] D. L. B. Jupp. Approximation to data by splines with free knots. *SIAM J. Numer. Anal.*, 15:328–343, 1978.

[18] A. J. Kearsley. Projections onto order simplexes and isotonic regression. *J. Res. Natl. Inst. Stand. Technol.*, 111:121–125, 2006.

[19] B. Kvasov. *Methods of Shape Preserving Spline Approximation*. World Scientific, Singapore, 2000.

[20] T. Lyche and L. L. Schumaker. Computation of smoothing and interpolating natural splines via local bases. *SIAM J. Numer. Anal.*, 10:1027–1038, 1973.

[21] D. F. McAllister and J. A. Roulier. An algorithm for computing a shape-preserving oscillatory quadratic spline. *ACM Trans. Math. Software*, 7:331–347, 1981.

[22] E. Passow and J. A. Roulier. Monotone and convex spline interpolation. *SIAM J. Numer. Anal.*, 14:904–909, 1977.

[23] T. Robertson, F. T. Wright, and R. L. Dykstra. *Order Restricted Statistical Inference*. Wiley, Chichester, New York, 1988.

[24] N. S. Sapidis and P. D. Kaklis. An algorithm for constructing convexity and monotonicity-preserving splines in tension. *Computer Aided Geometric Design*, 5:127–137, 1988.

[25] N. S. Sapidis, P. D. Kaklis, and T. A. Loukakis. A method for computing the tension parameters in convexity-preserving spline-in-tension interpolation. *Numer. Math.*, 54:179–192, 1988.

[26] L. L. Schumaker. *Spline Functions: Basic Theory*. Wiley, New York, 1981.

[27] L. L. Schumaker. On shape preserving quadratic spline interpolation. *SIAM Journal on Numerical Analysis*, 20:854–864, 1983.

[28] D. G. Schweikert. An interpolation curve using a spline in tension. *J. Math. Phys.*, 45:312–317, 1966.

[29] H. Späth. Exponential spline interpolation. *Computing*, 4:225–233, 1969.

[30] Y. Wang and J. Huang. Limiting distribution for monotone median regression. *J. Stat. Planning and Inference*, 108:281–287, 2002.

Chapter 13

Solving sparse linear systems with GMRES and CG methods on GPU clusters

Lilia Ziane Khodja, Raphaël Couturier, and Jacques Bahi
Femto-ST Institute, University of Franche-Comte, France

13.1	Introduction	311
13.2	Krylov iterative methods	312
	13.2.1 CG method	313
	13.2.2 GMRES method	314
13.3	Parallel implementation on a GPU cluster	317
	13.3.1 Data partitioning	317
	13.3.2 GPU computing	317
	13.3.3 Data communications	319
13.4	Experimental results	321
13.5	Conclusion	327
	Bibliography	328

13.1 Introduction

Sparse linear systems are used to model many scientific and industrial problems, such as the environmental simulations or the industrial processing of the complex or nonNewtonian fluids. Moreover, the resolution of these problems often involves the solving of such linear systems that are considered the most expensive process in terms of execution time and memory space. Therefore, solving sparse linear systems must be as efficient as possible in order to deal with problems of ever increasing size.

There are, in the jargon of numerical analysis, different methods of solving sparse linear systems that can be classified in two classes: direct and iterative methods. However, the iterative methods are often more suitable than their counterparts, direct methods, to solve these systems. Indeed, they are less memory-consuming and easier to parallelize on parallel computers than direct methods. Different computing platforms, sequential and parallel computers, are used to solve sparse linear systems with iterative solutions. Nowadays, graphics processing units (GPUs) have become attractive to solve these systems, due to their computing power and their ability to compute faster than traditional CPUs.

In Section 13.2, we describe the general principle of two well-known iterative methods: the conjugate gradient method and the generalized minimal residual method. In Section 13.3, we give the main key points of the parallel implementation of both methods on a cluster of GPUs. Finally, in Section 13.4, we present the experimental results, obtained on a CPU cluster and on a GPU cluster of solving large sparse linear systems.

13.2 Krylov iterative methods

Let us consider the following system of n linear equations in $\mathbb{R}$:

$$Ax = b, \qquad (13.1)$$

where $A \in \mathbb{R}^{n \times n}$ is a sparse nonsingular square matrix, $x \in \mathbb{R}^n$ is the solution vector, $b \in \mathbb{R}^n$ is the right-hand side, and $n \in \mathbb{N}$ is a large integer number.

The iterative methods for solving the large sparse linear system (13.1) proceed by successive iterations of a same block of elementary operations, during which an infinite number of approximate solutions $\{x_k\}_{k \geq 0}$ is computed. Indeed, from an initial guess x_0, an iterative method determines at each iteration $k > 0$ an approximate solution x_k which, gradually, converges to the exact solution x^* as follows:

$$x^* = \lim_{k \to \infty} x_k = A^{-1}b. \qquad (13.2)$$

The number of iterations necessary to reach the exact solution x^* is not known beforehand and can be infinite. In practice, an iterative method often finds an approximate solution $\tilde{x}$ after a fixed number of iterations and/or when a given convergence criterion is satisfied as follows:

$$\|b - A\tilde{x}\| < \varepsilon, \qquad (13.3)$$

where $\varepsilon < 1$ is the required convergence tolerance threshold.

Some of the most iterative methods that have proven their efficiency for solving large sparse linear systems are those called *Krylov subspace methods* [9]. In the present chapter, we describe two Krylov methods which are widely used: the CG method (conjugate gradient method) and the GMRES method (generalized minimal residual method). In practice, the Krylov subspace methods are usually used with preconditioners that allow the improvement of their convergence. So, in what follows, the CG and GMRES methods are used to solve the left-preconditioned sparse linear system:

$$M^{-1}Ax = M^{-1}b, \qquad (13.4)$$

where M is the preconditioning matrix.

13.2.1 CG method

The conjugate gradient method was initially developed by Hestenes and Stiefel in 1952 [7]. It is one of the well-known iterative methods to solve large sparse linear systems. In addition, it can be adapted to solve nonlinear equations and optimization problems. However, it can only be applied to problems with positive definite symmetric matrices.

The main idea of the CG method is the computation of a sequence of approximate solutions $\{x_k\}_{k \geq 0}$ in a Krylov subspace of order k as follows:

$$x_k \in x_0 + \mathcal{K}_k(A, r_0), \tag{13.5}$$

such that the Galerkin condition must be satisfied:

$$r_k \perp \mathcal{K}_k(A, r_0), \tag{13.6}$$

where x_0 is the initial guess, $r_k = b - Ax_k$ is the residual of the computed solution x_k, and $\mathcal{K}_k$ the Krylov subspace of order k:

$$\mathcal{K}_k(A, r_0) \equiv \text{span}\{r_0, Ar_0, A^2 r_0, \ldots, A^{k-1} r_0\}.$$

In fact, CG is based on the construction of a sequence $\{p_k\}_{k \in \mathbb{N}}$ of direction vectors in $\mathcal{K}_k$ which are pairwise A-conjugate (A-orthogonal):

$$p_i^T A p_j = 0, \quad i \neq j. \tag{13.7}$$

At each iteration k, an approximate solution x_k is computed by recurrence as follows:

$$x_k = x_{k-1} + \alpha_k p_k, \quad \alpha_k \in \mathbb{R}. \tag{13.8}$$

Consequently, the residuals r_k are computed in the same way:

$$r_k = r_{k-1} - \alpha_k A p_k. \tag{13.9}$$

In the case where all residuals are nonzero, the direction vectors p_k can be determined so that the following recurrence holds:

$$p_0 = r_0, \quad p_k = r_k + \beta_k p_{k-1}, \quad \beta_k \in \mathbb{R}. \tag{13.10}$$

Moreover, the scalars $\{\alpha_k\}_{k>0}$ are chosen so as to minimize the A-norm error $\|x^* - x_k\|_A$ over the Krylov subspace $\mathcal{K}_k$, and the scalars $\{\beta_k\}_{k>0}$ are chosen so as to ensure that the direction vectors are pairwise A-conjugate. So, the assumption that matrix A is symmetric and the recurrences (13.9) and (13.10) allow the deduction that:

$$\alpha_k = \frac{r_{k-1}^T r_{k-1}}{p_k^T A p_k}, \quad \beta_k = \frac{r_k^T r_k}{r_{k-1}^T r_{k-1}}. \tag{13.11}$$

Algorithm 15 shows the main key points of the preconditioned CG method. It allows the solving the left-preconditioned sparse linear system (13.4). In

Algorithm 15: left-preconditioned CG method

1 Choose an initial guess x_0;
2 $r_0 = b - Ax_0$;
3 $convergence =$ false;
4 $k = 1$;
5 **repeat**
6     $z_k = M^{-1} r_{k-1}$;
7     $\rho_k = (r_{k-1}, z_k)$;
8   **if** $k = 1$ **then**
9        $p_k = z_k$;
10   **else**
11        $\beta_k = \rho_k / \rho_{k-1}$;
12        $p_k = z_k + \beta_k \times p_{k-1}$;
13   **end**
14     $q_k = A \times p_k$;
15     $\alpha_k = \rho_k / (p_k, q_k)$;
16     $x_k = x_{k-1} + \alpha_k \times p_k$;
17     $r_k = r_{k-1} - \alpha_k \times q_k$;
18   **if** $(\rho_k < \varepsilon)$ **or** $(k \geq maxiter)$ **then**
19        $convergence =$ true;
20   **else**
21        $k = k + 1$;
22   **end**
23 **until** $convergence$;

this algorithm, ε is the convergence tolerance threshold, $maxiter$ is the maximum number of iterations, and $(\cdot,\cdot)$ defines the dot product between two vectors in $\mathbb{R}^n$. At every iteration, a direction vector p_k is determined, so that it is orthogonal to the preconditioned residual z_k and to the direction vectors $\{p_i\}_{i<k}$ previously determined (from line 8 to line 13). Then, at lines 16 and 17, the iterate x_k and the residual r_k are computed using formulas (13.8) and (13.9), respectively. The CG method converges after, at most, n iterations. In practice, the CG algorithm stops when the tolerance threshold ε and/or the maximum number of iterations $maxiter$ is reached.

13.2.2 GMRES method

The iterative GMRES method was developed by Saad and Schultz in 1986 [10] as a generalization of the minimum residual method MINRES [8]. Indeed, GMRES can be applied for solving symmetric or nonsymmetric linear systems.

The main principle of the GMRES method is to find an approximation minimizing at best the residual norm. In fact, GMRES computes a sequence of approximate solutions $\{x_k\}_{k>0}$ in a Krylov subspace $\mathcal{K}_k$ as follows:

$$x_k \in x_0 + \mathcal{K}_k(A, v_1), \quad v_1 = \frac{r_0}{\|r_0\|_2}, \tag{13.12}$$

so that the Petrov-Galerkin condition is satisfied:

$$r_k \perp A\mathcal{K}_k(A, v_1). \tag{13.13}$$

GMRES uses the Arnoldi iterations [2] to construct an orthonormal basis V_k for the Krylov subspace $\mathcal{K}_k$ and an upper Hessenberg matrix $\bar{H}_k$ of order $(k+1) \times k$:

$$V_k = \{v_1, v_2, \ldots, v_k\}, \quad \forall k > 1, v_k = A^{k-1}v_1, \tag{13.14}$$

and

$$AV_k = V_{k+1}\bar{H}_k. \tag{13.15}$$

Then, at each iteration k, an approximate solution x_k is computed in the Krylov subspace $\mathcal{K}_k$ spanned by V_k as follows:

$$x_k = x_0 + V_k y, \quad y \in \mathbb{R}^k. \tag{13.16}$$

From both formulas (13.15) and (13.16) and $r_k = b - Ax_k$, we can deduce that

$$\begin{aligned} r_k &= b - A(x_0 + V_k y) \\ &= r_0 - AV_k y \\ &= \beta v_1 - V_{k+1}\bar{H}_k y \\ &- V_{k+1}(\beta e_1 - \bar{H}_k y), \end{aligned} \tag{13.17}$$

such that $\beta = \|r_0\|_2$ and $e_1 = (1, 0, \cdots, 0)$ is the first vector of the canonical basis of $\mathbb{R}^k$. So, the vector y is chosen in $\mathbb{R}^k$ so as to minimize at best the Euclidean norm of the residual r_k. Consequently, a linear least-squares problem of size k is solved:

$$\min_{y \in \mathbb{R}^k} \|r_k\|_2 = \min_{y \in \mathbb{R}^k} \|\beta e_1 - \bar{H}_k y\|_2. \tag{13.18}$$

The QR factorization of matrix $\bar{H}_k$ is used (the decomposition of the matrix $\bar{H}$ into Q and R matrices) to compute the solution of this problem by using Givens rotations [9, 10], such that

$$\bar{H}_k = Q_k R_k, \quad Q_k \in \mathbb{R}^{(k+1) \times (k+1)}, \quad R_k \in \mathbb{R}^{(k+1) \times k}, \tag{13.19}$$

where Q_k is an orthogonal matrix and R_k is an upper triangular matrix.

The GMRES method computes an approximate solution with a sufficient precision after, at most, n iterations (n is the size of the sparse linear system to be solved). However, the GMRES algorithm must construct and store in the memory an orthonormal basis V_k whose size is proportional to the number of iterations required to achieve the convergence. Then, to avoid a huge memory storage, the GMRES method must be restarted at each m iterations, such

Algorithm 16: left-preconditioned GMRES method with restarts

1 Choose an initial guess x_0;
2 $convergence = $ false;
3 $k = 1$;
4 $r_0 = M^{-1}(b - Ax_0)$;
5 $\beta = \|r_0\|_2$;
6 **while** $\neg convergence$ **do**
7 $\quad v_1 = r_0/\beta$;
8 $\quad$ **for** $j = 1$ **to** m **do**
9 $\quad\quad w_j = M^{-1}Av_j$;
10 $\quad\quad$ **for** $i = 1$ **to** j **do**
11 $\quad\quad\quad h_{i,j} = (w_j, v_i)$;
12 $\quad\quad\quad w_j = w_j - h_{i,j}v_i$;
13 $\quad\quad$ **end**
14 $\quad\quad h_{j+1,j} = \|w_j\|_2$;
15 $\quad\quad v_{j+1} = w_j/h_{j+1,j}$;
16 $\quad$ **end**
17 $\quad$ Set $V_m = \{v_j\}_{1 \leq j \leq m}$ and $\bar{H}_m = (h_{i,j})$ is an upper Hessenberg matrix of size $(m+1) \times m$;
18 $\quad$ Solve a least-squares problem of size m: $min_{y \in \mathbb{R}^m}\|\beta e_1 - \bar{H}_m y\|_2$;
19 $\quad x_m = x_0 + V_m y_m$;
20 $\quad r_m = M^{-1}(b - Ax_m)$;
21 $\quad \beta = \|r_m\|_2$;
22 $\quad$ **if** $(\beta < \varepsilon)$ **or** $(k \geq maxiter)$ **then**
23 $\quad\quad convergence = $ true;
24 $\quad$ **else**
25 $\quad\quad x_0 = x_m$;
26 $\quad\quad r_0 = r_m$;
27 $\quad\quad k = k + 1$;
28 $\quad$ **end**
29 **end**

that m is very small ($m \ll n$), and with x_m as the initial guess to the next iteration. This allows the limitation of the size of the basis V to m orthogonal vectors.

Algorithm 16 shows the key points of the GMRES method with restarts. It solves the left-preconditioned sparse linear system (13.4), such that M is the preconditioning matrix. At each iteration k, GMRES uses the Arnoldi iterations (defined from line 7 to line 17) to construct a basis V_m of m orthogonal vectors and an upper Hessenberg matrix $\bar{H}_m$ of size $(m+1) \times m$. Then, it solves the linear least-squares problem of size m to find the vector $y \in \mathbb{R}^m$ which minimizes at best the residual norm (line 18). Finally, it computes an approximate solution x_m in the Krylov subspace spanned by V_m (line 19).

The GMRES algorithm is stopped when the residual norm is sufficiently small ($\|r_m\|_2 < \varepsilon$) and/or the maximum number of iterations ($maxiter$) is reached.

13.3 Parallel implementation on a GPU cluster

In this section, we present the parallel algorithms of both iterative CG and GMRES methods for GPU clusters. The implementation is performed on a GPU cluster composed of different computing nodes, such that each node is a CPU core managed by one MPI (message passing interface) process and equipped with a GPU card. The parallelization of these algorithms is carried out by using the MPI communication routines between the GPU computing nodes and the CUDA (compute unified device architecture) programming environment inside each node. In what follows, the algorithms of the iterative methods are called iterative solvers.

13.3.1 Data partitioning

The parallel solving of the large sparse linear system (13.4) requires a data partitioning between the computing nodes of the GPU cluster. Let p denote the number of the computing nodes on the GPU cluster. The partitioning operation consists of the decomposition of the vectors and matrices, involved in the iterative solver, in p portions. Indeed, this operation allows the assignment to each computing node i:

- a portion of size $\frac{n}{p}$ elements of each vector,
- a sparse rectangular submatrix A_i of size $(\frac{n}{p}, n)$, and
- a square preconditioning submatrix M_i of size $(\frac{n}{p}, \frac{n}{p})$,

where n is the size of the sparse linear system to be solved. In the first instance, we perform a naive row-wise partitioning (row-by-row decomposition) on the data of the sparse linear systems to be solved. Figure 13.1 shows an example of a row-wise data partitioning between four computing nodes of a sparse linear system (sparse matrix A, solution vector x, and right-hand side b) of size 16 unknown values.

13.3.2 GPU computing

After the partitioning operation, all the data involved from this operation must be transferred from the CPU memories to the GPU memories, in order to be processed by GPUs. We use two functions of the CUBLAS library (CUDA Basic Linear Algebra Subroutines) developed by

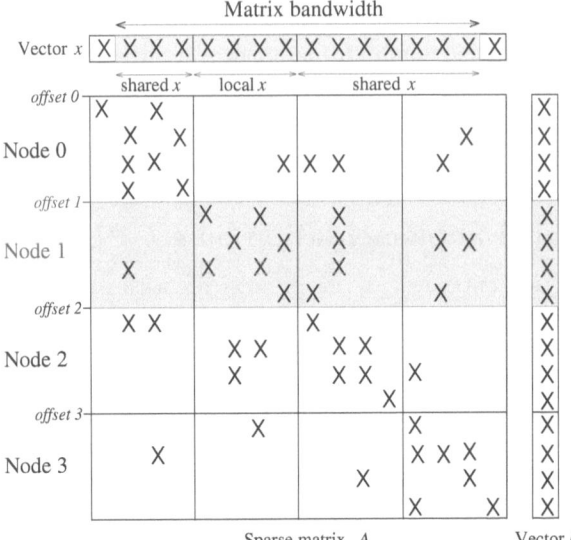

FIGURE 13.1. A data partitioning of the sparse matrix A, the solution vector x, and the right-hand side b into four portions.

NVIDIA [4]: cublasAlloc() for the memory allocations on GPUs and cublasSetVector() for the memory copies from the CPUs to the GPUs.

An efficient implementation of CG and GMRES solvers on a GPU cluster requires the determining of all parts of their codes that can be executed in parallel and, thus, takes advantage of the GPU acceleration. As many Krylov subspace methods, the CG and GMRES methods are mainly based on arithmetic operations dealing with vectors or matrices: sparse matrix-vector multiplications, scalar-vector multiplications, dot products, Euclidean norms, AXPY operations ($y \leftarrow ax + y$ where x and y are vectors and a is a scalar), and so on. These vector operations are often easy to parallelize and they are more efficient on parallel computers when they work on large vectors. Therefore, all the vector operations used in CG and GMRES solvers must be executed by the GPUs as kernels.

We use the kernels of the CUBLAS library to compute some vector operations of CG and GMRES solvers. The following kernels of CUBLAS (dealing with double floating point) are used: cublasDdot() for the dot products, cublasDnrm2() for the Euclidean norms, and cublasDaxpy() for the AXPY operations ($y \leftarrow ax + y$, compute a scalar-vector product and add the result to a vector). For the rest of the data-parallel operations, we code their kernels in CUDA. In the CG solver, we develop a kernel for the XPAY operation ($y \leftarrow x + ay$) used in line 12 in Algorithm 15. In the GMRES solver, we program a kernel for the scalar-vector multiplication (lines 7 and 15 in

Algorithm 16), a kernel to solve the least-squares problem, and a kernel to update the elements of the solution vector x.

The least-squares problem in the GMRES method is solved by performing a QR factorization on the Hessenberg matrix $\bar{H}_m$ with plane rotations and, then, solving the triangular system by backward substitutions to compute y. Consequently, solving the least-squares problem on the GPU is not efficient. Indeed, the triangular solves are not easy to parallelize and inefficient on GPUs. However, the least-squares problem to solve in the GMRES method with restarts has, generally, a very small size m. Therefore, we develop an inexpensive kernel which must be executed by a single CUDA thread.

The most important operation in CG and GMRES methods is the SpMV multiplication (sparse matrix-vector multiplication), because it is often an expensive operation in terms of execution time and memory space. Moreover, it requires taking care of the storage format of the sparse matrix in the memory. Indeed, the naive storage, row-by-row or column-by-column, of a sparse matrix can cause a significant waste of memory space and execution time. In addition, the sparse nature of the matrix often leads to irregular memory accesses to read the matrix nonzero values. So, the computation of the SpMV multiplication on GPUs can involve noncoalesced accesses to the global memory, which slows down its performances even more. One of the most efficient compressed storage formats of sparse matrices on GPUs is the HYB (hybrid) format [3]. It is a combination of ELLpack (ELL) and Coordinate (COO) formats. Indeed, it stores a typical number of nonzero values per row in ELL format and the remaining entries of exceptional rows in COO format. It combines the efficiency of ELL due to the regularity of its memory accesses and the flexibility of COO which is insensitive to the matrix structure. Consequently, we use the HYB kernel [1] developed by NVIDIA to implement the SpMV multiplication of CG and GMRES methods on GPUs. Moreover, to avoid the noncoalesced accesses to the high-latency global memory, we fill the elements of the iterate vector x in the cached texture memory.

13.3.3 Data communications

All the computing nodes of the GPU cluster execute in parallel the same iterative solver (Algorithm 15 or Algorithm 16) adapted to GPUs, but on their own portions of the sparse linear system: $M_i^{-1} A_i x_i = M_i^{-1} b_i$, $0 \leq i < p$. However, in order to solve the complete sparse linear system (13.4), synchronizations must be performed between the local computations of the computing nodes over the cluster. In what follows, two computing nodes sharing data are called neighboring nodes.

As already mentioned, the most important operation of CG and GMRES methods is the SpMV multiplication. In the parallel implementation of the iterative methods, each computing node i performs the SpMV multiplication on its own sparse rectangular submatrix A_i. Locally, it has only subvectors of size $\frac{n}{p}$ corresponding to rows of its submatrix A_i. However, it also requires

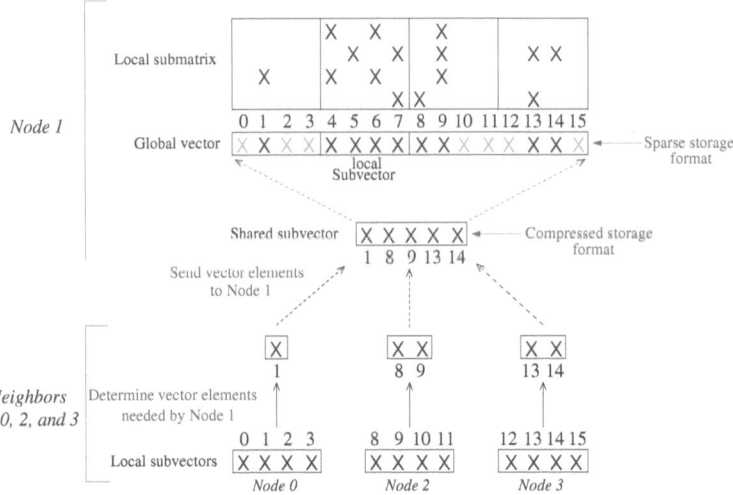

FIGURE 13.2. Data exchanges between *Node 1* and its neighbors *Node 0*, *Node 2*, and *Node 3*.

the vector elements of its neighbors, corresponding to the column indices on which its submatrix has nonzero values (see Figure 13.1). So, in addition to the local vectors, each node must also manage vector elements shared with neighbors and required to compute the SpMV multiplication. Therefore, the iterate vector x managed by each computing node is composed of a local subvector x^{local} of size $\frac{n}{p}$ and a subvector of shared elements x^{shared}. In the same way, the vector used to construct the orthonormal basis of the Krylov subspace (vectors p and v in CG and GMRES methods, respectively) is composed of a local subvector and a shared subvector.

Therefore, before computing the SpMV multiplication, the neighboring nodes over the GPU cluster must exchange between them the shared vector elements necessary to compute this multiplication. First, each computing node determines, in its local subvector, the vector elements needed by other nodes. Then, the neighboring nodes exchange between them these shared vector elements. The data exchanges are implemented by using the MPI point-to-point communication routines: blocking sends with MPI_Send() and nonblocking receives with MPI_Irecv(). Figure 13.2 shows an example of data exchanges between *Node 1* and its neighbors *Node 0*, *Node 2*, and *Node 3*. In this example, the iterate matrix A split between these four computing nodes is that presented in Figure 13.1.

After the synchronization operation, the computing nodes receive, from their respective neighbors, the shared elements in a subvector stored in a compressed format. However, in order to compute the SpMV multiplication, the computing nodes operate on sparse global vectors (see Figure 13.2). In this case, the received vector elements must be copied to the corresponding

Solving linear systems with GMRES and CG methods on GPU clusters   321

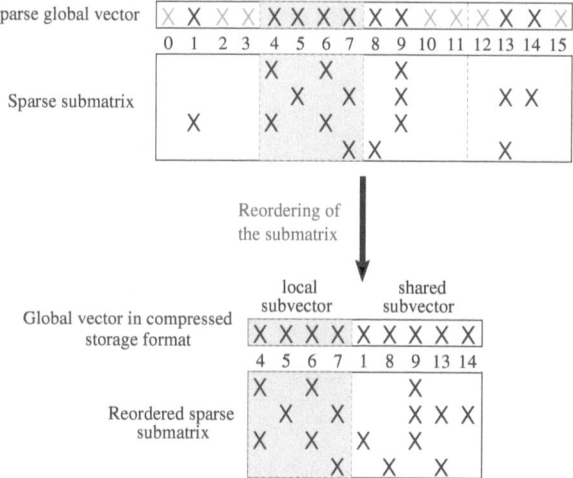

FIGURE 13.3. Columns reordering of a sparse submatrix.

indices in the global vector. So as not to need to perform this at each iteration, we propose to reorder the columns of each submatrix $\{A_i\}_{0\leq i<p}$, so that the shared subvectors could be used in their compressed storage formats. Figure 13.3 shows a reordering of a sparse submatrix (submatrix of *Node 1*).

A GPU cluster is a parallel platform with a distributed memory. So, the synchronizations and communication data between GPU nodes are carried out by passing messages. However, a GPU cannot exchange data with other GPUs in a direct way. Then, CPUs via MPI processes are in charge of the synchronizations within the GPU cluster. Consequently, the vector elements to be exchanged must be copied from the GPU memory to the CPU memory and vice versa before and after the synchronization operation between CPUs. We have used the CUBLAS communication subroutines to perform the data transfers between a CPU core and its GPU: cublasGetVector() and cublasSetVector(). Finally, in addition to the data exchanges, GPU nodes perform reduction operations to compute in parallel the dot products and Euclidean norms. This is implemented by using the MPI global communication MPI_Allreduce().

13.4 Experimental results

In this section, we present the performances of the parallel CG and GMRES linear solvers obtained on a cluster of 12 GPUs. Indeed, this GPU cluster of tests is composed of six machines connected by a 20GB/s InfiniBand net-

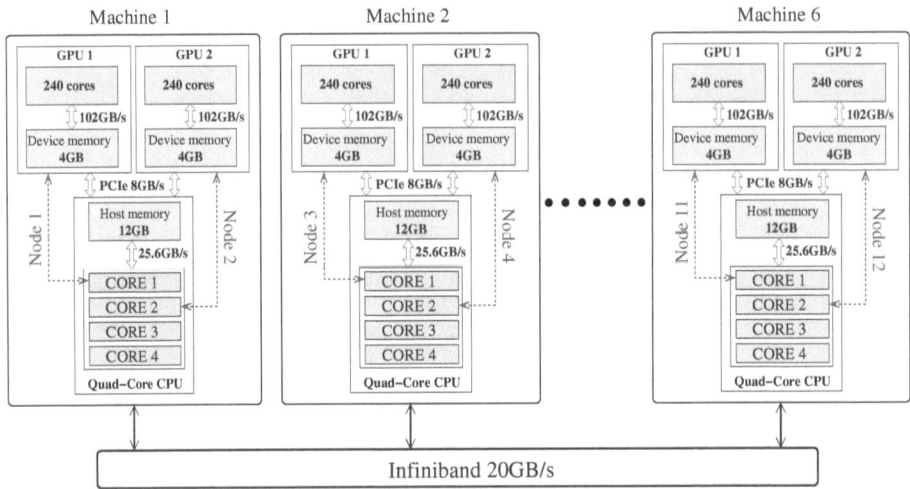

FIGURE 13.4. General scheme of the GPU cluster of tests composed of six machines, each with two GPUs.

work. Each machine is a Quad-Core Xeon E5530 CPU running at 2.4GHz and providing 12GB of RAM with a memory bandwidth of 25.6GB/s. In addition, two Tesla C1060 GPUs are connected to each machine via a PCI-Express 16x Gen 2.0 interface with a throughput of 8GB/s. A Tesla C1060 GPU contains 240 cores running at 1.3GHz and providing a global memory of 4GB with a memory bandwidth of 102GB/s. Figure 13.4 shows the general scheme of the GPU cluster that we used in the experimental tests.

Linux cluster version 2.6.39 OS is installed on CPUs. C programming language is used to code the parallel algorithms of both methods on the GPU cluster. CUDA version 4.0 [5] is used to program GPUs, using the CUBLAS library [4] to deal with vector operations in GPUs and, finally, MPI routines of OpenMPI 1.3.3 are used to carry out the communications between CPU cores. Indeed, the experiments are done on a cluster of 12 computing nodes, where each node is managed by one MPI process and is composed of one CPU core and one GPU card.

All tests are made on double-precision floating point operations. The parameters of both linear solvers are initialized as follows: the residual tolerance threshold $\varepsilon = 10^{-12}$, the maximum number of iterations $maxiter = 500$, the right-hand side b is filled with 1.0, and the initial guess x_0 is filled with 0.0. In addition, we limited the Arnoldi iterations used in the GMRES method to 16 iterations ($m = 16$). For the sake of simplicity, we have chosen the preconditioner M as the main diagonal of the sparse matrix A. Indeed, it allows us to easily compute the required inverse matrix M^{-1}, and it provides a relatively good preconditioning for not too ill-conditioned matrices. In the GPU computing, the size of thread blocks is fixed to 512 threads. Finally, the

Solving linear systems with GMRES and CG methods on GPU clusters

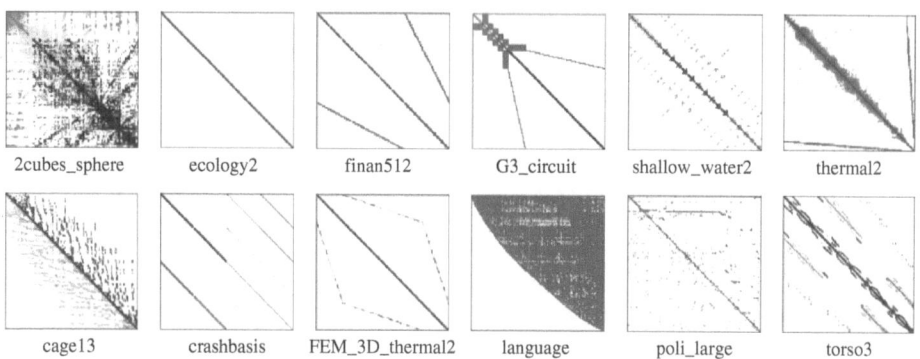

FIGURE 13.5. Sketches of sparse matrices chosen from the University of Florida collection.

Matrix Type	Matrix Name	# Rows	# Nonzeros	Bandwidth
Symmetric	2cubes_sphere	101,492	1,647,264	100,464
	ecology2	999,999	4,995,991	2,001
	finan512	74,752	596,992	74,725
	G3_circuit	1,585,478	7,660,826	1,219,059
	shallow_water2	81,920	327,680	58,710
	thermal2	1,228,045	8,580,313	1,226,629
Nonsymmetric	cage13	445,315	7,479,343	318,788
	crashbasis	160,000	1,750,416	120,202
	FEM_3D_thermal2	147,900	3,489.300	117,827
	language	399,130	1,216,334	398,622
	poli_large	15,575	33,074	15,575
	torso3	259,156	4,429,042	216,854

TABLE 13.1. Main characteristics of sparse matrices chosen from the University of Florida collection.

performance results, presented hereafter, are obtained from the mean value over 10 executions of the same parallel linear solver and for the same input data.

To get more realistic results, we have tested the CG and GMRES algorithms on sparse matrices of the University of Florida collection [6], that arise in a wide spectrum of real-world applications. We have chosen six symmetric sparse matrices and six nonsymmetric ones from this collection. In Figure 13.5, we show the structures of these matrices and in Table 13.1 we present their main characteristics which are the number of rows, the total number of nonzero values, and the maximal bandwidth. In the present chapter, the bandwidth of a sparse matrix is defined as the number of matrix columns separating the first and the last nonzero value on a matrix row.

Tables 13.2 and 13.3 show the performances of the parallel CG and GM-

Matrix	Time$_{cpu}$	Time$_{gpu}$	τ	# Iter.	Prec.	Δ
2cubes_sphere	0.132s	0.069s	1.93	12	1.14e-09	3.47e-18
ecology2	0.026s	0.017s	1.52	13	5.06e-09	8.33e-17
finan512	0.053s	0.036s	1.49	12	3.52e-09	1.66e-16
G3_circuit	0.704s	0.466s	1.51	16	4.16e-10	4.44e-16
shallow_water2	0.017s	0.010s	1.68	5	2.24e-14	3.88e-26
thermal2	1.172s	0.622s	1.88	15	5.11e-09	3.33e-16

TABLE 13.2. Performances of the parallel CG method on a cluster of 24 CPU cores vs. on a cluster of 12 GPUs.

Matrix	Time$_{cpu}$	Time$_{gpu}$	τ	# Iter.	Prec.	Δ
2cubes_sphere	0.234s	0.124s	1.88	21	2.10e-14	3.47e-18
ecology2	0.076s	0.035s	2.15	21	4.30e-13	4.38e-15
finan512	0.073s	0.052s	1.40	17	3.21e-12	5.00e-16
G3_circuit	1.016s	0.649s	1.56	22	1.04e-12	2.00e-15
shallow_water2	0.061s	0.044s	1.38	17	5.42e-22	2.71e-25
thermal2	1.666s	0.880s	1.89	21	6.58e-12	2.77e-16
cage13	0.721s	0.338s	2.13	26	3.37e-11	2.66e-15
crashbasis	1.349s	0.830s	1.62	121	9.10e-12	6.90e-12
FEM_3D_thermal2	0.797s	0.419s	1.90	64	3.87e-09	9.09e-13
language	2.252s	1.204s	1.87	90	1.18e-10	8.00e-11
poli_large	0.097s	0.095s	1.02	69	4.98e-11	1.14e-12
torso3	4.242s	2.030s	2.09	175	2.69e-10	1.78e-14

TABLE 13.3. Performances of the parallel GMRES method on a cluster 24 CPU cores vs. on cluster of 12 GPUs.

RES solvers, respectively, for solving linear systems associated to the sparse matrices presented in Table 13.1. They allow us to compare the performances obtained on a cluster of 24 CPU cores and on a cluster of 12 GPUs. However, Table 13.2 shows the performances of solving only symmetric sparse linear systems, due to the inability of the CG method to solve the nonsymmetric systems. In both tables, the second and third columns give, respectively, the execution times in seconds obtained on 24 CPU cores ($Time_{cpu}$) and that obtained on 12 GPUs ($Time_{gpu}$). Moreover, we take into account the relative gains τ of a solver implemented on the GPU cluster compared to the same solver implemented on the CPU cluster. The relative gains, presented in the fourth column, are computed as a ratio of the CPU execution time over the GPU execution time:

$$\tau = \frac{Time_{cpu}}{Time_{gpu}}. \tag{13.20}$$

In addition, Tables 13.2 and 13.3 give the number of iterations ($iter$), the precision (*prec*) of the solution computed on the GPU cluster, and the difference Δ between the solution computed on the CPU cluster and that computed on the GPU cluster. Both parameters *prec* and Δ allow us to validate and verify the accuracy of the solution computed on the GPU cluster. We have computed them as follows:

$$\Delta = max|x^{cpu} - x^{gpu}|, \tag{13.21}$$

$$prec = max|M^{-1}r^{gpu}|, \tag{13.22}$$

where Δ is the maximum vector element, in absolute value, of the difference between the two solutions x^{cpu} and x^{gpu} computed, respectively, on CPU and GPU clusters and *prec* is the maximum element, in absolute value, of the residual vector $r^{gpu} \in \mathbb{R}^n$ of the solution x^{gpu}. Thus, we can see that the solutions obtained on the GPU cluster were computed with a sufficient accuracy (about 10^{-10}) and they are, more or less, equivalent to those computed on the CPU cluster with a small difference ranging from 10^{-10} to 10^{-26}. However, we can notice from the relative gains τ that it is not efficient to use multiple GPUs for solving small sparse linear systems. In fact, a small sparse matrix does not allow us to maximize utilization of GPU cores. In addition, the communications required to synchronize the computations over the cluster increase the idle times of GPUs and slow down the parallel computations further.

Consequently, in order to test the performances of the parallel solvers, we developed in C programming language a generator of large sparse matrices. This generator takes a matrix from the University of Florida collection [6] as an initial matrix to build large sparse matrices exceeding ten million rows. It must be executed in parallel by the MPI processes of the computing nodes, so that each process can build its sparse submatrix. In the first experimental tests, we focused on sparse matrices having a banded structure, because they are those arising the most in the majority of numerical problems. So to generate the global sparse matrix, each MPI process constructs its submatrix by performing several copies of an initial sparse matrix chosen from the University of Florida collection. Then, it puts all these copies on the main diagonal of the global matrix (see Figure 13.6). Moreover, the empty spaces between two successive copies in the main diagonal are filled with subcopies (left-copy and right-copy in Figure 13.6) of the same initial matrix.

We have used the parallel CG and GMRES algorithms for solving sparse linear systems of 25 million unknown values. The sparse matrices associated to these linear systems are generated from those presented in Table 13.1. Their main characteristics are given in Table 13.4. Tables 13.5 and 13.6 show the performances of the parallel CG and GMRES solvers, respectively, obtained on a cluster of 24 CPU cores and on a cluster of 12 GPUs. Obviously, we can notice from these tables that solving large sparse linear systems on a GPU cluster is more efficient than on a CPU cluster (see relative gains τ). We can also notice that the execution times of the CG method, whether in a CPU cluster or in a GPU cluster, are better than those of the GMRES

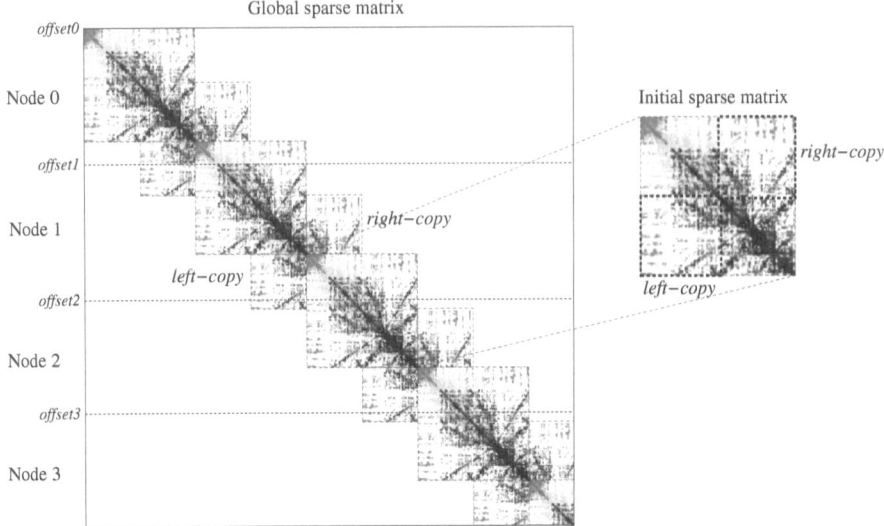

FIGURE 13.6. Parallel generation of a large sparse matrix by four computing nodes.

Matrix Type	Matrix Name	# Nonzeros	Bandwidth
Symmetric	2cubes_sphere	413, 703, 602	198, 836
	ecology2	124, 948, 019	2, 002
	finan512	278, 175, 945	123, 900
	G3_circuit	125, 262, 292	1, 891, 887
	shallow_water2	100, 235, 292	62, 806
	thermal2	175, 300, 284	2, 421, 285
Nonsymmetric	cage13	435, 770, 480	352, 566
	crashbasis	409, 291, 236	200, 203
	FEM_3D_thermal2	595, 266, 787	206, 029
	language	76, 912, 824	398, 626
	poli_large	53, 322, 580	15, 576
	torso3	433, 795, 264	328, 757

TABLE 13.4. Main characteristics of sparse banded matrices generated from those of the University of Florida collection.

method for solving large symmetric linear systems. In fact, the CG method is characterized by a better convergence rate and a shorter execution time of an iteration than those of the GMRES method. Moreover, an iteration of the parallel GMRES method requires more data exchanges between computing nodes compared to the parallel CG method.

Matrix	Time$_{cpu}$	Time$_{gpu}$	τ	# Iter.	Prec.	Δ
2cubes_sphere	1.625s	0.401s	4.05	14	5.73e-11	5.20e-18
ecology2	0.856s	0.103s	8.27	15	3.75e-10	1.11e-16
finan512	1.210s	0.354s	3.42	14	1.04e-10	2.77e-16
G3_circuit	1.346s	0.263s	5.12	17	1.10e-10	5.55e-16
shallow_water2	0.397s	0.055s	7.23	7	3.43e-15	5.17e-26
thermal2	1.411s	0.244s	5.78	16	1.67e-09	3.88e-16

TABLE 13.5. Performances of the parallel CG method for solving linear systems associated to sparse banded matrices on a cluster of 24 CPU cores vs. on a cluster of 12 GPUs.

Matrix	Time$_{cpu}$	Time$_{gpu}$	τ	# Iter.	Prec.	Δ
2cubes_sphere	3.597s	0.514s	6.99	21	2.11e-14	8.67e-18
ecology2	2.549s	0.288s	8.83	21	4.88e-13	2.08e-14
finan512	2.660s	0.377s	7.05	17	3.22e-12	8.82e-14
G3_circuit	3.139s	0.480s	6.53	22	1.04e-12	5.00e-15
shallow_water2	2.195s	0.253s	8.68	17	5.54e-21	7.92e-24
thermal2	3.206s	0.463s	6.93	21	8.89e-12	3.33e-16
cage13	5.560s	0.663s	8.39	26	3.29e-11	1.59e-14
crashbasis	25.802s	3.511s	7.35	135	6.81e-11	4.61e-15
FEM_3D_thermal2	13.281s	1.572s	8.45	64	3.88e-09	1.82e-12
language	12.553s	1.760s	7.13	89	2.11e-10	1.60e-10
poli_large	8.515s	1.053s	8.09	69	5.05e-11	6.59e-12
torso3	31.463s	3.681s	8.55	175	2.69e-10	2.66e-14

TABLE 13.6. Performances of the parallel GMRES method for solving linear systems associated to sparse banded matrices on a cluster of 24 CPU cores vs. on a cluster of 12 GPUs.

13.5 Conclusion

In this chapter, we have aimed at harnessing the computing power of a cluster of GPUs for solving large sparse linear systems. For this, we have used two Krylov subspace iterative methods: the CG and GMRES methods. The first method is well known for its efficiency to solve symmetric linear systems and the second one is used, particularly, to solve nonsymmetric linear systems.

We have presented the parallel implementation of both iterative methods on a GPU cluster. Particularly, the operations dealing with the vectors and/or matrices, of these methods, are parallelized between the different GPU computing nodes of the cluster. Indeed, the data-parallel vector operations are accelerated by GPUs, and the communications required to synchronize the parallel computations are carried out by CPU cores. For this, we have used heterogeneous CUDA/MPI programming to implement the parallel iterative algorithms.

In the experimental tests, we have shown that using a GPU cluster is efficient for solving linear systems associated to very large sparse matrices. The experimental results, discussed in the present chapter, show that a cluster of 12 GPUs is about 7 times faster than a cluster of 24 CPU cores for solving large sparse linear systems of 25 million unknown values. This is due to the GPUs ability to compute the data-parallel operations faster than the CPUs.

In our future works, we plan to test the parallel algorithms of CG and GMRES methods, adapted to GPUs, for solving large linear systems associated to sparse matrices of different structures. For example, the matrices having large bandwidths can lead to many data dependencies between the computing nodes and, thus, degrade the performances of both algorithms. So in this case, it would be interesting to study the different data partitioning techniques, in order to minimize the dependencies between the computing nodes and thus to reduce the total communication volume. This may improve the performances of both algorithms implemented on a GPU cluster. Moreover, in the recent GPU hardware and software architectures, the GPU-Direct system with CUDA version 5.0 is used so that two GPUs located on the same node or on distant nodes can communicate between each other directly without CPUs. This allows us to improve the data transfers between GPUs.

Bibliography

[1] CUSP Library. http://code.google.com/p/cusp-library/.

[2] W.E. Arnoldi. The principle of minimized iteration in the solution of the matrix eigenvalue problem. *Quarterly of Applied Mathematics*, 9(17):17–29, 1951.

[3] N. Bell and M. Garland. Efficient sparse matrix-vector multiplication on CUDA. NVIDIA Technical Report NVR-2008-004, NVIDIA Corporation, December 2008.

[4] NVIDIA Corporation. CUDA Toolkit 4.2 CUBLAS Library. 2012. http://developer.download.nvidia.com/compute/DevZone/docs/html/CUDALibraries/doc/CUBLAS_Library.pdf.

[5] NVIDIA Corporation. NVIDIA CUDA C Programming Guide. 2012. Version 4.2.

[6] T. Davis and Y. Hu. The University of Florida Sparse Matrix Collection. 1997. http://www.cise.ufl.edu/research/sparse/matrices/list_by_id.html.

[7] M.R. Hestenes and E. Stiefel. Methods of conjugate gradients for solving linear systems. *Journal of Research of the National Bureau of Standards*, 49(6):409–436, 1952.

[8] C.C. Paige and M.A. Saunders. Solution of sparse indefinite systems of linear equations. *SIAM Journal on Numerical Analysis*, 12(4):617–629, 1975.

[9] Y. Saad. *Iterative Methods for Sparse Linear Systems*. Society for Industrial and Applied Mathematics, second edition, 2003.

[10] Y. Saad and M.H. Schultz. GMRES: a generalized minimal residual algorithm for solving nonsymmetric linear systems. *SIAM Journal on Scientific and Statistical Computing*, 7(3):856–869, 1986.

Chapter 14

Solving sparse nonlinear systems of obstacle problems on GPU clusters

Lilia Ziane Khodja, Raphaël Couturier, and Jacques Bahi
Femto-ST Institute, University of Franche-Comte, France

Ming Chau
Advanced Solutions Accelerator, Castelnau Le Lez, France

Pierre Spitéri
ENSEEIHT-IRIT, Toulouse, France

14.1	Introduction	331
14.2	Obstacle problems	332
	14.2.1 Mathematical model	332
	14.2.2 Discretization	333
14.3	Parallel iterative method	334
14.4	Parallel implementation on a GPU cluster	337
14.5	Experimental tests on a GPU cluster	345
14.6	Red-black ordering technique	348
14.7	Conclusion	352
	Bibliography	353

14.1 Introduction

The obstacle problem is one kind of free boundary problem. It allows us to model, for example, an elastic membrane covering a solid obstacle. In this case, the objective is to find an equilibrium position of this membrane constrained to be above the obstacle and which tends to minimize its surface and/or its energy. The study of such problems occurs in many applications, for example, fluid mechanics, biomathematics (tumor growth process), or financial mathematics (American or European option pricing).

In this chapter, we focus on solutions of large obstacle problems defined in a three-dimensional domain. Particularly, the present study consists of solving large nonlinear systems derived from the spatial discretization of these problems. Owing to the great size of such systems, in order to reduce computation times, we proceed by solving them by parallel synchronous or asynchronous iterative algorithms. Moreover, we aim at harnessing the computing power of GPUs to accelerate computations of these parallel algorithms. For this, we use an iterative method involving a projection on a convex set, which is the projected Richardson method. We chose this method among other iterative methods because it is easy to implement on parallel computers and easy to adapt to GPU architectures.

In Section 14.2, we present the mathematical model of obstacle problems then, in Section 14.3, we describe the general principle of the parallel projected Richardson method. Next, in Section 14.4, we give the main key points of the parallel implementation of both synchronous and asynchronous algorithms of the projected Richardson method on a GPU cluster. In Section 14.5, we present the performances of both parallel algorithms obtained from simulations carried out on GPU clusters. Finally, in Section 14.6, we use the read-black ordering technique to improve the convergence and, thus, the execution times of the parallel projected Richardson algorithms on the GPU cluster.

14.2 Obstacle problems

In this section, we present the mathematical model of obstacle problems defined in a three-dimensional domain. This model is based on that presented in [5].

14.2.1 Mathematical model

An obstacle problem, arising for example in mechanics or financial derivatives, consists of solving a time-dependent nonlinear equation:

$$\begin{cases} \frac{\partial u}{\partial t} + b^t.\nabla u - \eta.\Delta u + c.u - f \geq 0,\ u \geq \phi,\ \text{a.e.w. in } [0,T] \times \Omega,\ \eta > 0, \\ (\frac{\partial u}{\partial t} + b^t.\nabla u - \eta.\Delta u + c.u - f)(u - \phi) = 0,\ \text{a.e.w. in } [0,T] \times \Omega, \\ u(0,x,y,z) = u_0(x,y,z), \\ \text{B.C. on } u(t,x,y,z) \text{ defined on } \partial\Omega, \end{cases}$$

(14.1)

where u_0 is the initial condition; $c \geq 0$, b, and η are physical parameters; T is the final time; $u = u(t,x,y,z)$ is an element of the solution vector U to compute; f is the right-hand side that could represent, for example, the external forces; B.C. describes the boundary conditions on the boundary $\partial\Omega$ of the domain Ω; ϕ models a constraint imposed to u; Δ is the Laplacian operator; ∇ is the gradient operator; a.e.w. means almost everywhere, and "." defines the products between two scalars, a scalar and a vector, or a matrix and a vector. In practice the boundary condition, generally considered, is the

Dirichlet condition (where u is fixed on $\partial\Omega$) or the Neumann condition (where the normal derivative of u is fixed on $\partial\Omega$).

The time-dependent equation (14.1) is numerically solved by considering an implicit or a semi-implicit time marching, where at each time step k a stationary nonlinear problem is solved:

$$\begin{cases} b^t.\nabla u - \eta.\Delta u + (c+\delta).u - g \geq 0,\ u \geq \phi, \text{ a.e.w. in } [0,T] \times \Omega,\ \eta > 0, \\ (b^t.\nabla u - \eta.\Delta u + (c+\delta).u - g)(u - \phi) = 0, \text{ a.e.w. in } [0,T] \times \Omega, \\ \text{B.C. on } u(t,x,y,z) \text{ defined on } \partial\Omega, \end{cases}$$
(14.2)

where $\delta = \frac{1}{k}$ is the inverse of the time step k, $g = f + \delta u^{prev}$ and u^{prev} is the solution computed at the previous time step.

14.2.2 Discretization

First, we note that the spatial discretization of the previous stationary problem (14.2) does not provide a symmetric matrix, because the convection-diffusion operator is not self-adjoint. Moreover, the fact that the operator is self-adjoint or not plays an important role in the choice of the appropriate algorithm for solving nonlinear systems derived from the discretization of the obstacle problem. Nevertheless, since the convection coefficients arising in the operator (14.2) are constant, we can formulate the same problem by self-adjoint operator by performing a classical change of variables. Then, we can replace the stationary convection-diffusion problem:

$$b^t.\nabla v - \eta.\Delta v + (c+\delta).v = g, \text{ a.e.w. in } [0,T] \times \Omega,\ c \geq 0,\ \delta \geq 0,\quad (14.3)$$

by the following stationary diffusion operator:

$$-\eta.\Delta u + (\frac{\|b\|_2^2}{4\eta} + c + \delta).u = e^{-a}g = f, \quad (14.4)$$

where $b = \{b_1, b_2, b_3\}$, $\|b\|_2$ denotes the Euclidean norm of b, and $v = e^{-a}.u$ represents the general change of variables such that $a = \frac{b^t(x,y,z)}{2\eta}$. Consequently, the numerical resolution of the diffusion problem (the self-adjoint operator (14.4)) is done by optimization algorithms, in contrast to that of the convection-diffusion problem (nonself-adjoint operator (14.3)) which is done by relaxation algorithms. In the case of our studied algorithm, the convergence is ensured by M-matrix property; then, the performance is linked to the magnitude of the spectral radius of the iteration matrix, which is independent of the condition number.

Next, the three-dimensional domain $\Omega \subset \mathbb{R}^3$ is set to $\Omega = [0,1]^3$ and discretized with a uniform Cartesian mesh constituted by $M = m^3$ discretization points, where m is related to the spatial discretization step by $h = \frac{1}{m+1}$. This is carried out by using a classical order 2 finite difference approximation of the Laplacian. So, the complete discretization of both stationary boundary value problems (14.3) and (14.4) leads to the solution of a large discrete complementary problem of the following form, when both Dirichlet or Neumann boundary conditions are used:

$$\begin{cases} \text{Find } U^* \in \mathbb{R}^M \text{ such that} \\ (A + \delta I)U^* - G \geq 0, U^* \geq \bar{\Phi}, \\ ((A + \delta I)U^* - G)^T(U^* - \bar{\Phi}) = 0, \end{cases} \qquad (14.5)$$

where A is a matrix obtained after the spatial discretization by a finite difference method, G is derived from the Euler first order implicit time marching scheme and from the discretized right-hand side of the obstacle problem, δ is the inverse of the time step k, and I is the identity matrix. The matrix A is symmetric when the self-adjoint operator is considered and nonsymmetric otherwise.

According to the chosen discretization scheme of the Laplacian, A is an M-matrix (irreducibly diagonal dominant, see [15]) and, consequently, the matrix $(A + \delta I)$ is also an M-matrix. This property is important to the convergence of iterative methods.

14.3 Parallel iterative method

Owing to the large size of the previous discrete complementary problem (14.5), we will solve it by parallel synchronous or asynchronous iterative algorithms (see [1–3]). In this chapter, we aim at harnessing the computing power of GPU clusters for solving these large nonlinear systems. Then, we choose to use the projected Richardson iterative method for solving the diffusion problem (14.4). Indeed, this method is based on the iterations of the Jacobi method, which are easy to parallelize on parallel computers and easy to adapt to GPU architectures. Then, according to the boundary value problem formulation with a self-adjoint operator (14.4), we can consider here the equivalent optimization problem and the fixed point mapping associated to its solution.

Assume that $E = \mathbb{R}^M$ is a Hilbert space, in which $\langle .,. \rangle$ is the scalar product and $\|.\|$ its associated norm. So, the general fixed point problem to be solved is defined as follows:

$$\begin{cases} \text{Find } U^* \in E \text{ such that} \\ U^* = F(U^*), \end{cases} \qquad (14.6)$$

where $U \mapsto F(U)$ is an application from E to E.

Let K be a closed convex set defined by

$$K = \{U | U \geq \Phi \text{ everywhere in } E\}, \qquad (14.7)$$

where Φ is the discrete obstacle function. In fact, the obstacle problem (14.5) is formulated as the following constrained optimization problem:

$$\begin{cases} \text{Find } U^* \in K \text{ such that} \\ \forall V \in K, J(U^*) \leq J(V), \end{cases} \quad (14.8)$$

where the cost function is given by

$$J(U) = \frac{1}{2}\langle \mathcal{A}.U, U\rangle - \langle G, U\rangle, \quad (14.9)$$

in which $\langle .,.\rangle$ denotes the scalar product in E, $\mathcal{A} = A + \delta I$ is a symmetric positive definite, and A is the discretization matrix associated with the self-adjoint operator (14.4) after change of variables.

For any $U \in E$; let $P_K(U)$ be the projection of U on K. For any $\gamma \in \mathbb{R}$, $\gamma > 0$, the fixed point mapping F_γ of the projected Richardson method is defined as follows:

$$U^* = F_\gamma(U^*) = P_K(U^* - \gamma(\mathcal{A}.U^* - G)). \quad (14.10)$$

In order to reduce the computation time, the large optimization problem is solved in a numerical way by using a parallel asynchronous algorithm of the projected Richardson method on the convex set K. Particularly, we will consider an asynchronous parallel adaptation of the projected Richardson method [12].

Let $\alpha \in \mathbb{N}$ be a positive integer. We consider that the space $E = \prod_{i=1}^{\alpha} E_i$ is a product of α subspaces E_i where $i \in \{1,\ldots,\alpha\}$. Note that $E_i = \mathbb{R}^{m_i}$, where $\sum_{i=1}^{\alpha} m_i = M$, is also a Hilbert space in which $\langle .,.\rangle_i$ denotes the scalar product and $|.|_i$ the associated norm, for all $i \in \{1,\ldots,\alpha\}$. Then, for all $u,v \in E$, $\langle u,v\rangle = \sum_{i=1}^{\alpha}\langle u_i, v_i\rangle_i$ is the scalar product on E.

Let $U \in E$, we consider the following decomposition of U and the corresponding decomposition of F_γ into α blocks:

$$\begin{aligned} U &= (U_1,\ldots,U_\alpha), \\ F_\gamma(U) &= (F_{1,\gamma}(U),\ldots,F_{\alpha,\gamma}(U)). \end{aligned} \quad (14.11)$$

Assume that the convex set $K = \prod_{i=1}^{\alpha} K_i$, such that $\forall i \in \{1,\ldots,\alpha\}, K_i \subset E_i$, and K_i is a closed convex set. Let also $G = (G_1,\ldots,G_\alpha) \in E$; for any $U \in E$, $P_K(U) = (P_{K_1}(U_1),\ldots,P_{K_\alpha}(U_\alpha))$ is the projection of U on K where $\forall i \in \{1,\ldots,\alpha\}, P_{K_i}$ is the projector from E_i onto K_i. So, the fixed point mapping of the projected Richardson method (14.10) can be written in the following way:

$$\forall U \in E, \forall i \in \{1,\ldots,\alpha\}, F_{i,\gamma}(U) = P_{K_i}(U_i - \gamma(\mathcal{A}_i.U - G_i)). \quad (14.12)$$

Note that $\mathcal{A}_i.U = \sum_{j=1}^{\alpha} \mathcal{A}_{i,j}.U_j$, where $\mathcal{A}_{i,j}$ denote block matrices of $\mathcal{A}$.

The parallel asynchronous iterations of the projected Richardson method for solving the obstacle problem (14.8) are defined as follows: let $U^0 \in E, U^0 = (U_1^0, \ldots, U_\alpha^0)$ be the initial solution, then for all $p \in \mathbb{N}$, the iterate $U^{p+1} = (U_1^{p+1}, \ldots, U_\alpha^{p+1})$ is recursively defined by

$$U_i^{p+1} = \begin{cases} F_{i,\gamma}(U_1^{\rho_1(p)}, \ldots, U_\alpha^{\rho_\alpha(p)}) & \text{if } i \in s(p), \\ U_i^p & \text{otherwise,} \end{cases} \quad (14.13)$$

where

$$\begin{cases} \forall p \in \mathbb{N}, s(p) \subset \{1, \ldots, \alpha\} \text{ and } s(p) \neq \emptyset, \\ \forall i \in \{1, \ldots, \alpha\}, \{p \mid i \in s(p)\} \text{ is enumerable,} \end{cases} \quad (14.14)$$

and $\forall j \in \{1, \ldots, \alpha\}$,

$$\begin{cases} \forall p \in \mathbb{N}, \rho_j(p) \in \mathbb{N}, 0 \leq \rho_j(p) \leq p \text{ and } \rho_j(p) = p \text{ if } j \in s(p), \\ \lim_{p \to \infty} \rho_j(p) = +\infty. \end{cases} \quad (14.15)$$

The previous asynchronous scheme of the projected Richardson method models computations that are carried out in parallel without order or synchronization (according to the behavior of the parallel iterative method) and describes a subdomain method without overlapping. It is a general model that takes into account all possible situations of parallel computations and nonblocking message passing. So, the synchronous iterative scheme is defined by

$$\forall j \in \{1, \ldots, \alpha\}, \forall p \in \mathbb{N}, \rho_j(p) = p. \quad (14.16)$$

The values of $s(p)$ and $\rho_j(p)$ are defined dynamically and not explicitly by the parallel asynchronous or synchronous execution of the algorithm. Particularly, They allow us to consider distributed computations whereby processors compute at their own pace according to their intrinsic characteristics and computational load. The parallelism between the processors is well described by the set $s(p)$ which contains at each step p the index of the components relaxed by each processor on a parallel way while the use of delayed components in (14.13) permits one to model nondeterministic behavior and does not imply inefficiency of the considered distributed scheme of computation. Note that, according to [11], theoretically, each component of the vector must be relaxed an infinite number of times. The choice of the relaxed components to be used in the computational process may be guided by any criterion, and in particular, a natural criterion is to pickup the most recently available values of the components computed by the processors. Furthermore, the asynchronous iterations are implemented by means of nonblocking MPI communication subroutines (asynchronous communications).

The important property ensuring the convergence of the parallel projected Richardson method, both synchronous and asynchronous algorithms, is the fact that $\mathcal{A}$ is an M-matrix. Moreover, the convergence proceeds from a result of [12]. Indeed, there exists a value $\gamma_0 > 0$, such that $\forall \gamma \in]0, \gamma_0[$, the parallel iterations (14.13), (14.14), and (14.15), associated to the fixed point mapping F_γ (14.12), converge to the unique solution U^* of the discretized problem.

14.4 Parallel implementation on a GPU cluster

In this section, we give the main key points of the parallel implementation of the projected Richardson method, both synchronous and asynchronous versions, on a GPU cluster, for solving the nonlinear systems derived from the discretization of large obstacle problems. More precisely, each nonlinear system is solved iteratively using the whole cluster. We use a heterogeneous CUDA and MPI programming. Indeed, the communication of data, at each iteration between the GPU computing nodes, can be either synchronous or asynchronous using the MPI communication subroutines, whereas inside each GPU node, a CUDA parallelization is performed.

Let S denote the number of computing nodes on the GPU cluster, where a computing node is composed of CPU core holding one MPI process and a GPU card. So, before starting computations, the obstacle problem of size $(NX \times NY \times NZ)$ is split into S parallelepipedic subproblems, each for a node (MPI process, GPU), as is shown in Figure 14.1. Indeed, the NY and NZ dimensions (according to the y and z axises) of the three-dimensional problem are split, respectively, into Sy and Sz parts, such that $S = Sy \times Sz$. In this case, each computing node has at most four neighboring nodes. This kind of data partitioning reduces the data exchanges at subdomain boundaries compared to a naive z-axis-wise partitioning.

All the computing nodes of the GPU cluster execute in parallel the Algorithm 17 on a three-dimensional subproblems of size $(NX \times ny \times nz)$. This algorithm gives the main key points for solving an obstacle problem defined in a three-dimensional domain, where A is the discretization matrix, G is the right-hand side, and U is the solution vector. After the initialization step, all the data generated from the partitioning operation are copied from the CPU memories to the GPU global memories to be processed on the GPUs. Next, the algorithm uses $NbSteps$ time steps to solve the global obstacle problem. In fact, it uses a parallel algorithm adapted to GPUs from the projected Richardson iterative method for solving the nonlinear systems of the obstacle problem. This function is defined by $Solve()$ in Algorithm 17. At every time step, the initial guess U^0 for the iterative algorithm is set to the solution found at the previous time step. Moreover, the right-hand side G is computed as follows:

$$G = \frac{1}{k}.U^{prev} + F$$

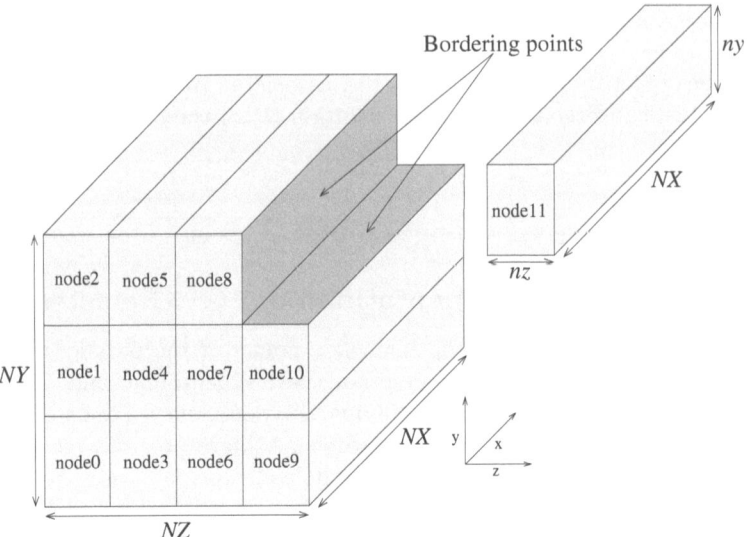

FIGURE 14.1. Data partitioning of a problem to be solved among $S = 3 \times 4$ computing nodes.

Algorithm 17: parallel solving of the obstacle problem on a GPU cluster

1 Initialization of the parameters of the subproblem;
2 Allocate and fill the data in the global memory GPU;
3 **for** $i = 1$ **to** $NbSteps$ **do**
4 $\quad G = \frac{1}{k}.U^0 + F$;
5 $\quad$ Solve(A, U^0, G, U, ε, $MaxRelax$);
6 $\quad U^0 = U$;
7 **end**
8 Copy the solution U back from GPU memory;

---

where k is the time step, U^{prev} is the solution computed in the previous time step, and each element $f(x, y, z)$ of the vector F is computed as follows:

$$f(x, y, z) = \cos(2\pi x) \cdot \cos(4\pi y) \cdot \cos(6\pi z). \qquad (14.17)$$

Finally, the solution U of the obstacle problem is copied back from the GPU global memories to the CPU memories. We use the communication subroutines of the CUBLAS (CUDA Basic Linear Algebra Subroutines) library [6] for the memory allocations in the GPU (cublasAlloc()) and the data transfers between the CPU and its GPU: cublasSetVector() and cublasGetVector().

As are many other iterative methods, the algorithm of the projected Richardson method is based on algebraic functions operating on vectors and/or matrices, which are more efficient on parallel computers when they

Algorithm 18: parallel iterative solving of the nonlinear systems on a GPU cluster (*Solve*() function)

1 $p = 0$;
2 $conv = false$;
3 $U = U^0$;
4 **repeat**
5   Determine_Bordering_Vector_Elements(U);
6   Compute_New_Vector_Elements(A, G, U);
7     $tmp = U^0 - U$;
8     $error = \|tmp\|_2$;
9     $U^0 = U$;
10     $p = p + 1$;
11     $conv = $ Convergence($error, p, \varepsilon, MaxRelax$);
12 **until** ($conv = true$);

work on large vectors. Its parallel implementation on the GPU cluster is carried out so that the GPUs execute the vector operations as kernels and the CPUs execute the serial codes, supervise the kernel executions and the data exchanges with the neighboring nodes, and supply the GPUs with data. Algorithm 18 shows the main key points of the parallel iterative algorithm (function *Solve*() in Algorithm 17). All the vector operations inside the main loop (**repeat ... until**) are executed by the GPU. We use the following functions of the CUBLAS library:

- `cublasDaxpy()` to compute the difference between the solution vectors U^p and U^{p+1} computed in two successive relaxations p and $p + 1$ (line 7 in Algorithm 18),

- `cublasDnrm2()` to perform the Euclidean norm (line 8), and

- `cublasDcpy()` for the data copy of a vector to another one in the GPU memory (lines 3 and 9).

The dimensions of the grid and blocks of threads that execute a given kernel depend on the resources of the GPU multiprocessor and the resource requirements of the kernel. So, if *block* defines the size of a thread block, which must not exceed the maximum size of a thread block, then the number of thread blocks in the grid, denoted by *grid*, can be computed according to the size of the local subproblem as follows:

$$grid = \frac{(NX \times ny \times nz) + block - 1}{block}.$$

However, when solving very large problems, the size of the thread grid can exceed the maximum number of thread blocks that can be executed on the GPUs (upto 65.535 thread blocks), and thus, the kernel will fail to launch.

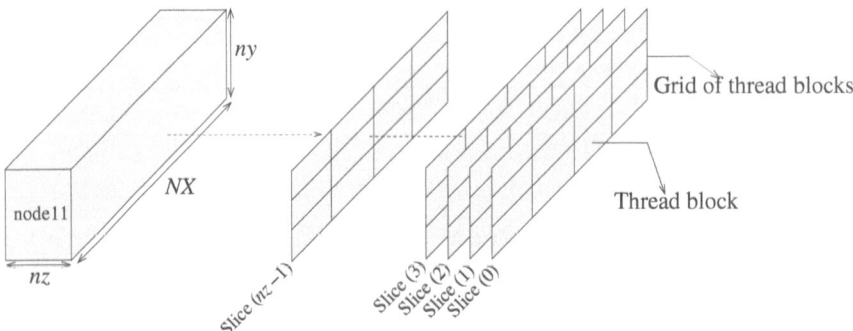

FIGURE 14.2. Decomposition of a subproblem in a GPU into nz slices.

Therefore, for each kernel, we decompose the three-dimensional subproblem into nz two-dimensional slices of size $(NX \times ny)$, as is shown in Figure 14.2. All slices of the same kernel are executed using a **for** loop by $NX \times ny$ parallel threads organized in a two-dimensional grid of two-dimensional thread blocks, as is shown in Listing 14.1. Each thread is in charge of nz discretization points (one from each slice), accessed in the GPU memory with a constant stride $(NX \times ny)$.

Listing 14.1. skeleton codes of a GPU kernel and a CPU function

```
/* GPU kernel */
__global__ void kernel(..., int n, int nx, int ny, int slices, int
    stride, ...)
{
    int tx = blockIdx.x * blockDim.x + threadIdx.x;//x-coordinate
    int ty = blockIdx.y * blockDim.y + threadIdx.y;//y-coordinate
    int tid = tx + ty * nx;                        //thread ID
    for(int i=0; i<slices; i++){
        if((tx<nx) && (ty<ny) && (tid<n)){
            ...
        }
        tid += stride;
    }
}

/* CPU function */
void Function(...)
{
    int n = NX * ny * nz; //size of the subproblem
    int slices = nz;
    int stride = NX * ny;
    int bx = 64, by = 4;
    int gx = (NX + bx - 1) / bx;
    int gy = (ny + by - 1) / by;
    dim3 block(bx,by);   //dimensions of a thread block
    dim3 grid(gx,gy);    //dimensions of the grid
    ...
    kernel<<<grid,block>>>(..., n, NX, ny, slices, stride, ...);
    ...
}
```

The function *Determine_Bordering_Vector_Elements*() (line 5 in Algorithm 18) determines the values of the vector elements shared at the boundaries with neighboring computing nodes. Its main operations are as follows:

1. define the values associated to the bordering points needed by the neighbors,

2. copy the values associated to the bordering points from the GPU to the CPU,

3. exchange the values associated to the bordering points between the neighboring CPUs, and

4. copy the received values associated to the bordering points from the CPU to the GPU.

The first operation of this function is implemented as kernels to be performed by the GPU:

- a kernel executed by $NX \times nz$ threads to define the values associated to the bordering vector elements along the y-axis, and

- a kernel executed by $NX \times ny$ threads to define the values associated to the bordering vector elements along the z-axis.

As mentioned previously, we develop the *synchronous* and *asynchronous* algorithms of the projected Richardson method. Obviously, in this scope, the synchronous or asynchronous communications refer to the communications between the CPU cores (MPI processes) on the GPU cluster, in order to exchange the vector elements associated to subdomain boundaries. For the memory copies between a CPU core and its GPU, we use the synchronous communication routines of the CUBLAS library: cublasSetVector() and cublasGetVector() in the synchronous algorithm and the asynchronous ones: cublasSetVectorAsync() and cublasGetVectorAsync() in the asynchronous algorithm. Moreover, we use the communication routines of the MPI library to carry out the data exchanges between the neighboring nodes. We use the following communication routines: MPI_Isend() and MPI_Irecv() to perform nonblocking sends and receives, respectively. For the synchronous algorithm, we use the MPI routine MPI_Waitall() which puts the MPI process of a computing node in blocking status until all data exchanges with neighboring nodes (sends and receives) are completed. In contrast, for the asynchronous algorithms, we use the MPI routine MPI_Test() which tests the completion of a data exchange (send or receives) without putting the MPI process in blocking status.

The function *Compute_New_Vector_Elements*() (line 6 in Algorithm 18) computes, at each iteration, the new elements of the iterate vector U. Its general code is presented in Listing 14.1 (CPU function). The iterations of

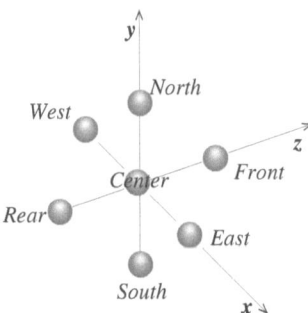

FIGURE 14.3. Matrix constant coefficients in a three-dimensional domain.

the projected Richardson method, based on those of the Jacobi method, are defined as follows:

$$u^{p+1}(x,y,z) = \tfrac{1}{Center}(g(x,y,z) - (Center \cdot u^p(x,y,z) + \\ West \cdot u^p(x-h,y,z) + East \cdot u^p(x+h,y,z) + \\ South \cdot u^p(x,y-h,z) + North \cdot u^p(x,y+h,z) + \\ Rear \cdot u^p(x,y,z-h) + Front \cdot u^p(x,y,z+h))), \quad (14.18)$$

where $u^p(x,y,z)$ is an element of the iterate vector U computed at the iteration p and $g(x,y,z)$ is a vector element of the right-hand side G. The scalars $Center$, $West$, $East$, $South$, $North$, $Rear$, and $Front$ define constant coefficients of the block matrix A. Figure 14.3 shows the positions of these coefficients in a three-dimensional domain.

The kernel implementations of the projected Richardson method on GPUs uses a perfect fine-grain multithreading parallelism. Since the projected Richardson algorithm is implemented as a fixed point method, each kernel is executed by a large number of GPU threads such that each thread is in charge of the computation of one element of the iterate vector U. Moreover, this method uses the vector elements updates of the Jacobi method, which means that each thread i computes the new value of its element u_i^{p+1} independently of the new values u_j^{p+1}, where $j \neq i$, of those computed in parallel by other threads at the same iteration $p+1$. Listing 14.2 shows the GPU implementations of the main kernels of the projected Richardson method, which are: the matrix-vector multiplication (`MV_Multiplication()`) and the vector elements updates (`Vector_Updates()`). The codes of these kernels are based on that presented in Listing 14.1.

Listing 14.2. GPU kernels of the projected Richardson method

```
/* Kernel of the matrix-vector multiplication */
__global__ void MV_Multiplication (..., double* U, double* Y)
{
    ...
    //Matrix coefficients filled in registers

    for(int tz=0; tz<slices; tz++){
        if((tx<NX) && (ty<ny) && (tid<n)){
            double sum = Center * fetch_double(U, tid);
            if(tx != 0)        sum += West  * fetch_double(U, tid-1);
            if(tx != (NX-1))   sum += East  * fetch_double(U, tid+1);
            if(ty != 0)        sum += South * fetch_double(U, tid-NX);
            if(ty != (ny-1))   sum += North * fetch_double(U, tid+NX);
            if(tz != 0)        sum += Rear  * fetch_double(U, tid-NX*ny);
            if(tz != (nz-1))   sum += Front * fetch_double(U, tid+NX*ny);
            Y[tid] = sum;
        }
        tid += stride;
    }
}

/* Kernel of the vector elements updates */
__global__ void Vector_Updates (..., double* G, double* Y, double* U)
{
    ...
    //Matrix coefficient filled in a register: Center

    for(int tz=0; tz<slices; tz++){
        if((tx<NX) && (ty<ny) && (tid<n)){
            double var = (G[tid]-Y[tid]) / Center + fetch_double(U, tid);
            if(var < 0.0) var = 0.0; //projection
            U[tid] = var;
        }
        tid += stride;
    }
}

/* Function to be executed by the CPU */
void Computation_New_Vector_Elements(double*A, double* G, double* U)
{
    double *Y;
    //Allocate a GPU memory space for the vector Y
    //Configure the kernel execution: grid and block
    //Elements of vector U filled in the texture memory

    MV_Multiplication<<<grid, block>>>(..., U, Y);
    Vector_Updates<<<grid, block>>>(..., G, Y, U);
}
```

Each kernel is executed by $NX \times ny$ GPU threads so that nz slices of $(NX \times ny)$ vector elements are computed in a **for** loop. In this case, each thread is in charge of one vector element from each slice (in total nz vector elements along the z-axis). We can notice from the formula (14.18) that the computation of a vector element $u^{p+1}(x,y,z)$, by a thread at iteration $p+1$, requires seven vector elements computed at the previous iteration p: two vector elements in each dimension plus the vector element at the intersection of the three axes x, y, and z (see Figure 14.4). So, to reduce the memory accesses to the high-latency global memory, the vector elements of the current slice can be stored in the low-latency shared memories of thread blocks, as is described in [10]. Nevertheless, the fact that the computation

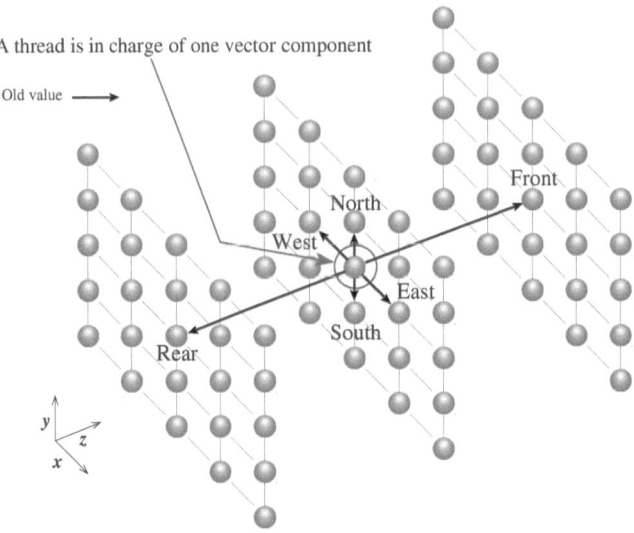

FIGURE 14.4. Computation of a vector element with the projected Richardson method.

of a vector element requires only two elements in each dimension does not allow us to maximize the data reuse from the shared memories. The computation of a slice involves in total $(bx + 2) \times (by + 2)$ accesses to the global memory per thread block, to fill the required vector elements in the shared memory where bx and by are the dimensions of a thread block. Then, in order to optimize the memory accesses on GPUs, the elements of the iterate vector U are filled in the cache texture memory (see [9]). In new GPU hardware and software as Fermi or Kepler, the global memory accesses are always cached in L1 and L2 caches. For example, for a given kernel, we can favor the use of the L1 cache to that of the shared memory by using the function `cudaFuncSetCacheConfig(Kernel, cudaFuncCachePreferL1)`. So, the initial access to the global memory loads the vector elements required by the threads of a block into the cache memory (texture or L1/L2 caches). Then, all the following memory accesses read from this cache memory. In Listing 14.2, the function `fetch_double(v,i)` is used to read from the texture memory the ith element of the double-precision vector v (see Listing 14.3). Moreover, the seven constant coefficients of matrix A can be stored in the constant memory but, since they are reused nz times by each thread, it is more efficient to fill them on the low-latency registers of each thread.

Listing 14.3. memory access to the cache texture memory

```
texture<int2,1> v;

static __inline__
__device__ double fetch_double(texture<int2,1> v, int i)
{
    int2 w = tex1Dfetch(v, i);
    return __hiloint2double(w.y, w.x);
}
```

The function $Convergence()$ (line 11 in Algorithm 18) allows us to detect the convergence of the parallel iterative algorithm and is based on the tolerance threshold ε and the maximum number of relaxations $MaxRelax$. We take into account the number of relaxations since that of iterations cannot be computed in the asynchronous case. Indeed, a relaxation is the update (14.13) of a local iterate vector U_i according to F_i. Then, counting the number of relaxations is possible in both synchronous and asynchronous cases. On the other hand, an iteration is the update of at least all vector components with F_i.

In the synchronous algorithm, the global convergence is detected when the maximal value of the absolute error, $error$, is sufficiently small and/or the maximum number of relaxations, $MaxRelax$, is reached, as follows:

$$error = \|U^p - U^{p+1}\|_2;$$
$$AllReduce(error, maxerror, MAX);$$
$$\text{if}((maxerror < \varepsilon) \text{ or } (p \geq MaxRelax))$$
$$conv \leftarrow true;$$

where the function $AllReduce()$ uses the MPI global reduction subroutine `MPI_Allreduce()` to compute the maximal value, $maxerror$, among the local absolute errors, $error$, of all computing nodes, and p (in Algorithm 18) is used as a counter of the local relaxations carried out by a computing node. In the asynchronous algorithms, the global convergence is detected when all computing nodes locally converge. For this, we use a token ring architecture around which a boolean token travels, in one direction, from a computing node to another. Starting from node 0, the boolean token is set to $true$ by node i if the local convergence is reached or to $false$ otherwise, and then, it is sent to node $i+1$. Finally, the global convergence is detected when node 0 receives from its neighbor node $S-1$, in the ring architecture, a token set to $true$. In this case, node 0 sends a stop message (end of parallel solving) to all computing nodes in the cluster.

14.5 Experimental tests on a GPU cluster

The GPU cluster of tests that we used in this chapter is an $20GB/s$ Infiniband network of six machines. Each machine is a Quad-Core Xeon E5530 CPU running at 2.4GHz. It provides a RAM memory of 12GB with a memory bandwidth of 25.6GB/s and it is equipped with two NVIDIA Tesla C1060 GPUs. A Tesla GPU contains in total 240 cores running at 1.3GHz. It provides 4GB of global memory with a memory bandwidth of 102GB/s, accessible by all its cores and also by the CPU through the PCI-Express 16x Gen 2.0 interface with a throughput of 8GB/s. Hence, the memory copy operations between the GPU and the CPU are about 12 times slower than those of the Tesla GPU memory. We have performed our simulations on a cluster of 24 CPU cores and on a cluster of 12 GPUs. Figure 13.4 describes the components of the GPU cluster of tests.

Linux cluster version 2.6.39 OS is installed on CPUs. C programming language is used for coding the parallel algorithms of the methods on both GPU cluster and CPU cluster. CUDA version 4.0 [13] is used for programming GPUs, using CUBLAS library [6] to deal with vector operations in GPUs, and finally, MPI functions of OpenMPI 1.3.3 are used to carry out the synchronous and asynchronous communications between CPU cores. Indeed, in our experiments, a computing node is managed by one MPI process and it is composed of one CPU core and one GPU card.

All experimental results of the parallel projected Richardson algorithms are obtained from simulations made in double precision data. The obstacle problems to be solved are defined in constant three-dimensional domain $\Omega \subset \mathbb{R}^3$. The numerical values of the parameters of the obstacle problems are $\eta = 0.2$, $c = 1.1$, f is computed by formula (14.17), and final time $T = 0.02$. Moreover, three time steps ($NbSteps = 3$) are computed with $k = 0.0066$. As the discretization matrix is constant along the time steps, the convergence properties of the iterative algorithms do not change. Thus, the performance characteristics obtained with three time steps will still be valid for more time steps. The initial function $u(0, x, y, z)$ of the obstacle problem (14.1) is set to 0, with a constraint $u \geq \phi = 0$. The relaxation parameter γ used by the projected Richardson method is computed automatically thanks to the diagonal entries of the discretization matrix. The formula and its proof can be found in [4]. The convergence tolerance threshold ε is set to 1e-04 and the maximum number of relaxations is limited to 10^6 relaxations. Finally, the number of threads per block is set to 256 threads, which gives, in general, good performances for most GPU applications. We have performed some tests for the execution configurations and have noticed that the best configuration of the 256 threads per block is an organization into two dimensions of sizes (64, 4).

The performance measures that we took into account are the execution times and the number of relaxations performed by the parallel iterative algorithms, both synchronous and asynchronous versions, on the GPU and CPU clusters. These algorithms are used for solving nonlinear systems derived from the discretization of obstacle problems of sizes 256^3, 512^3, 768^3, and 800^3. In Table 14.1 and Table 14.2, we show the performances of the parallel synchronous and asynchronous algorithms of the projected Richardson method implemented, respectively, on a cluster of 24 CPU cores and on a cluster of

Pb. size	Synchronous		Asynchronous		Gain%
	T_{cpu}	# Relax.	T_{cpu}	# Relax.	
256^3	575.22	198,288	539.25	198,613	6.25
512^3	19,250.25	750,912	18,237.14	769,611	5.26
768^3	206,159.44	1,635,264	183,582.60	1,577,004	10.95
800^3	222,108.09	1,769,232	188,790.04	1,701,735	15.00

TABLE 14.1. Execution times in seconds of the parallel projected Richardson method implemented on a cluster of 24 CPU cores.

Pb. size	Synchronous			Asynchronous			Gain%
	T_{gpu}	# Relax.	τ	T_{gpu}	# Relax.	τ	
256^3	29.67	100,692	19.39	18.00	94,215	29.96	39.33
512^3	521.83	381,300	36.89	425.15	347,279	42.89	18.53
768^3	4,112.68	831,144	50.13	3,313.87	750,232	55.40	19.42
800^3	3,950.87	899,088	56.22	3,636.57	834,900	51.91	7.95

TABLE 14.2. Execution times in seconds of the parallel projected Richardson method implemented on a cluster of 12 GPUs.

12 GPUs. In these tables, the execution time defines the time spent by the slowest computing node and the number of relaxations is computed as the summation of those carried out by all computing nodes.

In the sixth column of Table 14.1 and in the eighth column of Table 14.2, we give the gains in percentage obtained by using an asynchronous algorithm compared to a synchronous one. We can notice that the asynchronous version on CPU and GPU clusters is slightly faster than the synchronous one for both methods. Indeed, the cluster of tests is composed of local and homogeneous nodes communicating via low-latency connections. So, in the case of distant and/or heterogeneous nodes (or even with geographically distant clusters), the asynchronous version would be faster than the synchronous one. However, the gains obtained on the GPU cluster are better than those obtained on the CPU cluster. In fact, the computation times are reduced by accelerating the computations on GPUs while the communication times remain unchanged.

The fourth and seventh columns of Table 14.2 show the relative gains obtained by executing the parallel algorithms on the cluster of 12 GPUs instead on the cluster of 24 CPU cores. We compute the relative gain τ as a ratio between the execution time T_{cpu} spent on the CPU cluster over that T_{gpu} spent on the GPU cluster:

$$\tau = \frac{T_{cpu}}{T_{gpu}}.$$

We can see from these ratios that solving large obstacle problems is faster on the GPU cluster than on the CPU cluster. Indeed, the GPUs are more efficient than their counterpart CPUs at executing large data-parallel operations. In addition, the projected Richardson method is implemented as a fixed point based iteration and uses the Jacobi vector updates that allow a well-suited thread-parallelization on GPUs, such that each GPU thread is in charge of one vector component at a time without being dependent on other vector components computed by other threads. Then, this allows us to exploit at best the high performance computing of the GPUs by using all the GPU resources and avoiding the idle cores.

Finally, the number of relaxations performed by the parallel synchronous algorithm is different in the CPU and GPU versions, because the number of computing nodes involved in the GPU cluster and in the CPU cluster is different. In the CPU case, 24 computing nodes (24 CPU cores) are considered, whereas in the GPU case, 12 computing nodes (12 GPUs) are considered. As the number of relaxations depends on the domain decomposition, consequently it also depends on the number of computing nodes.

14.6 Red-black ordering technique

As is well known, the Jacobi method is characterized by a slow convergence rate compared to some iterative methods (for example, Gauss-Seidel method). So, in this section, we present some solutions to reduce the execution time and the number of relaxations and, more specifically, to speed up the convergence of the parallel projected Richardson method on the GPU cluster. We propose to use the point red-black ordering technique to accelerate the convergence. This technique is often used to increase the parallelism of iterative methods for solving linear systems [7, 8, 14]. We apply it to the projected Richardson method as a compromise between the Jacobi and Gauss-Seidel iterative methods.

The general principle of the red-black technique is as follows. Let t be the summation of the integer x-, y-, and z-coordinates of a vector element $u(x, y, z)$ on a three-dimensional domain: $t = x + y + z$. As is shown in Figure 14.5(a), the red-black ordering technique consists of the parallel computing of the red vector elements having even value t by using the values of the black ones, then the parallel computing of the black vector elements having odd values t by using the new values of the red ones.

This technique can be implemented on the GPU in two different manners:

- among all launched threads ($NX \times ny$ threads), only one thread out of two computes its red or black vector element at a time or

- all launched threads (on average half of $NX \times ny$ threads) compute the red vector elements first, and then the black ones.

Solving sparse nonlinear systems of obstacle problems on GPU clusters 349

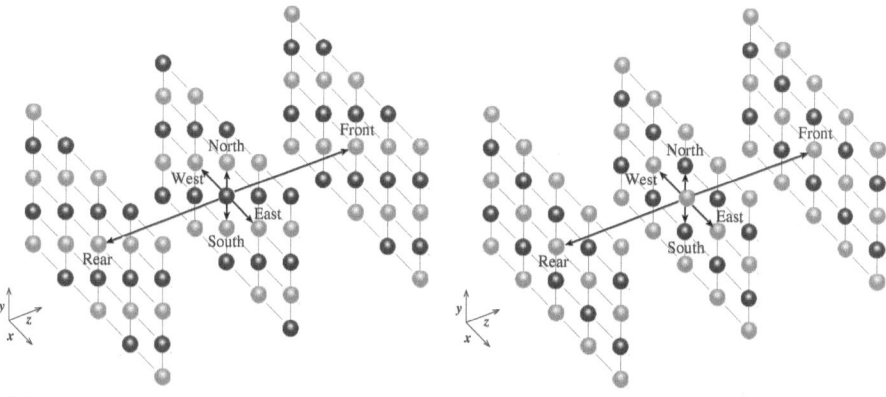

(a) Red-black ordering on x, y, and z axises   (b) Red-black ordering on y axis

FIGURE 14.5. Red-black ordering for computing the iterate vector elements in a three-dimensional space.

However, in both solutions, for each memory transaction, only half of the memory segment addressed by a half-warp is used. So, the computation of the red and black vector elements leads to using twice the initial number of memory transactions. Then, we apply the point red-black ordering accordingly to the y-coordinate, as is shown in Figure 14.5(b). In this case, the vector elements having even y-coordinate are computed in parallel using the values of those having odd y-coordinate and then vice-versa. Moreover, in the GPU implementation of the parallel projected Richardson method (Section 14.4), we have shown that a subproblem of size $(NX \times ny \times nz)$ is decomposed into nz grids of size $(NX \times ny)$. Then, each kernel is executed in parallel by $NX \times ny$ GPU threads, so that each thread is in charge of nz vector elements along the z-axis (one vector element in each grid of the subproblem). So, we propose to use the new values of the vector elements computed in grid i to compute those of the vector elements in grid $i + 1$. Listing 14.4 describes the kernel of the matrix-vector multiplication and the kernel of the vector elements updates of the parallel projected Richardson method using the red-black ordering technique.

Listing 14.4. GPU kernels of the projected Richardson method using the red-black technique

```
/* Kernel of the matrix-vector multiplication */
__global__ void MV_Multiplication(..., double*G, double*U, double*Y)
{
   ...
   //Matrix coefficients filled in registers:
   //Center, West, East, Front

   for(int tz=0; tz<slices; tz++){
      if((tx<NX) && (ty<ny) && (tid<n)){
         double sum = G[tid] - Center * fetch_double(U,tid);
         if(tx != 0)      sum -= West  * fetch_double(U,tid-1);
         if(tx != (NX-1)) sum -= East  * fetch_double(U,tid+1);
         if(tz != (nz-1)) sum -= Front * fetch_double(U,tid+NX*ny);
         Y[tid] = sum;
      }
      tid += stride;
   }
}

/* Kernel of the vector elements updates */
__global__ void Vector_Updates(..., int rb, double* Y, double* U)
{
   ...
   double val; //vector component computed in previous grid
   //Matrix coefficients filled in registers:
   //Center, South, North, Rear

   for(int tz=0; tz<slices; tz++){
      if((tx<NX) && (ty<ny) && (tid<n) && ((ty&1)==rb)){
         double sum = Y[tid] - South * fetch_double(U,tid-NX) -
                              North * fetch_double(U,tid+NX);
         //val: computed in previous grid
         if(tz != 0) sum -= Rear * val;
         sum = (sum / Center) + fetch_double(U,tid);
         if(sum < 0) sum = 0;  //projection
         U[tid] = val = sum;   //update of U
      }
      tid += stride;
   }
}

/* CPU Function */
void Computation_New_Vector_Components(double*A, double*G, double*U)
{
   double *Y;
   int red=0, black=1;
   //Allocate a GPU memory space for the vector Y
   //Configure the kernel execution: grid and block
   //Elements of vector U filled in the texture memory

   MV_Multiplication<<<grid,block>>>(..., G, U, Y);
   Vector_Updates<<<grid,block>>>(..., red,   Y, U);
   Vector_Updates<<<grid,block>>>(..., black, Y, U);
}
```

Finally, we exploit the concurrent executions between the host functions and the GPU kernels provided by the GPU hardware and software. In fact, the kernel launches are asynchronous (when this environment variable is not disabled on the GPUs), such that the control is returned to the host (MPI process) before the GPU has completed the requested task (kernel) [13]. Therefore, all the kernels necessary to update the local vector elements, $u(x,y,z)$

where $0 < y < (ny - 1)$ and $0 < z < (nz - 1)$, are executed first. Then, the values associated to the bordering vector elements are exchanged between the neighbors. Finally, the values of the vector elements associated to the bordering vector elements are updated. In this case, the computation of the local vector elements is performed concurrently with the data exchanges between neighboring CPUs and this in both synchronous and asynchronous cases.

In Table 14.3, we report the execution times and the number of relaxations performed on a cluster of 12 GPUs by the parallel projected Richardson algorithms; it can be noted that the performances of the projected Richardson algorithm are improved by using the point red-black ordering. We compare the performances of the parallel projected Richardson method with and without this later ordering (Tables 14.2 and 14.3). We can notice that both parallel synchronous and asynchronous algorithms are faster when they use the red-black ordering. Indeed, we can see in Table 14.3 that the execution times of these algorithms are reduced, on average, by 32% compared to those shown in Table 14.2.

Pb. size	Synchronous		Asynchronous		Gain%
	T_{gpu}	# Relax.	T_{gpu}	# Relax.	
256^3	18.37	71,988	12.58	67,638	31.52
512^3	349.23	271,188	289.41	246,036	17.13
768^3	2,773.65	590,652	2,222.22	532,806	19.88
800^3	2,748.23	638,916	2,502.61	592,525	8.92

TABLE 14.3. Execution times in seconds of the parallel projected Richardson method using red-black ordering technique implemented on a cluster of 12 GPUs.

In Figure 14.6, we study the ratio between the computation time and the communication time of the parallel projected Richardson algorithms on a GPU cluster. The experimental tests are carried out on a cluster composed of one to ten Tesla GPUs. We have focused on the weak scaling of both parallel, synchronous and asynchronous, algorithms using the red-black ordering technique. For this, we have fixed the size of a subproblem to 256^3 per computing node (a CPU core and a GPU). Then, Figure 14.6 shows the number of relaxations performed, on average, per second by a computing node. We can see from this figure that the efficiency of the asynchronous algorithm is almost stable, while that of the synchronous algorithm decreases (down to 81% in this example) with the increase in the number of computing nodes on the cluster. This is due to the fact that the ratio between the time of the computation over that of the communication is reduced when the computations are performed on GPUs. Indeed, GPUs compute faster than CPUs and communications are more time-consuming. In this context, asynchronous algorithms are more scal-

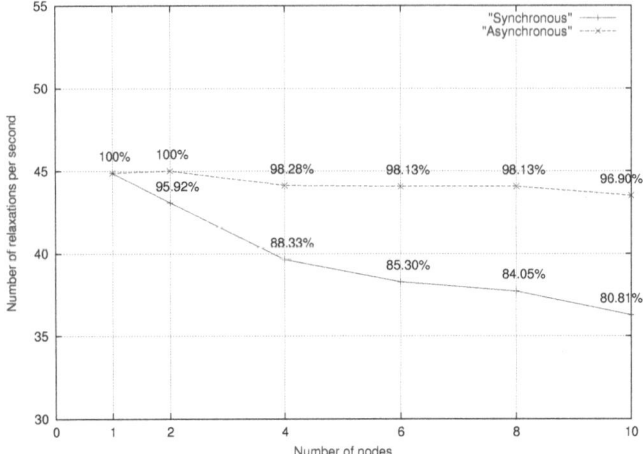

FIGURE 14.6. Weak scaling of both synchronous and asynchronous algorithms of the projected Richardson method using red-black ordering technique.

able than synchronous ones. So, with large scale GPU clusters, synchronous algorithms might be more penalized by communications, as can be deduced from Figure 14.6. That is why we think that asynchronous iterative algorithms are all the more interesting in this case.

14.7 Conclusion

Our main contribution, in this chapter, is the parallel implementation of an asynchronous iterative method on GPU clusters for solving large scale nonlinear systems derived from the spatial discretization of three-dimensional obstacle problems. For this, we have implemented both synchronous and asynchronous algorithms of the Richardson iterative method using a projection on a convex set. Indeed, this method uses point-based iterations of the Jacobi method that are very easy to parallelize on parallel computers. We have shown that its adapted parallel algorithms to GPU architectures allow us to exploit at best the computing power of the GPUs and to accelerate the resolution of large nonlinear systems. Consequently, the experimental results have shown that solving nonlinear systems of large obstacle problems with this method is about fifty times faster on a cluster of 12 GPUs than on a cluster of 24 CPU cores. Moreover, we have applied to this projected Richardson method the red-black ordering technique which allows it to improve its convergence rate. Thus, the execution times of both parallel algorithms performed on the cluster of 12 GPUs are reduced on average of 32%.

Afterwards, the experiments have shown that the asynchronous version is slightly more efficient than the synchronous one. In fact, the computations are accelerated by using GPUs while the communication times are still unchanged. In addition, we have studied the weak-scaling in the synchronous and asynchronous cases, which has confirmed that the ratio between the computations and the communications are reduced when using a cluster of GPUs. We highlight that asynchronous iterative algorithms are more scalable than synchronous ones. Therefore, we can conclude that asynchronous iterations are well suited to tackle scalability issues on GPU clusters.

In future works, we plan to perform experiments on large scale GPU clusters and on geographically distant GPU clusters, because we expect that asynchronous versions would be faster and more scalable on such architectures. Furthermore, we want to study the performance behavior and the scalability of other numerical algorithms which support, if possible, the model of asynchronous iterations.

Bibliography

[1] J.M. Bahi, S. Contassot-Vivier, and R. Couturier. *Parallel Iterative Algorithms: From Sequential to Grid Computing.* Numerical Analysis & Scientific Computing Series. Chapman & Hall/CRC, 2007.

[2] G.M. Baudet. Asynchronous iterative methods for multiprocessors. *Journal Assoc. Comput. Mach.*, 25:226–244, 1978.

[3] D.P. Bertsekas and J.N. Tsitsiklis. *Parallel and Distributed Computation: Numerical Methods.* Prentice-Hall, Inc., Upper Saddle River, NJ, USA, 1989.

[4] M. Chau, R. Couturier, J.M. Bahi, and P. Spiteri. Parallel solution of the obstacle problem in grid environments. *International Journal of High Performance Computing Applications*, 25(4):488–495, 2011.

[5] M. Chau, R. Couturier, J.M. Bahi, and P. Spiteri. Asynchronous grid computation for American options derivatives. *Advances in Engineering Software*, 60-61:136–144, 2012.

[6] NVIDIA Corporation. CUDA Toolkit 4.2 CUBLAS Library. 2012. http://developer.download.nvidia.com/compute/DevZone/docs/html/CUDALibraries/doc/CUBLAS_Library.pdf.

[7] D.J. Evans. Parallel S.O.R. iterative methods. *Parallel Computing*, 1(1):3–18, 1984.

[8] T. Iwashita and M. Shimasaki. Block red-black ordering method for parallel processing of ICCG solver. *High Performance Computing*, 2327(0), 2006.

[9] A. Leist, D.P. Playne, and K.A. Hawick. Exploiting graphical processing units for data-parallel scientific applications. *Concurrency and Computation: Practice and Experience*, 21(18):2400–2437, 2009.

[10] P. Micikevicius. 3D finite difference computation on GPUs using CUDA. In *Proceedings of 2nd Workshop on General Purpose Processing on Graphics Processing Units*, GPGPU-2, pages 79–84, New York, NY, USA, 2009. ACM.

[11] J.-C. Miellou. Algorithmes de relaxation chaotique à retards. *RAIRO Analyse numérique*, R1:55–82, 1975.

[12] J.-C. Miellou and P. Spiteri. Two criteria for the convergence of asynchronous iterations. *in P. Chenin et al. ed., Computers and Computing, Wiley Masson*, pages 91–95, 1985.

[13] NVIDIA. NVIDIA CUDA C Programming Guide. *Version 4.2*, 2012.

[14] Y. Saad. Iterative methods for sparse linear systems. *Society for Industrial and Applied Mathematics, second edition*, 2003.

[15] R.S. Varga. *Matrix Iterative Analysis*. Springer Series in Computational Mathematics. Springer, Dordrecht, 2009.

Chapter 15

Ludwig: multiple GPUs for a complex fluid lattice Boltzmann application

Alan Gray and Kevin Stratford

EPCC, The University of Edinburgh, United Kingdom

15.1	Introduction	355
15.2	Background	357
15.3	Single GPU implementation	360
15.4	Multiple GPU implementation	361
15.5	Moving solid particles	364
15.6	Summary	367
15.7	Acknowledgments	367
	Bibliography	368

15.1 Introduction

The lattice Boltzmann (LB) method (for an overview see, e.g., [18]) has become a popular approach to a variety of fluid dynamics problems. It provides a way to solve the incompressible isothermal Navier-Stokes equations and has the attractive features of being both explicit in time and local in space. This makes the LB method well suited to parallel computation. Many efficient parallel implementations of the LB method have been undertaken, typically using a combination of distributed domain decomposition and the Message Passing Interface (MPI). However, the potential performance benefits offered by GPUs has motivated a new "mixed-mode" approach to address very large problems. Here, fine-grained parallelism is implemented on the GPU, while MPI is reserved for larger scale parallelism. This mixed mode is of increasing interest to application programmers at a time when many supercomputing services are moving toward clusters of GPU accelerated nodes. The design questions which arise when developing a lattice Boltzmann code for this type of heterogeneous system are therefore worth studying. Furthermore, similar questions also recur in many other types of stencil-based algorithms.

The first applications of LB on GPUs were to achieve fluid-like effects in computer animation, rather than scientific applications per se. These early works include simple fluids [21], miscible two-component flow [27], and various image processing tasks based on the use of partial differential equations [26]. While these early works used relatively low level graphics APIs, the first CUDA runtime interface implementation was a two-dimensional simple fluid problem [19]. Following pioneering work on clusters of GPUs coupled via MPI to study air pollution [4], more recent work has included mixed OpenMP and CUDA [9], Posix threads and CUDA [12], and MPI and CUDA for increasingly large GPU clusters [2, 5, 25]. The heterogeneous nature of these systems has also spurred interest in approaches including automatic code generation [20] and auto-tuning [23] to aid application performance.

Many of these authors make use of another attractive feature of LB: the ability to include fixed solid-fluid boundary conditions as a straightforward addition to the algorithm to study, for example, flow in porous media. This points to an important application area for LB methods: that of complex fluids. Complex fluids include mixtures, surfactants, liquid crystals, and particle suspensions, and typically require additional physics beyond the bare Navier-Stokes equations to provide a full description [1]. The representation of this extra physics raises additional design questions for the application programmer. Here, we consider the *Ludwig* code [3], an LB application developed specifically for complex fluids (*Ludwig* was named for Boltzmann, 1844–1906). We will present the steps required to allow *Ludwig* to exploit efficiently both a single GPU and many GPUs in parallel. We show that *Ludwig* scales excellently to at least the one thousand GPU level (the largest resource available at the time of writing) with indications that the code will scale to much larger systems as they become available. In addition, we consider the steps required to represent fully-resolved moving solid particles (usually referred to as colloids in the context of complex fluids). Such particles need to have their surface resolved on the lattice scale in order to include relevant surface physics and must be able to move, e.g., to execute Brownian motion in response to random forces from the fluid. Standard methods are available to represent such particles (e.g., [7, 11]) which are amenable to effective domain decomposition and message passing [17]. We describe below the additional considerations which arise for such moving particles when developing an implementation on a GPU.

In the following section we provide a brief overview of the lattice Boltzmann method, and mention some of the general issues which can influence performance in the CPU context. In Section 15.3, we then describe the alterations which are required to exploit the GPU architecture effectively and highlight how the resulting code differs from the CPU version. In Section 15.4, we extend this description to include the steps required to allow exploitation of many GPUs in parallel while retaining effective scaling. We also present results for a typical benchmark for a fluid-only problem in Section 15.4 to demonstrate the success of the approach. In Section 15.5, we describe the design choices which are relevant to a GPU implementation of moving particles. Finally, we include a number of general observations on software engineering and maintenance which arise from our experience. A summary is provided in Section 15.6.

15.2 Background

For a general complex fluid problem the starting point is the fluid velocity field $\mathbf{u}(\mathbf{r})$, whose evolution obeys the Navier-Stokes equations describing the conservation of mass (or density ρ), and momentum:

$$\rho[\partial_t \mathbf{u} + (\mathbf{u}.\nabla)\mathbf{u}] = -\nabla p + \eta \nabla^2 \mathbf{u} + \mathbf{f}(\mathbf{r}), \qquad (15.1)$$

where p is the isotropic pressure and η is the viscosity. A local force $\mathbf{f}(\mathbf{r})$ provides a means for coupling to other complex fluid constituents, e.g., it might represent the force exerted on the fluid by a curved interface between different phases or components.

The LB approach makes use of a regular three-dimensional lattice (see Figure 15.1) with discrete spacing Δr. It also makes use of a discrete velocity space $\mathbf{c}_i$, where the $\mathbf{c}_i$ are chosen to capture the correct symmetries of the Navier-Stokes equations. A typical choice, used here, is the so-called D3Q19 basis in three dimensions where there is one velocity such that $\mathbf{c}\Delta t$ is zero, along with six extending to the nearest neighbor lattice sites, and twelve extending to the next-nearest neighbor sites (Δt being the discrete time step). The fundamental object in LB is then the distribution function $f_i(\mathbf{r}; t)$ whose moments are related to the local hydrodynamic quantities: the fluid density, momentum, and stress. The time evolution of the distribution function is described by a discrete Boltzmann equation

$$f_i(\mathbf{r} + \mathbf{c}_i\Delta t; t) - f_i(\mathbf{r}; t) = -\mathcal{L}_{ij} f_j(\mathbf{r}; t). \qquad (15.2)$$

It is convenient to think of this in two stages. First, the right-hand side represents the action of a collision operator $\mathcal{L}_{ij}$, which is local to each lattice site and relaxes the distribution toward a local equilibrium at a rate ultimately related to the fluid viscosity. Second, the left-hand side represents a propagation step (sometimes referred to as streaming step), in which each element i of the distribution is displaced $\mathbf{c}_i \Delta t$, i.e., one lattice spacing in the appropriate direction per discrete time step.

More specifically, we store a vector of 19 double-precision floating point values at each lattice site for the distribution function $f_i(\mathbf{r}; t)$. The collision operation $\mathcal{L}_{ij}$, which is local at each lattice site, may be thought of as follows. A matrix-vector multiplication $\mathcal{M}_{ij} f_j$ is used to transform the distributions into the hydrodynamic quantities, where $\mathcal{M}_{ij}$ is a constant 19x19 matrix related to the choice of $\mathbf{c}_i$. The nonconserved hydrodynamic quantities are then relaxed toward their (known) equilibrium values and are transformed back to new

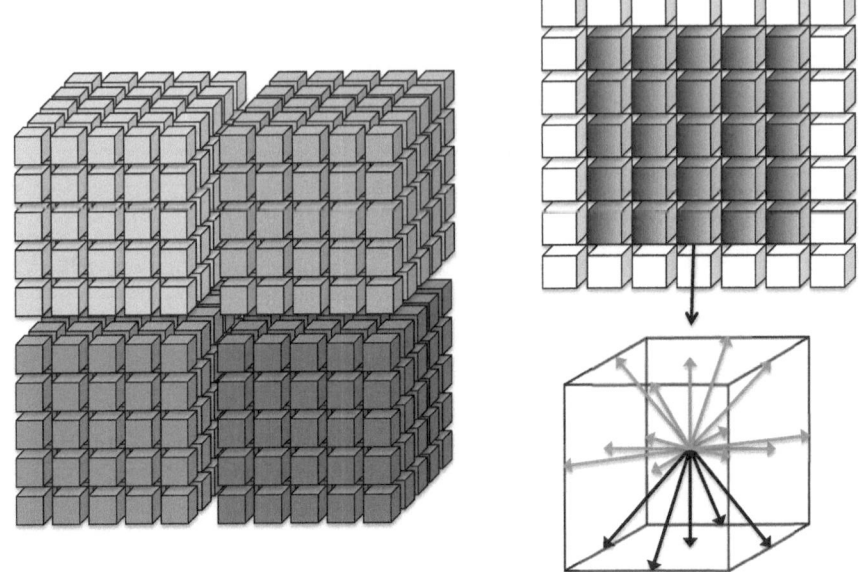

FIGURE 15.1. Left: the lattice is decomposed between MPI tasks. For clarity we show a 2D decomposition of a 3D lattice, but in practice we decompose in all 3 dimensions. Halo cells are added to each subdomain (as shown on the upper right for a single slice) which store data retrieved from remote neighbors in the halo exchange. Lower right: the D3Q19 velocity set resident on a lattice site; highlighted are the 5 "outgoing" elements to be transferred in a specific direction.

post-collision distributions via the inverse transformation $\mathcal{M}_{ij}^{-1}$. This gives rise to the need for a minimum of 2×19^2 floating point multiplications per lattice site. (Additional operations are required to implement, for example, the force $\mathbf{f}(\mathbf{r})$.) In contrast, the propagation stage consists solely of moving the distribution values between lattice sites and involves no floating point operations.

In the CPU version, the *Ludwig* implementation stores one time level of distribution values $f_i(\mathbf{r};t)$. This distribution is stored as a flat array of C data type `double` and laid out so that all elements of the velocity are contiguous at a given site (often referred to as "array-of-structures" order). This is appropriate for sums over the distribution required locally at the collision stage as illustrated schematically in Listing 15.1: the fact that consecutive loads are from consecutive memory addresses allows the prefetcher to engage fully. (The temporary scalar `a_tmp` allows caching of the intermediate accumulated

Listing 15.1. collision schematic for CPU.

```
/* loop over lattice sites */
for (is = 0; is < nsites; is++) {
  ...
  /* load distribution from ``array-of-structures'' */
  for (i = 0; i < 19; i++)
    f[i] = f_aos[19*is + i]
  ...
  /* perform matrix-vector multiplication */
  for (i = 0; i < 19; i++) {
    a_tmp = 0.0;
    for (j = 0; j < 19; j++) {
      a_tmp += f[j]*M[i][j];
    }
    a[i] = a_tmp;
  }
  ...
}
```

value in the innermost loop.) A variety of standard sequential optimizations are relevant for the collision stage (loop unrolling, inclusion of SIMD operations, and so on [22]). For example, the collision stage in *Ludwig* includes explicit SIMD code, which is useful if the compiler on a given platform cannot identify the appropriate vectorization. The propagation stage is separate and is organized as a "pull" (again, various optimizations have been considered, e.g., [8, 15, 24]). No further optimization is done here beyond ensuring that the ordering of the discrete velocities allows memory access to be as efficient as possible. While these optimizations are important, it should be remembered that for some complex fluid problems, the hydrodynamics embodied in the LB calculation is a relatively small part of the total computational cost (at the 10% level in some cases). This means optimization effort may be better concentrated elsewhere.

The regular 3D decomposition is illustrated in Fig. 15.1. Each local subdomain is surrounded by a halo, or ghost, region of width one lattice site. While the collision is local, elements of the distribution must be exchanged at the edges of the domains to facilitate the propagation. To achieve the full 3D halo exchange, the standard approach of shifting the relevant data in each coordinate direction in turn is adopted. This requires appropriate synchronization, i.e., a receive in the the first coordinate direction must be complete before a send in the second direction involving relevant data can take place, and so on. We note that only "outgoing" elements of the distribution need to be sent at each edge. For the D3Q19 model, this reduces the volume of data traffic from 19 to 5 of the $f_i(\mathbf{r}; t)$ per lattice site at each edge. In the CPU version, the necessary transfers are implemented in place using a vector of MPI datatypes with appropriate stride for each direction.

15.3 Single GPU implementation

In this section we describe the steps taken to enable *Ludwig* for the GPU. There are a number of crucial issues: first, the minimization of data traffic between host and device; second, the optimal mapping of available parallelism onto the architecture; and third, the issue of memory coalescing. We discuss each of these in turn.

While the most important section of the LB in terms of floating-point performance is the collision stage, this cannot be the only consideration for a GPU implementation. It is essential to offload all computational activity which involves the main data structures (such as the distribution) to the GPU. This includes kernels with relatively low computational demand, such as the propagation stage. All relevant data then remain resident on the GPU, to avoid expensive host-device data transfer at each iteration of the algorithm. Such transfers would negate any benefit of GPU acceleration. We note that for a complex fluid code, this requirement can extend to a considerable number of kernels, although we limit the discussion to collision and propagation for brevity.

To achieve optimal performance, it is vital to exploit fully the parallelism inherent in the GPU architecture, particularly for those matrix-vector operations within the collision kernel. The GPU architecture features a hierarchy of parallelism. At the lowest level, groups of 32 threads (warps) operate in lockstep on different data elements: this is SIMD-style vector-level parallelism. Multiple warps are combined into a thread block (in which communication and synchronization are possible), and multiple blocks can run concurrently across the streaming multiprocessors in the GPU (with no communication or synchronization possible across blocks). To decompose on the GPU, we must choose which part of the collision to assign to which level of parallelism.

While there exists parallelism within the matrix-vector operation, the length of each vector (here, 19) is smaller than the warp size and typical thread block sizes. So we simply decompose the loop over lattice sites to all levels of parallelism, i.e., we use a separate CUDA thread for every lattice site, and each thread performs the full matrix-vector operation. We find that a block size of 256 performs well on current devices: we therefore decompose the lattice into groups of 256 sites and assign each group to a block of CUDA threads. As the matrix $\mathcal{M}_{ij}$ is constant, it is assigned to the fast *constant* on-chip device memory.

For the propagation stage, the GPU implementation adds a second time level of distribution values. The data dependencies inherent in the propagation mean that the in-place propagation of the CPU version cannot be parallelized effectively without the additional time level. As both time levels may remain resident on the GPU, this is not a significant overhead.

An architectural constraint of GPUs means that optimal global memory

```
Listing 15.2. collision schematic for GPU.
```
```
/* compute current site index 'is' from CUDA thread and block */
/* indices and load distribution from "structure-of-arrays" */
for (i = 0; i < 19; i++)
  f[i] = f_soa[nsites*i + is]

/* perform matrix-vector multiplication as before */
...
```

bandwidth is only achieved when data are structured such that threads within a *half-warp* (a group of 16 threads) load data from the same memory segment in a single transaction: this is memory coalescing. It can be seen that the array-of-structures ordering used for the distribution in the CPU code would not be suitable for coalescing; in fact, it would result in serialized memory accesses and relative poor performance. To meet the coalescing criteria and allow consecutive threads to read consecutive memory addresses on the GPU, we transpose the layout of the distribution so that, for each velocity component, consecutive sites are contiguous in memory ("structure-of-arrays" order). A schematic of the GPU collision code is given in Listing 15.2.

Ludwig was modified to allow a choice of distribution data layout at compilation time depending on the target architecture: CPU or GPU. We defer some further comments on software engineering aspects of the code to the summary.

15.4 Multiple GPU implementation

To execute on multiple GPUs, we use the same domain decomposition and message passing framework as the CPU version. Within each subdomain (allocated to one MPI task) the GPU implementation proceeds as described in the previous section. The only additional complication is that halo transfers between GPUs must be staged through the host (in the future, direct GPU-to-GPU data transfers via MPI may be possible, obviating the need for these steps). This means host MPI sends must be preceded by appropriate device-to-host transfers and host MPI receives must be followed by corresponding host-to-device transfers.

In practice, this data movement requires additional GPU kernels to pack and unpack the relevant data before and after corresponding MPI calls. However, the standard shift algorithm, in which each coordinate direction is treated in turn, does provide some scope for the overlapping of different operations. For example, after the data for the first coordinate direction have been retrieved by the host, these can be exchanged using MPI between hosts at the same time as kernels for packing and retrieving of data for the second coordinate direction are executed. This overlapping must respect the synchronization required to ensure that data values at the corners of the subdomain are transferred correctly. We use a separate CUDA stream for each coordinate direction: this allows some of the host-device communication time to be effectively "hidden" behind the host-host MPI communication, resulting in an overall speedup. The improvement is more pronounced for the smaller local lattice size, perhaps because of less CPU memory bandwidth contention. The overlapping is then a particularly valuable aid to strong scaling, where the local system size decreases as the number of GPUs increases.

To demonstrate the effectiveness of our approach, we compare the performance of both CPU and GPU versions of *Ludwig*. To test the complex fluid nature of the code, the problem is actually an immiscible fluid mixture which uses a second distribution function to introduce a composition variable. The interested reader is referred to [16] for further details. The largest total problem size used is $2548 \times 1764 \times 1568$. The CPU system features a Cray XE6 architecture with 2 16-core AMD Opteron CPUs per node, and with nodes interconnected using Cray Gemini technology. For GPU results, a Cray XK6 system is used: this is very similar to the XE6, but has one CPU per node replaced with an NVIDIA X2090 GPU. Each node in the GPU system therefore features a single Opteron CPU acting as a host to a single GPU. The internode interconnect architecture is the same as for the Cray XE6. The GPU performance tests use a prototype Cray XK6 system with 936 nodes (the largest available at the time of writing). To provide a fair comparison, we compare scaling on a *per node* basis. That is, we compare 1 fully occupied 32-core CPU node (running 32 MPI tasks) with 1 GPU node (host running 1 MPI task, and 1 device). We believe this is representative of the true "cost" of a simulation in terms of accounting budgets, or electricity.

Figure 15.2 shows the results of both weak and strong scaling tests (top and bottom panels, respectively). For weak scaling, where the local subdomain size is kept fixed (here a relatively large 196^3 lattice), the time taken by an ideal code would remain constant. It can be seen that while scaling of the CPU version shows a pronounced upward slope, the GPU version scales almost perfectly. The advantage of the GPU version over the CPU version is the fact that the GPU version has a factor of 32 fewer MPI tasks per node, so communications require a significantly smaller number of larger data transfers. The performance advantage of the GPU version ranges from a factor of around 1.5 to around 1.8. Careful studies by other authors [23] have found the absolute performance (in terms of floating-point operations per second, or floating-point operations per Watt) to be remarkably similar between architectures.

Perhaps of more interest is the strong scaling picture (lower panel in Figure 15.2) where the performance as a function of the number of nodes is measured for fixed problem size. We consider four different fixed problem sizes on both CPU (up to 512 nodes shown) and GPU (up to 768 nodes). To allow compari-

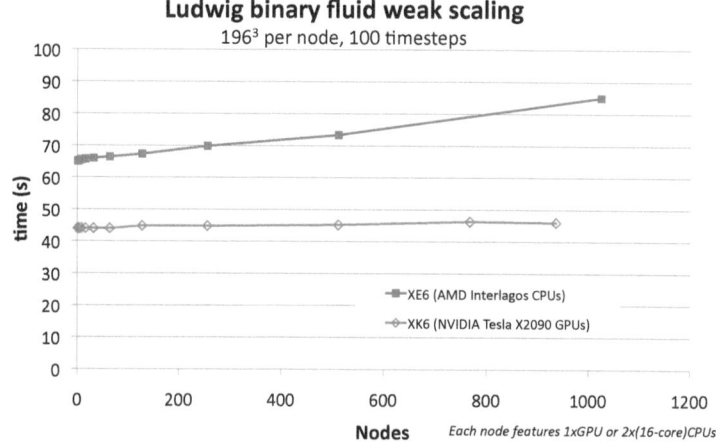

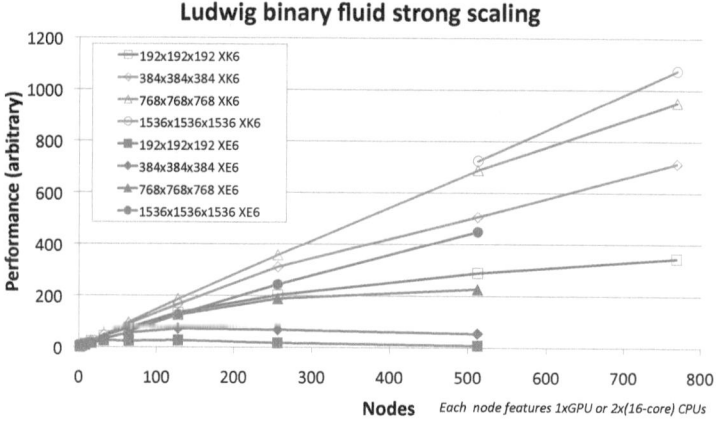

FIGURE 15.2. The weak (top) and strong (bottom) scaling of *Ludwig*. Closed shapes denote results using the CPU version run on the Cray XE6 (using two 16-core AMD Interlagos CPUs per node), while open shapes denote results using the GPU version on the Cray XK6 (using a single NVIDIA X2090 GPU per node). Top: the benchmark time is shown where the problem size per node is constant. Bottom: performance is shown where, for each of the four problem sizes presented, the results are scaled by the lattice volume, and all results are normalized by the same arbitrary constant.

son, the results are scaled by total system size in each case. For strong scaling, the disparity in the number of MPI tasks is clearly revealed in the failure of the CPU version to provide any significant benefit beyond a modest number of nodes as communication overheads dominate. In contrast, the GPU version

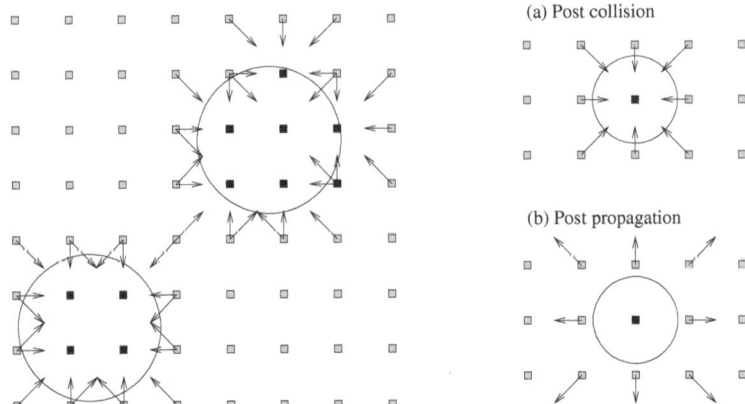

FIGURE 15.3. A two-dimensional schematic picture of spherical particles on the lattice. Left: a particle is allowed to move continuously across the lattice, and the position of the surface defines fluid lattice sites (gray) and solid lattice sites (black). The discrete surface is defined by links where propagation would intersect the surface (arrows). Note the discrete shape of the two particles is different. Right: post-collision distributions are reversed at the surface by the process of bounce-back on links, which replaces the propagation.

shows reasonably robust scaling even for the smaller system sizes and good scaling for the larger systems.

15.5 Moving solid particles

The introduction of moving solid particles poses an additional hurdle to efficient GPU implementation of an LB code such as *Ludwig*. In this section, we give a brief overview of the essential features of the CPU implementation, and how the considerations raised in the previous sections—maximizing parallelism and minimising both host-to-device and GPU-to-GPU data movement—shape the design decisions for a GPU implementation. We restrict our discussion to a simple fluid in this section; additional solid-fluid boundary conditions (e.g., wetting at a fluid-fluid-solid contact line) usually arise elsewhere in the calculation and are broadly independent of the hydrodynamic boundary conditions which are described in what follows.

Moving solid particles (here, spheres) are defined by a center position which is allowed to move continuously across the space of the lattice, and a fixed radius which is typically a few lattice spacings Δr. The surface of the particle is defined by a series of *links* where a discrete velocity propagation $\mathbf{c}_i \Delta t$ would intercept or cut the spherical shell (see Fig. 15.3). Hydrodynamic boundary conditions are then implemented via the standard approach of bounce-back on links [7, 11], where the relevant post-collision distribution values are reversed at the propagation stage with an appropriate correction to allow for the solid body motion. The exchange of momentum at each link must then be accumulated around the entire particle surface to provide the net hydrodynamic force and torque on the sphere. The particle motion can then be updated in a molecular dynamics-like step.

For the CPU implementation, a number of additional MPI communications are required: (1) to exchange the center position, radius, etc. of each particle as it moves and (2) to allow the accumulation of the force and torque for each particle (the links of which may be distributed between up to 8 MPI tasks in three dimensions). Appropriate marshaling of these data can provide an effective parallelization for relatively dense particle suspensions of up to 40% solid volume fraction [17]. As a final consideration, fluid distributions must be removed and replaced consistently as a given particle moves across the lattice and changes its discrete shape.

With these features in mind, we can identify a number of competing concerns which are relevant to a GPU implementation:

1. Minimization of host-device data transfer would argue for moving the entire particle code to the GPU. However, the code in question involves largely conditional logic (e.g., identifying cut surface links) and irregular memory accesses (e.g., access to distribution elements around a spherical particle). These operations seem poorly suited to effective parallelization on the GPU. As an additional complication, the sums required over the particle surface would involve potentially tricky and inefficient reductions in GPU memory.

2. The alternative is to retain the relevant code on the CPU. While the transfer of the entire distribution $f_i(\mathbf{r}; t)$ between host and device at each time step is unconscionable owing to PCIe bus bandwidth considerations, the transfer of only relevant distribution information to allow bounce-back on links is possible. At modest solid volumes only a very small fraction of the distribution data is involved in bounce-back (e.g., a 2% solid volume fraction of particles radius $2.3\Delta r$ would involve approximately 1% of the distribution data). This option also has the advantage that no further host-device data transfers are necessary to allow the MPI exchanges required for particle information.

We have implemented the second option as follows. For each subdomain, a list of boundary-cutting links is assembled on the CPU; the list includes the identity of the relevant element of the distribution in each case. This list, together with the particle information required to compute the correct bounce-back term, is transferred to the GPU. The updates to the relevant elements of the distribution can then take place on the GPU. The corresponding information to compute the update of the particle dynamics is returned to the CPU, where the reduction over the surface links is computed. The change of particle shape may be dealt with in a similar manner: the relatively small number of updates required at any one time step (or however frequently the particle position is updated) can be marshaled to the GPU as necessary. This preliminary implementation is found to be effective on problems involving up to 4% solid volume fraction.

We note that the CPU version actually avoids the collision calculation at solid lattice points by consulting a look-up table of solid/fluid status. On the GPU, it is perhaps preferable to perform the collision stage everywhere instead of moving the look-up table to the GPU and introducing the associated logic. Ultimately, the GPU might favor other boundary methods which treat solid and fluid on a somewhat more equal basis, for example, the immersed boundary method [6, 13, 14] or smoothed profile method [10]. However, the approach adopted here allows us to exploit the GPU for the intensive fluid simulation while maintaining the complex code required for particles on the CPU. Overheads of CPU-GPU transfer are minimized by transferring only those data relevant to the hydrodynamic interaction implemented via bounce-back on links.

It is perhaps interesting at this point to make some more general observations on the software engineering challenge presented when extending an existing CPU code to the GPU. The already complex task of maintaining the code in a portable fashion while also maintaining performance is currently formidable. To help this process, we have followed a number of basic principles. First, in order to port to the GPU in an incremental fashion, we have tried to maintain the modular structure of the CPU where possible. For each data structure, such as the distribution, a separate analogue is maintained in both the CPU and GPU memory spaces. However, the GPU copy does not include the complete CPU structure: in particular, nonintrinsic datatypes such as MPI datatypes are not required on the GPU. Functions to marshal data between CPU and GPU are provided for each data structure, abstracting the underlying CUDA implementation. (This reasonably lightweight abstraction layer could also allow an OpenCL implementation to be developed.) This makes it easy to switch between the CPU and GPU for different components in the code, which is useful in development and testing. GPU functionality can be added incrementally while retaining a code that runs correctly (albeit slowly due to data transfer overheads). Once all relevant components are moved to the GPU, it becomes possible to remove such data transfers and keep the entire problem resident on the device.

15.6 Summary

We have described the steps taken to implement, on both single and multiple GPUs, the *Ludwig* code which was originally designed for complex fluid problems on a conventional CPU architecture. We have added the necessary functionality using NVIDIA CUDA and discussed the important changes to the main data structures. By retaining domain decomposition and message passing via MPI, we have demonstrated it is possible to scale complex fluid problems to large numbers of GPUs in parallel. By following the key design criteria of maximizing parallelism and minimizing host-device data transfers, we have confirmed that the mixed MPI-GPU approach is viable for scientific applications. For the more intricate problem of moving solid particles, we find it is possible to retain the more serial elements related to particle link operations on the CPU, while offloading only the parallel lattice-based operations involving the LB distribution to the GPU. Again, this minimizes host-device movement of data.

From the software engineering viewpoint, some duplication of code to allow efficient implementation on both host and device is currently required. This issue might be addressed by approaches such as automatic kernel generation, but may also be addressed naturally in time as GPU and CPU hardware converge. Portable abstractions and APIs, perhaps based on approaches such as OpenCL, will also facilitate the development and maintenance of portable codes which also exhibit portable performance (perhaps in conjunction with automatic tuning approaches).

So, while the challenges in designing portable and efficient scientific applications remain very real, this work provides some hope that large clusters of GPU machines can be used effectively for a wide range of complex fluid problems.

15.7 Acknowledgments

AG was supported by the CRESTA project, which has received funding from the European Community's Seventh Framework Programme (ICT-2011.9.13) under Grant Agreement no. 287703. KS was supported by UK EPSRC under grant EV/J007404/1.

Bibliography

[1] C. K. Aidun and J. R. Clausen. Lattice Boltzmann method for complex flows. *Ann. Rev. Fluid Mech.*, 42:439–472, 2010.

[2] M. Bernaschi, M. Fatica, S. Melchionna, S. Succi, and E. Kaxiras. A flexible high-performance lattice Boltzmann GPU code for the simulations of fluid flow in complex geometries. *Concurrency Computat.: Pract. Exper.*, 22:1–14, 2010.

[3] J.-C. Desplat, I. Pagonabarraga, and Bladon P. Ludwig: A parallel lattice-Boltzmann code for complex fluids. *Comput. Phys. Comms.*, 134:273–290, 2001.

[4] Z. Fan, F. Qiu, A. Kaufman, and S. Yoakum-Stover. GPU cluster for high performance computing. In IEEE Computer Society Press, editor, *Proceedings of ACM/IEEE Supercomputing Conference*, pages 47–59, Pittsburgh, PA, 2004.

[5] C. Feichtinger, J. Habich, H. Köstler, G. Hager, U. Rüde, and G. Wellein. A flexible patch-based lattice Boltzmann parallelization approach for heterogeneous GPU-CPU clusters. *Parallel Computing*, 37:536–549, 2011.

[6] Z.-G. Feng and E. E Michaelides. The immersed boundary-lattice Boltzmann method for solving fluid-particles interaction problem. *J. Comp. Phys.*, 195:602–628, 2004.

[7] A. J. C. Ladd. Numerical simulations of particle suspensions via a discretized Boltzmann equation. Part 1. theoretical foundation and Part ii. Numerical results. *J. Fluid Mech.*, 271:285–339, 1994.

[8] K. Mattila, J. Hyväluoma, T. Rossi, Aspnäs M., and J. Westerholm. An efficient swap algorithm for the lattice Boltzmann method. *Comput. Phys. Comms.*, 176:200–210, 2007.

[9] J. Myre, S. D. C. Walsh, D. Lilja, and M. O. Saar. Performance analysis of single-phase, multiphase, and multicomponent lattice Boltzmann fluid flow simulations on GPU clusters. *Concurrency Computat.: Pract. Exper.*, 23:332–350, 2011.

[10] Y. Nakayama and R. Yammamoto. Simulation method to resolve hydrodynamic interactions in colloidal dispersions. *Phys. Rev. E*, 71:036707, 2005.

[11] N.-Q. Nguyen and A. J. C. Ladd. Lubrication corrections for lattice Boltzmann simulations of particle suspensions. *Phys. Rev. E*, 66:046708, 2002.

[12] C. Obrecht, F. Kuznik, B. Tourancheau, and J.-J. Roux. Multi-GPU implementation of the lattice Boltzmann method. *Comput. Math. with Applications*, 65:252–261, 2013.

[13] C. S. Peskin. Flow patterns around heart valves; a numerical method. *J. Comp. Phys.*, 10:252–271, 1972.

[14] C. S. Peskin. The immersed boundary method. *Acta Nummerica*, 11:479–517, 2002.

[15] T. Pohl, M. Kowarschik, J. Wilke, K. Igelberger, and U. Rüde. Optimization and profiling of the cache performance of parallel lattice Boltzmann code. *Parallel Process Lett.*, 13:549–560, 2003.

[16] K. Stratford, R. Adhikari, I. Pagonabarraga, and J.-C. Desplat. Lattice Boltzmann for binary fluids with suspended colloids. *J. Stat. Phys.*, 121:163–178, 2005.

[17] K. Stratford and I. Pagonabarraga. Parallel domain decomposition for lattice Boltzmann with moving particles. *Comput. Math. with Applications*, 55:1585–1593, 2008.

[18] S. Succi. *The Lattice Boltzmann Equation and Beyond*. Oxford University Press, Oxford, 2001.

[19] J. Tölke. Implementation of a lattice Boltzmann kernel using the compute unified device architecture developed by NVIDIA. *Comput. Visual Sci.*, 13:29–39, 2010.

[20] S. D. C. Walsh and M. O. Saar. Developing extensible lattice Boltzmann simulators for general-purpose graphics-processing units. *Comm. Comput. Phys.*, 13:867–879, 2013.

[21] X. Wei, W. Li, K. Müller, and A. E. Kaufman. The lattice Boltzmann method for simulating gaseous phenomena. *IEEE Transactions on Visualization and Computer Graphics*, 10:164–176, 2004.

[22] G. Wellein, T. Zeiser, G. Hager, and S. Donath. On the single processor performance of simple lattice Boltzmann kernels. *Computers and Fluids*, 35:910–919, 2006.

[23] S. Williams, L. Oliker, J. Carter, and J. Shalf. Extracting ultra-scale lattice Boltzmann performance via hierarchical and distributed auto-tuning. In *International Conference for High Performance Computing, Networking, Storage and Analysis (SC)*, pages 1–12, Berkeley, CA, 2011. IEEE Computer Society Press.

[24] M. Wittmann, T. Zeiser, G. Hager, and G. Wellein. Comparison of different propagation steps for lattice Boltzmann methods. *Comput. Math with Appl.*, 2012. doi:10.1016/j.camwa.2012.05.002.

[25] W. Xian and A. Takayuki. Multi-GPU performance of incompressible flow computation by lattice Boltzmann method on GPU cluster. *Parallel Comput.*, 37:521–535, 2011.

[26] Y. Zhao. Lattice Boltzmann based PDE solver on the GPU. *Visual Comput.*, 24:323–333, 2008.

[27] H. Zhu, X. Liu, Y. Liu, and E. Wu. Simulation of miscible binary mixtures based on lattice Boltzmann method. *Comp. Anim. Virtual Worlds*, 17:403–410, 2006.

Chapter 16

Numerical validation and performance optimization on GPUs of an application in atomic physics

Rachid Habel
Télécom SudParis, France

Pierre Fortin, Fabienne Jézéquel, and Jean-Luc Lamotte
Laboratoire d'Informatique de Paris 6, Université Pierre et Marie Curie, France

Stan Scott
School of Electronics, Electrical Engineering & Computer Science, The Queen's University of Belfast, United Kingdom

16.1	Introduction	372
16.2	2DRMP and the PROP program	373
	16.2.1 Principles of R-matrix propagation	373
	16.2.2 Description of the PROP program	375
	16.2.3 CAPS implementation	376
16.3	Numerical validation of PROP in single precision	377
	16.3.1 Medium case study	378
	16.3.2 Huge case study	380
16.4	Towards a complete deployment of PROP on GPUs	381
	16.4.1 Computing the output R-matrix on GPU	382
	16.4.2 Constructing the local R-matrices on GPU	384
	16.4.3 Scaling amplitude arrays on GPU	385
	16.4.4 Using double-buffering to overlap I/O and computation	385
	16.4.5 Matrix padding	386
16.5	Performance results	387
	16.5.1 PROP deployment on GPU	387
	16.5.2 PROP execution profile	389
16.6	Propagation of multiple concurrent energies on GPU	391
16.7	Conclusion and future work	392
	Bibliography	393

16.1 Introduction

As described in Chapter 1, GPUs are characterized by hundreds of cores and theoretically perform one order of magnitude better than CPUs. An important factor to consider when programming on GPUs is the cost of data transfers between CPU memory and GPU memory. Thus, to have good performance on GPUs, applications should be coarse-grained and have a high arithmetic intensity (i.e., the ratio of arithmetic operations to memory operations). Another important aspect of GPU programming is that floating-point operations are preferably performed in single precision, if the validity of results is not impacted by that format. The GPU compute power for floating-point operations is indeed greater in single precision than in double precision. The peak performance ratio between single precision and double precision varies, for example, for NVIDIA GPUs from 12 for the first Tesla GPUs (C1060), to 2 for the Fermi GPUs (C2050 and C2070), and to 3 for the latest Kepler architecture (K20/K20X). As far as AMD GPUs are concerned, the latest AMD GPU (Tahiti HD 7970) presents a ratio of 4. Moreover, GPU internal memory accesses and CPU-GPU data transfers are faster in single precision than in double precision because of the different format lengths.

This chapter describes the deployment on GPUs of PROP, a program of the 2DRMP [2, 10] suite which models electron collisions with H-like atoms and ions at intermediate energies. 2DRMP operates successfully on serial computers, high performance clusters, and supercomputers. The primary purpose of the PROP program is to propagate a global R-matrix [1], $\Re$, in the two-electron configuration space. The propagation needs to be performed for all collision energies, for instance, hundreds of energies, which are independent. Propagation equations are dominated by matrix multiplications involving submatrices of $\Re$. However, the matrix multiplications are not straightforward in the sense that $\Re$ dynamically changes the designation of its rows and columns and increases in size as the propagation proceeds [11].

In a preliminary investigation PROP was selected by GENCI[1] and CAPS,[2] following their first call for projects in 2009–2010 aimed at deploying applications on hybrid systems based on GPUs. First CAPS recast the propagation equations with larger matrices. For matrix products the GPU performance gain over CPU increases indeed with the matrix size, since the CPU-GPU transfer overhead becomes less significant and since CPUs are still more efficient for fine computation grains. Then, using HMPP,[3] a commercial hybrid and parallel compiler, CAPS developed a version of PROP in which matrix

---

[1]GENCI: Grand Equipement National de Calcul Intensif, www.genci.fr

[2]CAPS is a software company providing products and solutions for many-core application programming and deployment, www.caps-entreprise.com

[3]HMPP (Hybrid Multicore Parallel Programming) or CAPS compiler, see: www.caps-entreprise.com/hmpp.html

multiplications are performed on the GPU or the CPU, depending on the matrix size. Unfortunately this partial GPU implementation of PROP does not offer significant acceleration.

The work described in this chapter, which is based on a study presented in [3], aims at improving PROP performance on GPUs by exploring two directions. First, because the original version of PROP is written in double precision, we study the numerical stability of PROP in single precision. Second, we deploy the whole computation code of PROP on GPUs to avoid the overhead generated by data transfers and we propose successive improvements (including one specific to the Fermi architecture) in order to optimize the GPU code.

16.2 2DRMP and the PROP program

16.2.1 Principles of R-matrix propagation

2DRMP [2, 10] is part of the CPC library.[4] It is a suite of seven programs aimed at creating virtual experiments on high performance and grid architectures to enable the study of electron scattering from H-like atoms and ions at intermediate energies. The 2DRMP suite uses the two-dimensional R-matrix propagation approach [1]. In 2DRMP the two-electron configuration space (r_1, r_2) is divided into sectors. Figure 16.1 shows the division of the two-electron configuration space (r_1, r_2) into 4 vertical *strips* representing 10 *sectors*. The key computation in 2DRMP, performed by the PROP program, is the propagation of a global R-matrix, $\Re$, from sector to sector across the internal region, as shown in Fig. 16.1.

We consider the general situation in Fig. 16.2 where we assume that we already know the global R-matrix, $\Re^I$, associated with the boundary defined by edges 5, 2, 1, and 6 in domain D and we wish to evaluate the new global R-matrix, $\Re^O$, associated with edges 5, 3, 4, and 6 in domain D' following propagation across subregion d. Input edges are denoted by I (edges 1 and 2), output edges by O (edges 3 and 4) and common edges by X (edges 5 and 6). Because of symmetry, only the lower half of domains D and D' has to be considered. The global R-matrices, $\Re^I$ in domain D and $\Re^O$ in domain D', can be written as

$$\Re^I = \begin{pmatrix} \Re^I_{II} & \Re^I_{IX} \\ \Re^I_{XI} & \Re^I_{XX} \end{pmatrix} \quad \Re^O = \begin{pmatrix} \Re^O_{OO} & \Re^O_{OX} \\ \Re^O_{XO} & \Re^O_{XX} \end{pmatrix}. \quad (16.1)$$

From the set of local R-matrices, $\mathbf{R}_{ij}$ ($i, j \in \{1, 2, 3, 4\}$), associated with

---
[4]CPC: Computer Physics Communications, http://cpc.cs.qub.ac.uk/

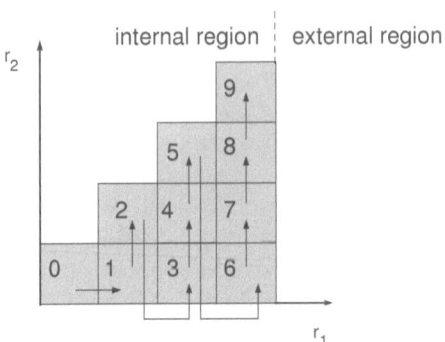

FIGURE 16.1. Subdivision of the configuration space (r_1, r_2) into a set of connected sectors.

subregion d, we can define

$$\mathbf{r}_{II} = \begin{pmatrix} \mathbf{R}_{11} & \mathbf{R}_{12} \\ \mathbf{R}_{21} & \mathbf{R}_{22} \end{pmatrix}, \quad \mathbf{r}_{IO} = \begin{pmatrix} \mathbf{R}_{13} & \mathbf{R}_{14} \\ \mathbf{R}_{23} & \mathbf{R}_{24} \end{pmatrix}, \quad (16.2a)$$

$$\mathbf{r}_{OI} = \begin{pmatrix} \mathbf{R}_{31} & \mathbf{R}_{32} \\ \mathbf{R}_{41} & \mathbf{R}_{42} \end{pmatrix}, \quad \mathbf{r}_{OO} = \begin{pmatrix} \mathbf{R}_{33} & \mathbf{R}_{34} \\ \mathbf{R}_{43} & \mathbf{R}_{44} \end{pmatrix}, \quad (16.2b)$$

where I represents the input edges 1 and 2, and O represents the output edges 3 and 4 (see Fig. 16.2). The propagation across each sector is characterized by equations (16.3a) to (16.3d).

$$\mathfrak{R}^O_{OO} = \mathbf{r}_{OO} - \mathbf{r}^T_{IO}(\mathbf{r}_{II} + \mathfrak{R}^I_{II})^{-1}\mathbf{r}_{IO}, \quad (16.3a)$$
$$\mathfrak{R}^O_{OX} = \mathbf{r}^T_{IO}(\mathbf{r}_{II} + \mathfrak{R}^I_{II})^{-1}\mathfrak{R}^I_{IX}, \quad (16.3b)$$
$$\mathfrak{R}^O_{XO} = \mathfrak{R}^I_{XI}(\mathbf{r}_{II} + \mathfrak{R}^I_{II})^{-1}\mathbf{r}_{IO}, \quad (16.3c)$$
$$\mathfrak{R}^O_{XX} = \mathfrak{R}^I_{XX} - \mathfrak{R}^I_{XI}(\mathbf{r}_{II} + \mathfrak{R}^I_{II})^{-1}\mathfrak{R}^I_{IX}. \quad (16.3d)$$

The matrix inversions are not explicitly performed. To compute $(\mathbf{r}_{II} + \mathfrak{R}^I_{II})^{-1}\mathbf{r}_{IO}$ and $(\mathbf{r}_{II} + \mathfrak{R}^I_{II})^{-1}\mathfrak{R}^I_{IX}$, two linear systems are solved.

While equations (16.3a)–(16.3d) can be applied to the propagation across a general subregion two special situations should be noted: propagation across a diagonal subregion and propagation across a subregion bounded by the r_1-axis at the beginning of a new strip.

In the case of a diagonal subregion, from symmetry considerations, edge 2 is identical to edge 1 and edge 3 is identical to edge 4. Accordingly, with only one input edge and one output edge equations (16.2a)–(16.2b) become

$$\mathbf{r}_{II} = 2\mathbf{R}_{11}, \quad \mathbf{r}_{IO} = 2\mathbf{R}_{14}, \quad (16.4a)$$
$$\mathbf{r}_{OI} = 2\mathbf{R}_{41}, \quad \mathbf{r}_{OO} = 2\mathbf{R}_{44}. \quad (16.4b)$$

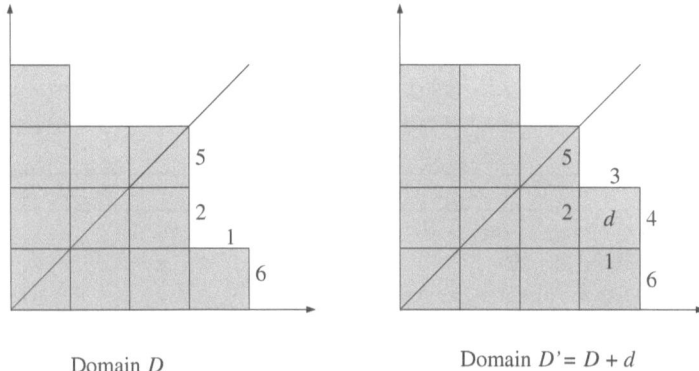

FIGURE 16.2. Propagation of the R-matrix from domain D to domain D'.

Data Set	Local R-Matrix Size	Strips	Sectors	Final Global R-Matrix Size	Scattering Energies
Small	90 × 90	4	10	360 × 360	6
Medium	90 × 90	4	10	360 × 360	64
Large	383 × 383	20	210	7660 × 7660	6
Huge	383 × 383	20	210	7660 × 7660	64

TABLE 16.1. Characteristics of four data sets.

In the case of a subregion bounded by the r_1-axis at the beginning of a new strip, we note that the input boundary I consists of only one edge. When propagating across the first subregion in the second strip there is no common boundary X: in this case only equation (16.3a) needs to be solved.

Having obtained the global R-matrix $\Re$ on the boundary of the innermost subregion (labeled 0 in Fig. 16.1), $\Re$ is propagated across each subregion in the order indicated in Fig. 16.1, working systematically from the r_1-axis at the bottom of each strip across all subregions to the diagonal.

16.2.2 Description of the PROP program

The PROP program computes the propagation of the R-matrix across the sectors of the internal region. Table 16.1 shows four different data sets used in this study and highlights the principal parameters of the PROP program. PROP execution can be described by Algorithm 19. First, amplitude arrays and correction data are read from data files generated by the preceding program of the 2DRMP suite. Then, local R-matrices are constructed from amplitude arrays. Correction data is used to compute correction vectors added to the diagonal of the local R-matrices. The local R-matrices, together with the input R-matrix, $\Re^I$, computed on the previous sector, are used to compute the current output R-matrix, $\Re^O$. At the end of a sector evaluation, the output R-matrix becomes the input R-matrix for the next evaluation.

Algorithm 19: PROP algorithm

1 **for** *all scattering energies* **do**
2   **for** *all sectors* **do**
3   Read amplitude arrays;
4   Read correction data;
5   Construct local R-matrices;
6   From $\Re^I$ and local R-matrices, compute $\Re^O$;
7        $\Re^O$ becomes $\Re^I$ for the next sector;
8   **end**
9   Compute physical R-Matrix;
10 **end**

On the first sector, there is no input R-matrix yet. To bootstrap the propagation, the first output R-matrix is constructed using only one local R-matrix. On the last sector, that is, on the boundary of the inner region, a physical R-matrix corresponding to the output R-matrix is computed and stored into an output file.

In the PROP program, sectors are characterized into four types, depending on the computation performed:

- the starting sector (labeled 0 in Fig. 16.1);
- the axis sectors (labeled 1, 3, and 6 in Fig. 16.1);
- the diagonal sectors (labeled 2, 5, and 9 in Fig. 16.1);
- the off-diagonal sectors (labeled 4, 7, and 8 in Fig. 16.1).

The serial version of PROP is implemented in Fortran 90 and for linear algebra operations uses BLAS and LAPACK routines which are fully optimized for x86 architecture. This program serially propagates the R-matrix for all scattering energies. Since the propagations for these different energies are independent, there also exists an embarrassingly parallel version of PROP that spreads the computations of several energies among multiple CPU nodes via MPI.

16.2.3 CAPS implementation

In order to handle larger matrices, and thus obtain better GPU speedup, CAPS recast equations (16.3a) to (16.3d) into one equation. The output R-

matrix $\Re^O$ defined by equation (16.1) is now computed as follows:

$$\Re^O = \Re^{O\,'} + UA^{-1}V, \qquad (16.5)$$

$$\text{with } \Re^{O\,'} = \begin{pmatrix} \mathbf{r}_{OO} & 0 \\ 0 & \Re^I_{XX} \end{pmatrix}, \; U = \begin{pmatrix} -r^T_{IO} \\ \Re^I_{XI} \end{pmatrix}, \qquad (16.6)$$

$$A = \mathbf{r}_{II} + \Re^I_{II} \text{ and } V = (\mathbf{r}_{IO} \quad -\Re^I_{IX}). \qquad (16.7)$$

To compute $W = A^{-1}V$, no matrix inversion is performed. The matrix system $AW = V$ is solved. This reimplementation of PROP reduces the number of equations to be solved and the number of matrix copies for evaluating each sector. For instance, for an off-diagonal sector, copies fall from 22 to 5, matrix multiplications from 4 to 1, and calls to a linear equation solver from 2 to 1.

To implement this version, CAPS used HMPP, a commercial hybrid and parallel compiler, based on compiler directives such as the new OpenACC standard.[5] If the matrices are large enough (the limit sizes are experimental parameters), they are multiplied on the GPU, otherwise on the CPU. CAPS used Intel's MKL (Math Kernel Library) BLAS implementation on an Intel Xeon x5560 quad core CPU (2.8 GHz) and the CUBLAS library (CUDA 2.2) on one Tesla C1060 GPU. On the large data set (see Table 16.1), CAPS obtained a speedup of 1.15 for the GPU version over the CPU one (with multi-threaded MKL calls on the four CPU cores). This limited gain in performance is mainly due to the use of double precision computation and to the small or medium sizes of most matrices. For these matrices, the computation gain on the GPU is indeed strongly affected by the overhead generated by transferring these matrices from the CPU memory to the GPU memory to perform each matrix multiplication and then transferring the result back to the CPU memory.

Our goal is to speed up PROP more significantly by porting the whole code to the GPU and therefore avoiding the intermediate data transfers between the host (CPU) and the GPU. We will also study the stability of PROP in single precision because single-precision computation is faster on the GPU and CPU-GPU data transfers are twice as fast as those performed in double precision.

16.3 Numerical validation of PROP in single precision

Floating-point input data, computation, and output data of PROP are originally in double-precision format. PROP produces a standard R-matrix

---

[5]See: www.openacc-standard.org

H-file [2] and a collection of Rmat00X files (where X ranges from 0 to the number of scattering energies − 1) holding the physical R-matrix for each energy. The H-file and the Rmat00X files are binary input files of the FARM program [2] (last program of the 2DRMP suite). Their text equivalents are the prop.out and the prop00X.out files. To study the validity of PROP results in single precision, first, reference results are generated by running the serial version of PROP in double precision. Data used in the most costly computation parts are read from input files in double precision format and then cast to single precision format. PROP results (input of FARM) are computed in single precision and written into files in double precision.

16.3.1 Medium case study

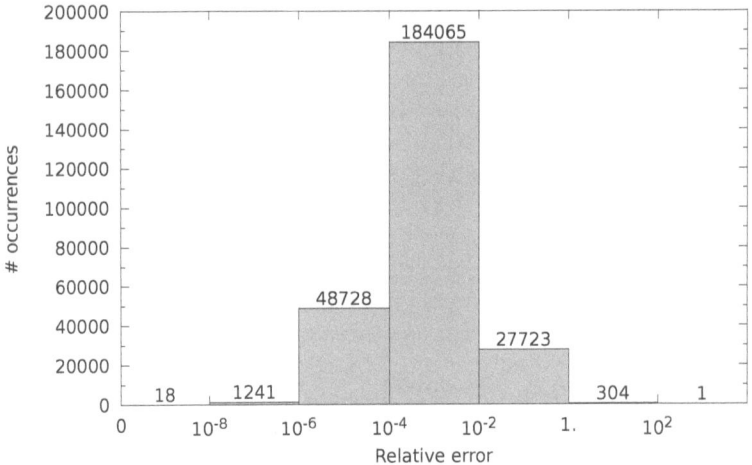

FIGURE 16.3. Error distribution for medium case in single precision.

The physical R-matrices, in the prop00X.out files, are compared to the reference ones for the medium case (see Table 16.1). The relative error distribution is given in Fig. 16.3. We focus on the largest errors.

- Errors greater than 10^2: the only impacted value is of order 10^{-6} and is negligible compared to the other ones in the same prop00X.out file.

- Errors between 1 and 10^2: the values corresponding to the largest errors are of order 10^{-3} and are negligible compared to the majority of the other values which range between 10^{-2} and 10^{-1}.

- Errors between 10^{-2} and 1: the largest errors ($\geq 6\%$) impact values the order of magnitude of which is at most 10^{-1}. These values are negligible. Relative errors of approximately 5% impact values the order of

File	Largest Relative Error	File	Largest Relative Error
1s1s	0.02	1s3p	0.11
1s2s	0.06	1s3d	0.22
1s2p	0.08	1s4s	0.20
1s3s	0.17	2p4d	1.60

TABLE 16.2. Impact on FARM of the single precision version of PROP.

magnitude of which is at most 10^2. For instance, the value 164 produced by the reference version of PROP becomes 172 in the single precision version.

To study the impact of the single precision version of PROP on the FARM program, the cross-section results files corresponding to transitions 1s1s, 1s2s, 1s2p, 1s3s, 1s3p, 1s3d, 1s4s, 2p4d are compared to the reference ones. Table 16.2 shows that all cross-section files are impacted by errors. Indeed in the 2p4d file, four relative errors are greater than one and the maximum relative error is 1.60. However, the largest errors impact negligible values. For example, the maximum error (1.60) impacts a reference value which is 4.5 10^{-4}. The largest values are impacted by low errors. For instance, the maximum value (1.16) is impacted by a relative error of the order 10^{-3}.

To examine in more detail the impact of PROP on FARM, cross-sections above the ionization threshold (1 Ryd) are compared in single and double precision for transitions among the 1s, ...4s, 2p, ...4p, 3d, 4d target states. This comparison is carried out by generating 45 plots. In all the plots, results in single and double precision match except for a few scattering energies which are very close to pseudo-state thresholds. For example, Fig. 16.4(a) and 16.5(a) present the scattering energies corresponding to the 1s2p and 1s4d cross-sections computed in single and double precision. For some cross-sections, increasing a threshold parameter from 10^{-4} to 10^{-3} in the FARM program results in energies close to threshold being avoided, and therefore, the cross-sections in double and single precision match more accurately. This is the case for instance for cross-section 1s2p (see Fig. 16.4(b)). However for other cross-sections (such as 1s4d) some problematic energies remain even if the threshold parameter in the FARM program is increased to 10^{-3} (see Fig. 16.5(b)). A higher threshold parameter would be required for such cross-sections.

As a conclusion, the medium case study shows that the execution of PROP in single precision leads to a few inexact scattering energies to be computed by the FARM program for some cross-sections. Thanks to a suitable threshold parameter in the FARM program these problematic energies may possibly be skipped. Instead of investigating more deeply the choice of such a parameter for the medium case, we analyze the single-precision computation in a more realistic case in Sect. 16.3.2.

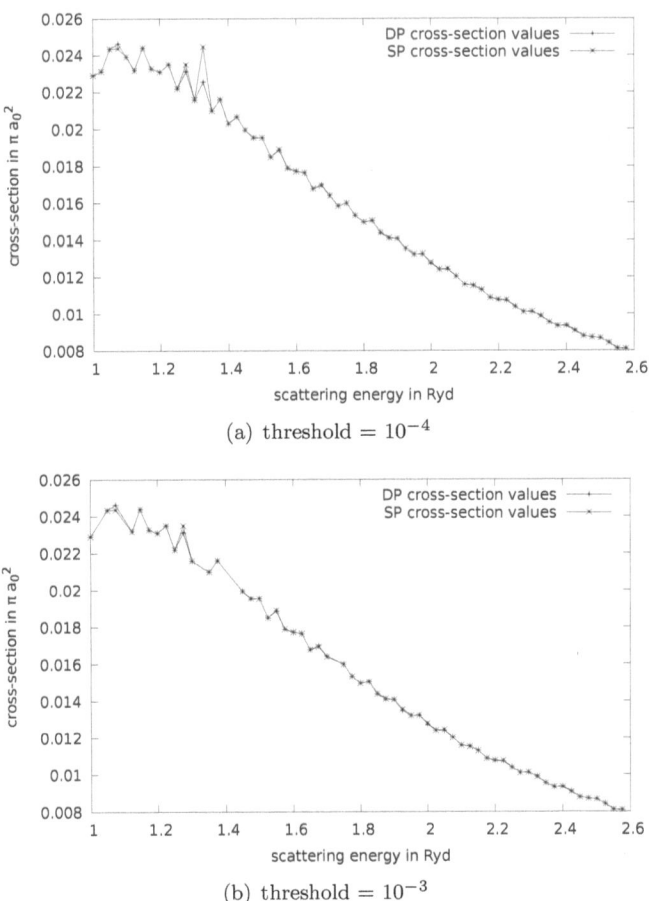

FIGURE 16.4. 1s2p cross-section, 10 sectors.

16.3.2 Huge case study

We study here the impact on FARM of the PROP program run in single precision for the huge case (see Table 16.1). The cross-sections corresponding to all atomic target states 1s...7i are explored, which leads to 406 comparison plots. It should be noted that in this case, over the same energy range above the ionization threshold, the density of pseudo-state thresholds is significantly increased compared to the medium case. As expected, all the plots exhibit large differences between single and double precision cross-sections. For example Fig. 16.6 and 16.7 present the 1s2p and 1s4d cross-sections computed in single—and double—precision for the huge case. We can conclude that PROP in single precision gives invalid results for realistic simulation cases

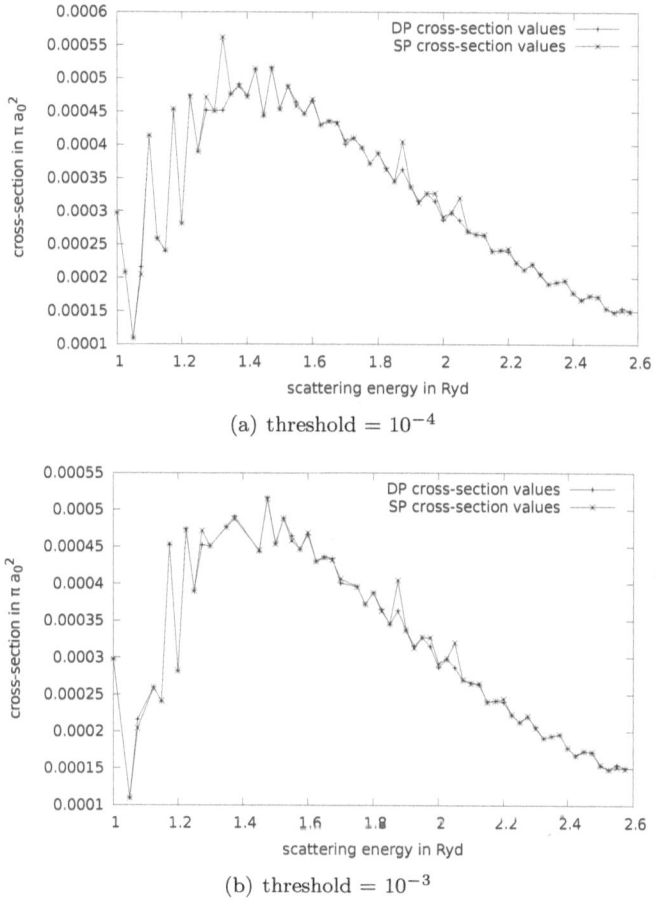

FIGURE 16.5. 1s4d cross-section, 10 sectors.

above the ionization threshold. Therefore, the deployment of PROP on GPU, described in Sect. 16.4, has been carried out in double precision.

16.4 Towards a complete deployment of PROP on GPUs

We now detail how PROP has been progressively deployed on GPUs in double precision in order to avoid the expensive memory transfers between the host and the GPU. Different versions with successive improvements and optimizations are presented. We use CUDA [8] for GPU programming, as well

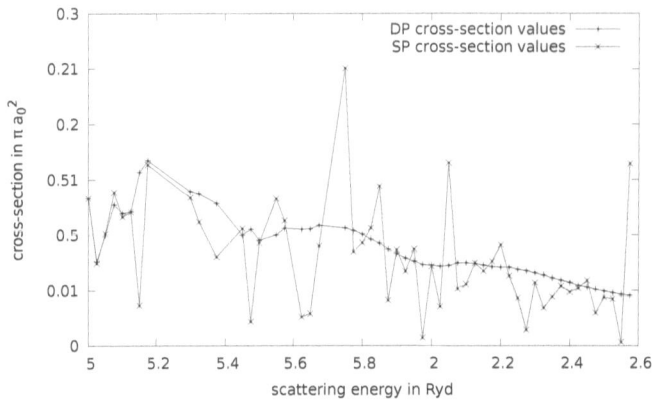

FIGURE 16.6. 1s2p cross-section, threshold = 10^{-4}, 210 sectors.

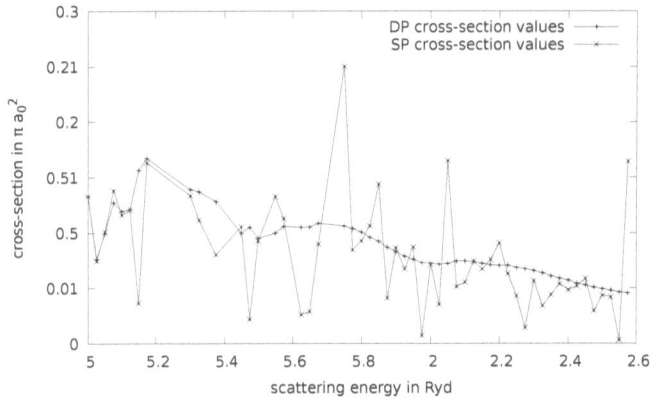

FIGURE 16.7. 1s4d cross-section, threshold = 10^{-4}, 210 sectors.

as the CUBLAS [7] and MAGMA [4] libraries for linear algebra operations. Since PROP is written in Fortran 90, *wrappers* in C are used to enable calls to CUDA kernels from PROP routines.

16.4.1 Computing the output R-matrix on GPU

As mentioned in Algorithm 19, evaluating a sector mainly consists of constructing local R-matrices and computing one output R-matrix, $\Re^O$. In this first step of the porting process, referred to as GPU V1, we consider only the computation of $\Re^O$ on the GPU. We distinguish the following six steps, related to equations (16.5), (16.6), and (16.7), and illustrated in Fig. 16.8 for an off-diagonal sector.

Step 1 ("Input copies"): data are copied from $\Re^I$ to temporary arrays (A,

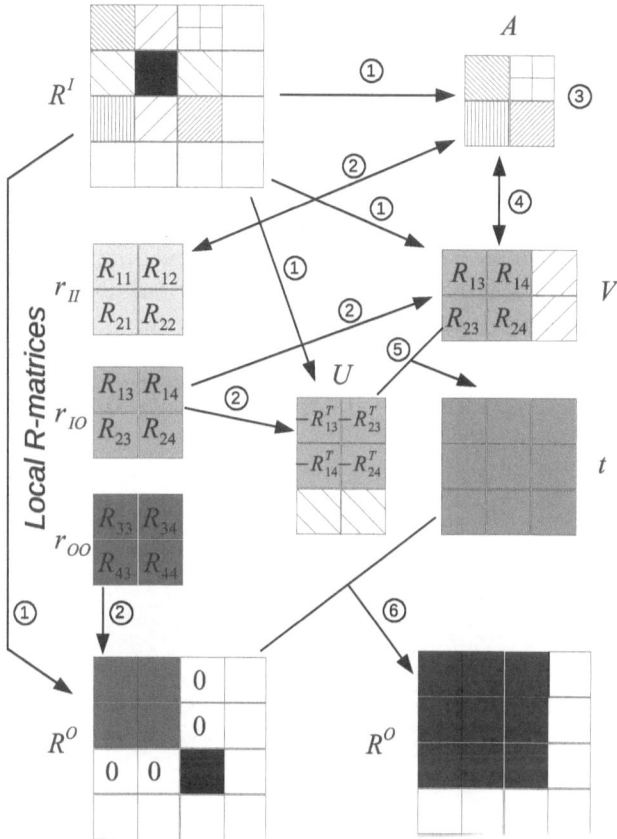

FIGURE 16.8. The six steps of an off-diagonal sector evaluation.

U, V) and to $\Re^O$. These copies, along with possible scalings or transpositions, are implemented as CUDA kernels which can be applied to two matrices of any size starting at any offset. Memory accesses are coalesced [8] in order to provide the best performance for such memory-bound kernels.

Step 2 ("Local copies"): data are copied from local R-matrices to temporary arrays (U, V) and to $\Re^O$. Moreover data from local R-matrix $\mathbf{r}_{II}$ is added to matrix A (via a CUDA kernel) and zeroes are written in $\Re^O$ where required.

Step 3 ("Linear system solving"): matrix A is factorized using the MAGMA DGETRF routine and the result is stored in-place.

Step 4 ("Linear system solving," cont.): the matrix system of linear equations $AW = V$ is solved using the MAGMA DGETRS routine. The solution is stored in matrix V.

Step 5 ("Output matrix product"): matrix U is multiplied by matrix V using the CUBLAS DGEMM routine. The result is stored in a temporary matrix t.

Step 6 ("Output add"): t is added to $\Re^O$ (CUDA kernel).

All the involved matrices are stored in the GPU memory. Only the local R-matrices are first constructed on the host and then sent to the GPU memory, since these matrices vary from sector to sector. The evaluation of the axis and diagonal sectors is similar. However, fewer operations and copies are required because of symmetry considerations [10].

16.4.2 Constructing the local R-matrices on GPU

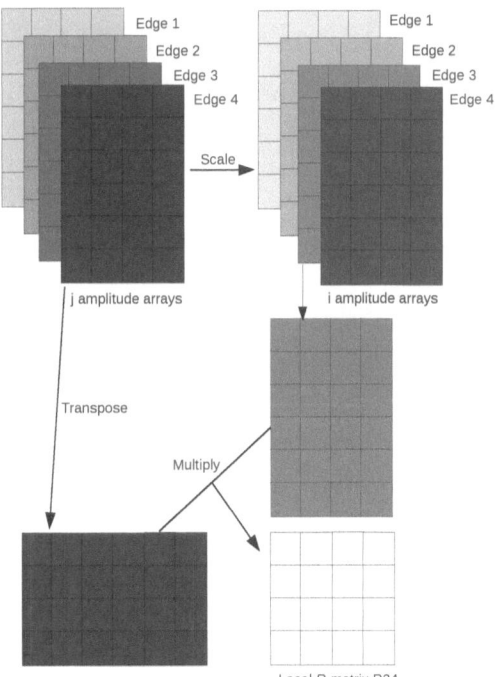

FIGURE 16.9. Constructing the local R-matrix R34 from the j amplitude array associated with edge 4 and the i amplitude array associated with edge 3.

Local R-matrices are constructed using two three-dimensional arrays, i and j. Each three-dimensional array contains four matrices corresponding to the surface amplitudes associated with the four edges of a sector. Those matrices are named *amplitude arrays*. j amplitude arrays are read from data files and i amplitude arrays are obtained by scaling each row of the j amplitude arrays. The main part of the construction of a local R-matrix, presented in Fig. 16.9, is a matrix product between one i amplitude array and one transposed j amplitude array which is performed by a single DGEMM BLAS call. In this version, hereafter referred to as GPU V2, i and j amplitude arrays are transferred to the GPU memory and the required matrix multiplications are performed on the GPU (via CUBLAS routines).

The involved matrices have medium sizes (either 3066 × 383 or 5997 × 383), and performing these matrix multiplications on the GPU is expected to be faster than on the CPU. However, this implies a greater communication volume between the CPU and the GPU since the i and j amplitude arrays are larger than the local R-matrices. It can be noticed that correction data are also used in the construction of a local R-matrix, but this is a minor part in the computation. However, these correction data also have to be transferred from the CPU to the GPU for each sector.

16.4.3 Scaling amplitude arrays on GPU

It should be worthwhile to try to reduce the CPU-GPU data transfers of the GPU V2, where the i and j amplitude arrays are constructed on the host and then sent to the GPU memory for each sector. In this new version, hereafter referred to as GPU V3, we transfer only the j amplitude arrays and the required scaling factors (stored in one 1D array) to the GPU memory, so that the i amplitude arrays are then directly computed on the GPU by multiplying the j amplitude arrays by these scaling factors (via a CUDA kernel). Therefore, we save the transfer of four i amplitude arrays on each sector by transferring only this 1D array of scaling factors. Moreover, scaling j amplitude arrays is expected to be faster on the GPU than on the CPU, thanks to the massively parallel architecture of the GPU and its higher internal memory bandwidth.

16.4.4 Using double-buffering to overlap I/O and computation

As described in Algorithm 19, there are two main steps in the propagation across a sector: reading amplitude arrays and correction data from I/O files and evaluating the current sector. Fig. 16.10 shows the I/O times and the evaluation times of each sector for the huge case execution (210 sectors, 20 strips) of the GPU V3 on one C1060. Whereas the times required by the off-diagonal sectors are similar within each of the 20 strips, the times for diagonal sectors of each strip are the shortest ones, the times for the axis sectors being intermediate. The I/O times are roughly constant among all strips. The evaluation time is equivalent to the I/O time for the first sectors.

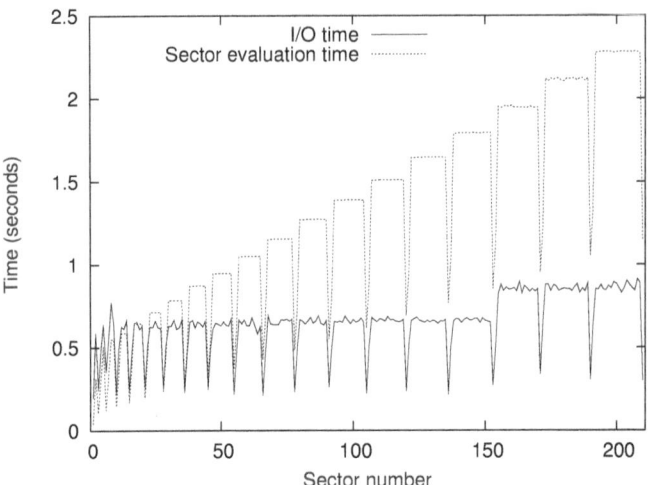

FIGURE 16.10. Compute and I/O times for the GPU V3 on one C1060.

But this evaluation time grows linearly with the strip number and rapidly exceeds the I/O time.

It is thus interesting to use a double-buffering technique to overlap the I/O time with the evaluation time: for each sector, the evaluation of sector n is performed (on GPU) simultaneously with the reading of data for sector $n+1$ (on CPU). This requires the duplication in the CPU memory of all the data structures used for storing data read from I/O files for each sector. This version, hereafter referred to as GPU V4, uses POSIX threads. Two threads are executed concurrently: an I/O thread that reads data from I/O files for each sector, and a computation thread, dedicated to the propagation of the global R-matrix, that performs successively for each sector all necessary computations on GPU, as well as all required CPU-GPU data transfers. The evaluation of a sector uses the data read for this sector as well as the global R-matrix computed on the previous sector. This dependency requires synchronizations between the I/O thread and the computation thread which are implemented through standard POSIX thread mechanisms.

16.4.5 Matrix padding

The MAGMA DGETRF/DGETRS performance and the CUBLAS DGEMM performance are reduced when the sizes (or the leading dimensions) of the matrix are not multiples of the inner blocking size [5]. This inner blocking size can be 32 or 64, depending on the computation and on the underlying GPU architecture [4,6]. In this version (GPU V5), the matrices are therefore padded with 0.0 (and 1.0 on the diagonal for the linear systems) so that their sizes are multiples of 64. This corresponds indeed to the optimal size for the

Numerical validation and GPU performance in atomic physics

matrix product on the Fermi architecture [6]. And as far as linear system solving is concerned, all the matrices have sizes which are multiples of 383: we therefore use padding to obtain multiples of 384 (which are again multiples of 64). It can be noticed that this padding has to be performed dynamically as the matrices increase in size during the propagation (when possible the maximum required storage space is, however, allocated only once in the GPU memory).

16.5 Performance results

16.5.1 PROP deployment on GPU

PROP Version	Execution Time	
CPU Version: 1 Core	201m32s	
CPU Version: 4 Cores	113m28s	
GPU Version	C1060	C2050
GPU V1 (§ 16.4.1)	79m25s	66m22s
GPU V2 (§ 16.4.2)	47m58s	29m52s
GPU V3 (§ 16.4.3)	41m28s	23m46s
GPU V4 (§ 16.4.4)	27m21s	13m55s
GPU V5 (§ 16.4.5)	24m27s	12m39s

TABLE 16.3. Execution time of PROP on CPU and GPU.

Table 16.3 presents the execution times of PROP on CPUs and GPUs, each version solves the propagation equations in the form (16.5-16.7) as proposed by CAPS. Figure 16.11(a) (respectively, 16.11(b)) shows the speedup of the successive GPU versions over one CPU core (respectively, four CPU cores). We use here Intel Q8200 quad-core CPUs (2.33 GHz), one C1060 GPU, and one C2050 (Fermi) GPU, located at UPMC (Université Pierre et Marie Curie, Paris, France). As a remark, the execution times measured on the C2050 would be the same on the C2070 and on the C2075, the only difference between these GPUs being their memory size and their TDP (Thermal Design Power). We emphasize that the execution times correspond to the complete propagation for all six energies of the large case (see Table 16.1), that is to say to the complete execution of the PROP program. Since energies are independent, execution times for more energies (e.g. the huge case) should be proportional to those reported in Table 16.3.

These tests, which have been performed with CUDA 3.2, CUBLAS 3.2, and MAGMA 0.2, show that the successive GPU versions of PROP offer increasing,

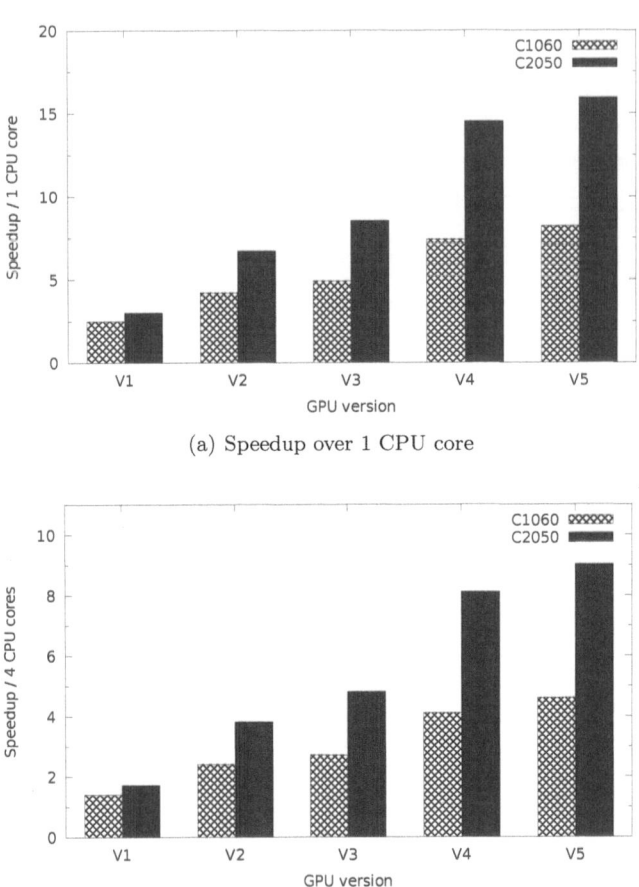

FIGURE 16.11. Speedup of the successive GPU versions.

and at the end interesting, speedups. More precisely, V2 shows that it is worth increasing slightly the CPU-GPU communication volume in order to perform large enough matrix products on the GPU. This communication volume can fortunately be reduced thanks to V3, which also accelerates the computation of amplitude arrays thanks to the GPU. The double-buffering technique implemented in V4 effectively enables the overlapping of I/O operations with computation, while the padding implemented in V5 also improves the computation times. It is noticed that the padding does offer much more performance gain with, for example, CUDA 3.1 and the MAGMA DGEMM [6]: the speedup with respect to one CPU core was increased from 6.3 to 8.1 on C1060 and from 9.5 to 14.3 on C2050. Indeed CUBLAS 3.2 performance has been improved through MAGMA code for matrix sizes which are not multiples of the

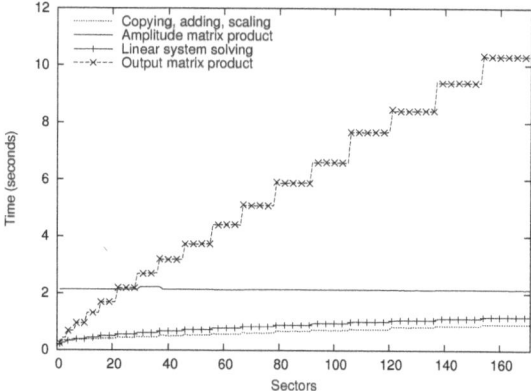

FIGURE 16.12. CPU (1 core) execution times for the off-diagonal sectors of the large case.

inner blocking size [5]. Although for all versions the C2050 (with its improved double precision performance) offers up to almost double speedup compared to the C1060, the performance obtained with both architectures justifies the use of the GPU for such an application.

16.5.2 PROP execution profile

We detail here the execution profile on the CPU and the GPU for the evaluation of all off-diagonal sectors (the most representative ones) for a complete energy propagation. Figures 16.12 and 16.13 show CPU and GPU execution times for the 171 off-diagonal sectors of the large case (see Table 16.1). "Copying, adding, scaling" corresponds to the amplitude array construction (scaling) as well as to Steps 1, 2, and 6 in Sect. 16.4.1, all implemented via CUDA kernels. "Amplitude matrix product" corresponds to the DGEMM call to construct the local R-matrices from the i and j amplitude arrays. "Linear system solving" and "Output matrix product" correspond respectively to steps 3-4 and to step 5 in Sect. 16.4.1. "CPU-GPU transfers" in Fig. 16.13 aggregate transfer times for the j amplitude arrays and the scaling factors, as well as for the correction data.

On one CPU core (see Fig. 16.12), matrix products for the construction of the local R-matrices require the same computation time during the whole propagation. Likewise the CPU time required by matrix products for the output R-matrix is constant within each strip. But as the global R-matrix is propagated from strip to strip, the sizes of the matrices U and V increase, and so does their multiplication time. The time required to solve the linear system increases slightly during the propagation. These three operations ("Amplitude matrix product," "Output matrix product," and "Linear system solving") are clearly dominant in terms of computation time compared to the

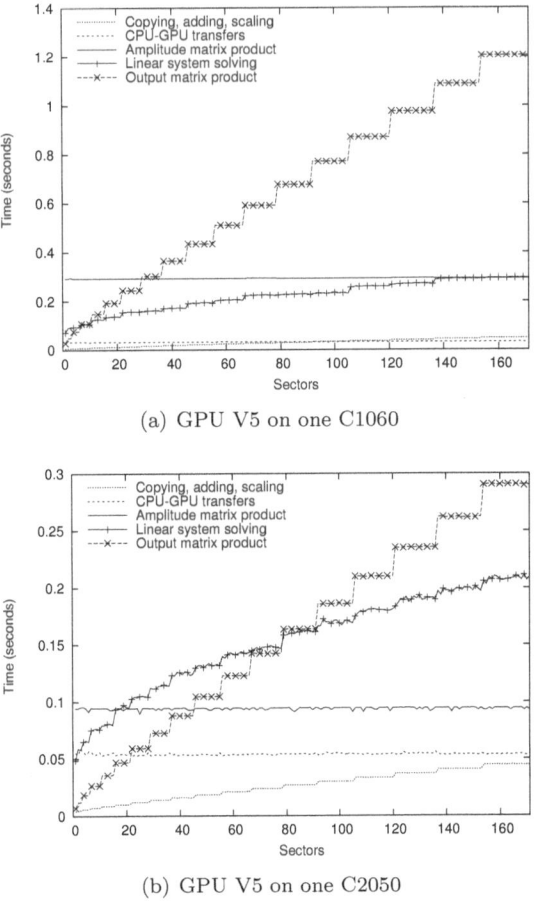

FIGURE 16.13. GPU execution times for the off-diagonal sectors of the large case.

other remaining operations, which justify our primary focus on these three linear algebra operations.

On the C1060 (see Fig. 16.13(a)), we have generally managed to obtain a similar speedup for all operations (around 8, which corresponds to Fig. 16.11(a)). Only the linear system solving presents a lower speedup (around 4). The CUDA kernels and the remaining CPU-GPU transfers make a minor contribution to the overall computation time and do not require additional improvements.

On the C2050 (see Fig. 16.13(b)), additional speedup is obtained for all operations, except for the CPU-GPU transfers and the linear system solving. The CPU-GPU transfers are mainly due to the j amplitude arrays, and currently still correspond to minor times. When required, the double-buffering technique may also be used to overlap such transfers with computation on the GPU.

16.6 Propagation of multiple concurrent energies on GPU

Finally, we present here an improvement that can benefit from the Fermi architecture, as well as from the newest Kepler architecture, both of which enable the concurrent execution of multiple CUDA kernels, thus offering additional speedup on GPUs for small or medium computation grain kernels. In our case, the performance gain on the GPU is indeed limited since most matrices have small or medium sizes. By using multiple streams within one CUDA context [8], we can propagate multiple energies concurrently on the Fermi GPU. It can be noticed that all GPU computations for all streams are launched by the same host thread. We therefore rely here on the *legacy API* of CUBLAS [7] (like MAGMA) without thread safety problems. A *breadth first* issue order is used for kernel launches [9]: for a given GPU kernel, all kernel launchs are indeed issued together in the host thread, using one stream for each concurrent energy, in order to maximize concurrent kernel execution. Of course, the memory available on the GPU must be large enough to store all data structures required by each energy. Moreover, multiple streams are also used within the propagation of each single energy in order to enable concurrent executions among the required kernels.

	Number of Energies	1	2	4	8	16
Medium Case	Time (s)	11.18	6.87	5.32	4.96	4.76
	Speedup	-	1.63	2.10	2.26	2.35
Large Case	Number of Energies	1	2		3	
	Time (s)	509.51	451.49		436.72	
	Speedup	-	1.13		1.17	

TABLE 16.4. Performance results with multiple concurrent energies on one C2070 GPU. GPU initialization times are not considered here.

In order to have enough GPU memory to run two or three concurrent energies for the large case, we use one C2070 GPU (featuring 6GB of memory) with one Intel X5650 hex-core CPU, CUDA 4.1 and CUBLAS 4.1, as well as the latest MAGMA release (version 1.3.0). Substantial changes have been required in the MAGMA calls with respect to the previous versions of PROP that were using MAGMA 0.2. Table 16.4 presents the speedups obtained on the Fermi GPU for multiple concurrent energies (up to sixteen since this is the maximum number of concurrent kernel launches currently supported [8]). With the medium case, speedups greater than 2 are obtained with four concurrent energies or more. With the more realistic large case, the performance gain is lower mainly because of the increase in matrix sizes, which implies a better GPU usage with only one energy on the GPU. The concurrent execution of multiple kernels is also limited by other operations on the GPU [8, 9] and by the current MAGMA code which prevents concurrent MAGMA calls in different streams. Better speedups can be expected on the latest Kepler GPUs which offer additional compute power, and whose *Hyper-Q* feature may help improve further the GPU utilization with concurrent energies. To the contrary, the same code on the C1060 shows no speedup since the concurrent kernel launches are serialized on this previous GPU architecture.

16.7 Conclusion and future work

In this chapter, we have presented our methodology and our results in the deployment on a GPU of an application (the PROP program) in atomic physics.

We have started by studying the numerical stability of PROP using single precision arithmetic. This has shown that PROP using single precision, while relatively stable for some small cases, gives unsatisfactory results for realistic simulation cases above the ionization threshold where there is a significant density of pseudo-states. It is expected that this may not be the case below the ionization threshold where the actual target states are less dense. This requires further investigation.

We have therefore deployed the PROP code in double precision on a GPU, with successive improvements. The different GPU versions each offer increasing speedups over the CPU version. Compared to the single (respectively, four) core(s) CPU version, the optimal GPU implementation gives a speedup of 8.2 (resp., 4.6) on one C1060 GPU, and a speedup of 15.9 (resp., 9.0) on one C2050 GPU with improved double-precision performance. An additional gain of around 15% can also be obtained on one Fermi GPU with large memory (C2070) thanks to concurrent kernel execution.

Such speedups cannot be directly compared with the 1.15 speedup obtained with the HMPP version, since in our tests the CPUs are different and the CUBLAS versions are more recent. However, the programming effort required progressively to deploy PROP on GPUs clearly offers improved and interesting speedups for this real-life application in double precision with varying-sized matrices.

We are currently working on a hybrid CPU-GPU version that spreads the computations of the independent energies on both the CPU and the GPU. This will enable multiple energy execution on the CPU, with one or several core(s) dedicated to each energy (via multithreaded BLAS libraries). Multiple concurrent energies may also be propagated on each Fermi GPU. By merging this work with the current MPI PROP program, we will obtain a scalable hybrid CPU-GPU version. This final version will offer an additional level of parallelism, thanks to the MPI standard, in order to exploit multiple nodes with multiple CPU cores and possibly multiple GPU cards.

Bibliography

[1] P. G. Burke, C. J. Noble, and M. P. Scott. R-matrix theory of electron scattering at intermediate energies. *Proceedings of the Royal Society of London A*, 410:287–310, 1987.

[2] V. M. Burke, C. J. Noble, V. Faro-Maza, A. Maniopoulou, and N. S. Scott. FARM_2DRMP: a version of FARM for use with 2DRMP. *Computer Physics Communications*, 180:2450–2451, 2009.

[3] P. Fortin, R. Habel, F. Jézéquel, J.-L. Lamotte, and N. S. Scott. Deployment on GPUs of an application in computational atomic physics. In *12th IEEE International Workshop on Parallel and Distributed Scientific and Engineering Computing (PDSEC) in Conjunction with the 25th International Parallel and Distributed Processing Symposium (IPDPS)*, 2011.

[4] MAGMA (Matrix Algebra on GPU and Multicore Architectures). Available at http://icl.cs.utk.edu/magma.

[5] R. Nath, S. Tomov, and J. Dongarra. Accelerating GPU kernels for dense linear algebra. In *Proc. of VECPAR'10*, pages 83–92, Berkeley, CA (USA), June 2010.

[6] R. Nath, S. Tomov, and J. Dongarra. An improved MAGMA GEMM for Fermi graphics processing units. *International Journal of High Performance Computing Applications*, 24(4):511–515, 2010.

[7] *NVIDIA CUDA Toolkit* 4.1, CUBLAS Library, January 2012.

[8] *NVIDIA CUDA C Programming Guide*, version 4.1, November 2011.

[9] S. Rennich. *CUDA C/C++ Streams and Concurrency, NVIDIA*, January 2012.

[10] N. S. Scott, M. P. Scott, P. G. Burke, T. Stitt, V. Faro-Maza, C. Denis, and A. Maniopoulou. 2DRMP: A suite of two-dimensional R-matrix propagation codes. *Computer Physics Communications*, 180:2424–2449, 2009.

[11] T. Stitt, N. S. Scott, M. P. Scott, and P. G. Burke. 2-D R-matrix propagation: a large scale electron scattering simulation dominated by the multiplication of dynamically changing matrices. In *Proc. of VECPAR'02*, pages 354–367, Porto, Portugal, June 2002.

Chapter 17

A GPU-accelerated envelope-following method for switching power converter simulation

Xuexin Liu, Sheldon Xiang-Dong Tan
Department of Electrical Engineering, University of California, Riverside, CA, USA

Hai Wang
University of Electronics Science and Technology of China, Chengdu, Sichuan, China

Hao Yu
School of Electrical & Electronic Engineering, Nanyang Technological University, Singapore

17.1	Introduction	395
17.2	The envelope-following method in a nutshell	398
17.3	New parallel envelope following method	400
	17.3.1 GMRES solver for Newton update equation	400
	17.3.2 Parallelization on GPU platforms	402
	17.3.3 Gear-2 based sensitivity calculation	404
17.4	Numerical examples	406
17.5	Summary	410
17.6	Glossary	410
	Bibliography	410

17.1 Introduction

Power converters have seen a surge of new trends and novel applications due to their widespread use in renewable energy systems and emerging hybrid and purely-electric vehicles. More efficient simulation techniques for power converters are urgently needed to meet more design constraints. In this chapter, we present a novel envelope-following parallel transient analysis method for the general switching power converters. The new method first exploits the parallelisim in the envelope-following method and parallelize the Newton update solving part, which is the most computational expensive, in GPU platforms to boost the simulation performance. To further speed up the iterative GMRES solving for Newton update equation in the envelope-following method, we apply the matrix-free Krylov subspace basis generation technique, which was previously used for RF simulation. Last, the new method also applies more robust Gear-2 integration to compute the sensitivity matrix instead of traditional integration methods.

Over the past decades, power electronics and especially switching power converters have seen a surge of new trends and novel applications — from the growing significance of PWM (pulse-width modulation) rectifiers and multilevel inverters to their widespread use in renewable (like solar and wind) energy systems, smart grids and emerging electric and hybrid vehicles [20].

This requires more efficient simulation techniques for power electronics to meet the new application and demanding design constraints. To facilitate the design of typical power electronics circuits, many special-purpose simulation algorithms and tools were developed. Among them is the envelope-following method [5,14,22], which is able to calculate the slowly changing contour, or envelope, of a carrier waveform with a much higher switching frequency. This is the case for switching power converters, which have fast switching currents to convert powers from one level to another level. In those switching power converters, it is the envelope, which is the power voltage delivered, not the fast switching waves in every cycle, that is of interest to the designers. As shown in Figure 17.1(a), the solid line is the waveform of the output node in a Buck converter [13], the dots are the simulation points of SPICE, and the appended dash line is the envelope.

Obviously, traditional SPICE will incur an extraordinarily high simulation time for this task, since it has to integrate the circuit's differential equation with many time points in every clock cycle to get the accurate details of the carrier. For switching power converters, the waveform of the carrier in consequent cycles does not change much, envelope-following method is an approximation analysis method, which skips over several cycles (the dash line in Figure 17.1(b)), the so called envelope step, without simulating them, and then carries out a correction, which usually contains a sensitivity-based Newton iteration or shooting until convergence, in order to begin the next envelope step.

Also, iterative GMRES solver is typically used in the envelope-following method to compute the solution of Newton update due to its efficiency compared to direct LU method. However, as the Jacobian matrix or the sensitivity matrix in the equation to be solved is dense, explicit computing of the Jacobian is a very expense process. Recently, the matrix-free GMRES was proposed [19] for RF shooting based simulation. The new method leads to significant savings due to its implicit calculation of new basis vectors without the explicit formulation of the sensitivity matrix.

Modern computer architecture has shifted towards designs that employ so called multi-core processor or chip-multiprocessors (CMP) [3,9]. The family

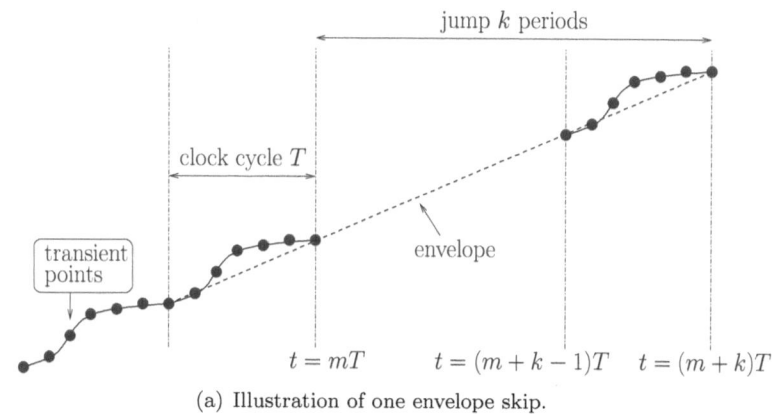

(a) Illustration of one envelope skip.

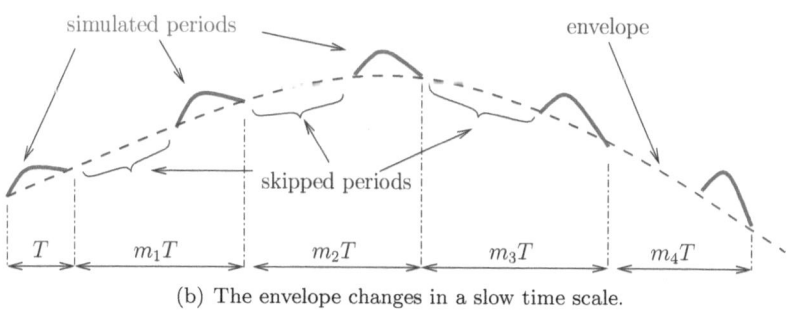

(b) The envelope changes in a slow time scale.

FIGURE 17.1. Transient envelope-following analysis. (Both figures reflect backward Euler style envelope-following.)

of graphic processing units (GPU) are among the most powerful many-core computing systems in mass-market use [17]. For instance, the state-of-the-art NVIDIA Tesla T10 chip has a peak performance of over 1 TFLOPS versus about 80–100 GFLOPS of Intel i5 series Quad-core CPUs [12]. In addition to the primary use of GPUs in accelerating graphics rendering operations, there has been considerable interest in exploiting GPUs for general purpose computation (GPGPU) [7]. The introduction of new parallel programming interfaces for general purpose computation, such as Computer Unified Device Architecture (CUDA), Stream SDK, and OpenCL [4, 11, 18], has made GPUs an attractive choice for developing high-performance scientific computation tools and solving practical engineering problems. Hence, the applications with GPGPUs are rapidly growing in a broad variety of parallel numerical computation works [1].

After the basic algorithm of envelope-following is briefly reviewed and Newton update equation is derived in Section 17.2. In Section 17.3 presents the new parallel envelope method where the CPU parallelization, matrix-free GMRES and Gear-2 integration will be discussed. We will show the parallelization of the Newton update, which is the most time consuming step, in the envelop-follow method, in GPU platforms. We will also present the new GMRES solver using matrix-free Jacobian-vector multiplication and the Gear-2 integration for sensitivity based Newton update equation. Numerical examples are shown in Section 17.4, and finally, this chapter is summarized in Section 17.5.

17.2 The envelope-following method in a nutshell

In transient analysis, a nonlinear circuit's behavior can be represented by a coupled set of nonlinear first-order differential algebraic equations (DAE) of the form

$$\frac{\mathrm{d}}{\mathrm{d}t} q(x(t)) + f(x(t)) = b(t), \qquad (17.1)$$

where $x(t) \in \mathbb{R}^N$ is the vector of circuit variables, usually comprising node voltages and possibly branch currents, $f(\cdot)$ is a nonlinear function that maps $x(t)$ to a vector of N entries most of which are sums of resistive currents at a node, $q(\cdot)$ maps $x(t)$ to a vector of N entries of capacitor charges or inductor fluxes, and $b(t)$ contains the input source values. For many power electronics circuits, the input switching signal is known and is periodic with clock period T.

Assume the time inside one period T has been discretised into M time steps, $0 = t_0 < t_1 < t_2 < \cdots < t_M = T$, and the i-th step length is $h_i = t_i - t_{i-1}$. In practice, the discretisation is nonuniform to control truncation error and convergence. Then, given an initial condition $x(0)$ at $t = 0$, numerical integration is applied to find the time domain solution of circuit state $x(t)$ at each time step till $t = T$.

For those circuits whose carrier waveforms vary slowly in a large number of periods, the envelope-following method only integrates the DAE in several selected periods and then jumps over to a new time point. By repeating this "simulate and skip" action, envelope-following method attains its efficiency compared to conventional transient analysis, but still can accurately obtain the slow varying envelope.

For example, if the clock period is T, and the suitable jump interval for the envelope is k periods, then the envelope step is kT. Suppose the state at time $t = mT$ is known as $x(mT)$, and the envelope-following wants to obtain the state at the next envelope step, $x((m+k)T)$. If the envelope step is much larger than the clock period (k is much bigger than one), envelope-following will lead to significant savings in simulation time.

To estimate $x((m + k)T)$, a forward Euler style jumping relies only on $x(mT)$ and $x((m-1)T)$, i.e.,

$$x((m+k)T) = x(mT) + k\left[x(mT) - x((m-1)T)\right].$$

However, this approach is inefficient due to its restriction on envelope step k, since larger k usually causes instability. Instead, backward Euler jumping,

$$x((m+k)T) - x(mT) = k\left[x((m+k)T) - x((m+k-1)T)\right],$$

allows larger envelope steps. Here $x((m+k-1)T)$ is the unknown variable to be solved by Newton iteration, and $x((m+k)T)$ is dependent on $x((m+k-1)T)$ in each iteration. The Newton update equation in each iteration is thus expressed as

$$\Delta x((m+k-1)T) = A^{-1}\left[(k-1)x^j((m+k)T)\right.$$
$$\left. - kx^j((m+k-1)T) + x(mT)\right], \quad (17.2)$$

where the Jacobian matrix A is computed as a combination of circuit sensitivity matrix and identity matrix, as

$$A = (k-1)\frac{\mathrm{d}x((m+k)T)}{\mathrm{d}x((m+k-1)T)} - kI = (k-1)J - kI. \quad (17.3)$$

In each Newton iteration, $x((m+k-1)T)$ is used as initial condition to calculate $x((m+k)T)$ by integrating the DAE in one clock period. Meanwhile, the conductance matrix $G_i = \mathrm{d}f(x(t_i))/\mathrm{d}x(t_i)$ and the capacitance matrix $C_i = \mathrm{d}q(x(t_i))/\mathrm{d}x(t_i)$ at each time step are used to derive the sensitivity matrix $J = \mathrm{d}x((m+k)T)/\mathrm{d}x((m+k-1)T)$.

Different integration rules can be applied to the computation of sensitivity matrix. It can be easily derived that, if the DAE is integrated with backward

Euler rule, the sensitivity is

$$J = \frac{\mathrm{d}x_M}{\mathrm{d}x_0} = \prod_{i=1}^{M} \left(\frac{1}{h_i}C_i + G_i\right)^{-1} \frac{1}{h_i}C_{i-1}$$

In summary, the envelope-following method is fundamentally a boosted version of traditional transient analysis, with certain skips over several periods and a Newton iteration to update or correct the errors brought by the skips, as is exhibited by Figure 17.2.

17.3 New parallel envelope-following method

In this section, we explain how to efficiently use matrix-free GMRES to solve the Newton update problems with implicit sensitivity calculation, i.e., the steps enclosed by the double dashed block in Figure 17.2. Then implementation issues of GPU acceleration will be discussed in detail. Finally, the Gear-2 integration is briefly introduced.

17.3.1 GMRES solver for Newton update equation

<u>G</u>eneralized <u>M</u>inimum <u>Res</u>idual, or GMRES method is an iterative method for solving systems of linear equations ($Ax = b$) with dense matrix A. The standard GMRES is given in Algorithm 20. It constructs a Krylov subspace with order m,

$$\mathcal{K}_m = \mathrm{span}(b, Ab, A^2 b, \ldots, A^{m-1}b),$$

where the approximate solution x_m resides. In practice, an orthonormal basis V_m that spans the subspace $\mathcal{K}_m$ can be generated by the Arnoldi iterations. The goal of GMRES is to search for an optimal coefficient y such that the linear combination $x_m = V_m y$ will minimize its residual $\|b - Ax_m\|_2$. The Arnoldi iteration also creates a by-product, an upper Hessenberg matrix $\tilde{H}_m \in R^{(m+1) \times m}$. Thus, the projection of A on the orthonormal basis V_m is described by the Arnoldi decomposition $AV_m = V_{m+1}\tilde{H}_m$, which is useful to check the residual at each iteration without forming x_m, and to solve for coefficient y when residual is smaller than a preset tolerance [8].

At a first glance, the cost of using standard GMRES directly to solve the Newton update in Eq. (17.2) seems to come mainly from two parts: the formulation of the sensitivity matrix $J = \mathrm{d}x_M/\mathrm{d}x_0$ in Eq. (17.9) in Section 17.3.3, and the iteration inside the standard GMRES, especially the matrix-vector multiplication and the orthonormal basis construction (Line 5 through Line 11 in Algorithm 20). Based on the observation that only the matrix-vector product is required in GMRES, the work in [19] introduces an efficient matrix-free

GPU-accelerated envelope-following method

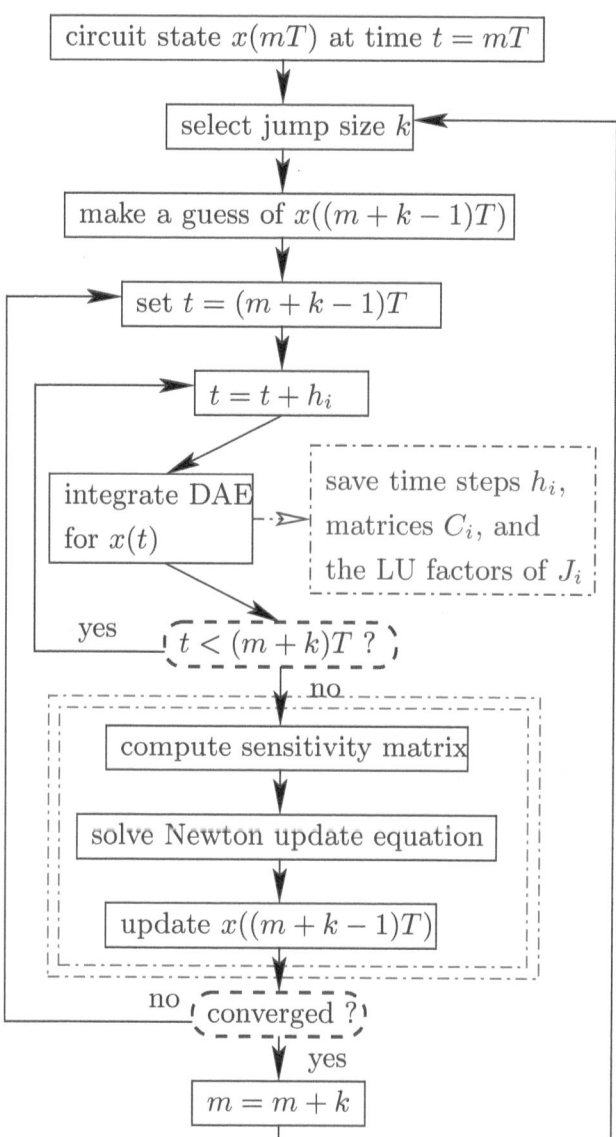

FIGURE 17.2. The flow of envelope-following method.

algorithm in the shooting-Newton method, where the equation solving part also involves with a sensitivity matrix. The matrix-free method does not take an explicit matrix as input, but directly passes the saved capacitance matrices C_i and the LU factorizations of J_i, $i = 0, \ldots, M$, into the Arnoldi iteration for Krylov subspace generation. Therefore, it avoids the cost of forming the dense sensitivity matrix and focuses on subspace construction. Briefly speak-

Algorithm 20: standard GMRES algorithm

Input: $A \in \mathbb{R}^{N \times N}$, $b \in \mathbb{R}^N$, and initial guess $x_0 \in \mathbb{R}^N$
Output: $x \in \mathbb{R}^N$: $\|b - Ax\|_2 < tol$

1 $r = b - Ax_0$;
2 $h_{1,0} = \|r\|_2$;
3 $m = 0$;
4 **while** $m < max\_iter$ **do**
5 $\quad m = m + 1$; $v_m = r/h_{m,m-1}$;
6 $\quad r = Av_m$;
7 $\quad$ **for** $i = 1 \ldots m$ **do**
8 $\quad\quad h_{i,m} = \langle v_i, r \rangle$;
9 $\quad\quad r = r - h_{i,m} v_i$;
10 $\quad$ **end**
11 $\quad h_{m+1,m} = \|r\|_2$;
12 $\quad$ Compute the residual ϵ;
13 $\quad$ **if** $\epsilon < tol$ **then**
14 $\quad\quad$ Solve the problem: minimize $\|b - Ax_m\|_2$;
15 $\quad\quad$ Return $x_m = x_0 + V_m y_m$;
16 $\quad$ **end**
17 **end**

ing, Line 5 will be replaced by a procedure without explict A, and we will talk about the flow of matrix-free generation of new basis vectors in later sections.

17.3.2 Parallelization on GPU platforms

There exist many GPU-friendly computing operations in GMRES, such as the vector addition (axpy), 2-norm of vector (nrm2), and sparse matrix-vector multiplication (csrmv), which have been parallelized in the CUDA Basic Linear Algebra Subroutine (CUBLAS) Library and the CUDA Sparse Linear Algebra Library (CUSPARSE) [16].

GPU programming is typically limited by the data transfer bandwidth as GPU favors computationally intensive algorithms [12]. So how to efficiently transfer the data and wisely partition the data between CPU memory and GPU memory is crucial for GPU programming. In the following, we discuss these two issues in our implementation.

As noted in Section 17.2, the envelope-following method requires the matrices gathered from all the time steps in a period in order to solve a Newton update. At each time step, SPICE has to linearize device models, stamp matrix elements into MNA (short for modified nodal analysis) matrices, and solve circuit equations in its inner Newton iteration. When convergence is attained, circuit states are saved and then next time step begins. This is also the time when we store the needed matrices for the envelope-following computation.

Since these data are involved in the calculation of Gear-2 sensitivity matrix in the generation of Krylov subspace vectors in Algorithm 21, it is desirable that all of these matrices are transferred to GPU for its data parallel capability.

To efficiently transfer those large data, we explore asynchronous memory copy between host and device in the recent GPUs (Fermi architecture), so that the copying overlaps with the host's computing of the next time step's circuit state. The implementation of asynchronous matrices copy includes two parts: allocating page-locked memory, also known as pinned memory, where we save matrices for one time step, and using asynchronous memory transfer to copy these data to GPU memory. While it is known that page-locked host memory is a scarce resource and should not be overused, the demand of memory size of the data generated at one time step can be generously accommodated by today's mainstream server configurations.

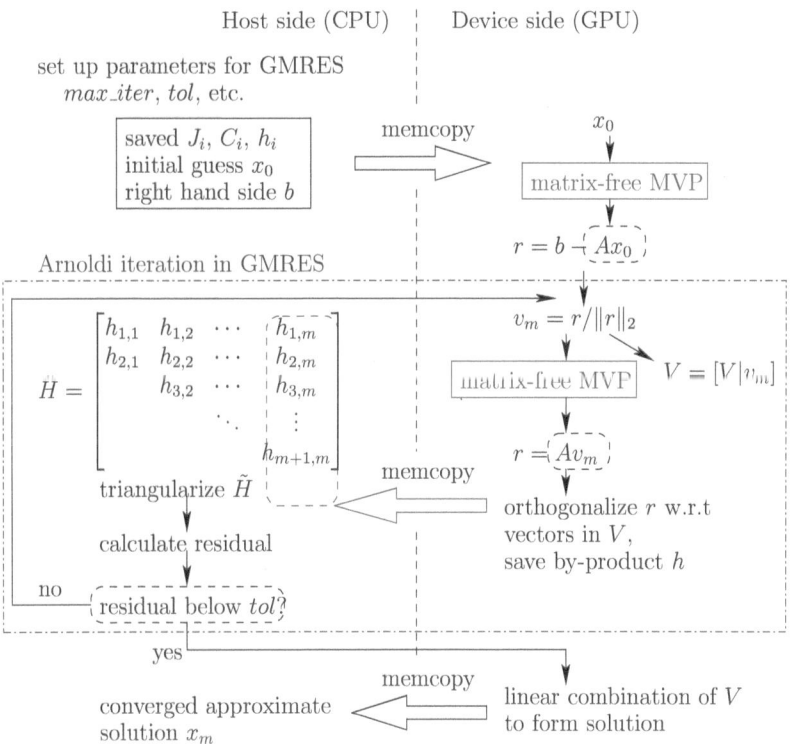

FIGURE 17.3. GPU parallel solver for envelope-following update.

The second issue is to decide the location of data between CPU and GPU memories. Therefore let us first make a rough sketch of the quantities in the GMRES Algorithm 20. Although GMRES tends to converge quickly for most circuit examples, i.e., the iteration number $m \ll N$, the space for storing all the subspace basis V_m of N-by-m, i.e., m column vectors with N-length, is still big. In addition, every newly generated matrix-vector product needs to be orthogonalized with respect to all its previous basis vectors in the Arnoldi iterations. Hence, keeping all the vectors of V_m in GPU global memory allows GPU to handle those operations, such as inner-product of basis vectors (dot) and vector subtraction (axpy), in parallel.

On the other hand, it is better to keep the Hessenberg matrix $\tilde{H}$, where intermediate results of the orthogonalization are stored, at the host side. This comes with the following reasons. First, its size is $(m+1)$-by-m at most, rather small if compared to circuit matrices and Krylov basis. Besides, it is also necessary to triangularize $\tilde{H}$ and check the residual regularly in each iteration so the GMRES can return the approximate solution as soon as the residual is below a preset tolerance. Hence, in consideration of the serial nature of the trianularization, the small size of Hessenberg matrix, and the frequent inspection of values by host, it is preferable to allocate $\tilde{H}$ in CPU (host) memory. As shown in Figure 17.3, the memory copy from device to host is called each time when Arnoldi iteration generates a new vector and the orthogonalization produces the vector h.

17.3.3 Gear-2 based sensitivity calculation

The Gear-2 integration method is a backward differentiation formula (BDF), which approximates the derivative of a function using information from past few steps. Gear-2 method has been shown to be more suitable for many practical problems such as stiff problems where circuit behavior is affected by different time constants (fast ones with large poles and slow ones with small poles) [21]. Switching power converters and RF systems are typically stiff systems as waveforms changing in two different time scales are involved.

Given the DAE (17.1), at the i-th time step, Gear-2 integration approximates the derivative $\mathrm{d}q(x(t))/\mathrm{d}t$ by a two-step backward finite difference,

$$\frac{\mathrm{d}q(x(t_i))}{\mathrm{d}t} = \frac{1}{h_i}\left[q(x(t_i)) - \alpha_1^i q(x(t_{i-1})) - \alpha_2^i q(x(t_{i-2}))\right],$$

where the coefficients α's are chosen to satisfy Gear's backward differentiation formula [6]. For uniformly discretised time steps, the index i in h_i, α_1^i and α_2^i can be omitted.

For the first step t_1, only the initial condition at t_0 is available. Therefore backward Euler is used, i.e.,

$$\tfrac{1}{h_1}[q(x_1) - q(x_0)] + f(x_1) = b_1. \tag{17.4}$$

Since x_0 directly affects the solution of x_1, the sensitivity matrix up to now is

$$\frac{\mathrm{d}x_1}{\mathrm{d}x_0} = \left[\frac{1}{h_1}C_1 + G_1\right]^{-1}\frac{C_0}{h_1} = J_1^{-1}\frac{C_0}{h_1}. \tag{17.5}$$

Let J_i denote $(1/h_i)C_i + G_i$ in the remaining part of this chapter. In addition, the LU factorizations of J_i are already calculated since they are used to solve the circuit state at each time step before we calculate the sensitivity.

Next, for the solution at t_2, with the previous two steps information available, Gear-2 integration can be applied,

$$\frac{1}{h_2}[q(x_2) - \alpha_1 q(x_1) - \alpha_2 q(x_0)] + f(x_2) = b_2. \tag{17.6}$$

In view of the sensitivity of x_2 with respect to the changes of x_0, (17.6) indicates that x_2's perturbation can be traced back to x_0 along two paths: x_2 is directly affected by x_0, and x_2 is also affected indirectly by x_0 via x_1. That is,

$$\frac{dx_2}{dx_0} = \frac{\partial x_2}{\partial x_1}\frac{dx_1}{dx_0} + \frac{\partial x_2}{\partial x_0},$$

where the two partial derivatives are

$$\frac{\partial x_2}{\partial x_1} = J_2^{-1}\frac{\alpha_1}{h_2}C_1, \qquad \frac{\partial x_2}{\partial x_0} = J_2^{-1}\frac{\alpha_2}{h_2}C_0,$$

and dx_1/dx_0 is calculated previously in (17.5). Thus, the sensitivity matrix deduced from (17.6) is

$$\frac{dx_2}{dx_0} = J_2^{-1}\frac{\alpha_1}{h_2}C_1 \cdot J_1^{-1}\frac{C_0}{h_1} + J_2^{-1}\frac{\alpha_2}{h_2}C_0. \tag{17.7}$$

Likewise, for the third time step t_3, apply the Gear-2 integration formula to DAE (17.1),

$$\frac{1}{h_3}[q(x_3) - \alpha_1 q(x_2) - \alpha_2 q(x_1)] + f(x_3) = b_3, \tag{17.8}$$

and the chain rule for sensitivity is

$$\frac{dx_3}{dx_0} = \frac{\partial x_3}{\partial x_2}\frac{dx_2}{dx_0} + \frac{\partial x_3}{\partial x_1}\frac{dx_1}{dx_0} = J_3^{-1}\left[\frac{\alpha_1}{h_3}C_2\frac{dx_2}{dx_0} + \frac{\alpha_2}{h_3}C_1\frac{dx_1}{dx_0}\right],$$

where both dx_2/dx_0 and dx_1/dx_0 are ready from the previous two time steps, i.e., Eqs. (17.7) and (17.5). Follow this chain rule in the aforementioned recursive style, the sensitivity matrix for Gear-2 integration is computed along the remaining time steps up to the M-th step,

$$J = \frac{dx_M}{dx_0} = J_M^{-1}\left[\frac{\alpha_1}{h_M}C_{M-1}\frac{dx_{M-1}}{dx_0} + \frac{\alpha_2}{h_M}C_{M-2}\frac{dx_{M-2}}{dx_0}\right]. \tag{17.9}$$

Note that as the matrix-free GMRES method is applied, which only requires matrix-vector multiplication and no explicit J is required, as explained in Algorithm 21.

With matrix-free method, the matrix-vector multiplication in Line 5 of Algorithm 20 is replaced by the iteration shown in Algorithm 21, which calculates the multiplication product of the Gear-2 sensitivity we encounter in

Algorithm 21: the matrix-free method for Krylov subspace construction

Input: current Krylov subspace basis vector v, time step lengths h_i, saved C_i matrices and LU factors of J_i, $i = 0, \ldots, M$
Output: matrix-vector product r, such that $r = Av$

1 solve $J_1 p_2 = h_1^{-1} C_0 v$ for p_2;
2 solve $J_2 p_1 = h_2^{-1} (\alpha_1 C_1 p_2 + \alpha_2 C_0 v)$ for p_1;
3 **for** $i = 3 \ldots M$ **do**
4 $\quad$ solve $J_i p_0 = \alpha_1 h_i^{-1} C_{i-1} p_1 + \alpha_2 h_i^{-1} C_{i-2} p_2$ for p_0 ;
5 $\quad$ $p_2 = p_1$;
6 $\quad$ $p_1 = p_0$;
7 **end**
8 $r = (k-1) p_0 - kv$;

envelope-following and a basis vector in the Krylov subspace. Note that the scaling and shift of J in $A = (k-1)J - kI$, as described in Eq. (17.3), is taken into consideration by Line 8 of Algorithm 21.

For the matrix-free generation of new basis vectors, it is straightforward to apply CUBLAS and CUSPARSE routines, and some customized GPU kernel functions to implement Algorithm 21 with the stored LU matrices of J_i and C_i mentioned earlier. For example in Line 4, as the iteration index i traverses all the M time points' matrix information, sparse matrix-vector multiplication `csrmv` is first called to calculate $C_{i-1} p_1$ and $C_{i-2} p_2$. And after the two resulted vectors are combined by `axpy` of CUBLAS, p_0 is solved for by `trsv`, since as we noted before, J_i is already in LU factorization form when transient simulation at step i finished.

17.4 Numerical examples

The presented algorithm has been prototyped and numerical experiments have been carried out on a server which has an Intel Xeon quad-core CPU with 2.0 GHz clock speed, and 24 GBytes memory. The GPU card mounted on this server is NVIDIA's Tesla C2070 (Fermi), which contains 448 cores (14 MPs × 32 cores per MP) running at a 1.30 GHz and has 4 GBytes on-chip memory. Some initial results have been published in [15].

The envelope-following method with the proposed Gear-2 sensitivity matrix computation is added to an open-source SPICE, implemented in C [2]. Our envelope-following program is implemented by following the algorithm mentioned in [10]. To solve the Newton update equation, different methods are used to compare the computation time, such as direct LU, GMRES with

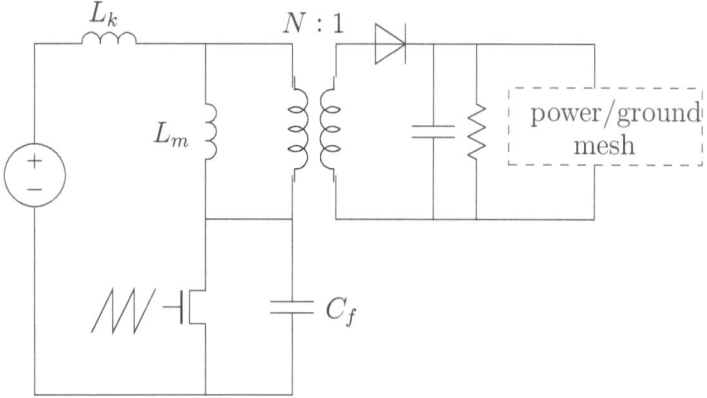

FIGURE 17.4. Diagram of a zero-voltage quasi-resonant flyback converter.

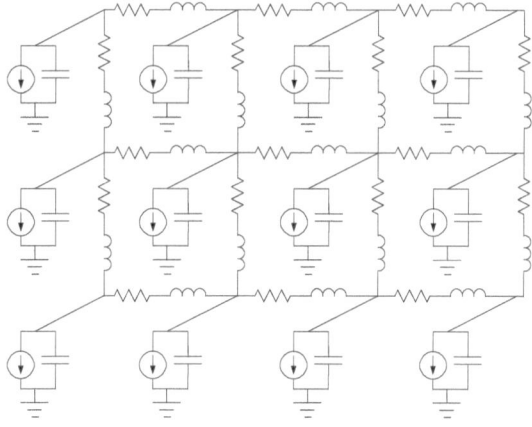

FIGURE 17.5. Illustration of power/ground network model.

explicitly formed matrix, and GMRES with implicit matrix-vector multiplication (matrix-free). Moreover, the matrix-free method is also incorporated to the same SPICE simulator using CUDA C programming interface, as described in Section 17.3.2.

We use several integrated on-chip converters as simulation examples to measure running time and speedup. They include a Buck converter, a quasi-resonant flyback converter (shown in Figure 17.4), and two boost converters. Each converter is directly integrated with on-chip power grid networks, since the performance of the converters should be studied with their loads and we can easily observe the waveforms at different nodes in a power grid (see Figure 17.5 for a simplified power grid structure).

Figure 17.6 and Figure 17.7 show the waveform at output node of the resonant flyback converter and the Buck converter. Note that on the enve-

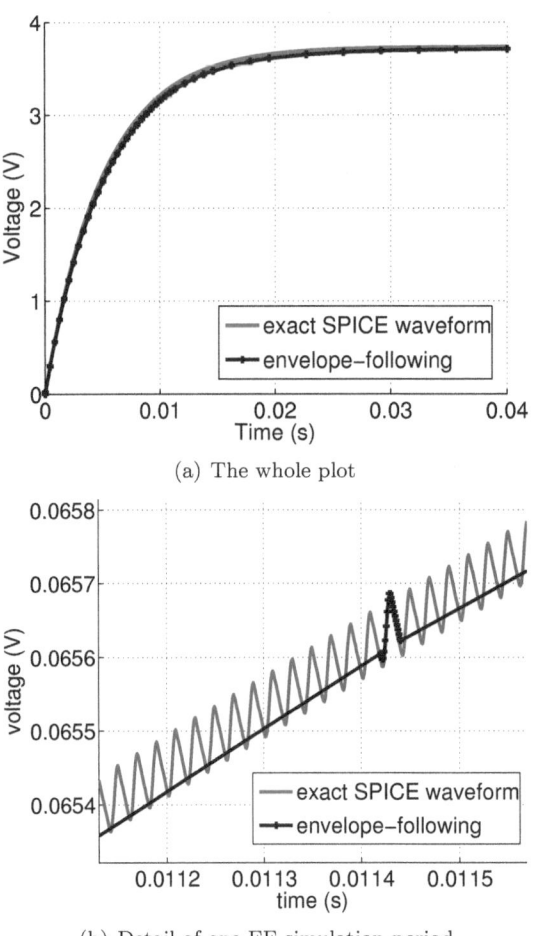

(a) The whole plot

(b) Detail of one EF simulation period

FIGURE 17.6. Flyback converter solution calculated by envelope-following. The red curve is traditional SPICE simulation result, and the back curve is the envelope-following output with simulation points marked.

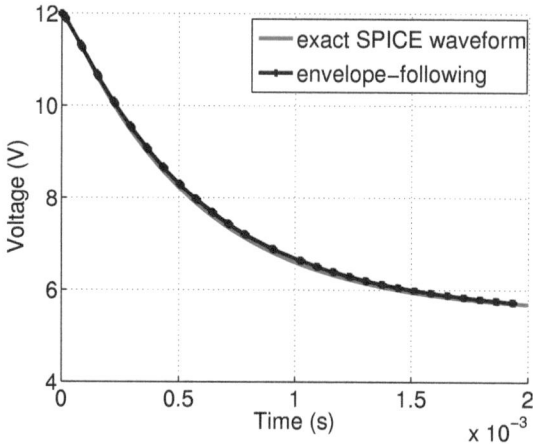

FIGURE 17.7. Buck converter solution calculated by envelope-following.

Circuit	Nodes	Direct LU	Explicit GMRES	Implicit GMRES CPU	Implicit GMRES GPU	X
Buck	910	423.8	420.3	36.8	3.9	9.4×
Flyback	941	462.4	459.6	64.5	7.4	8.7×
Boost-1	976	695.1	687.7	73.2	6.2	11.8×
Boost-2	1093	729.5	720.8	71.0	8.5	9.9×

TABLE 17.1. CPU and GPU time comparisons (in seconds) for solving Newton update equation with the proposed Gear-2 sensitivity.

lope curve, the darker dots in separated segments indicate the real simulation points that were calculated in those cycles, and the segments without dots are the envelope jumps where no simulation were done. It can be verified that the proposed Gear-2 envelope-following method produces a envelope matching the original waveform well.

For the comparison of running time spent in solving Newton update equation, Table 17.1 lists the time cost by direct method, explicit GMRES, matrix-free GMRES, and GPU matrix-free GMRES. All methods carry out the Gear-2 based envelope-following method, but they handle the sensitivity and equation solving in different implementation steps. It is obvious that as long as the sensitivity matrix is explicitly formed, such as in the cases of direct method and explicit GMRES, the cost is much higher than the implicit methods. When the matrix-free technique is applied to generate matrix-vector products implicitly, the computation cost is greatly reduced. Thus, for the same example, implicit GMRES would be one order of magnitude faster than explicit GMRES. Furthermore, our GPU parallel implementation of implicit GMRES makes this method even faster, with a further 10× speedup.

17.5 Summary

In this chapter, we have presented a new envelope-following method for transient analysis of switching power converters. First, the computationally expensive step, the solving of the Newton update equation, has been parallelized on CUDA-enabled GPU platforms with iterative GMRES solver to boost performance of the analysis method. To further speed up the GMRES solving for the Newton update equation, we have employed the matrix-free Krylov basis generation technique. The proposed method also applies the more robust Gear-2 integration to compute the sensitivity matrix. Experimental results from several integrated on-chip power converters have shown that the proposed GPU envelope-following algorithm can lead to about 10× speedup compared to its CPU counterpart, and 100× faster than the traditional envelope-following methods while still keep the similar accuracy.

17.6 Glossary

Envelope-Following: In transient simulation of switching power circuits, nodal voltage waveforms in neighboring high frequency clock cycles are similar, but not exactly duplicates. Envelope-following technique approximates the slowly changing transient trend over a lot of clock cycles without calculating waveforms in all cycles.

Bibliography

[1] CUDA community showcase. http://www.nvidia.com/.

[2] NGSPICE. http://ngspice.sourceforge.net/.

[3] AMD Inc. Multi-core processors—the next evolution in computing (White Paper), 2006. http://multicore.amd.com.

[4] AMD Inc. AMD Steam SDK. http://developer.amd.com/gpu/ATIStreamSDK, 2011.

[5] P. Feldmann and J. Roychowdhury. Computation of circuit waveform envelopes using an efficient, matrix-decomposed harmonic balance algorithm. In *Proc. ICCAD*, pages 295–300, Nov. 1996.

[6] C. W. Gear. *Numerical Initial Value Problems in Ordinary Differential Equations*. Prentice-Hall, Englewood Cliffs, NJ, 1971.

[7] D. Göddeke. General-purpose computation using graphics hardware. http://www.gpgpu.org/, 2011.

[8] G. H. Golub and C. Van Loan. *Matrix Computations*. The Johns Hopkins University Press, 3rd edition, 1996.

[9] Intel Corporation. Intel multi-core processors, making the move to quad-core and beyond (White Paper), 2006. http://www.intel.com/multi-core.

[10] T. Kato et al. Envelope following analysis of an autonomous power electronic system. In *IEEE COMPEL'06*, pages 29–33, July 2006.

[11] Khronos Group. Open Computing Language (OpenCL). http://www.khronos.org/opencl, 2011.

[12] D. B. Kirk and W.-M. Hwu. *Programming Massively Parallel Processors: A Hands-on Approach*. Morgan Kaufmann Publishers Inc., San Francisco, CA, 2010.

[13] P. Krein. *Elements of Power Electronics*. Oxford University Press, 1997.

[14] K. Kundert, J. White, and A. Sangiovanni-Vincentelli. An envelope following method for the efficient transient simulation of switching power and filter circuits. In *Proc. ICCAD*, pages 446–449, Oct. 1988.

[15] X.-X. Liu, S. X.-D. Tan, H. Wang, and H. Yu. A GPU-accelerated envelope-following method for switching power converter simulation. pages 1349–1354, March 2012.

[16] NVIDIA, 2011. http://developer.nvidia.com/cuda-toolkit-40.

[17] NVIDIA Corporation, 2011. http://www.nvidia.com.

[18] NVIDIA Corporation. CUDA (Compute Unified Device Architecture), 2011. http://www.nvidia.com/object/cuda_home.html.

[19] R. Telichevesky, K. Kundert, and J. White. Efficient steady-state analysis based on matrix-free Krylov-subspace methods. pages 480–484, 1995.

[20] A. M. Trzynadlowski. *Introduction to Modern Power Electronics*, second edition, Wiley, 2010.

[21] J. Vlach and K. Singhal. *Computer Methods for Circuit Analysis and Design*. Van Nostrand Reinhold, New York, NY, 1995.

[22] J. White and S. Leeb. An envelope-following approach to switching power converter simulation. *IEEE Trans. Power Electron.*, 6(2):303–307, Apr. 1991.

Part VI

Other

Chapter 18

Implementing multi-agent systems on GPU

Guillaume Laville, Christophe Lang, Bénédicte Herrmann, and Laurent Philippe

Femto-ST Institute, University of Franche-Comte, France

Kamel Mazouzi

Franche-Comte Computing Center, University of Franche-Comte, France

Nicolas Marilleau

UMMISCO, Institut de Recherche pour le Developpement (IRD), France

18.1	Introduction		416
18.2	Running agent-based simulations		417
	18.2.1	Multi-agent systems and parallelism	417
	18.2.2	MAS implementation on GPU	418
18.3	A first practical example		420
	18.3.1	The Collembola model	420
	18.3.2	Collembola implementation	421
	18.3.3	Collembola performance	423
18.4	Second example		424
	18.4.1	The MIOR model	424
	18.4.2	MIOR implementation	425
		18.4.2.1 Execution mapping on GPU	426
		18.4.2.2 Data structures translation	427
		18.4.2.3 Critical resources access management	428
		18.4.2.4 Termination detection	430
	18.4.3	Performance of MIOR implementations	430
18.5	Analysis and recommendations		434
	18.5.1	Analysis	434
	18.5.2	MAS execution workflow	435
	18.5.3	Implementation challenges	435
	18.5.4	MCSMA	436
18.6	Conclusion		437
	Bibliography		438

415

18.1 Introduction

In this chapter we introduce the use of Graphical Processing Units (GPU) for multi-agents-based systems as an example of a not-so-regular application that could benefit from the GPU computing power. Multi-Agent Systems (MAS) are a simulation paradigm used to study the behavior of dynamic systems. Dynamic systems as physical systems are often modeled by mathematical representations and their dynamic behavior is simulated by differential equations. The simulation of the system thus often relies on the resolution of a linear system that can be efficiently computed on a graphical processing unit as shown in the preceding chapters. But when the behavior of the system elements is not uniformly driven by the same law, when these elements have their own behavior, the modeling process is too complex to rely on formal expressions. In this context MAS is a recognized approach to model and simulate systems where individuals have an autonomous behavior that cannot be simulated by the evolution of a set of variables driven by mathematical laws. MAS are often used to simulate natural or collective phenomena whose individuals are too numerous or various to provide a unified algorithm describing the system evolution. The agent-based approach is to divide these complex systems into individual self-contained entities with their smaller set of attributes and functions. But, as for mathematical simulations, when the size of the MAS increases, the need of computing power and memory also increases. For this reason, multi-agent systems should benefit from the use of distributed computing architectures. Clusters and grids are often identified as the main solution to increase simulation performance but GPUs are also a promising technology with an attractive performance/cost ratio.

Conceptually a MAS is a distributed system as it favors the definition and description of large sets of individuals, the agents, that can be run in parallel. As a large set of agents could have the same behavior, a Single Instruction Multiple Data (SIMD) execution architecture should fit the simulation execution. Most of the agent-based simulators are, however, designed with a sequential scheme in mind, and these simulators seldom use more than one core for their execution. Due to simulation scheduling constraints, data sharing and exchange between agents and the huge amount of interactions between agents and their environment, it is indeed rather difficult to distribute an agent based simulator, for instance, to take advantage of new multithreaded computer architectures. Thus, guidelines and tools dedicated to MAS paradigm and High Performance Computing (HPC) are now a need for other complex system communities. Note that, from the described structure (large number of agents sharing data), we can conclude that MAS would more easily benefit from many-core architectures than from other kinds of parallelism.

Another key point that advocates for the use of many-core in MAS is the growing need for multiscale simulations. Multiscale simulations explore problems with interactions between several scales. The different scales use different granularity of the structure and potentially different models. Most of the time the lower scale simulations provide results to higher scale simulations. In that case the execution of the simulations can easily be distributed between the local cores and a many-core architecture, i.e., a GPU device.

We explore in this chapter the use of many-core architectures to execute agent-based simulations. We illustrate our reflexion with two cases: the Collembola simulator designed to simulate the diffusion of Collembola between plots of land and the MIOR (MIcro ORganism) simulator that reproduces effects of earthworms on bacteria dynamics in a bulked soil. In Section 18.2 we present the work related to MAS and parallelization with a special focus on many-core use. In sections 18.3 and 18.4 we present in detail two multi-agent models, their GPU implementations, the conducted experiments, and their performance results. The first model, given in Section 18.3, illustrates the use of a GPU architecture to speed up the execution of some computation-intensive functions while the main model is still executed on the central processing unit. The second model, given in Section 18.4, illustrates the use of a GPU architecture to implement the whole model on the GPU processor which implies deeper changes in the initial algorithm. In Section 18.5 we propose a more general reflexion on these implementations and provide some guidelines. Then, we conclude in Section 18.6 on the possible generalization of our work.

18.2 Running agent-based simulations

In this section, we present the context of MAS, their parallelization, and we report several existing works on using GPU to simulate multi-agent systems.

18.2.1 Multi-agent systems and parallelism

Agent-based systems are often used to simulate natural or collective phenomena whose actors are too numerous or various to provide a simple unified algorithm describing the studied system dynamic [21]. The implementation of an agent based simulation usually starts by designing the underlying agent-based model (ABM). Most ABM are based around a few types of entities such as agents, one environment, or an interaction organization [9]. In the complex system domain, the environment often describes a real space, its structure (e.g. soil textures and porosities), and its dynamics (e.g., organic matter decomposition). It is a virtual world in which agents represent studied entities (e.g., biotic organisms) evolution. The actual agent is animated by a behavior that can range between reactivity (only reacts to external stimuli) and cognition (makes complex decisions based on environmental and internal factors). Interaction and organization define functions, types, and patterns of communications of their member agents in the system [20]. Note that, depending on the MAS, agents can communicate either directly through special primitives or indirectly through the information stored in the environment.

Agent-based simulations have been used for more than a decade to reproduce, understand and even predict complex system dynamics. They have proved their usefulness in various scientific communities. Nowadays generic agent based frameworks such as Repast [19] or NetLogo [24] are promoted to implement simulators. Many ABMs such as the crown model representing a city wide scale [25] tend however to require a large number of agents to provide a realistic behavior and reliable global statistics. Moreover, an achieved model analysis needs to resort to an experiment plan, consisting of multiple simulation runs, to obtain enough confidence in a simulation. In this case the available computing power often limits the simulation size, and the resulting range thus requires the use of parallelism to explore bigger configurations.

For that, three major approaches can be identified:

1. parallelizing experiments execution on a cluster or a grid (one or a few simulations are submitted to each core) [3,6],

2. parallelizing the simulator on a cluster (the environment of the MAS is split and run on several distributed nodes) [8,18],

3. optimizing the simulator by taking advantage of computer resources (multi-threading, GPU, and so on) [2].

In the first case, experiments are run independent of each other and only simulation parameters are changed between two runs so that a simple version of an existing simulator can be used. This approach does not, however, allow to run larger models. In the second and the third case, model and code modifications are necessary. Only a few frameworks, however, introduce distribution in agent simulation (Madkit [12], MASON [22], repastHPC [7]), and parallel implementations are often based on the explicit use of threads using shared memory [13] or cluster libraries such as MPI [14].

Parallelizing a multi-agent simulation is, however, complex due to space and time constraints. Multi-agent simulations are usually based on a synchronous execution: at each time step, numerous events (space data modification, agent motion) and interactions between agents happen. Distributing the simulation on several computers or grid nodes to guarantee a distributed synchronous execution and coherency. This often leads to poor performance or complex synchronization problems. Multicore execution or delegating part of this execution to others processors such as GPUs [4] is usually easier to implement since all the threads share the data and the local clock.

18.2.2 MAS implementation on GPU

The last few years have seen the appearance of new generations of graphic cards based on more general purpose execution units which are promising for large systems such as MAS. Using matrix-based data representations and SIMD computations is however not always straightforward in MAS, where data structures and algorithms are tightly coupled to the described simulation. However, works from existing literature show that MAS can benefit from these performance gains on various simulation types, such as traffic simulation [25], cellular automata [10], mobile-agent based path-finding [23] or genetic algorithms [17]. Note that an application-specific adaptation process was required in the case of these MAS: some of the previous examples are driven by mathematical laws (path-finding) or use a natural mapping between a discrete environment (cellular automaton) and GPU cores. Unfortunately, this mapping often requires algorithmic adaptations in other models but experience shows that the more reactive a MAS is the more adapted its implementation is to GPU.

The first step in the adaptation of an ABM to GPU platforms is the choice of language. On the one hand, the Java programming language is often used for the implementation of MAS due to its availability on numerous platforms or frameworks and its focus on high-level, object-oriented programming. On the other hand, GPU platforms can only run specific languages such as OpenCL or CUDA. OpenCL (supported on AMD, Intel, and NVIDIA hardware) better suits the portability concerns across a wide range of hardware needed the agent simulators, as opposed to CUDA which is an NVIDIA-specific library.

OpenCL is a C library which provides access to the underlying CPU or GPU threads using an asynchronous interface. Various OpenCL functions allow the compilation and the execution of programs on these execution resources, the copying of data buffers between devices, and the collection of profiling information.

This library is based around three main concepts:

- the *kernel* (similar to a CUDA kernel), which represents a runnable program containing instructions to be executed on the GPU;

- the *work-item* (equivalent to a CUDA thread), which is analogous to the concept of thread on GPU, in that it represents one running instance of a GPU kernel; and

- the *work-group* (or execution block) which is a set of work-items sharing some memory to speed up data accesses and computations. Synchronization operations such as barrier can only be used across the same work-group.

Running an OpenCL computation consists of launching numerous work-items that execute the same kernel. The work-items are submitted to a submission queue to optimize the available cores usage. A calculus is achieved once all these kernel instances have terminated.

The number of work-items used in each work-group is an important implementation choice which determines how many tasks will share the same cache memory. Data used by the work-items can be stored as N-dimensions matrices in local or global GPU memory. Since the size of this memory is often limited to a few hundred kilobytes, choosing this number often implies a compromise between the model synchronization or data requirements and the available resources.

In the case of agent-based simulations, each agent can be naturally mapped to a work-item. Work-groups can then be used to represent groups of agents or simulations sharing common data (such as the environment) or algorithms (such as the background evolution process).

In the following examples a binding named JOCL [1] is used to access the OpenCL platform from the Java programming language.

In the next sections we present two practical cases that will be studied in detail, from the model to its implementation and performance.

18.3 A first practical example

The first model, the Collembola model, simulates the propagation of collembolas in fields and forests. It is based on a diffusion algorithm which illustrates the case of agents with a simple behavior and few synchronization problems.

18.3.1 The Collembola model

The Collembola model is an example of multi-agent system using GIS (Geographical Information System) and survey data (population count) to model the evolution of the biodiversity across land plots. A first version of this model has been developed with the Netlogo framework by Bioemco and UMMISCO researchers. In this model, the biodiversity is modeled by populations of athropod individuals, the Collembola, which can reproduce and diffuse to favorable new habitats. The simulator allows us to study the diffusion of Collembola, between plots of land depending on their use (artifical soil, crop, forest, etc.) In this model the environment is composed of the studied land, and collembola are used as agents. Every land plot is divided into several cells, each cell representing a surface unit (16x16 meters). A number of individuals per collembola species is associated to each cell. The model evolution is then based on a common diffusion model that diffuses individuals between cells. Each step of the simulation is based on four stages, as shown on Figure 18.1:

The algorithm is quite simple but includes two costly operations, the reproduction and the diffusion, that must be parallelized to improve the model performances.

The **reproduction** stage consists in updating the total population of each plot by taking the individuals arrived at the preceding computation step. This

Implementing multi-agent systems on GPU 421

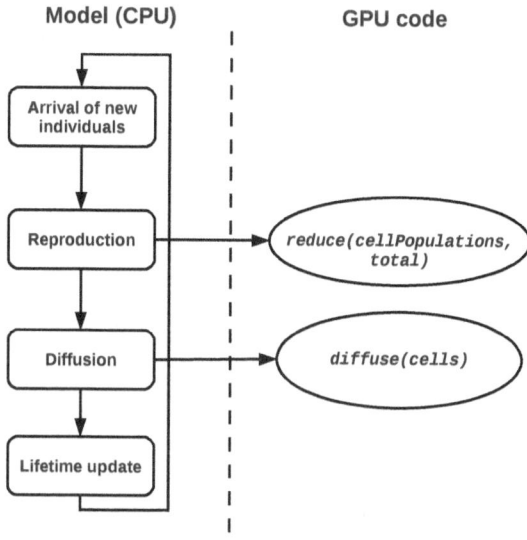

FIGURE 18.1. Evolution algorithm of the Collembola model.

stage involves processing the whole set of cells of the environment to sum their population. The computed value is recorded in the plot associated to each cell. This process can be assimilated to a reduction operation on all the population cells associated to one plot to obtain its population.

The **diffusion** stage simulates the natural behavior of the Collembola that tends toward occupying the whole space over time. This stage consists in computing a new value for each cell depending on the population of the neighbor cells. This process can be assimilated to a linear diffusion at each iteration of the population of the cells across their neighbors.

These two processes are quite common in numerical computations so that the Collembola model can be adapted to a GPU execution without much difficulty.

18.3.2 Collembola implementation

In the collembola simulator biodiversity is modeled by populations of collembola individuals, which can reproduce and diffuse to favorable new habitats. This is implemented as a fixed reproduction factor, applied to the size of each population, followed by a linear diffusion of each cell population to its eight neighbors. These reproduction and diffusion processes are followed by two more steps on the GPU implementation. The first one consist of culling of populations in an inhospitable environment, by checking each cell value and terrain type, and setting its population to zero if necessary. The final simulation step is the reduction of the cell populations for each plot, to obtain an updated plot population for statistic purposes. This separate computation step, done while updating each cell population in the reference sequential algorithm, is motivated by synchronization problems and allows the reduction of the total number of memory accesses needed to updated those populations.

Listing 18.1. collembola openCL diffusion kernel

```
#include "collem_structures.h"

kernel void diffuse(
    global CollemWorld *world,
    global int *populations,
    global int *overflows,
    global int *newPopulations)
{
    const int i = get_global_id(0);
    const int j = get_global_id(1);

    // Compute the population to diffuse for this cell
    OVERFLOW(world, overflows, i, j) = CELL(world, populations, i, j)
        * world->diffusionRate;

    // _syncthreads() in CUDA
    barrier(CLK_GLOBAL_MEM_FENCE);

    // Retrieve neighbors surplus and add them to the current cell
    // population
    int surplus = OVERFLOW(world, overflows, i + 1, j) + OVERFLOW(
        world, overflows, i - 1, j) + OVERFLOW(world, overflows, i, j
        - 1) + OVERFLOW(world, overflows, i, j + 1);

    CELL(world, newPopulations, i, j) = CELL(world, populations, i, j)
        + surplus / 8.0 - OVERFLOW(world, overflows, i, j);
}
```

The reproduction, diffusion and culling steps are implemented on GPU (Figure 18.1) as a straight mapping of each cell to an OpenCL work-item (GPU thread). Listing 18.1 gives the kernel for the diffusion implementation. To prevent data coherency problems, the diffusion step is split into two phases, separated by an execution *barrier*. In the first phase each cell diffusion overflow is calculated and divided by the number of neighbors. Note that, on the border of the grid, populations can also overflow outside the environment grid but we do not manage those external populations, since there are no reason to assume our model to be isolated of its surroundings. The overflow by neighbors value is stored for each cell before encountering the barrier. After the barrier is met, each cell reads the overflows stored by all its neighbors at the previous step and applies them to its own population. In this manner, only one barrier is required to ensure the consistency of population numbers, since no cell ever modify a value other than its own.

Listing 18.2 gives the kernel for the reduction implementation. The only step requiring numerous synchronized accesses is the reduction one: in this first approach, we chose to use *atomic_add* operation to implement this process, but more efficient implementations using partial reduction and local GPU memory could be implemented.

Listing 18.2. collembola OpenCL reduction kernel

```
#include "collem_structures.h"

kernel void reduce(
    global CollemWorld *world,
    global int *populations,
    global int *patchesOwners,
    global int *plotsPopulations)
{
    const int i = get_global_id(0);
    const int j = get_global_id(1);

    // Retrieve the owner parcel
    const int plotId = CELL(world, patchesOwners, i, j);
    if (plotId == -1) return;

    // Add cell population to the owner parcel population
    atomic_add(plotsPopulations + plotId, CELL(world, populations, i,
        j));
}
```

18.3.3 Collembola performance

In this part we present the performance of the collembola model on various CPU and GPU execution platforms. Figure 18.2 shows that the number of cores and the processor architecture as a strong influence on the obtained results

- Older GPU cards can be slower than modern processors. This can be explained by the new cache and memory access optimizations implemented in newer generations of GPU devices. These optimizations reduce the penalties associated with irregular and frequent global memory accesses. They are not available on our Tesla nodes.

- GPU curves exhibit an odd-even pattern in their performance results. Since this phenomenon is visible on two distinct manufacturer hardware, driver, and OpenCL implementation, it is likely the result of the decomposition process based on warp of fixed, power-of-two sizes.

- The number of cores is not the only determining factor: an Intel Core i7 2600K processor, even with only four cores, can provide better performance than a Phenom one.

Both graphs show that using the GPU to parallelize part of the simulator results in tangible performance gains over a CPU execution on modern hardware. These gains are more mixed on older GPU platforms due to the limitations when dealing with irregular memory or execution patterns often encountered in MAS systems. This can be closely linked to the availability of caching facilities on the GPU hardware and its dramatic effects depend on the locality and frequency of data accesses. In this case, even if the Tesla architecture offers more execution cores and is the far costlier solution, more

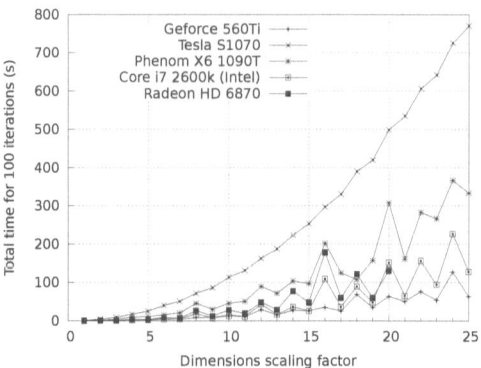

FIGURE 18.2. Performance of the Collembola model on CPU and GPU.

recent, cheaper, solutions such as high-end GPU provide better performance when the execution is not constrained by memory size.

18.4 Second example

The second model, the MIOR model, simulates the behavior of microbian colonies. Its execution model is more complex so that it requires changes in the initial algorithm and the use of synchronization to benefit from the GPU architecture.

18.4.1 The MIOR model

The MIOR [5] model was developed to simulate local interactions in soil between microbial colonies and organic matters. It reproduces each small cubic unit ($0.002m^3$) of soil as a MAS.

Multiple implementations of the MIOR model have already been realized, in Smalltalk and Netlogo, in 2 or 3 dimensions. The last implementation, used in our work and referenced as MIOR in the rest of the chapter, is freely accessible online as WebSimMior[1].

MIOR is based around two types of agents: (1) the Meta-Mior (MM), which represents microbial colonies consuming carbon and (2) the Organic Matter (OM) which represents carbon deposits occurring in soil.

The Meta-Mior agents are characterized by two distinct behaviors:

---

[1] http://www.IRD.fr/websimmior/

- *breath*: this action converts mineral carbon from the soil to carbon dioxide (CO_2) that is released into the soil and

- *growth*: by this action each microbial colony fixes the carbon present in the environment to reproduce itself (augment its size). This action is only possible if the colony breathing needs are covered, i.e., enough mineral carbon is available.

These behaviors are described in Algorithm 22.

Algorithm 22: evolution step of each Meta-Mior (microbial colony) agent

Input: A static array *mmList* of MM agents
1   A static array *omList* of OM agents
2   A MIOR environment *world*
3 $breathNeed \leftarrow world.respirationRate \times mm.carbon$;
4 $growthNeed \leftarrow world.growthRate \times mm.carbon$;
5 $availableCarbon \leftarrow totalAccessibleCarbon(mm)$;
6 **if** $availableCarbon > breathNeed$ **then**
    /* Breath */
7     $mm.active \leftarrow true$;
8     $availableCarbon \leftarrow availableCarbon - consumCarbon(mm, breathNeed)$;
9     $world.CO2 \leftarrow world.CO2 + breathNeed$;
10   **if** $availableCarbon > 0$ **then**
        /* Growth */
11         $growthConsum \leftarrow max(totalAccessCarbon(mm), growthNeed)$;
12         $consumCarbon(mm, growthConsum)$;
13         $mm.carbon \leftarrow mm.carbon + growthConsum$;
14   **end**
15 **else**
16     $mm.active \leftarrow false$
17 **end**

Since this simulation takes place at a microscopic scale, a large number of these simulations must be executed for each macroscopic simulation step to model a realistic-sized unit of soil. This leads to large computing needs despite the small computation cost of each individual simulation.

18.4.2 MIOR implementation

As pointed out previously, the MIOR implementation implied more changes for the initial code to be run on GPU. As a first attempt, we tried a simple GPU implementation of the MIOR simulator, with only minimal

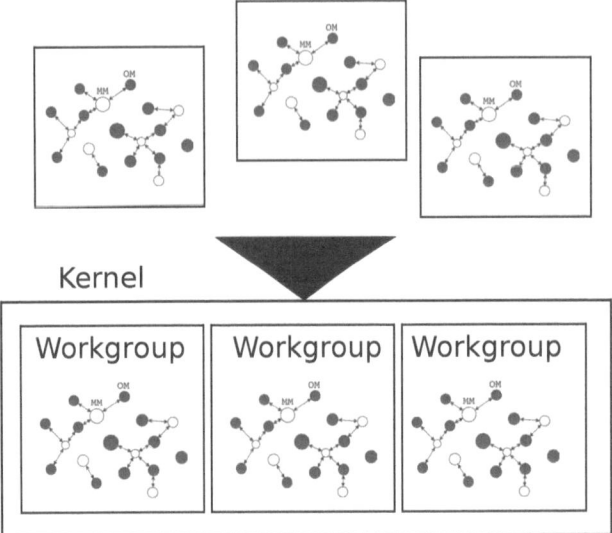

FIGURE 18.3. Consolidation of multiple simulations in one OpenCL kernel execution.

changes to the CPU algorithm. Execution times showed the inefficiency of this approach and highlighted the necessity of adapting the simulator to take advantage of the GPU execution capabilities [15]. In this part, we show the main changes which were realized to adapt the MIOR simulator on GPU architectures.

18.4.2.1 Execution mapping on GPU

Each MIOR simulation is represented by a work-group, and each agent by a work-item. A kernel is in charge of the life cycle process for each agent of the model. This kernel is executed by all the work-items of the simulation each on its own GPU core.

The usage of one work-group for each simulation allows the easy execution of multiple simulations in parallel, as shown on Figure 18.3. By taking advantage of the execution overlap possibilities provided by OpenCL, it then becomes possible to exploit all the cores at the same time, even if an unique simulation is too small to use all the available GPU cores. However, the maximum size of a work-group is limited (to 512), which allows us to execute only one simulation per work-group when using 310 threads (number of OM in the reference model) to execute the simulation.

The usage of the GPU to execute multiple simulations is initiated by the CPU. The CPU keeps total control of the simulator execution flow. Thus, optimized scheduling policies (such as submitting kernels in batch, limiting the number of kernels, or asynchronously retrieving the simulation results) can be defined to minimize the cost related to data transfers between CPU and GPUs.

18.4.2.2 Data structures translation

The adaptation of the MIOR model to GPU requires the mapping of the data model to OpenCL data structures. The environment and the agents are represented by arrays of structures, where each structure describes the state of one entity. The behaviors of these entities are implemented as OpenCL functions to be called from the kernels during execution.

Since the main program is written in Java, JOCL is responsible for the allocation and mapping of the object data structures to OpenCL ones before execution.

Four main data structures are defined: (1) an array of MM agents, representing the state of the microbial colonies. (2) an array of OM agents, representing the state of the carbon deposits. (3) a topology matrix, which stores accessibility information between the two types of agents of the model (4) a world structure, which contains all the global input data (metabolism rate, numbers of agents) and output data (quantity of CO_2 produced) of the simulation. The C-like OpenCL structures used to represent each type of to agent and the environment are illustrated in (Figure 18.3). These data structures are initialized by the CPU and then copied on the GPU.

Listing 18.3. main data structures used in a MIOR simulation

```
// Microbial colony
typedef struct MM {
    float    x;
    float    y;
    int      carbon;
    int      dormancy;
} MM;

// Carbon deposit
typedef struct OM {
    float    x;
    float    y;
    int      carbon;
    int      lock;
} OM;

// MIOR Environment
typedef struct World {
    int      nbMM;
    int      nbOM;
    int      RA;        // Action radius
    float    RR;        // Breathing rate
    float    GR;        // Growth rate
    float    K;         // Decomposition rate
    int      width;     // Size of the model
    int      minSize;   // Minimal size for a colony
    int      CO2;       // Total CO2 in the model
    int      lock;
} World;
```

The world topology is stored as a two-dimension matrix which represents OM indexes on the abscissa and MM indexes on the ordinate. Each agent walks through its line/column of the matrix at each iteration to determinate which agents can be accessed during the simulation. Since many agents are not connected, this matrix is sparse, which introduces a big number of useless memory accesses. To reduce the impact of these memory accesses we use a compacted, optimized representation of this matrix based on [11], as illustrated in Figure 18.4. This compact representation considers each line of the matrix as an index list, and only stores accessible agents compactly, to reduce the number of nonproductive accesses.

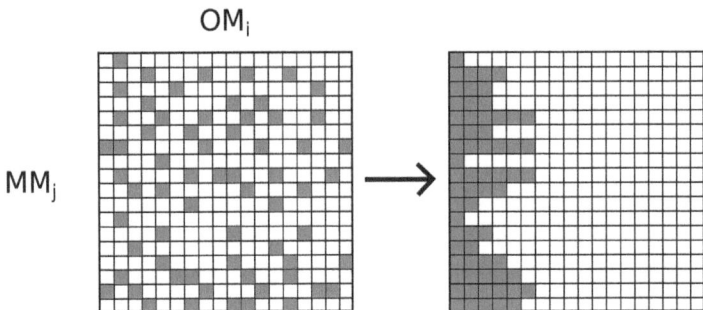

FIGURE 18.4. Compact representation of the topology of a MIOR simulation.

Since dynamic memory allocation is not possible yet in OpenCL and is only provided in the latest revisions of the CUDA standard, these matrices are statically allocated. The allocation is based on the worst-case scenario where all OM and MM are linked since the real occupation of the matrix cells cannot be deduced without some kind of preprocessing computations.

18.4.2.3 Critical resources access management

One of the main concers in the MIOR model is to ensure that all the microbial colonies will have an equitable access to carbon resources, when multiple colonies share the same deposits. Access synchronizations are mandatory in these cases to prevent conflicting updates on the same data that may lead to calculation error (e.g. loss of matter).

On massively parallel architectures such as GPUs, these kind of synchronization conflicts can lead to an inefficient implementation by enforcing a quasi-sequential execution. It is necessary, in the case of MIOR as well as for other ABM, to ensure that each work-item is not too constrained in its execution.

Listing 18.4. main MIOR kernel

```
kernel void simulate(
    global          MM       allMMList[],
    global          OM       allOMList[],
    global          World    allWorlds[],
    constant        int      allMMOffsets[],
    constant        int      allOMOffsets[],
    global const    int      allMMCSR[],
    global const    int      allOMCSR[],
    global          int      allParts[])
{
    const int iSim = get_group_id(0);
    const int i    = get_local_id(0);

    global World *       world    = allWorlds + iSim;
    global MM *          mmList   = allMMList + iSim * NB_MM;
    global OM *          omList   = allOMList + iSim * NB_OM;
    constant int *       mmOffsets = allMMOffsets + iSim * NB_MM;
    constant int *       omOffsets = allOMOffsets + iSim * NB_OM;
    global const int *   mmCSR    = allMMCSR;
    global const int *   omCSR    = allOMCSR;
    global int *         parts    = allParts + iSim * PARTS_SIZE;

    if (i < world->nbOM) {
        scatter(omList, omOffsets, omCSR, parts, world);
    }

    barrier(CLK_GLOBAL_MEM_FENCE);

    if (i < world->nbMM) {
        live(mmList, mmOffsets, mmCSR, parts, world);
    }

    barrier(CLK_GLOBAL_MEM_FENCE);

    if (i < world->nbOM) {
        gather(omList, omOffsets, omCSR, parts, world);
    }

    barrier(CLK_GLOBAL_MEM_FENCE);
}
```

From the sequential algorithm (Algorithm 22) where all the agents share the same data, we have developed a parallel algorithm composed of three sequential stages separated by synchronization barriers. This new algorithm is based on the distribution of the available OM carbon deposits into parts at the beginning of each execution step. The three stages, illustrated in Listing 18.4, are the following:

1. *scattering*: the available carbon in each carbon deposit (OM) is equitably dispatched among all accessible MM in the form of parts,

2. *live*: each MM consumes carbon in its allocated parts for its breathing and growing processes, and

3. *gathering*: unconsumed carbon in parts is gathered back into the carbon deposits.

This solution suppresses the data synchronization needed by the first algorithm, thus the need for synchronization barriers, and requires only one kernel launch from Java as described on Listing 18.5.

Listing 18.5. MIOR simulation launcher

```
@Override
protected void doLiveImpl() {
    simulateKernel.setArguments(mmMem, omMem, worldsMem, mmOffsetsMem,
        omOffsetsMem, mmCSRMem, omCSRMem, partsMem);

    if (blockSize < Math.max(nbOM, nbMM)) {
        throw new RuntimeException("blockSize (" + blockSize +
            ") too small to execute the simulation");
    }

    OCLEvent event = queue.enqueue1DKernel(simulateKernel, nbSim *
        blockSize, blockSize);

    OCLEvent.waitFor(event);

    if (!isBatchModeEnabled()) {
        queue.blockingReadBuffer(mmMem, mmList, 0, mmMem.getSize());
        queue.blockingReadBuffer(omMem, omList, 0, omMem.getSize());
        queue.blockingReadBuffer(worldsMem, worlds, 0, worldsMem.
            getSize());
    }
}
```

18.4.2.4 Termination detection

The termination of a MIOR simulation is reached when the model stabilizes and no more CO_2 is produced. This termination detection can be done on either the CPU or the GPU but it requires a global view on the system execution.

In the first case, when the CPU controls the GPU simulation process, the detection is done in two steps: (1) the CPU starts the execution of a simulation step on the GPU and (2) the CPU retrieves the GPU data and determines if another iteration must be launched or if the simulation has terminated. This approach allows a fine-grain control over the GPU execution, but it requires many costly data transfers as each iteration result must be sent from the GPU to the CPU. In the case of the MIOR model these costs are mainly due to the inherent PCI-express port latencies rather than to bandwidth limitation since data sizes remains rather small, on the order of few dozens of Megabytes.

In the second case the termination detection is directly implemented on the GPU by checking the amount of available carbon between two iterations. The CPU does not have any feedback while the simulation is running, but retrieves the results once the kernel execution is finished. This approach minimizes the number of transfers between the CPU and the GPU.

18.4.3 Performance of MIOR implementations

In this part we present several MIOR GPU implementations using the distribution/gathering process described in the previous section and compare their performance on two distinct hardware platform, i.e., two different GPU devices. Five incremental MIOR implementations were realized with an increasing level of adaptation for the algorithm: in all cases, we choose the average time over 50 executions as a performance indicator.

- The **GPU 1.0** implementation is a direct implementation of the existing algorithm and its data structures where data dependencies were removed, and it uses the noncompact topology representation described in Section 18.4.2.2

- The **GPU 2.0** implementation uses the previously described compact representation of the topology and remains otherwise identical to the GPU 1.0 implementation.

- The **GPU 3.0** implementation introduces the manual copy into local (private) memory of often-accessed global data, such as carbon parts or topology information.

- The **GPU 4.0** implementation is a variant of the GPU 1.0 implementation but allows the execution of multiple simulations for each kernel execution.

- The **GPU 5.0** implementation is a multisimulation version of the GPU 2.0 implementation, using the execution of multiple simulations for each kernel execution as for GPU 4.0.

The two last implementations **GPU 4.0** and **GPU 5.0** illustrate the gain provided by a better usage of the hardware resources, thanks to the driver execution overlapping capabilities. A sequential version of the MIOR algorithm, labeled **CPU**, is included for comparison purpose. This sequential version was developed in Java, the same language used for GPU implementations.

For these performance evaluations, two platforms are used. The first one is representative of the kind of hardware which is available on HPC clusters. It is a cluster node dedicated to GPU computations with two Intel X5550 processors running at 2.67GHz and one Tesla C1060 GPU device running at 1.3GHz and composed of 240 cores (30 multiprocessors). The second platform illustrates what can be expected from a personal desktop computer built a few years ago. It uses an Intel Q9300 CPU, running at 2.5GHz, and a Geforce 8800GT GPU running at 1.5GHz and composed of 112 cores (14 multiprocessors). The purpose of these two platforms is to assess the benefit that could be obtained when a scientist has access either to specialized hardware as a cluster or tries to take advantage of its own personal computer.

Figures 18.5 and 18.6 show the execution time for 50 simulations on the two hardware platforms. A size factor is applied to the problem: at scale 1, the model contains 38 MM and 310 OM, while at the scale 6 these numbers are multiplied by six. The size of the environment is modified as well to maintain

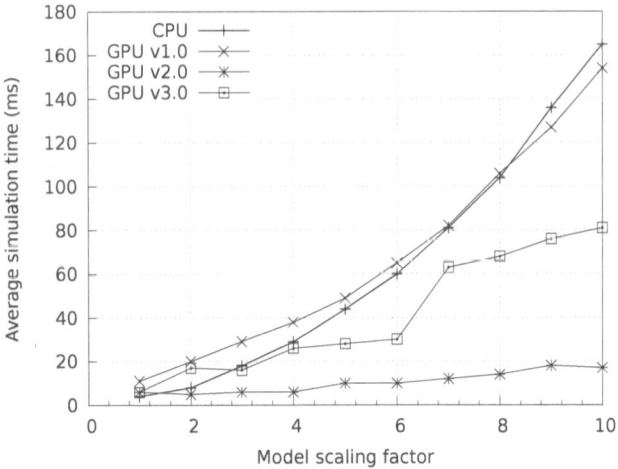

FIGURE 18.5. CPU and GPU performance on a Tesla C1060 node.

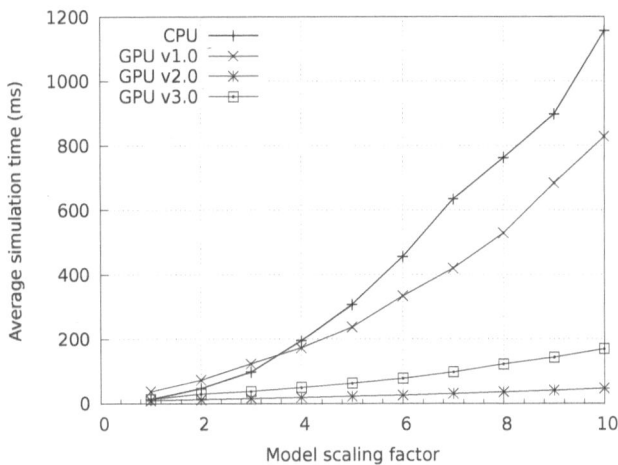

FIGURE 18.6. CPU and GPU performance on a personal computer with a Geforce 8800GT.

the same average agent density in the model. This scaling factor displays the impact of the chosen size of simulation on performance.

The charts show that for small problems the execution times of all the implementations are very close. This is because the GPU execution does not have enough threads (representing agents) for an optimal usage of GPU resources. This trend changes around scale 5 where GPU 2.0 and GPU 3.0 take the advantage over the GPU 1.0 and CPU implementations. This advantage continues to grow with the scaling factor, and reaches a speedup of 10 at the scale 10 between the fastest single-simulation GPU implementation and the first, naive one GPU 1.0.

Multiple trends can be observed in these results. First, optimizations for the GPU hardware show a large, positive impact on performance, illustrating the strong requirements on the algorithm properties to reach execution efficiency. These charts also show that despite the vast difference in numbers of cores between the two GPU platforms, the same trends can be observed in both cases. We can therefore expect similar results on other GPU cards, without the need for more adaptations.

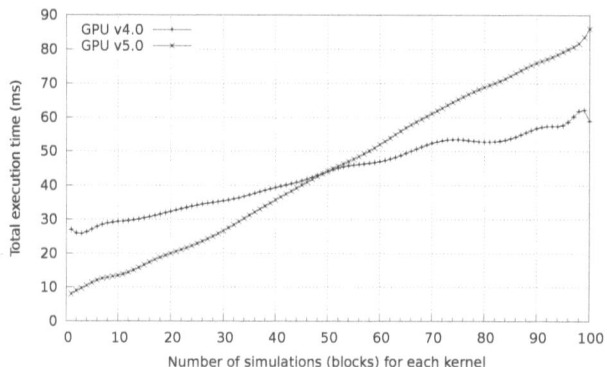

FIGURE 18.7. Execution time of one multi-simulation kernel on the Tesla platform.

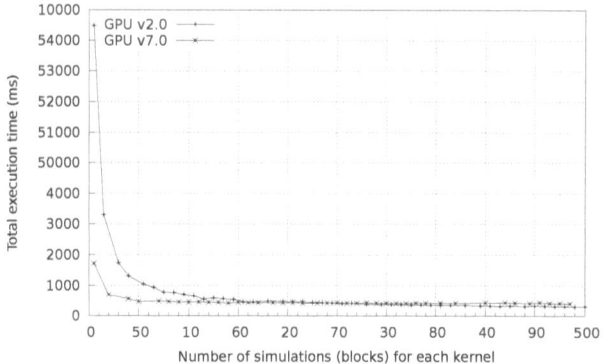

FIGURE 18.8. Total execution time for 1000 simulations on the Tesla platform, while varying the number of simulations for each kernel.

There are two ways to measure simulations performance: (1) by executing only one kernel, and varying its size (the number of simulations executed in parallel), as shown in Figure 18.7, to test the costs linked to the parallelization process or (2) by executing a fixed number of simulations and varying the size of each kernel, as shown in Figure 18.8.

Figure 18.7 illustrates the execution time for only one kernel. It shows that for small numbers of simulations run in parallel, the compact implementation of the model topology is faster than the two-dimension matrix representation. This trends reverse with more than 50 simulations in parallel, which can be explained either by the nonlinear progression of the synchronization costs or by the additional memory required for the access-efficient representation.

Figure 18.8 illustrates the execution time of a fixed number of simulations. It shows that for a small number of simulations run in parallel, the costs resulting from program setup, data copies, and launch on GPU are very detrimental to performance. Once the number of simulations executed for each kernel grows, these costs are counterbalanced by computation costs. This trend is more marked in the case of the sparse implementation (GPU 4.0) than in the compact one but appears on both curves. With more than 30 simulations for each kernel, execution times stall, since hardware limits are reached. This indicates that the cost of preparing and launching kernels become negligible compared to the computing time once a good GPU occupancy rate is achieved.

18.5 Analysis and recommendations

In this section we synthesize the observations done on the two models and identify some recommendations for implementing complex systems on GPU platforms.

18.5.1 Analysis

In both the collembola and the MIOR model, a critical problematic is the determination of the parts of the simulation that are to be run on GPU and which are to remain on the CPU. The determination of these parts is a classic step of any algorithm parallelization and must take into account considerations such as the cost of the different parts of the algorithm and the expected gains.

In the case of the collembola model two steps of the algorithm were ported to GPU. Both steps use straightforward, easily parallelizable operations where a direct gain can be expected by using more execution cores without important modifications to the algorithm.

In the MIOR model case, however, no such inherently parallelizable parts are evident in the original sequential algorithm. This is mainly explained by the rate of interactions between agents in this model in the form of two operations (breathing, growth) using heavily-shared carbon resources. In this case the algorithm had to be more profoundly modified while keeping in mind the need to remain true to the original model, to synchronize the main execution step of all agents in the model, to ensure equity, and to minimize the numbers of synchronizations. The minimization is done by factoring the distribution of carbon in the model in two separated steps at the beginning and the end of each iteration rather than at multiple points of the execution.

18.5.2 MAS execution workflow

Many MAS simulations decompose their execution process into discrete evolution steps where each step represents a quantum of time (minimal unit of time described). At the end of each step many global data, graphical displays or output files are updated. This execution model may not correctly fit on GPU platforms as they assume more or less a batch-like workflow model. The execution model must be split into the following ever repeating steps:

- Allocation of GPU data buffers
- Copy of data from CPU to GPU
- GPU kernels execution
- Copy of results from GPU to CPU

This workflow works well if the considered data transfer time is negligible compared to GPU execution or can be done in parallel, thanks to the asynchronous nature of OpenCL. If we are to update the MAS model after each iteration then performance risks being degraded. This is illustrated in the MIOR model by the fact that the speedup observed on GPU is much more significant for bigger simulations, which imply longer GPU execution times. Our solution to this problem is to desynchronize the execution of the MAS model and its GPU parts by requesting the execution of multiple steps of the GPU simulations for each launch.

Another related prerequisite of GPU execution is the ability to have many threads executed, to allow an efficient exploitation of the superior number of cores provided by the architecture. In the case of MAS models, this means that executing one agent at a time on GPU is meaningless in regard to GPU usage, copying cost, and actual gain in execution time, if the agent computations are not complex enough. In the MIOR and the collembola models, this is solved by executing the computations for all agents of the model at the same time. If the model has only chronic needs for intensive computations, then some kind of batching mechanism is required to store waiting treatments in a queue, until the total sum of waiting computations justifies the transfer cost to the GPU platform.

18.5.3 Implementation challenges

Besides the execution strategy challenges described above, some implementation challenges also occur when implementing an OpenCL version of a MAS model, mainly related to the underlying limitations of the execution platform.

The first one is the impossibility (except in latest CUDA versions) to dynamically allocate memory during execution. This is a problem in the case of models where the number of agents can vary during the simulation, such as prey-predator models. In this case, the only solution is to overestimate the size of arrays or data structures to accommodate these additional individuals, or to use the CPU to resize data structures when these situations occur. Both approaches require trending either memory or performance and are not always practical.

Another limitation is the impossibility to store pointers in data structures, since OpenCL only allows one-dimension static arrays. This precludes the usage of structures such as linked-list, graphs or sparse matrices not represented by some combination of static arrays, and can be another source of memory or performance losses.

In the case of MIOR, this problem is especially exacerbated in the case of neighboring storage: both representations consume much more memory than is actually required, since the worst case (all agents have access to all others agents) has to be taken into account when dimensioning the data structure.

The existence of different generations of GPU hardware is also a challenge. Older implementations, such as the four year old Tesla C1060 cards, have very strong constraints in term of memory accesses and requires very regular access patterns to perform efficiently. MAS having many random accesses, such as MIOR, or many small global memory accesses, such as Collembola, are penalized on these older cards. Fortunately, these requirements are less present is modern cards, which offer caches and other facilities traditionally present on CPU to offset these kinds of penalties.

The final concern is related to the previous ones and often results in more memory consumption. The amount of memory available on GPU cards is much more limited and adding new memory capabilities is more costly compared to expending a CPU RAM. On computing clusters, hardwares nodes with 128GB of memory or more have become affordable, whereas newer Tesla architecture remains limited to 16GB of memory. This can be a concern in the case of big MAS models, or small ones which can only use memory-inefficient OpenCL structures.

18.5.4 MCSMA

As shown in the previous sections, many data representation choices are common to entire classes of MAS. The paradigm of grid, for example, is often encountered in models where each cell constitutes either the elementary unit of simulation (SugarScape [10]) or a discretization of the environment space (Pathfinding [13]). These grids can be considered as two- or three-dimensional matrices, whose processing can be directly distributed.

Another common data representation encountered in MAS system is the usage of 2D or 3D coordinates to store the position of each agent of the model. In this case, even if the environment is no longer discrete, the location information still imply computations (Euclidean distances) which can be parallelized.

MCSMA [16] is a framework developed to provide to the MAS designer those basic data structures and the associated operations, to facilitate the portage of existing MAS on GPU. Two levels of utilization are provided to the developer, depending on its usage profile:

- A high-level library, composed of modules regrouping classes of operations. Such operation can distance computations in 1D, 2D or 3D grids, diffusion or reduction operations on matrices...

- A low-level API which allows the developer direct access to the GPU and the inner working of MCSMA, to develop new modules in the case where the required operations are not yet provided by the platform.

Both usage levels were illustrated in the above two practical cases. In MIOR, the whole algorithm (baring initialization) is ported on GPU as a specific plugin which allows executing n MIOR simulations and retrieving their results. This first approach requires extensive adaptations to the original algorithm. In collembola, to the contrary, the main steps of the algorithm remain executed on the CPU, and only specific operations are delegated to generic, already existing diffusion and reduction kernels. The fundamental algorithm is not modified and GPU execution is only applied to specific parts of the execution which may benefit from it. These two programming approaches allow incremental adaptations of existing Java MAS to accelerate their execution on GPU, while retaining the option to develop their own reusable or more efficient module to supplement the already existing ones.

18.6 Conclusion

This chapter has addressed the issue of complex system simulation by using agent-based paradigms and GPU hardware. From the experiments on two existing agent-based models of soil science we have provided useful information on the architecture, the algorithm design, and the data management to run agent-based simulations on GPU, and more generally to run computationally intensive applications that are not based on purely-matricial models. The first result of this work is that adapting the algorithm to a GPU architecture is possible and suitable to speed up agent based simulations as illustrated by the MIOR model. Coupling CPU with GPU seems to be an interesting way to take better advantage of the power given by computers and clusters as illustrated by the collembola model. From our point of view the adaptation process is less costy in time than a traditional parallelization on distributed nodes and not much difficult than a standard multithreaded parallelization, since all the data remains on the same host and can be shared in central memory. The usage of OpenCL also enables a portable simulator that can be run on different graphical units. Even using a mainstream card such as the GPU card of a standard computer can lead to significant performance improvements. This is an interesting result as it opens up the field of inexpensive HPC to a large community.

In this perspective, we are working on MCSMA, a development platform that would facilitate the use of GPU or many-core architectures for multi-agent simulations. Our first work has been the definition of common, efficient, reusable data structures, such as grids or lists. Another goal is to provide easier means to control the distribution of specific processes between CPU or GPU, to allow the easy exploitation of the strengths of each platform in the same multi-agent simulation. We think that the same approach, i.e., developing specific environments that facilitate the developer access to the GPU power, can be applied in many domains with computationally intensive needs to open the GPU use to a larger community.

Bibliography

[1] JOCL: Java Bindings for OpenCL. http://www.jocl.org.

[2] B. G. Aaby, K. S. Perumalla, and S. K. Seal. Efficient simulation of agent-based models on multi-GPU and multi-core clusters. In *Proceedings of the 3rd International ICST Conference on Simulation Tools and Techniques*, SIMUTools '10, pages 29:1–29:10, ICST, Brussels, Belgium, 2010. ICST (Institute for Computer Sciences, Social-Informatics and Telecommunications Engineering).

[3] E. Blanchart, C. Cambier, C. Canape, B. Gaudou, T.-N. Ho, T.-V. Ho, C. Lang, F. Michel, N. Marilleau, and L. Philippe. EPIS: A grid platform to ease and optimize multi-agent simulators running. In *PAAMS*, volume 88 of *Advances in Intelligent and Soft Computing*, pages 129–134, Salamanca, Spain, 2011. Springer.

[4] A. Bleiweiss. Multi agent navigation on the GPU. *GDC09 Game Developers Conference*, 2099.

[5] C. Cambier, D. Masse, M. Bousso, and E. Perrier. An offer versus demand modelling approach to assess the impact of micro-organisms spatiotemporal dynamics on soil organic matter decomposition rates. *Ecological Modelling*, pages 301–313, 2007.

[6] F. Chuffart, N. Dumoulin, T. Faure, and G. Deffuant. Simexplorer: Programming experimental designs on models and managing quality of modelling process. *IJAEIS*, 1(1):55–68, 2010.

[7] N.T. Collier and M.J North. *Repast SC++: A Platform for Large-scale Agent-based Modeling*, chapter Large-Scale Computing Techniques for Complex System Simulation. Wiley (In Press), 2011.

[8] B. Cosenza, G. Cordasco, R. De Chiara, and V. Scarano. Distributed load balancing for parallel agent-based simulations. In *19th Euromicro International Conference on Parallel, Distributed and Network-Based Computing*, pages 62–69, Ayia Napa, Cyprus, 2011.

[9] T. Da Silva Joao Luis and Y. Demazeau. Vowels co-ordination model. In *AAMAS*, pages 1129–1136, Bologna, Italy, 2002.

[10] Rahmani K. D'Souza R. M., Lysenko M. SugarScape on steroids: Simulating over a million agents at interactive rates. In *Proceedings of the Agent 2007 Conference*, Chicago, USA, 2007.

[11] J. Gómez-Luna, J.-M. González-Linares, J.-I. Benavides, and N. Guil. Parallelization of a video segmentation algorithm on CUDA-enabled graphics processing units. In *15th Euro-Par Conference*, pages 924–935, Berlin, Heidelberg, 2009. Springer-Verlag.

[12] O. Gutknecht and J. Ferber. Madkit: a generic multi-agent platform. In *Proceedings of the Fourth International Conference on Autonomous Agents*, AGENTS '00, pages 78–79, New York, NY, USA, 2000. ACM.

[13] S. J. Guy, J. Chhugani, C. Kim, N. Satish, M. C. Lin, D. Manocha, and P. Dubey. Clearpath: highly parallel collision avoidance for multi-agent simulation. In *ACM Siggraph/Eurographics Symposium on Computer Animation*, pages 177–187, New Orleans, LA, USA, 2009. ACM.

[14] M. Kiran, P. Richmond, M. Holcombe, L. S. Chin, D. Worth, and C. Greenough. Flame: simulating large populations of agents on parallel hardware architectures. In *Proceedings of the 9th International Conference on Autonomous Agents and Multiagent Systems, volume 1*, AAMAS, pages 1633–1636, Richland, SC, 2010. International Foundation for Autonomous Agents and Multiagent Systems.

[15] G. Laville, K. Mazouzi, C. Lang, N. Marilleau, and L. Philippe. Using GPU for multi-agent multi-scale simulations. In *DCAI'12, 9-th Int. Conf. on Advances in Intelligent and Soft Computing, volume 151 of Advances in Intelligent and Soft Computing*, pages 197–204, Salamanca, Spain, March 2012. Springer.

[16] G. Laville, K. Mazouzi, C. Lang, N. Marilleau, and L. Philippe. Using GPU for multi-agent soil simulation. In *PDP 2013, 21st Euromicro International Conference on Parallel, Distributed and Network-based Computing*, pages 392–399, Belfast, Ireland, February 2013. IEEE Computer Society Press.

[17] O. Maitre, N. Lachiche, P. Clauss, L. Baumes, A. Corma, and P. Collet. Efficient parallel implementation of evolutionary algorithms on GPGPU cards. In Henk Sips, Dick Epema, and Hai-Xiang Lin, editors, *Euro-Par 2009 Parallel Processing*, volume 5704 of *Lecture Notes in Computer Science*, pages 974–985. Springer Berlin / Heidelberg, 2009. 10.1007/978-3-642-03869-3

[18] N. Marilleau, C. Lang, P. Chatonnay, and L. Philippe. An agent based framework for urban mobility simulation. In *PDP*, pages 355–361, Montbéliard, France, 2006. IEEE Computer Society.

[19] M. J. North, T. R. Howe, N. T. Collier, and J. R. Vos. A declarative model assembly infrastructure for verification and validation. In Springer, editor, *Advancing Social Simulation: The First World Congress*, Heidelberg, FRG, 2007.

[20] J. J. Odell, H. Van Dyke Parunak, and M. Fleischer. The role of roles in designing effective agent organizations. In Alessandro Garcia, Carlos Lucena, Franco Zambonelli, Andrea Omicini, and Jaelson Castro, editors, *Software engineering for large-scale multi-agent systems*, pages 27–38. Springer-Verlag, Berlin, Heidelberg, 2003.

[21] F. Schweitzer and J. D. Farmer. *Brownian Agents and Active Particles: Collective Dynamics in the Natural and Social Sciences*. Complexity. Springer, 2003.

[22] L. Sean, C. Cioffi-Revilla, L. Panait, K. Sullivan, and G. Balan. MASON: A multi-agent simulation environment. *Simulation: Transactions of the society for Modeling and Simulation International*, 82(7):517–527, 2005.

[23] R. Silveira, L. Fischer, J. A. S. Ferreira, E. Prestes, and L. Nedel. Path-planning for RTS games based on potential fields. In *Proceedings of the Third International Conference on Motion in Games*, MIG'10, pages 410–421, Berlin, Heidelberg, 2010. Springer-Verlag.

[24] E. Sklar. NetLogo, a multi-agent simulation environment. *Artificial Life*, 13(3):303–311, 2011.

[25] David Strippgen and Kai Nagel. Multi-agent traffic simulation with CUDA. *International Conference on High Performance Computing Simulation*, pages 106–114, 2009.

Chapter 19

Pseudorandom number generator on GPU

Raphaël Couturier and Christophe Guyeux

Femto-ST Institute, University of Franche-Comte, France

19.1	Introduction		441
19.2	Basic reminders		443
	19.2.1	A short presentation of chaos	443
	19.2.2	On Devaney's definition of chaos	443
	19.2.3	Chaotic iterations	444
19.3	Toward efficiency and improvement for CI PRNG		445
	19.3.1	First efficient implementation of a PRNG based on chaotic iterations	445
	19.3.2	Efficient PRNGs based on chaotic iterations on GPU	446
	19.3.3	Naive version for GPU	446
	19.3.4	Improved version for GPU	447
	19.3.5	Chaos evaluation of the improved version	448
19.4	Experiments		449
19.5	Summary		450
	Bibliography		450

19.1 Introduction

Randomness is of importance in many fields such as scientific simulations or cryptography. Random numbers can mainly be generated by either a deterministic and reproducible algorithm called a pseudorandom number generator (PRNG), or by a physical nondeterministic process having all the characteristics of a random noise, called a truly random number generator (TRNG). In this chapter, we focus on reproducible generators, useful for instance in Monte Carlo-based simulators. These domains need PRNGs that are statistically irreproachable. In some fields such as in numerical simulations, speed is a strong requirement that is usually attained by using parallel architectures. In that case, a recurrent problem is that a deflation of the statistical qualities is often reported, when the parallelization of a good PRNG is realized. This is why ad hoc PRNGs for each possible architecture must be found to achieve both speed and randomness. On the other hand, speed is not the main requirement in cryptography: the most important point is to define *secure* generators able to withstand malicious attacks. Roughly speaking, an attacker should not be able in practice to make the distinction between numbers obtained with the secure generator and a true random sequence. Or, in an equivalent formulation, he or she should not be able (in practice) to predict the next bit of the generator, having the knowledge of all the binary digits that have been already released [9]. "Being able in practice" refers here to the possibility to achieve this attack in polynomial time and to the exponential growth of the difficulty of this challenge when the size of the parameters of the PRNG increases.

Finally, a small part of the community working in this domain focuses on a third requirement: to define chaotic generators [8, 11, 17]. The main idea is to benefit from a chaotic dynamical system to obtain a generator that is unpredictable, disordered, sensible to its seed, or in other words, chaotic. These scientists' desire is to map a given chaotic dynamics into a sequence that seems random and unassailable due to chaos. However, the chaotic maps used as a pattern are defined in the real line whereas computers deal with finite precision numbers. This distortion leads to a deflation of both chaotic properties and speed. Furthermore, authors of such chaotic generators often claim their PRNG as secure due to their chaos properties, but there is no obvious relation between chaos and security as it is understood in cryptography. This is why the use of chaos for PRNG still remains marginal and disputable. However, we have established in previous researches that these flaws can be circumvented by using a tool called choatic iterations. Such investigations have led to the definition of a new family of PRNGs that are chaotic while being fast and statistically perfect, or cryptographically secure [1, 2, 5, 6]. This family is improved and adapted for GPU architectures in this chapter.

Let us finish this introduction by noticing that, in this chapter, statistical perfection refers to the ability to pass the whole *BigCrush* battery of tests, which is widely considered as the most stringent statistical evaluation of a sequence claimed as random. This battery can be found in the well-known TestU01 package [12]. More precisely, each time we performed a test on a PRNG, we ran it twice in order to observe if all p-values were inside [0.01, 0.99]. In fact, we observed that few p-values (fewer than 10 out of 160) are sometimes outside this interval but inside [0.001, 0.999], so that is why a second run has allowed us to confirm that the values outside are not for the same test. With this approach all our PRNGs pass the *BigCrush* successfully and all p-values are at least once inside [0.01, 0.99]. Chaos, for its part, refers to the well-established definition of a chaotic dynamical system defined by Devaney [7].

The remainder of this chapter is organized as follows. Basic definitions and terminologies about both topological chaos and chaotic iterations are provided in the next section. Various chaotic iterations based on pseudorandom number generators are then presented in Section 19.3. They encompass naive and improved efficient generators for CPU and for GPU. These generators are finally experimented in Section 19.4.

19.2 Basic reminders

This section is devoted to basic definitions and terminologies in the fields of topological chaos and chaotic iterations. We assume the reader is familiar with basic notions on topology (see for instance [7]).

19.2.1 A short presentation of chaos

Chaos theory studies the behavior of dynamical systems that are perfectly predictable, yet appear to be wildly amorphous and meaningless. Chaotic systems are highly sensitive to initial conditions, which is popularly referred to as the butterfly effect. In other words, small differences in initial conditions (such as those due to rounding errors in numerical computation) yield widely diverging outcomes, in general rendering long-term prediction impossible [11]. This happens even though these systems are deterministic, meaning that their future behavior is fully determined by their initial conditions, with no random elements involved [11]. That is, the deterministic nature of these systems does not make them predictable [11, 16]. This behavior is known as deterministic chaos, or simply chaos. It has been well-studied in mathematics and physics, leading among other things to the well-established definition of Devaney which can be found next.

19.2.2 On Devaney's definition of chaos

Consider a metric space $(\mathcal{X}, d)$ and a continuous function $f : \mathcal{X} \longrightarrow \mathcal{X}$, for one-dimensional dynamical systems of the form:

$$x^0 \in \mathcal{X} \text{ and } \forall n \in \mathbb{N}^*, x^n = f(x^{n-1}), \tag{19.1}$$

the following definition of chaotic behavior, formulated by Devaney [7], is widely accepted.

Definition 1 *A dynamical system of Form (19.1) is said to be chaotic if the following conditions hold.*

- *Topological transitivity:*

$$\forall U, V \text{ open sets of } \mathcal{X}, \exists k > 0, f^k(U) \cap V \neq \varnothing. \tag{19.2}$$

Intuitively, a topologically transitive map has points that eventually move under iteration from one arbitrarily small neighborhood to any other. Consequently, the dynamical system cannot be decomposed into two disjoint open sets that are invariant under the map. Note that if a map possesses a dense orbit, then it is clearly topologically transitive.

- Density of periodic points in $\mathcal{X}$:

  Let $P = \{p \in \mathcal{X} | \exists n \in \mathbb{N}^* : f^n(p) = p\}$ the set of periodic points of f. Then P is dense in $\mathcal{X}$:

  $$\overline{P} = \mathcal{X}. \tag{19.3}$$

  The density of periodic orbits means that every point in space is closely approached by periodic orbits in an arbitrary way. Topologically mixing systems failing this condition may not display sensitivity to initial conditions presented below and, hence, may not be chaotic.

- Sensitive dependence on initial conditions:

  $\exists \varepsilon > 0, \forall x \in \mathcal{X}, \forall \delta > 0, \exists y \in \mathcal{X}, \exists n \in \mathbb{N}, d(x,y) < \delta$ and $d\left(f^n(x), f^n(y)\right) \geqslant \varepsilon$.

  Intuitively, a map possesses sensitive dependence on initial conditions if there exist points arbitrarily close to x that eventually separate from x by at least ε under the iteration of f. Not all points near x need to be eventually separate from x under iteration, but there must be at least one such point in every neighborhood of x. If a map possesses sensitive dependence on initial conditions, then for all practical purposes, the dynamics of the map defy numerical computation. Small errors in computation that are introduced by round-off may become magnified upon iteration. The results of numerical computation of an orbit, no matter how accurate, may bear no resemblance whatsoever with the real orbit.

When f is chaotic, then the system $(\mathcal{X}, f)$ is chaotic and quoting Devaney [7, p. 50]: "it is unpredictable because of the sensitive dependence on initial conditions. It cannot be broken down or decomposed into two subsystems which do not interact because of topological transitivity. And, in the midst of this random behavior, we nevertheless have an element of regularity." Fundamentally different behaviors are consequently possible and occur in an unpredictable way.

19.2.3 Chaotic iterations

Let us now introduce an example of a dynamical systems family that has the potentiality to become chaotic, depending on the choice of the iteration function. This family is the basis of the PRNGs we will develop during this chapter.

Definition 2 *The set $\mathbb{B}$ denoting $\{0,1\}$, let $f : \mathbb{B}^\mathsf{N} \longrightarrow \mathbb{B}^\mathsf{N}$ be an "iteration" function and S be a sequence of subsets of $[\![1, \mathsf{N}]\!]$ called a chaotic strategy. Then, the so-called* chaotic iterations *are defined by [15]:*

$$\begin{cases} x^0 \in \mathbb{B}^\mathsf{N}, \\ \forall n \in \mathbb{N}^*, \forall i \in [\![1; \mathsf{N}]\!], x_i^n = \begin{cases} x_i^{n-1} & \text{if } i \notin S^n \\ f(x^{n-1})_i & \text{if } i \in S^n. \end{cases} \end{cases} \quad (19.4)$$

In other words, at the nth iteration, only the cells of S^n are iterated. Chaotic iterations generate a set of vectors; they are defined by an initial state x^0, an iteration function f, and a chaotic strategy S [3]. These "chaotic iterations" can behave chaotically as defined by Devaney, depending on the choice of f [3]. For instance, chaos is obtained when f is the vectorial negation. Note that, with this example of function, chaotic iterations defined above can be rewritten as

$$x^0 \in [\![0, 2^\mathsf{N} - 1]\!], \ \mathcal{S}^n \in \mathcal{P}\left([\![1, 2^\mathsf{N} - 1]\!]\right)^\mathbb{N}, \ x^{n+1} = x^n \oplus \mathcal{S}^n, \quad (19.5)$$

where $\mathcal{P}(X)$ stands for the set of subsets of X, whereas $a \oplus b$ is the bitwise exclusive or operation between the binary representation of the integers a and b. Note that the term $\mathcal{S}^n$ is directly and obviously linked to the S^n of Eq. 19.4. Such an iterative sequence satisfies the Devaney's definition of chaos.

19.3 Toward efficiency and improvement for CI PRNG

19.3.1 First efficient implementation of a PRNG based on chaotic iterations

Listing 19.1. C code of the sequential PRNG based on chaotic iterations

```
{
unsigned int CIPRNG() {
  static unsigned int x = 123123123;
  unsigned long t1 = xorshift();
  unsigned long t2 = xor128();
  unsigned long t3 = xorwow();
  x = x^(unsigned int)t1;
  x = x^(unsigned int)(t2>>32);
  x = x^(unsigned int)(t3>>32);
  x = x^(unsigned int)t2;
  x = x^(unsigned int)(t1>>32);
  x = x^(unsigned int)t3;
  return x;
}
```

In Listing 19.1 a sequential version of the proposed PRNG based on chaotic iterations is presented, which extends the generator family formerly presented in [4,5]. The xor operator is represented by ^. This function uses three classical 64-bit PRNGs, namely the xorshift, the xor128, and the xorwow [13]. In the following, we call them "xor-like PRNGs". As each xor-like PRNG uses 64-bits whereas our proposed generator works with 32-bits, we use the command (unsigned int), which selects the 32 least significant bits of a given integer, and the code (unsigned int)(t>>32) in order to obtain the 32 most significant bits of t.

Thus producing a pseudorandom number needs 6 xor operations with 6 32-bit numbers that are provided by 3 64-bit PRNGs. This version successfully passes the stringent *BigCrush* battery of tests [12]. At this point, we have defined an efficient and statistically unbiased generator. Its speed is directly related to the use of linear operations, but for the same reason, this fast generator cannot be proven as secure.

19.3.2 Efficient PRNGs based on chaotic iterations on GPU

In order to benefit from the computing power of GPU, a program needs to have independent blocks of threads that can be computed simultaneously. In general, the larger the number of threads is, the more local memory is used, and the less branching instructions are used (if, while, etc.) and so, the better the performances on GPU are. Obviously, having these requirements in mind, it is possible to build a program similar to the one presented in Listing 19.1, which computes pseudorandom numbers on GPU. To do so, we must first recall that in the CUDA [14] environment, threads have a local identifier called ThreadIdx, which is relative to the block containing them. Furthermore, in CUDA, parts of the code that are executed by the GPU are called *kernels*.

19.3.3 Naive version for GPU

It is possible to deduce from the CPU version a quite similar version adapted to GPU. The simple principle consists of making each thread of the GPU compute the CPU version of our PRNG. Of course, the three xor-like PRNGs used in these computations must have different parameters. In a given thread, these parameters are randomly picked from another PRNGs. The initialization stage is performed by the CPU. To do this, the Indirection, Shift, Accumulate, Add, and Count (ISAAC) PRNG [10] is used to set all the parameters embedded into each thread.

The implementation of the three xor-like PRNGs is straightforward when their parameters have been allocated in the GPU memory. Each xor-like works with an internal number x that saves the last generated pseudorandom number. Additionally, the implementation of the *xor128*, the *xorshift*, and the *xorwow*, respectively, require 4, 5, and 6 unsigned long as internal variables.

Algorithm 23: main kernel of the GPU "naive" version of the PRNG based on chaotic iterations

Input: InternalVarXorLikeArray: array with internal variables of the 3 xor-like PRNGs in global memory;
NumThreads: number of threads;
Output: NewNb: array containing random numbers in global memory

1 **if** *threadIdx is concerned by the computation* **then**
2   retrieve data from InternalVarXorLikeArray[threadIdx] in local variables;
3   **for** *i=1 to n* **do**
4   compute a new PRNG as in Listing 19.1;
5   store the new PRNG in NewNb[NumThreads*threadIdx+i];
6   **end**
7   store internal variables in InternalVarXorLikeArray[threadIdx];
8 **end**

Algorithm 23 presents a naive implementation of the proposed PRNG on GPU. Due to the available memory in the GPU and the number of threads used simultaneously, the number of random numbers that a thread can generate inside a kernel is limited (i.e., the variable n in Algorithm 23). For instance, if 100,000 threads are used and if $n = 100$[1], then the memory required to store all of the internal variables of both the xor-like PRNGs[2] and the pseudorandom numbers generated by our PRNG, is equal to $100,000 \times ((4 + 5 + 6) \times 2 + (1 + 100)) = 1,310,000$ 32-bit numbers, that is, approximately 52Mb.

This generator is able to pass the whole *BigCrush* battery of tests, for all the versions that have been tested depend on their number of threads (called NumThreads in our algorithm, tested up to 5 million).

19.3.4 Improved version for GPU

As GPU cards using CUDA have shared memory between threads of the same block, it is possible to use this feature in order to simplify the previous algorithm, i.e., to use fewer than 3 xor-like PRNGs. The solution consists in computing only one xor-like PRNG by thread, saving it into the shared memory, and then using the results of some other threads in the same block of threads. In order to define which thread uses the result of which other one, we can use a combination array that contains the indexes of all threads and for which a combination has been performed.

In Algorithm 24, two combination arrays are used. The variable offset is computed using the value of combination_size. Then we can compute o1 and o2 representing the indexes of the other threads whose results are used by the current one. In this algorithm, we consider that a 32-bit xor-like PRNG has

---
[1] In fact, we need to add the initial seed (a 32-bit number).
[2] We multiply this number by 2 in order to count 32-bit numbers.

been chosen. In practice, we use the xor128 proposed in [13] in which unsigned longs (64 bits) have been replaced by unsigned integers (32 bits).

This version can also pass the whole *BigCrush* battery of tests.

Algorithm 24: main kernel for the chaotic iterations based PRNG GPU efficient version

Input: InternalVarXorLikeArray: array with internal variables of 1 xor-like PRNGs in global memory;
NumThreads: Number of threads;
array_comb1, array_comb2: Arrays containing combinations of size combination_size;
Output: NewNb: array containing random numbers in global memory

1 **if** *threadId is concerned* **then**
2   retrieve data from InternalVarXorLikeArray[threadIdx] in local variables including shared memory and x;
3   offset = threadIdx%combination_size;
4   o1 = threadIdx-offset+array_comb1[offset];
5   o2 = threadIdx-offset+array_comb2[offset];
6   **for** *i=1 to n* **do**
7   t=xor-like();
8   t=t^shmem[o1]^shmem[o2];
9   shared_mem[threadId]=t;
10   x = x^t;
11   store the new PRNG in NewNb[NumThreads*threadIdx+i];
12   **end**
13   store internal variables in InternalVarXorLikeArray[threadIdx];
14 **end**

19.3.5 Chaos evaluation of the improved version

A run of Algorithm 24 consists of an operation ($x = x \oplus t$) having the form of Equation 19.5, which is equivalent to the iterative system of Eq. 19.4. That is, an iteration of the general chaotic iterations is realized between the last stored value x of the thread and a strategy t (obtained by a bitwise exclusive or between a value provided by a xor-like() call and two values previously obtained by two other threads). To be certain that such iterations correspond to the chaotic one recalled at the end of Section 19.2.2, we must guarantee that this dynamical system iterates on the space $\mathcal{X} = \mathbb{B}^N \times \mathcal{P}\left(\llbracket 1, 2^N \rrbracket\right)^\mathbb{N}$. The left term x obviously belongs to $\mathbb{B}^N$. To prevent any flaws of chaotic properties, we must check that the right term (the last t in Algorithm 24), corresponding to the strategies, can possibly be equal to any integer of $\llbracket 1, 2^N \rrbracket$.

Such a result is obvious; as for the xor-like(), all the integers belonging to its interval of definition can occur at each iteration, and thus the last t respects the requirement. Furthermore, it is possible to prove by an immediate mathematical induction that, supposing that the initial x is uniformly distributed, the two other stored values shmem[o1] and shmem[o2] are uniformly distributed too (this is the induction hypothesis), and thus the next x is finally uniformly distributed.

Thus Algorithm 24 is a concrete realization of the general chaotic iterations presented previously, and for this reason, it satisfies the Devaney's formulation of chaotic behavior.

19.4 Experiments

Different experiments have been performed in order to measure the generation speed. We have used one computer equipped with a Tesla C1060 NVIDIA GPU card and an Intel Xeon E5530 cadenced at 2.40 GHz, and a second computer equipped with a smaller CPU and a GeForce GTX 280. All the cards have 240 cores.

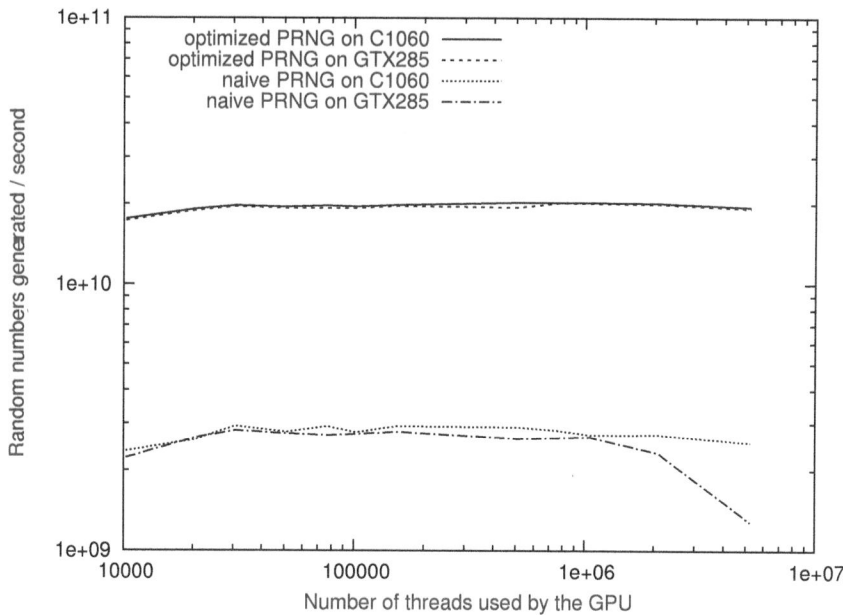

FIGURE 19.1. Quantity of pseudorandom numbers generated per second with the xorlike-based PRNG.

In Figure 19.1 we compare the quantity of pseudorandom numbers generated per second with various xor-like based PRNGs. In this figure, the optimized versions use the xor64 described in [13], whereas the naive versions embed the three xor-like PRNGs described in Listing 19.1. In order to obtain the optimal performances, the storage of pseudorandom numbers to the GPU memory has been removed. This step is time-consuming and slows down the numbers generation. Moreover this storage is completely useless, in case of applications that consume the pseudorandom numbers directly after they have been generated. We can see that when the number of threads is greater than approximately 30,000 and less than 5 million, the number of pseudorandom numbers generated per second is almost constant. With the naive version, this value ranges from 2.5 to 3GSamples/s. With the optimized version, it is approximately equal to 20GSamples/s. Finally we can remark that both GPU cards are quite similar, but in practice, the Tesla C1060 has more memory than the GTX 280, and this memory should be of better quality. As a comparison, Listing 19.1 leads to the generation of about 138MSample/s when using one core of the Xeon E5530.

These experiments allow us to conclude that it is possible to generate a very large quantity of pseudorandom numbers statistically perfect with the xor-like version.

19.5 Summary

In this chapter, a PRNG based on chaotic iterations is presented. It is proven to be chaotic according to Devaney. Efficient implementations on GPU using xor-like PRNGs as input generators have shown that a very large quantity of pseudorandom numbers can be generated per second (about 20Gsamples/s on a Tesla C1060), and that these proposed PRNGs succeed in passing the hardest battery in TestU01, namely, the *BigCrush*.

Bibliography

[1] J. Bahi, X. Fang, and C. Guyeux. An optimization technique on pseudorandom generators based on chaotic iterations. In *INTERNET'2012, 4th Int. Conf. on Evolving Internet*, pages 31–36, Venice, Italy, June 2012.

[2] J. Bahi, X. Fang, C. Guyeux, and Q. Wang. Evaluating quality of chaotic pseudo-random generators. application to information hiding. *IJAS, International Journal On Advances in Security*, 4(1-2):118–130, 2011.

[3] J. Bahi and C. Guyeux. Hash functions using chaotic iterations. *Journal of Algorithms and Computational Technology*, 4(2):167–181, 2010.

[4] J. Bahi and C. Guyeux. Topological chaos and chaotic iterations, application to hash functions. *Neural Networks (IJCNN2010)*, pages 1–7, July 2010.

[5] J Bahi, C. Guyeux, and Q. Wang. A novel pseudo-random generator based on discrete chaotic iterations. In *INTERNET'09, 1-st Int. Conf. on Evolving Internet*, pages 71–76, Cannes, France, August 2009.

[6] J. Bahi, C. Guyeux, and Q. Wang. A pseudo random numbers generator based on chaotic iterations. application to watermarking. In *WISM 2010, Int. Conf. on Web Information Systems and Mining*, volume 6318 of *LNCS*, pages 202–211, Sanya, China, October 2010.

[7] R. L. Devaney. *An Introduction to Chaotic Dynamical Systems*. Addison-Wesley, Redwood City, CA, 2nd edition, 1989.

[8] J. Gleick. *Chaos: Making a New Science*. Open Road, 2011.

[9] O. Goldreich. *Foundations of Cryptography: Basic Tools*. Cambridge University Press, 2007.

[10] R. J. Jenkins. ISAAC. In *IWFSE: International Workshop on Fast Software Encryption*, volume 1039 of *Lecture Notes in Computer Science*, pages 41–49. Springer Verlag, 1996.

[11] S. H. Kellert. *In the Wake of Chaos: Unpredictable Order in Dynamical Systems*. Science and Its Conceptual Foundations. University of Chicago Press, 1994.

[12] P. L'Ecuyer and R. J. Simard. TestU01: A C library for empirical testing of random number generators. *ACM Trans. Math. Softw*, 33(4):22–40, 2007.

[13] G. Marsaglia. Xorshift rngs. *Journal of Statistical Software*, 8(14):1–6, 2003.

[14] NVIDIA. *CUDA CUBLAS library*, 2011. Version 4.0.

[15] F. Robert. *Discrete Iterations: A Metric Study*, volume 6 of *Springer Series in Computational Mathematics*. 1986.

[16] C. Werndl. What are the new implications of chaos for unpredictability? *The British Journal for the Philosophy of Science*, 60(1):195–220, 2009.

[17] C. Wu, Z. Xu, W. Lin, and J. Ruan. Stochastic Properties in Devaney's Chaos. *Chaos, Solitons and Fractals*, 23(4):1195 – 1199, 2005.

Chapter 20

Solving large sparse linear systems for integer factorization on GPUs

Bertil Schmidt and Hoang-Vu Dang

Johannes Gutenberg University of Mainz, Germany

20.1	Introduction	453
20.2	Block Wiedemann algorithm	454
20.3	SpMV over GF(2) for NFS matrices using existing formats on GPUs	455
20.4	A hybrid format for SpMV on GPUs	459
	20.4.1 Dense format	459
	20.4.2 Sliced COO	460
	20.4.3 Determining the cut-off point of each format	464
20.5	SCOO for single-precision floating-point matrices	465
20.6	Performance evaluation	466
	20.6.1 Experimental setup	466
	20.6.2 Experimental results of SpMV on NFS matrix	466
	20.6.3 Experimental results of SpMV on floating-point matrices	467
	20.6.3.1 Performance comparison to existing GPU formats	467
	20.6.3.2 Performance comparison to a CPU implementation	469
20.7	Conclusion	470
	Bibliography	471

20.1 Introduction

The Number Field Sieve (NFS) is the current state-of-the-art integer factorization method. It requires the solution of a large sparse linear system over Galois Field GF(2) (called the linear algebra step). The Block Wiedemann (BW) [8] algorithm can be used to solve such a large sparse linear system efficiently using iterative sparse matrix vector multiplication (SpMV).

Recent integer factorization efforts have been using CPU clusters to solve the large sparse linear system [1, 17]. The RSA-768 factorization [17], for example, reported a runtime of 3 months for the linear algebra step on a cluster with 48 AMD dual hex-core CPUs. Previous work on parallelizing the linear algebra step focused on using CPU clusters and grids [2, 11, 12, 18]. In this chapter, we present a CUDA approach that can be used to accelerate the costly iterative SpMV operation for matrices derived from NFS.

The memory access pattern in the SpMV operation generally consists of regular access patterns over the matrix and irregular access patterns over the vector. The irregular access pattern over the vector is a challenge that is more pronounced on the GPU than on the CPU, because of the smaller cache and the restrictive memory access pattern requirement to achieve maximum performance. However, a high-end GPU has an order-of-magnitude higher bandwidth than a high-end CPU; e.g., a GeForce GTX 580 has 192.4 GB/s memory bandwidth, while an Intel Core-i7 has a maximum of 25.6 GB/s memory bandwidth.

SpMV on the GPU has been explored previously in several papers [4,6,7,14] for matrices derived from scientific computing applications. However, sparse matrices derived from NFS have generally different properties, i.e. they are larger, have a few dense rows, and have many extremely sparse rows. The large size of the matrix causes the BL and BW algorithms to require a large number of SpMV iterations. This means that the time spent for matrix preprocessing and matrix data transfer to the GPU memory is negligible compared to the total runtime. Thus, approaches to SpMV on GPUs for NFS matrices may be different from previously published GPU SpMV approaches. We further present an extension of our SpMV method for binary-valued NFS matrices to single-precision floating-point matrices.

20.2 Block Wiedemann algorithm

The BW algorithm heuristically finds n vectors in the kernel space of a $d \times d$ binary matrix B; n is one of two parameters m, n, called blocking factors. BW consists of the following steps:

- **Step 1 (BW1):** Compute the matrix sequence

$$A_i = x \cdot B^i \cdot y, \forall i = 1, ..., \frac{d}{m} + \frac{d}{n} + O(1), \quad (20.1)$$

where x, y are randomly chosen binary matrices of size $m \times d$ and $d \times n$, respectively.

- **Step 2 (BW2):** The Berlekamp-Massey algorithm [19] is used to compute a generating polynomial of the matrix sequence A from BW1 in the form

$$F(X) = \sum_{i=1}^{\frac{d}{n}+O(1)} C_i \cdot X^i, \qquad (20.2)$$

where C_i is an $n \times n$ binary matrix

- **Step 3 (BW3):** Compute the sequence of matrices S of size $d \times n$ such that

$$S_i = B^i \cdot y \cdot C_i, \forall i = 1, ..., \frac{d}{n} + O(1). \qquad (20.3)$$

With high probability, $B \cdot \sum S_i = 0$, for which $\sum S_i$ is output as a solution.

We can treat x, y, and S_i as vectors of block width m or n. Assuming B is a sparse matrix with γ nonzeros per row on average and ignoring m and n as constant, the complexity of BW1 and BW3 is $O(\gamma d^2)$. Using the subquadratic algorithm by Thomé [19], BW2 has a complexity of $O(d \log^2 d)$. Thus our approach to accelerate BW is based on accelerating BW1 and BW3 on a GPU while BW2 is still done on a CPU.

20.3 SpMV over GF(2) for NFS matrices using existing formats on GPUs

In this section, we review a few relevant previously published sparse matrix formats on GPUs and study their performance when applied to sparse matrices over GF(2) derived from integer factorization with NFS.

In Algorithm 25, we consider the sparse binary matrix B of size $d \times d$ and a dense vector X of size $d \times n$ bit, where n is called the *blocking factor*. Modern processors generally support 64-bit operations. Thus, typical blocking factors are of the form of $64 \cdot k$, $k \in \mathbb{N}$. Note that doubling the blocking factor roughly halves the number of SpMV iterations required but doubles the input vector size.

For all $0 \leq i \leq d-1$ let $c\_index[i]$ column index array of $B[i]$ which contains the indices of the nonzero entries of row i. Then, the following pseudocode shows a single SpMV iteration of B with input vector X and result vector Y.

The costly operations in the SpMV pseudocode are the memory accesses for loading ind, $X[ind]$, $Y[i]$ and storing $Y[i]$. To speed up those operations on any architecture a common approach is to design a cache-friendly order of accessing the memory. The order is especially important for the vectors X and Y, since their memory locations might be accessed multiple times. Accesses to Y can be minimized by storing the intermediate results in fast memory and only write the accumulated result to Y.

Algorithm 25: SpMV-column-index

Input: $c\_index[i]$: column array of rows i of B
Input: X: input vector
Output: $Y = B \cdot X$

1 **for** $i = 0$ *to* $d - 1$ **do**
2    $Y[i] = 0$;
3   **forall the** $ind \in c\_index[i]$ **do**
4       $Y[i] = Y[i] \oplus X[ind]$;
5       $\oplus$: bitwise XOR operation
6   **end**
7 **end**

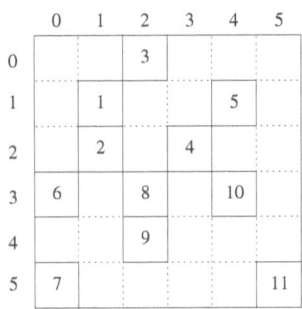

FIGURE 20.1. An example square matrix of size 6×6 (zero values are not shown).

CUDA implementations of SpMV generally store both matrix and vectors in *global memory*. Memory accesses to B and Y can usually be coalesced. Memory accesses to X are random and noncoalesced, thus having higher latency, but *texture memory* can be used to take advantage of the *texture cache*. Intermediate results could utilize *shared memory*, which has low latency but is very limited in size and only shared between threads of the same block. Shared memory is further divided into *banks*. When threads in a warp access the shared memory, *bank conflicts* should be avoided, otherwise the access will be serialized.

We now briefly review the CUDA implementation of a number of SpMV matrix formats published in previous papers [4,14]. We also describe the data structure in which the matrix in Figure 20.1 is stored using different formats as examples. Please note that when storing a binary matrix, the *value* array can be removed.

Coordinate list (COO)

For each nonzero, both its column and row indices are explicitly stored. The Cusp implementation [3] stores elements in sorted order of row indices ensuring that entries with the same row index are stored contiguously.

```
coo.row_index = {0, 1, 1, 2, 2, 3, 3, 3, 4, 5, 5}
coo.col_index = {2, 1, 4, 1, 3, 0, 2, 4, 2, 0, 5}
coo.value     = {3, 1, 5, 2, 4, 6, 8, 10, 9, 7, 11}
```

Compressed sparse row (CSR)

Nonzeros are sorted by the row index, and only their column indices are explicitly stored in a column array. Additionally, the vector *row_start* stores indices of the first nonzero element of each row in the column array.

```
csr.row_start = {0, 1, 3, 5, 8, 9, 12}
csr.col_index = {2, 1, 4, 1, 3, 0, 2, 4, 2, 0, 5}
csr.value     = {3, 1, 5, 2, 4, 6, 8, 10, 9, 7, 11}
```

Ellpack (ELL)

Let K be the maximum number of nonzero elements in any row of the matrix. Then, for each row, ELL stores exactly K elements (extra padding is required for rows that contain fewer than K nonzero elements). Only column indices are required to store in an array, the row index can be implied since exactly K elements are stored per row. The Cusp implementation stores the column indices in a transposed manner so that consecutive threads can access consecutive memory addresses.

```
ell.col_index = {
                 2, 1, 1, 0, 2, 0,
                 *, 4, 3, 2, *, 5,
                 *, *, *, 4, *, *}
ell.value     = {
                 3, 1, 2, 6, 9, 7,
                 *, 5, 4, 8, *, 11,
                 *, *, *, 10, *, *}
```

Hybrid (HYB)

The HYB format heuristically computes a value K and stores K nonzeros per rows in the ELL format. When a row has more than K nonzeros, the trailing nonzeros are stored in COO. This design decreases the storage overhead due to ELL padding elements and thus improves the overall performance.

```
hyb.nnz_per_row  = 2
hyb.ell.col_index = {2, 1, 1, 0, 2, 0, *, 4, 3, 2, *, 5}
hyb.ell.value    = {3, 1, 2, 6, 9, 7, *, 5, 4, 8, *, 11}
hyb.coo.row_index = {3}
hyb.coo.col_index = {4}
hyb.coo.value    = {10}
```

Sliced Ellpack (SLE)

This format partitions the matrix into horizontal slices of S adjacent rows [14]. Each slice is stored in ELLPACK format. The maximum number of nonzeros may be different for each slice. An additional array *slice_start* is used to index the first element in each slice. The matrix rows are usually sorted by the number of nonzeros per row in order to move rows with similar number of nonzeros together.

```
sle.slice_size  = 2
sle.col_index   = {
                  2, 1, *, 4,
                  1, 0, 3, 2, *, 4,
                  2, 0, *, 5}
sle.value       = {
                  3, 1, *, 5,
                  2, 6, 4, 8, *, 10
                  9, 7, *, 11}
sle.slice_start = {0, 4, 10, 14}
```

	RSA-170	RSA-190	KILOBIT	RSA-768
Max dimension	10.4M	26.1M	66.7M	192M
Nonzeros	994.7M	2,451M	9,538M	27,797M
Max row weight	5.5M	14M	28.2M	82.6M
Min row weight	3	3	2	2
Avg row weight	95.08	93.6	143	144

TABLE 20.1. Properties of some NFS matrices.

The existing formats do not achieve good performance due to the special structure of NFS matrices. Row weights of an NFS matrix have a very wide range (see Table 20.1). Hence using one warp or one thread per row (as in CSR or ELLPACK) results in an unbalanced workload. Moreover, the NFS matrices are highly unstructured, which does not facilitate cache-friendly vector access patterns using existing formats. These reasons have motivated our design of a new format.

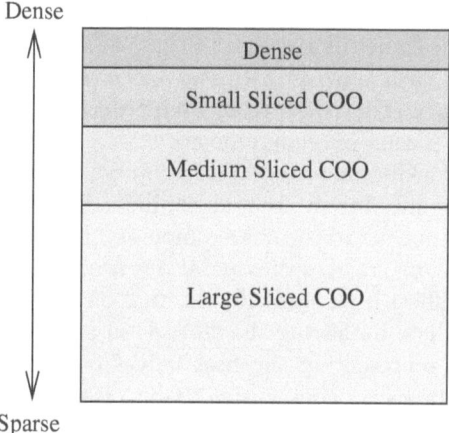

FIGURE 20.2. Partitioning of a row-sorted NFS matrix into four formats.

20.4 A hybrid format for SpMV on GPUs

As a preprocessing step, we reorder the rows of the matrix by their *row weight*, in nonincreasing order. The row weight of row j of B is defined as the total number of nonzero elements in row j. We then partition the sorted matrix rows into at most four consecutive parts. Each part uses a different format. The different formats are optimized for the sparseness properties of each partition as shown in Figure 20.2. For the densest part, we use a dense format. When the matrix gets less dense, we switch to another format which we call Sliced COO (SCOO). SCOO has three variants, small, medium, and large. Our formats are now described in more detail.

20.4.1 Dense format

The dense format is used for the dense part of the matrix. This format uses 1 bit per matrix entry. Within a column, 32 matrix entries are stored as a 32-bit integer. Thus, 32 rows are stored as N consecutive integers.

Each CUDA thread works on a column. Each thread fetches one element from the input vector in coalesced fashion. Then, each thread checks the 32 matrix entries one by one. When the matrix entry is a nonzero, the thread performs an XOR operation between the element from the input vector and the partial result for the row. This means each thread accesses the input vector only once to do work on up to 32 nonzeros. The partial result from each thread needs to be stored and combined to get the final result for the 32 rows. These operations are performed in CUDA shared memory.

The 32 threads in a warp share 32 shared memory entries to store the partial results from the 32 rows. Since all threads in a warp execute a common instruction at the same time, access to these 32 entries can be made exclusive. The result from each warp in a thread block is combined using p-reduction on shared memory. The result from each thread block is combined using an atomic XOR operation on global memory.

When the blocking factor is larger than 64, access to the shared memory needs to be reorganized to avoid bank conflicts. Each thread can read/write up to 64-bit data at a time to the shared memory. If a thread is accessing 128-bit data for example, two read/write operations need to be performed. Thus, there will be bank conflicts if we store 128-bit data on contiguous addresses. We can avoid bank conflicts by having the threads in a warp first access consecutive 64-bit elements representing the first halves of the 128-bit elements. Then, the threads again access consecutive 64-bit elements representing the second halves of the 128-bit elements. The same modification can be applied to other formats as well.

20.4.2 Sliced COO

The SCOO format is adapted from the CADO-NFS software for CPUs [15]. The aim is to reduce irregular accesses to the input vector and increase the texture cache hit rate. SCOO stores the column index and the row index of each nonzero. A number of consecutive rows form a slice. Nonzeros within a slice are sorted by their column index. We give an example of SCOO in Figure 20.3, threads working on a slice can access contiguous elements from the input vector. In the figure, S1, S2 are two slices. Entries in the input vector denote the corresponding nonzero elements with which the input vector element is multiplied.

For each nonzero, two bytes are used to store the column index. However, two bytes is not enough for large RSA matrices. Thus, we further divide a slice into groups. Group i contains nonzeros with column index between $i \times 2^{16}$ and $(i+1) \times 2^{16} - 1$. An additional array stores the starting position of each group in the slice.

Solving large sparse linear systems for integer factorization on GPUs 461

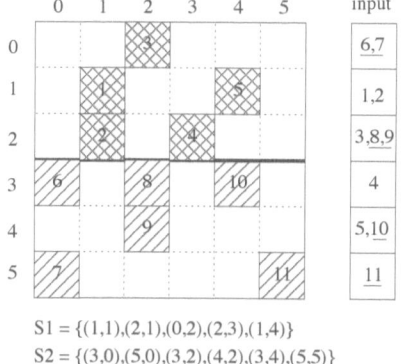

S1 = {(1,1),(2,1),(0,2),(2,3),(1,4)}
S2 = {(3,0),(5,0),(3,2),(4,2),(3,4),(5,5)}

FIGURE 20.3. Example of the memory access pattern for a 6×6 matrix stored in sliced COO format (slice size = 3 rows).

Listing 20.1. SpMV with SCOO format

```
// compute y = B*x (B is stored in SCOO formats [ cols, rows, values,
//offsets, numPacks, numRows ])
// LANE_SIZE = 2^k
// NUM_ROWS_PER_SLICE is computed based on sparsity
template <const uint32_t THREADS_PER_BLOCK, const uint32_t
    NUM_ROWS_PER_SLICE, const uint32_t LANE_SIZE>
__global__ void
sliced_coo_kernel(
                const uint32_t numRows,
                const uint32_t numPacks,
                const uint32_t * cols,
                const uint16_t * rows,
                const float * values,
                const uint32_t * offsets,
                const float * x,
                float * y)
{
    // ~ threadIdx.x % LANE_SIZE
    const int thread_lane = threadIdx.x & (LANE_SIZE-1);

    const int row_lane = threadIdx.x/(LANE_SIZE);

    __shared__ float sdata[NUM_ROWS_PER_SLICE][LANE_SIZE];

    const uint32_t packNo=blockIdx.x;
    const uint32_t limit = ( (packNo==numPacks-1)?((numRows-1)%
        NUM_ROWS_PER_SLICE)+1:NUM_ROWS_PER_SLICE );

    const uint32_t begin = offsets[packNo];
    const uint32_t end = offsets[packNo+1];
    for(int i=row_lane; i<limit; i+=THREADS_PER_BLOCK/LANE_SIZE)
        sdata[i][thread_lane] = 0;

    __syncthreads();

    for(int32_t index=begin+threadIdx.x; index<end; index+=
        THREADS_PER_BLOCK){
        const uint32_t col = cols[index];
        const uint16_t row = rows[index];
        const float value = values[index];
```

```
            // fetch x[col] from texture cache
40          const float input = fetch_x(col, x) * value;
            atomic_add(&sdata[row][thread_lane], input);
        }

        __syncthreads();
45
        // Parallel reduction
        // Sum of row i is available in column i%LANE_SIZE
        for (uint32_t i=row_lane; i<limit; i+=THREADS_PER_BLOCK/LANE_SIZE)
        {
            float *p = sdata[i];
50          int des = (thread_lane+i) & (LANE_SIZE-1);

            // assuming lane size <= 32
            if (LANE_SIZE>16 && thread_lane<16)
                p[des]+=p[(des+16)&(LANE_SIZE-1)]; __syncthreads();
55          if (LANE_SIZE>8 && thread_lane<8)
                p[des]+=p[(des+8)&(LANE_SIZE-1)]; __syncthreads();
            if (LANE_SIZE>4 && thread_lane<4)
                p[des]+=p[(des+4)&(LANE_SIZE-1)]; __syncthreads();
            if (LANE_SIZE>2 && thread_lane<2)
60              p[des]+=p[(des+2)&(LANE_SIZE-1)]; __syncthreads();
            if (LANE_SIZE>1 && thread_lane<1)
                p[des]+=p[(des+1)&(LANE_SIZE-1)]; __syncthreads();
        }

65      __syncthreads();
        const uint32_t actualRow = packNo * NUM_ROWS_PER_SLICE;

        for(int r = threadIdx.x; r < limit; r+=THREADS_PER_BLOCK)
            y[actualRow+r] = sdata[r][thread_lane];
70  }
```

One thread block works on a slice, one group at a time. Neighboring threads work on neighboring nonzeros in the group in parallel. Each thread works on more than one row of the matrix. Thus, each thread needs some storage to store the partial results and combine them with the results from the other threads to generate the final output. Each entry costs equally as the element of the vector, i.e., the blocking factor in bytes. Shared memory is used as intermediate storage as it has low latency however it is limited to 48 KB per SM in Fermi. The global memory is accessed only once at the end to write the final output. The CUDA kernel for SpMV with Sliced COO format is shown in Listing 20.1.

Since neighboring nonzeros may or may not come from the same row (recall that we sorted them by column index), many threads may access the same entry in the shared memory if it is used to store result of the same rows. Thus, shared memory entries are either assigned exclusively to a single thread or shared by using atomic XOR operations. Based on the way we allocate the shared memory, we further divide the Sliced COO format into three different subformats: small, medium, and large.

Small sliced (SS) COO

In this subformat, each thread has one exclusive entry in shared memory to store the partial result for each row. The assignment of the shared memory is organized such that each thread in a warp accesses only one bank and there

Sliced COO subformats	Small	Medium	Large
Memory sharing	No sharing	Among warp	Among block
Access method	Direct	Atomic XOR	Atomic XOR
Bank conflict	No	No	Yes
# Rows per Slice	12	192	6144

TABLE 20.2. Sliced COO subformat comparison (# rows per slices is based on $n = 64$).

is no bank accessed by more than one thread. Thus, there is no bank conflict. A p-reduction operation on shared memory is required to combine partial results from each thread.

The maximum number of rows per slice is calculated as *size of shared memory per SM in bits / (number of threads per block * blocking factor)*. We use 512 threads per block for 64-bit blocking factor which gives 12 rows, and 256 threads per block for 128 and 256-bit blocking factor, which gives 12 and 6 rows, respectively. Hence, one byte per row index is sufficient for this subformat.

Medium sliced (MS) COO

In this subformat, each thread in a warp gets an entry in the shared memory to store the partial result for each row. However, this entry is shared with the threads in other warps. Access to the shared memory uses an atomic XOR operation. Each thread in a warp accesses only one bank, avoiding bank conflicts. A p reduction operation on shared memory is required to combine the 32 partial results.

The maximum number of rows per slice is calculated as *size of shared memory per SM in bits / (32 * blocking factor)* where 32 is the number of threads in a warp. This translates to 192, 96, and 48 rows per slice for blocking factors of 64, 128, and 256-bits, respectively. Hence, one byte per row index is sufficient for this format.

Large sliced (LS) COO

In this subformat, the result for each row gets one entry in shared memory, which is shared among all threads in the thread block. Access to shared memory uses an atomic XOR operation. Thus, there will be bank conflicts. However, this drawback can be compensated for by a higher texture cache hit rate. There is no p-reduction on shared memory required.

The maximum number of rows per slice is calculated as *size of shared memory per SM in bits / blocking factor*. This translates to 6144, 3072, and 1536 rows per slice for blocking factors of 64, 128, and 256-bits, respectively. We need two bytes for the row index.

Table 20.2 summarizes the differences between the three subformats.

a) Each thread in a block is allocated one memory entry for the result of a row and works on consecutive 12 rows. A p-reduction to Thread#1 is needed to get the final result of one row.

b) Each warp in a block is allocated one memory entry per row and works on 192 consecutive rows, threads in each warp use atomic XOR to update the value in a given entry. A p-reduction for each row to memory in Warp#1 is needed to obtain the final result of that row.

c) A thread block works on 6144 consecutive rows and shares the memory entry of each row. Any update to the memory has to use atomic XOR.

20.4.3 Determining the cut-off point of each format

To determine which format to use, we compare the performance of two consecutive formats in terms of giga nonzeros ($gnnz$) per second for a given matrix, starting with the dense format and the SS-COO format. The two formats start from the same row (the first row) and work on the minimum number of rows possible. For the dense format, the minimum number of rows is 32. For the SS-COO format (and its variants), the minimum number of rows is *the number of rows in a slice* times *the number of multiprocessors in the GPU*, since one thread block works on one slice and one thread block is assigned to one multiprocessor.

The next comparison depends on the result of the current comparison. If the dense format performs better, we decide to use it for rows 1 to 32, and we continue comparing the dense format and the SS-COO format starting from row 33. However, if the SS-COO format performs better, we compare its performance with the next format, MS-COO, starting from the same row, and so on. The idea is to stop considering the denser format once the sparser format outperforms it. Once we get to the comparison between MS-COO and LS-COO, and LS-COO performs better, we don't need to do any further comparisons. LS-COO should be used for the rest of the matrix.

For Sliced COO format, it is essential to note that when one slice is assigned to each multiprocessor, the load for one multiprocessor may be much higher than that of the other multiprocessors. This is because the matrix rows have been reordered by their weight in a nonincreasing order, so the first slice contains more nonzero entries than the rest. Thus, we need to further reorder the rows so that each multiprocessor gets the same level of load.

The cut-off point determination is performed once per matrix and as a preprocessing step. Its runtime is equivalent to less than five iterations of SpMV.

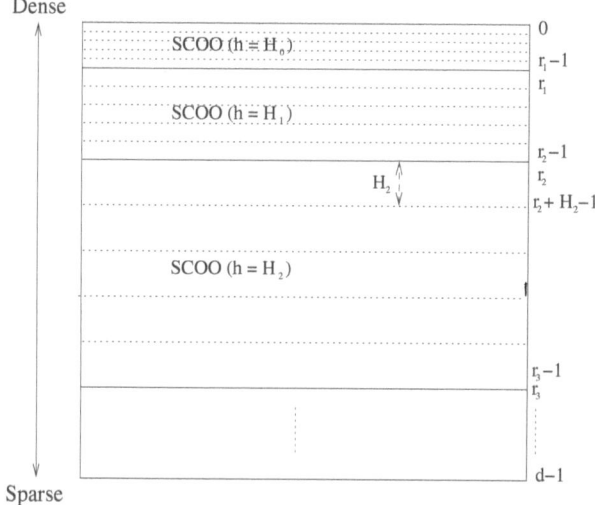

FIGURE 20.4. Partitioning of a row-sorted floating-point matrix into SCOO format.

20.5 SCOO for single-precision floating-point matrices

The design of SCOO depends on performing atomic operations to combine the partial results. Atomic operations for double-precision are not available on current CUDA-enabled devices. Thus, we present the SCOO extension for only single-precision floating-point matrices.

To extend SCOO for general matrices in real numbers, we first adapt to the variety of sparsity in real number matrices. Unlike NFS matrices, the sparsity is not predictable. Hence, instead of using three subformats (SS-COO, MS-COO, LS-COO) with a fixed slice size (12, 192, 6144), we make the slice size a variable. Let h denote the slice size, S denote the size in bytes of shared memory per thread block ($S = 49152$ for Fermi), and b denote the size of each matrix value in bytes ($b = 4$ for single-precision floating point). h must be a positive integer value of the form of $h = \frac{S}{2^i b}$, $\forall 0 \le i \le T$ where $T = \log_2$ (maximum number of thread per block).

For example in Fermi GPUs, single-precision floating point, $h \in \{12288, 6144, 3072, 1536, 768, 384, 192, 96, 48, 24, 12\}$ for $S = 49152, b = 4, T = 10$. The algorithm described in Section 20.4.3 can be used to determine the cut-off point for each subformat. Figure 20.4 illustrates the final matrix in SCOO format assuming the number of multiprocessors in the GPU is 5 and the row index ranges from $0..d-1$. H_i and r_i are the slice size and starting row index of the sub-matrix i.

Hardware	C2075	GTX-580	Core-i7 2700K
# Cores	448	512	4
Clock speed (Ghz)	1.15	1.57	3.5
Memory type	GDDR5	GDDR5	DDR3-1600
Memory size (GB)	6	3	16
Max Memory bandwidth (GB/s)	144	192	21

TABLE 20.3. Overview of hardware used in the experiments.

More details of the SCOO format for floating-point matrices can be found in [9].

20.6 Performance evaluation

20.6.1 Experimental setup

Table 20.3 specifies the GPUs and the CPU workstation used for performance evaluation. The performance is measured in terms of $gnnz$ and $Gflop/s$ for binary and floating-point matrices. Measured GPU performance includes neither PCIe data transfers nor matrix preprocessing. These are realistic assumptions since SpMV applications usually consist of a large number of iterations where the sparse matrix is iteratively multiplied by the input/output vectors.

Our source code is compiled and executed using CUDA toolkit and driver version 4.2 under Linux Ubuntu 12.04.

20.6.2 Experimental results of SpMV on NFS matrix

We have evaluated our implementation on an NVIDIA Tesla C2075 (ECC disabled) with 6 GB RAM. We have compared the GPU performance with the open-source CADO-NFS [15] program running on Intel Core i7-920 CPU with 12 GB DDR3-1066 memory. The RSA-170 matrix (see Table 20.1) is used for performance evaluation.

The speedups compared to the multithreaded CADO-NFS bucket implementation on an Intel Core i7-920 are given in parentheses in Table 20.3. The GPU memory required to store the sparse matrix and the corresponding bytes per nonzero (nnz) are also reported. CADO-NFS contains several CPU optimized SpMV implementations using multithreading and SSE instructions. In this experiment, we compare the performance of our implementation to the CPU cache-optimized bucket format of CADO-NFS using 8 threads.

Solving large sparse linear systems for integer factorization on GPUs 467

Blocking factor	C2075 (speedup)	2 x C2075 (speedup)	Core i7-920 CADO-NFS	GPU memory (bytes/nnz)
64	4.28 (4.9)	8.38 (9.5)	0.88	2748 MB (2.90)
128	2.78 (5.6)	5.37 (10.7)	0.5	2967 MB (3.13)
256	1.88 (7.0)	3.68 (13.6)	0.27	3000 MB (3.16)

TABLE 20.4. Performance of SpMV on RSA-170 matrix.

Table 20.4 shows the result. The dual-GPU implementation achieves speedups between 1.93 and 1.96 compared to the single-GPU performance.

Table 20.5 shows the individual performance of each subformat (in $gnnz/s$), the percentage of nonzeros are included in parentheses. The results show that the performance decreases when the matrix gets sparser. The MS-COO and LS-COO performance degrade when the blocking factor is increased from 64 to 128 and 256-bit. This is caused by the increased number of bank conflicts and serialization of atomic XOR operations on larger blocking factors. Thus, the SS-COO format gets a higher percentage of nonzeros with 128 and 256-bit blocking factor.

More detail results of our full block Wiedemann CUDA implementation as well as a multi-GPU implementation can be found in [16]

Blocking factor	Dense	SS-COO	MS-COO	LS-COO
64	13.66 (24%)	9.66 (11%)	8.13 (13%)	2.77 (52%)
128	9.66 (15%)	7.53 (24%)	3.23 (6%)	1.86 (55%)
256	7.00 (15%)	4.21 (21%)	2.34 (6%)	1.33 (58%)

TABLE 20.5. Performance for each of the four subformat partitions of the RSA-170 matrix on a C2075.

20.6.3 Experimental results of SpMV on floating-point matrices

20.6.3.1 Performance comparison to existing GPU formats

We compare the performance of our SCOO format to available SpMV implementations on both GPU and CPU. The set of selected test matrices are collected from the University of Florida Sparse Matrix Collection [10]. We have chosen the biggest matrices from different areas that with their corresponding input and output vectors can still fit into the 6GB global memory of a C2075 GPU. Table 20.6 gives an overview of those matrices.

We compare the SCOO format to the CSR, COO, and HYB format of Cusp 0.3.0. Other Cusp formats are not able to run on the large tested matrices that we selected. The results are shown in Figure 20.5. The performances are

Name	row	column	nz/row	Description
GL7d19	1,911,130	1,955,309	19,53	combinatorial problem
relat9	12,360,060	549,336	3,15	combinatorial problem
wikipedia-20070206	3,566,907	3,566,907	12,62	directed graph
wb-edu	9,845,725	9,845,725	5,81	directed graph
road_usa	23,947,347	23,947,347	2,41	undirected graph
hugebubbles-00010	19,458,087	19,458,087	3,00	undirected graph
circuit5M	5,558,326	5,558,326	10,71	circuit simulation
nlpkkt120	3,542,400	3,542,400	26,85	optimization problem
cage15	5,154,859	5,154,859	19,24	directed weighted
kron_g500-logn21	2,097,152	2,097,152	86,82	undirected multigraph
indochina-2004	7,414,866	7,414,866	26,18	directed graph
nlpkkt160	8,345,600	8,345,600	27,01	optimization problem
rgg_n_2_24_s0	16,777,216	16,777,216	15,80	undirected random
uk-2002	18,520,486	18,520,486	16,10	directed graph

TABLE 20.6. Overview of sparse matrices used for performance evaluation.

in terms of Gflop/s which is based on the assumption of two flops per nonzero entry of the matrix [5, 7].

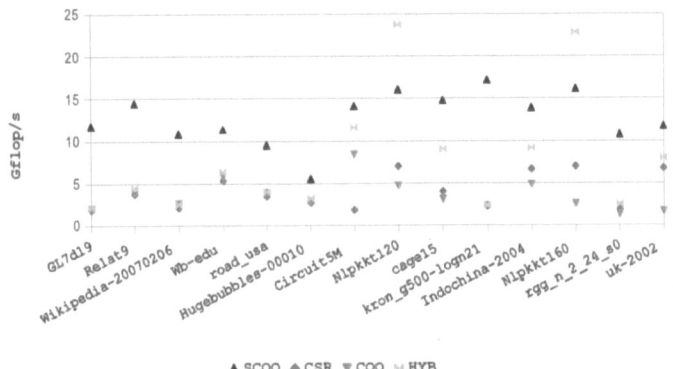

FIGURE 20.5. Performance comparison of SCOO and other GPU formats for each test matrix on a Fermi Tesla C2075 (ECC disabled).

The SCOO format achieves a stable performance for different matrices in single-precision mode. In most cases a performance of over 10 Gflop/s can be sustained. For some highly unstructured matrices such as *GL7d19*, *wikipedia-20070206*, *rgg_n_2_24_s0*, and *kron_g500-logn21*, SCOO achieves high speedups ranging from 3 to 6 compared to the best performaning Cusp format.

For most matrices, HYB produces the best performance among the tested Cusp formats. HYB is able to outperform SCOO for only two matrices: *nlpkkt120* and *nlpkkt160*. Both matrices have a similar structure, i.e., they consist of consecutive rows that have a very similar number of nonzero coefficients which are suitable to be stored in the ELL section of the HYB format. Moreover the nonzeros are close to each other facilitating coaleasing and cache-friendly access patterns by nature. SCOO is able to outperform COO and CSR for all tested matrices.

In matrix *Relat9* we observe some patterns but the matrix is still generally unstructured, thus SCOO is able to achieve about 2 times speed up compared to HYB which is the best among tested Cusp formats in this case. The average speedup of SCOO for the tested matrices is 3.0 compared to CSR, 5.02 compared to COO, 2.15 compared to HYB.

We show the visualization of sparse matrices *nlpkkt120*, *relat9*, *GL7d19* in Figure 20.6(a), 20.6(b), 20.6(c) using MatView [13]. The white color represents zero entries, gray color represents nonzero entries.

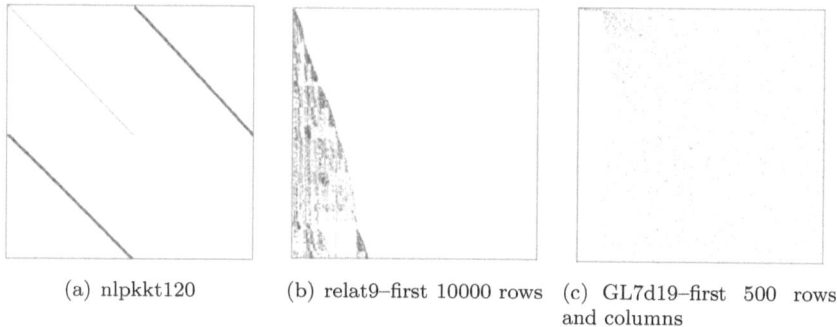

(a) nlpkkt120   (b) relat9–first 10000 rows   (c) GL7d19–first 500 rows and columns

FIGURE 20.6. Visualization of *nlpkkt120*, *relat9*, and *GL7d19* matrix.

20.6.3.2 Performance comparison to a CPU implementation

We used the Intel MKL library 10.3 in order to compare SCOO performance to an optimized CPU implementation. MKL SpMV receives the input matrices in CSR format. The results are shown in Figure 20.7. Using a GTX-580, we achieved speedups ranging between 5.5 and 18 over MKL on a 4-core CPU with hyper-threading using 8 threads. Also note that the SCOO performance on a GTX-580 is around 1.5 times faster than on the C2075 due to the increased memory bandwidth and clock speed. The storage requirement for the *rgg_n_2_24_s0* and *uk-2002* matrices and associated input/output vectors slightly exceeds the 3 GB global memory of the GTX-580 and thus are not included.

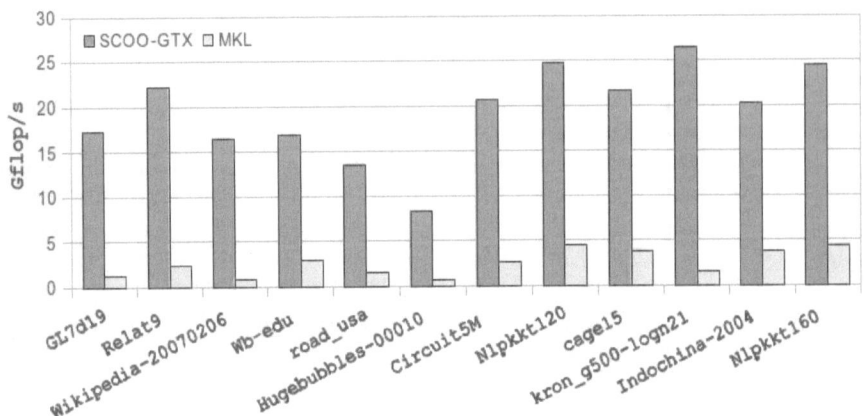

FIGURE 20.7. Performance of the SCOO on a GTX-580 and a CPU implementation using MKL performed on a Core-i7 2700K using 8 threads.

20.7 Conclusion

In this chapter, we have presented our implementation of iterative SpMV for NFS matrices on GPUs with the CUDA programming language. Our GPU implementation takes advantage of the variety of sparseness properties in NFS matrices to produce suitable formats for different parts. The GPU implementation shows promising improvement over an optimized CPU implementation. As the size of integers in factorization projects is expected to increase further, the linear algrebra step of NFS will become an even bigger bottleneck. The size and sparseness of matrices generated by the NFS sieving step are growing significantly with the size of the integer to be factored. Thus, a large GPU cluster is required to accelerate the linear algebra step. However, in order to achieve scalability for larger problem sizes, the amounts of GPU RAM and data transfer bandwidth need to be increased in addition to the number of GPUs.

We further adapted the proposed Sliced COO format to single-precision floating-point numbers and evaluated it with large and sparse matrices derived from other computational science applications. We have published our code at https://github.com/danghvu/cudaSpmv

Bibliography

[1] K. Aoki, J. Franke, T. Kleinjung, A. K. Lenstra, and D. A. Osvik. A kilobit special number field sieve factorization. In *ASIACRYPT*, Kuching, Malaysia, 2007.

[2] K. Aoki, T. Shimoyama, and H. Ueda. Experiments on the linear algebra step in the number field sieve. In *Proceedings of the Security 2nd International Conference on Advances in Information and Computer Security*, IWSEC'07, pages 58–73, Berlin, Heidelberg, 2007. Springer-Verlag.

[3] N. Bell and M. Garland. Cusp: Generic parallel algorithms for sparse matrix and graph computations. http://cusplibrary.github.io/.

[4] N. Bell and M. Garland. Efficient sparse matrix-vector multiplication on CUDA. NVIDIA Technical Report NVR-2008-004, NVIDIA Corporation, December 2008.

[5] N. Bell and M. Garland. Implementing sparse matrix-vector multiplication on throughput-oriented processors. In *SC '09: Proceedings of the Conference on High Performance Computing Networking, Storage and Analysis*, pages 1–11, New York, NY, USA, 2009. ACM.

[6] B. Boyer, J.-G. Dumas, and P. Giorgi. Exact sparse matrix-vector multiplication on gpu's and multicore architectures. *CoRR*, abs/1004.3719, 2010.

[7] J. W. Choi, A. Singh, and R. W. Vuduc. Model-driven autotuning of sparse matrix-vector multiply on GPUs. *ACM SIGPLAN Notices - PPoPP '10*, 45:115–126, January 2010.

[8] D. Coppersmith. Solving homogeneous linear equations over GF(2) via block Wiedemann algorithm. *Mathematics of Computation*, 62, 1994.

[9] H.-V. Dang and B. Schmidt. The sliced COO format for sparse matrix-vector multiplication on CUDA-enabled GPUs. In *International Conference on Computational Science ICCS, Procedia Vol. 9*, pages 57–66, 2012.

[10] I. S. Duff, R. G. Grimes, and J. G. Lewis. Sparse matrix test problems. *ACM Trans. Math. Softw.*, 15:1–14, March 1989.

[11] W. Hwang and D. Kim. Load Balanced Block Lanczos Algorithm over GF(2) for Factorization of Large Keys. In *High Performance Computing, HiPC*, pages 375–386, Bangalore, India, 2006.

[12] T. Kleinjung, L. Nussbaum, and E. Thomé. Using a grid platform for slving large sparse linear systems over GF(2). In *11th ACM/IEEE International Conference on Grid Computing (Grid 2010)*, Brussels Belgique, 10 2010.

[13] J. Kohl. MatView: Scalable sparse matrix viewer. http://www.csm.ornl.gov/ kohl/MatView/.

[14] A. Monakov, A. Lokhmotov, and A. Avetisyan. Automatically tuning sparse matrix-vector multiplication for GPU architectures. In *International conference on High-Performance Embedded Architectures and Compilers,HiPEAC*, pages 111–125, Pisa, Italy, 2010.

[15] P. Gaudry et al. CADO-NFS. http://cado-nfs.gforge.inria.fr/.

[16] B. Schmidt, H. Aribowo, and H.-V. Dang. Iterative Sparse Matrix-Vector Multiplication for accelerating the Block Wiedemann Algorithm over GF(2) on Multi-graphics Processing Unit Systems. *Concurrency and Computation: Practice and Experience*, 25:586–603, 2013.

[17] T. Kleinjung et al. Factorization of a 768-bit RSA modulus. In *International Crytology Conference*, pages 333–350, 2010.

[18] T. Kleinjung et al. A heterogeneous computing environment to solve the 768-bit RSA challenge. *Cluster Computing*, 15(1):53–68, 2012.

[19] E. Thomé. Subquadratic computation of vector generating polynomials and improvement of the block Wiedemann algorithm. *J. Symb. Comput.*, 33:757–775, May 2002.

Index

AIAC, 120
Arnoldi iterations, 404
asynchronous iterations, 120, 336, 341, 345, 352

BLAS, 376, 377, 385, 393
boundary condition, 82, 258, 260
boundary volume problem, 87
Boussinesq models, 255
branch-and-bound, 156, 224
    branching, 225
    cutting-plane, 227
    node selection, 226
BSP parallel scheme, 106

carrier waveforms, 399
chaos, 443
chaotic
    iterations, 444
    systems, 443
collembola model, 420
combinatorial optimization, 184
compressed storage format, 319
    COO, 319, 457
    CSR, 457
    ELL, 319, 457
    HYB, 319, 457
    SLE, 458
    sliced COO, 459
computation auxiliary, 134
computing node, 317, 337
concept, 77
concurrent kernel execution, 391, 392
constrained splines, 295, 298
convergence, 89, 268, 312, 326, 333, 337, 348
    global, 126

    local, 126, 128
    maximum number of iterations, 314, 317
    maximum number of relaxations, 345
    tolerance threshold, 312, 314, 345
convergence detection, 122
    global, 126
convolution, 53
CUBLAS, 16, 317, 321, 338, 339, 341, 377, 382, 384, 385, 387, 388, 391, 392
CUDA, 108
    data transfer, 108
    stream, 108, 112, 115
CUDA functions
    cudaMalloc, 13
CUDA functions
    cudaMemcpy, 15
    timer, 15
CUDA keywords
    __shared__, 15
    blockDim, 15
    blockIdx, 15
    thread index, 15
    threadIdx, 15
Cutil library, 26
    cutLoadPGMi, 26
    cutSavePGMi, 26
    timer usage, 28

defect correction, 88, 90, 267, 268
    iteration, 88
density of periodic points, 444
differential algebraic equations, 398
dispersion, 276

474   Index

domain decomposition, 93, 269
double-buffering, 385, 386, 388, 390

envelope-following, 397, 399
Euler
    backward Euler, 397, 399
    forward Euler, 83, 266, 399

filtering, 280
finite difference, 77, 83, 88, 251, 261, 262
flexible order, 79, 268
flowshop scheduling problem, 154
fluid, 256

Galerkin condition, 313
Gear-2 integration, 404
GPU
    architecture of a, 5
    based local search, 203
    cluster, 321, 322, 345
    coalesced memory accesses, 383
    CPU/GPU communication, 188, 190
    data placement, 188
    memory management, 197, 207
    sequence, 110
    streamed sequence, 113
    thread divergence, 164, 189, 197, 202
    thread mapping, 199, 200
    threads synchronization, 189
GPU-based frameworks
    libCUDAOptimize, 201, 202
    ParadisEO-MO-GPU, 201–203
    PUGACE, 201

heat conduction, 82
Hermite splines, 297, 302
Hessenberg matrix, 315, 316, 319, 400
Hilbert space, 334, 335
HMPP, 372, 377, 392

initial value problem, 82
instruction-level parallelism, 37, 40

integer linear programming, 224
isotone regression, 297, 303
iterative method, 88, 312, 348
    Arnoldi iterations, 315, 316, 322, 400
    CG, 313, 317, 319
    Gauss-Seidel, 348
    GMRES, 314, 317, 319, 396, 400, 402
    Jacobi, 334, 342, 348
    Krylov subspace, 312, 315, 400, 406
    Krylov subspace, 313
    MINRES, 314
    Newton iteration, 399, 402
    number field sieve, 453
    projected Richardson, 334, 335, 338, 342
    red-black ordering, 348, 349

Jacobian, 263
Jacobian matrix, 396, 399

kinematic, 276
Krylov subspace, 396

LAPACK, 376
Laplace operator, 82
Laplace problem, 261
lattice Boltzmann method, 355
library, *see* software library
linear programming, 217
lower bound, 160
LU, 396, 405, 406
Ludwig code, 356

MAGMA functions
    DGEMM, 388
    DGETRF, 383, 386
    DGETRS, 383, 386
mass conservation, 256
matrix multiplication
    block cyclic, 118, 142
matrix-free, 77, 396, 405, 406
memory access optimizations, 168
memory hierarchy, 9

cache memory, 9
global memory, 9, 35
local memory, 9, 33
pinned memory, 26
registers, 9, 35
shared memory, 9, 26, 35, 62
texture memory, 25, 26
message
    loss/miss, 124
    stamping, 131
metaheuristics
    algorithmic-level parallelism, 186
    ant colony optimization, 189, 190, 197
    evolutionary algorithm, 187
    iterated local search, 185, 203
    iteration-level parallelism, 186
    parallel metaheuristics, 184, 186, 187
    particle swarm optimization, 193, 203
    simulated annealing, 185
    solution-level parallelism, 187
    tabu search, 187, 191
method of lines, 82, 263
MLS (minimum lower sets) algorithm, 304
modified nodal analysis, or MNA, 402
momentum conservation, 257
monotonicity, 295, 298, 302, 303
MPI, 107
    blocking, 122, 320, 341
    global, 321, 345
    nonblocking, 320, 336, 341
    synchronous, 122
multi-agent system, 416
multi-GPU, 91, 237, 321
multigrid, 89, 268
multiple output per thread, 55

neighboring node, 319, 320, 339
nonlinear, 332–334, 337
number field sieve
    Berlekamp-Massey, 454

block Wiedemann, 453
blocking factors, 454
kernel space, 454
numerical validation, 377

obstacle problem, 332, 333, 335, 337
OpenACC, 377
OpenMP, 107
    barrier, 112, 118
    mutex, 123
    parallel region, 114
    thread creation, 112
ordered access, 142
ORWL, 140
    handle, 142
    location, 141
    operation, 145
    resource, 141
        data, 141
        device, 141
    schedule, 145
    task, 145
overlap
    computation and communication, 106, 122
    computation and computation, 107
    GPU data transfers with GPU computation, 112

padding, 386–388
page-locked data, 114
parallel
    evaluation of bounds, 158
    reduction, 228
    tree exploration, 158
parareal, 95, 282
PAV algorithm, 303
PDE example, 137
performance model, 237
Petrov-Galerkin condition, 315
Poisson equation, 87
POSIX threads, 386
potential flow, 255
power converters, 395

power electronics, 396
precision, 277
    double precision, 372, 373, 377–381, 389, 392
    mixed, 278
    single precision, 372, 373, 377–380, 392
preconditioning, 89, 93, 267
    left-preconditioning, 267
prefetching, 26, 36, 62, 67
PRNG, 441

real-time simulation, 286
reduction operation, 16
register count, 35, 37, 39
relative gain, 324, 347
relaxation, 266
    zone, 266
residual, 124

sensitive dependence on initial conditions, 444
sensitivity matrix, 396, 399, 404
simplex
    revised method, 219
    standard method, 217
software library, 74
sparse linear system, 312, 319
    preconditioned, 312, 313, 316
SPICE, 396, 402, 406
spline, 295
SpMV multiplication, 319, 320
stability, 263
stencil, 79, 261
switching frequency, 396
synchronous iterations, 336, 341, 345, 352

TDP (thermal design power), 387
templates, 76
termination, 122
TestU01, 442
time integration, 83, 263
topological transitivity, 443
topology, 93
transient analysis, 398

truncation error, 398
type definitions, 85

wrapper, 382